Banach and Locally Convex Algebras

Banach and Locally Convex Algebras

A. Ya. Helemskii

Translated by A. West

CLARENDON PRESS · OXFORD
1993

Oxford University Press, Walton Street, Oxford OX2 6DP
Oxford New York Toronto
Delhi Bombay Calcutta Madras Karachi
Kuala Lumpur Singapore Hong Kong Tokyo
Nairobi Dar es Salaam Cape Town
Melbourne Auckland Madrid
and associated companies in
Berlin Ibadan

Oxford is a trade mark of Oxford University Press

Published in the United States by
Oxford University Press Inc., New York

A catalogue record for this book is available from the British Library

Library of Congress Cataloging in Publication Data
(Data available)
ISBN 0–19–853578–3

Typeset by the translator
Printed in Great Britain by
Bookcraft (Bath) Ltd
Midsomer Norton, Avon

Preface to the English edition

I am very glad that this book will now be available to the English reader. I would like to use this opportunity to express my sincere gratitude to Professor H. G. Dales, who was the initiator of its translation on behalf of OUP, and to Dr A. West, who did the translation work with his customary highly professional skill.

Especially I would like to emphasize the excellent and really creative work of the English editor of the book, Dr J. Rennison. Besides eradicating numerous small mistakes from the book and adapting its terminology to the standards of the English reader, he made many wise and inventive comments and suggestions, which were gladly and gratefully accepted. If this book is indeed read by English-speaking experts and students with interest and pleasure, this will be due to a great extent to his taste and professional judgement.

Moscow A. Ya. H.
October 1992

Preface to the Russian edition

Almost half a century has passed since the appearance of a short report by Gel'fand [112], announcing to the mathematical world the creation of a new theory. Since that time quite a few books have been written about Banach algebras and their generalizations, including some monographs of an instructional character. We would now like to recall some of these with gratitude.

The author and many of his colleagues had occasion to study Banach algebras using two remarkable books from the pen of the initiator of the theory and his close collaborators: these were *Commutative normed rings* by I. M. Gel'fand, D. A. Raikov, and G. E. Shilov [20] and *Normed rings* by M. A. Naimark [53]. *The general theory of Banach algebras* by C. E. Rickart [64] with its masterly account of the algebraic approach to the subject has played an important role in promoting this discipline abroad. Of the later books, *Complete normed algebras* by F. F. Bonsall and J. Duncan [9] deserves, in our opinion, special mention. In this book, a careful account of the origins of the theory is integrated with an account of the latest achievements at that time—up to the year 1972—(they even managed to include the first results on cohomology and amenable algebras which had just appeared).

All the books mentioned are destined to have a long life.

But as the second half of the eighties has already passed, we hope that this new book on 'the algebras of analysis' will also be of use. It was thought of primarily as a textbook and intended as a first acquaintance with the subject; it is not surprising that it does not include many of the varied results in this large domain of functional analysis which has such an extensive boundary layer. At the same time we have tried, in some measure, as far as the size and proportions of the book allowed, to reflect some of the contemporary trends in the development of the theory.

The detailed table of contents gives the specialist a fairly complete indication of the material in the book. We therefore limit ourselves to a few remarks about the contents.

We have curtailed the account of the more traditional section—the theory of commutative Banach algebras—at a fairly early stage, having proved the basic theorem about representing these as function algebras and having explained what this says in specific situations. This is the boundary at which, relatively speaking, the 'general' results, among which this theorem is of great merit, finish and the 'special' results begin, for which this theorem becomes part of the fundamentals and serves as an initial principle and a basic working method. If we have left a lot of interesting and important questions about the 'commutative' theory outside the scope of the book, we do not suffer any pangs of remorse. There is an excellent book available in Russian, *Uniform algebras* by T. W. Gamelin [17], where all these useful and beautiful things—Shilov boundaries, analytic discs, Gleason parts, etc.—are set out very thoroughly and in great detail.

Then we have given greater consideration than in previous monographs to Banach and polynormed modules with their various interrelations and applications. In particular, in the context of the 'modules of analysis' we have considered such concepts as 'freeness', projectivity, injectivity, and flatness, together with their applications to derivations, extensions, and other questions. There is some discussion, in the light of these concepts, of several classes of algebras introduced in the seventies and actively studied since then; among these, the best known are the amenable algebras mentioned above.

Perhaps Chapter 0, which contains the indispensable background material, deserves some explanation. It should not be thought that its balance exactly reflects the comparative degree to which the corresponding concepts and results figure in the basic text. Priority is given here to those questions contained in the literature that are difficult to access or which are set out—by algebraists for algebraists, say—in a way not quite as we would have wished. Primarily we have in mind Banach and topological tensor products. But what has been said applies, although to a lesser degree, to elementary category theory and homology theory and also to polynormed spaces. This material, we are convinced, should be included as the compulsory minimum for a contemporary textbook on the subject we are interested in. But it is also important because it is of independent interest and its applications go far beyond the scope of this discipline.

At the same time we have restricted ourselves to a comparatively short mention of a lot of other preliminary material which is of no less importance in an account of the subject. For the most part this concerns questions in analysis and topology that are included in a university course or are, at least, in university textbooks. For the latter we have to a large extent been guided by the books of A. N. Kolmogorov and S. V. Fomin [42] and A. A. Kirillov and A. D. Gvishiani [41].

* * *

Let us mention some details on the technical side of the exposition. As usual, in the text, theorems, definitions, and other 'mathematical statements' have different shades of meaning. A single numbering system is used which is independent within each section (it does not matter whether a section is divided into subsections or not) or within a chapter if that chapter is not divided into sections. A reference of the kind 'according to IV.2.14' refers to the fourteenth 'statement' of §2, Chapter IV. A reference of the kind 'see the beginning of §1.1' is to something said at the beginning of the first subsection of the first section. Here, as in references to 'statements', the number of the chapter is omitted if reference is made to something within the same chapter.

Theorems, lemmas, propositions are given with complete proofs (it is better to say that the proofs claim to be complete). The latter begin with ▶ and end with ◼;

when placed side by side (▶■) these signs mean 'obviously' or 'can be verified immediately'. These signs are not placed after corollaries, which are always considered to be obvious, since they proceed from what has been said earlier.

There are many exercises among the enumerated 'statements'; they are incorporated in the text and are considered as an integral part of it. In the light of the hints that are sometimes given, they should prove fairly simple. If these exercises (and 'assertions' which we shall speak of later) were removed, then the account would still be complete: we emphasize that they are not used between the signs ▶ and ■. Nevertheless, in our opinion, it would be unwise to omit them: they are fairly conceptual and in solving them the reader verifies how well he has understood the definition or the theorem that the given exercise accompanies.

Another kind of fact given without proof—but for another reason—is an 'assertion'. There are not so many of them. Mostly they are results of fundamental importance having a simple and effective formulation; however, their proofs are either fairly complicated or hinge on theorems in analysis or algebra which go beyond the scope of our minimum background (it may be one or the other or both). Therefore, knowing about the existence of such a result is desirable even on a first acquaintance with the subject, but spending time and effort on analyzing its proof is not worthwhile for the present: at the given stage of study it is wiser to move on further. Typical 'assertions' in our text are theorems about multidimensional holomorphic calculus (Shilov–Arens–Calderon and Taylor) or, let us say, Kadison's theorem that algebraic and topological irreducibility coincide for representations of C^* algebras.

The sign $\Leftrightarrow$ replaces the words 'if and only if'. An equation representing a definition is often denoted by the sign :=. The tailed arrow $\mapsto$ is used when we are explaining how a given map acts on the elements or—more rarely—how a functor acts on objects and morphisms. So we say: 'we consider the function $w : \mathbb{C} \to \mathbb{C} : z \mapsto z^2$'.

Sections which are bracketed like this paragraph are for supplementary information; often this is something we wish to talk about without being unnecessarily formal. Of course, the main text does not depend on the 'bracketed text'.

The list of references consists mostly, although not always, of monographs and articles to which we refer directly in the text. Needless to say, it is far from complete; in particular many old papers are excluded, as these are sufficiently well 'catered for' in subsequent books. A comprehensive list of references on 'algebras of analysis' and their applications could form a separate book comparable in size with ours.

Contents

Chapter 0
Foundations

Strictly speaking, we shall only start on the subject of this book after we have introduced an additional structure — multiplication — into the objects of linear functional analysis — Banach spaces and some more general spaces; this is precisely what we shall talk about in the first definition of the first chapter. But first, in Chapter 0, it is necessary to indicate the stock of background material which is important for us as a foundation.

Some of this material is widely known and is contained in standard textbooks but there is some which can be found only in specialized monographs and articles in journals. Therefore the amount of detail given in different places in this chapter is very varied.

§0. REMINDERS FROM SET THEORY, LINEAR ALGEBRA, AND TOPOLOGY

This section is a sort of reference: mostly we simply define precisely the terms and notations to be used later.

0.1°. We assume that a minimal amount of set theory is known, as contained, for example, in [42, Ch.I].

As usual, we use the standard "open-face" notation: $\mathbb{C}$, $\mathbb{R}$ and so on, for the classic sets of numbers. We define only some of them precisely; $\mathbb{N} := \{1, 2, 3, \cdots,\}$, $\mathbb{T} := \{z \in \mathbb{C} : |z| = 1\}$, $\mathbb{D} := \{z \in \mathbb{C} : |z| < 1\}$, $\overline{\mathbb{D}} := \mathbb{D} \cup \mathbb{T}$; $\mathbb{R}^{[+)} := \{t \in \mathbb{R} : t \geq 0\}$, $\mathbb{R}^{(+)} := \mathbb{R}^{[+)} \setminus \{0\}$, finally, $\mathbb{C}^{(+)} := \{z \in \mathbb{C} : z = |z| e^{i\varphi}, -\pi < \varphi < \pi\}$.

Maps from the set M into the set N (or "between M and N") are denoted, as a rule, by $\varphi : M \longrightarrow N$ or $M \xrightarrow{\varphi} N$; moreover M is often called the *domain* and N the *codomain* of the map φ. When we explain how the map is determined on elements, we use notation of the type $\varphi : M \longrightarrow N : x \longmapsto y$ or simply $\varphi : x \longmapsto y$ if it is clear what the domain and the codomain are. For example, the notation $1_M : M \longrightarrow M : x \longmapsto x$ explains how the identity map acts (and how it is denoted); we write 1 instead of 1_M when M is fixed.

The words "injective", "surjective" and "bijective" when applied to maps have the usual set-theoretic sense; the same applies to the term "image of the map φ" (notation $\operatorname{Im}\varphi$).

Let $\varphi : M \longrightarrow N$ be given. Then for $M_0 \subseteq M$ we can consider the map $\varphi|_{M_0} : M_0 \longrightarrow N : x \longmapsto \varphi(x)$, called the *restriction* of φ to M_0, and for $\operatorname{Im}\varphi \subseteq N_0 \subseteq N$ we can consider the map $\varphi|^{N_0} : M \longrightarrow N_0 : x \longmapsto \varphi(x)$, called the *corestriction* of φ to N_0. The phrase "φ is an *extension* of ψ" means that ψ is a restriction of φ. The restriction of the identity map of M to $M_0 \subseteq M$ is called the *natural embedding* of M_0 in M and is denoted by **in**.

The symbol $\times$ in expressions of the type $M \times M$ or $\mathop{\Large\times}\limits_{\nu\in\Lambda} M_\nu$ denotes the Cartesian product. The totality of all maps between M and N (it is also the Cartesian product of copies of the set N, indexed by the set M) is denoted by N^M. The totality of n-tuples from the set M is denoted, as usual, by M^n. The disjoint union (that is, the union of non-intersecting sets) is denoted by the sign $\coprod$.

A *diagram* (of sets) is any collection of sets together with maps connecting some of them. For example, one of the simplest diagrams is the "triangle"

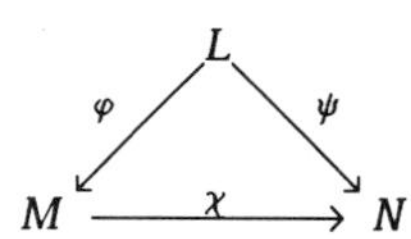

and the "square"

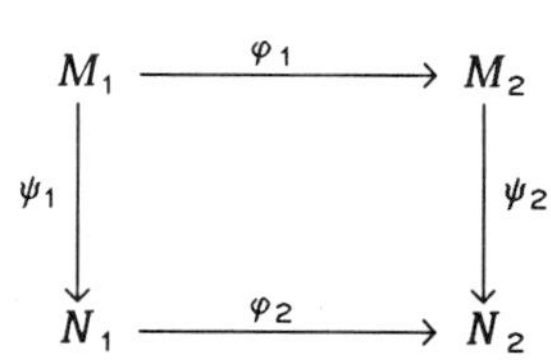

A diagram is said to be *commutative* if the following is satisfied: given two collections, $\varphi_1, \cdots, \varphi_n$ and $\psi_1, \cdots, \psi_m$, of its maps, as long as the compositions $\varphi_m \circ \cdots \circ \varphi_1$ and $\psi_n \circ \cdots \circ \psi_1$ make sense and have the same domain and the same codomain, then these compositions coincide. For example, the triangle being commutative simply means that $\chi \circ \varphi = \psi$, and the square being commutative means that $\varphi_2 \circ \psi_1 = \psi_2 \circ \varphi_1$.

0.2°. The reader needs to have enough linear algebra to cover a standard university course [19], [43].

Let us stress that, throughout the whole book, when we say "linear space" we mean that the scalars are complex numbers; there are a few exceptions which will be specially mentioned. The words "operator" and "bioperator", "functional" and "bifunctional" always means linear operator and bilinear operator, linear functional and bilinear functional respectively. A *linear isomorphism* is an isomorphism in the sense of linear algebra (= an operator having an inverse). The linear hull of a set M in a linear space is denoted by $l.h.(M)$. The *kernel* of the operator $\varphi : E \longrightarrow F$ is denoted by $\operatorname{Ker}\varphi$ (= the preimage of zero), and its natural embedding in E is denoted by $\ker\varphi : \operatorname{Ker}\varphi \longrightarrow E$.

The linear space of the operators from E into F is denoted by $\mathscr{L}(E, F)$; we write $\mathscr{L}(E)$ instead of $\mathscr{L}(E, E)$ and $E^{\#}$ instead of $\mathscr{L}(E, \mathbb{C})$. This last space (= the space of functionals) is called the *algebraic dual of E*.

The value of a functional on an element will often be denoted by the symbols $\langle f, x\rangle$ or $\langle x, f\rangle$ rather than $f(x)$.

If E_0 is a subspace of E, then the coset of E_0 in E that contains x will be denoted by $x + E_0$ or $\dot{x}$. The operator $x \longmapsto \dot{x}$ sending E to the quotient space E/E_0 is called the *natural projection* and is denoted by **pr**.

Quite often, when considering linear spaces (possibly with additional structures), we shall come across the following situation. Let E_0 be a subspace in E, and F_0 a subspace of F, and let $\varphi : E \longrightarrow F$ be an operator such that $\varphi(E_0) \subseteq F_0$. Then it is easy to see that there exists a unique operator ψ for which the diagram

$$\begin{array}{ccc} E & \xrightarrow{\ \varphi\ } & F \\ {\scriptstyle \mathbf{pr}}\downarrow & & \downarrow{\scriptstyle \mathbf{pr}} \\ E/E_0 & \xrightarrow{\ \psi\ } & F/F_0 \end{array}$$

is commutative. In this case we shall say that ψ *is induced by the operator* φ. In particular, when $\operatorname{Ker}\varphi \supseteq E_0$, we may, by taking $F_0 = \{0\}$, talk about the operator $\psi : E/E_0 \longrightarrow F$ induced by φ. We note that, in the case when φ is surjective and $\operatorname{Ker}\varphi = E_0$, this last operator is a linear isomorphism.

The symbol $\oplus$ in expressions of the type $E \oplus F$ or $\bigoplus_{\nu\in\Lambda} E_\nu$ always means the direct sum of linear spaces and does not take into account, unless explicitly stated otherwise, the possible presence of some additional structure. If $E = F \oplus G$ is satisfied, then we say that G is a *linear complement* of F in E.

0.3°. Almost all the material we need in the theory of metric and topological spaces is set out in the textbooks [42], [41]; we make only a few refinements and additions.

First of all we shall also use the concept of a *subbase* as well as the concept of a base for a topological space (or its topology): this is a system of open sets whose finite intersections form a base. Obviously, any system of subsets of a given set may be considered as a subbase of a topology that it determines uniquely. A subbase of neighbourhoods for a fixed point of a topological space can be defined analogously: this is a system of open subsets whose finite intersections form a base (= defining system) for the neighbourhoods of this point.

A map between two topological spaces is said to be *open* if the image of every open set is open (in this way a *homeomorphism* may be defined as a bijective map that is at the same time both continuous and open). A continuous map φ is said to be *topologically injective* if its corestriction to $\operatorname{Im}\varphi$ is a homeomorphism. A continuous open surjective map is also said to be *topologically surjective*.

The Cartesian product $\bigtimes_{\nu\in\Lambda}\Omega_\nu$ of a family of topological spaces is itself provided with a standard topology. Its subbase is formed by sets indexed by all possible pairs $(\nu\in\Lambda, U)$, where U is an open set in Ω_ν and consisting of all those elements ("rows"), whose ν-coordinate belongs to U. The disjoint union $\coprod_{\nu\in\Lambda}\Omega_\nu$ is also furnished with a topology whose base is formed by all the open sets in all the Ω_ν.

As in many textbooks, we say that a topological space for which "Borel's lemma is satisfied" is *quasicompact*, that is, from any open cover one can select a finite subcover, and we say that a Hausdorff quasicompact space is *compact* (or a *compactum*). To facilitate future references we recall three classic theorems about compact spaces.

Alexandrov's Theorem. *Every continuous bijective map of a compact space onto a Hausdorff space is a homeomorphism.*

Urysohn's Theorem. *Let* Ω *be a normal space* (in particular, a compact space) *and let* Δ_1 *and* Δ_2 *be closed non-intersecting subsets. Then there are continuous functions* $g_i:\Omega\longrightarrow[0,1]$; $i=1,2$, *such that* $g_i|_{\Delta_i}\equiv 1$ *and* $g_1g_2=0$. *As a consequence, if* $g_1g_2\neq 0$ *for every pair of continuous functions* $g_1, g_2\neq 0$, *then* Ω *consists of a single point.*

Tietze–Urysohn Theorem. *Any continuous complex-valued function given on a closed subset of a normal topological space* (in particular, a compact space) *may be extended to a continuous function on the whole of that space.*

A (necessarily Hausdorff) topological space is said to be *locally compact* if each point has a neighbourhood with a compact closure. To every such space one may add one other point "∞" (in the given context this is said to be *at infinity*)[1] so that one obtains a compact space; the latter is called the *one-point, or Alexandrov, compactification of* Ω and is denoted by Ω_+. For this, the open sets in $\Omega_+ = \Omega \cup \{\infty\}$ are declared to be the sets in the topology of Ω and, besides these, the complements in Ω_+ of compact subsets Ω. (We recommend the reader to verify that Ω_+ has the required properties and explain where the local compactness of the original space is actually used.) As examples we note that, up to a homeomorphism, $\mathbb{N}_+ = \{1, 1/2, 1/3, \cdots\} \cup \{0\}$, $\mathbb{R}_+ = \mathbb{T}$, and $\mathbb{C}_+$ is the two-dimensional sphere. (In general the construction of the one-point compactification is a natural generalization of the classic construction of the extended complex plane.)

§1. REMINDERS AND ADDITIONAL MATERIAL FROM LINEAR FUNCTIONAL ANALYSIS

In choosing the material from functional analysis that we shall consider to be already known, we turned to the university syllabus covered by the textbooks [42] and [41].

§1.1. Normed, seminormed and polynormed spaces

The reader is familiar with what a norm is; a *seminorm* is a map $x \longmapsto \|x\|$, satisfying the same axioms, except that $\|x\| = 0$, generally speaking, does not imply $x = 0$. By analogy with a normed space, a space furnished with a seminorm is said to be *seminormed.* When we say "a seminormed space $(E; \|\cdot\|)$" we are defining more precisely which seminorm we are talking about; if we consider several spaces we shall often denote the seminorm in E by $\|\cdot\|_E$. The (closed) unit ball in a seminormed space is defined exactly as in a normed space although this set may now contain a non-trivial subspace; it is denoted by $\underline{S}_E$ or simply by $\underline{S}$.

Example (0.1.1). Every $f \in E^\#$ gives a seminorm $\|\cdot\|_f$ in E by the rule $\|x\|_f = |f(x)|$. If $f \neq 0$ and $\dim E > 1$, such a seminorm is never a norm. In a similar way every $x \in E$ gives the seminorm $\|\cdot\|_x$ in $E^\#$ according to the rule $\|f\|_x = |f(x)|$.

Example (0.1.2). Let E consist of some set of bounded complex functions on the set M (for example, in the case $M = \mathbb{N}$ these will be sequences). In this situation we shall often consider

---

[1] We are talking about an arbitrary element that does not belong to Ω.

the so-called *uniform norm* (or sup-norm) $\|\cdot\|_0$ in E, putting $\|x\|_0 := \sup\{|x(t)| : t \in M\}$ for $x = x(t) \in E$.

A linear space furnished with a scalar product (denoted, as a rule, by $\langle\cdot, \cdot\rangle$), is called a *pre-Hilbert* space. Recall that such a space can always be considered to have a norm $\|x\| := \sqrt{\langle x, x\rangle}$.

In certain important questions (although not in all) we have to deal with linear spaces furnished not with one but with a whole family of seminorms. Such spaces are called *polynormed spaces*[2] (and also, taking a slightly different approach, *locally convex spaces*; for the equivalence of both concepts see, for example, [41, III, §1]). We feel we must look at these in more detail.

For polynormed spaces, as also for normed spaces, what is important is not so much the specific form of the system of seminorms but the topology that it gives. If $(E; \|\cdot\|_\nu; \nu \in \Lambda)$ is our space (the meaning of such a notation is obvious), then a subbase of this topology is described by the collection of sets of the form

$$\{x \in E : \|x - x_0\|_\nu < \varepsilon\} \quad \text{for all} \quad x_0 \in E, \quad \nu \in \Lambda \quad \text{and} \quad \varepsilon > 0.$$

(In the case of a normed space we get the standard topology.) The closure (in this topology) of the set $l.h.(M)$ in E is denoted by $\overline{l.h.}(M)$.

We give two facts that are useful and easy to verify.

Proposition (0.1.3). *A given net in* E *converges to* $x \in E$ $\Longleftrightarrow$ *it converges to* x *in the seminormed space* $(E; \|\cdot\|_{\nu_0})$ *for every* $\nu_0 \in \Lambda$. ▸■

Proposition (0.1.4). E *is Hausdorff* $\Longleftrightarrow$ *for every* $x \in E\setminus\{0\}$ *there is some* $\nu \in \Lambda$ *such that* $\|x\|_\nu > 0$. ▸■

If E is a polynormed space, then a *defining system of seminorms for* E is any system that is equivalent to the original system, that is, one giving the same topology. Obviously, given a defining system of seminorms, one also has its so-called *saturation*: this is the system obtained from the original by adding in the maxima of all possible finite collections of its seminorms (instead of maxima one could alternatively take sums). Therefore we are justified, when convenient, to assume beforehand that the given system is *saturated* (= coincides with its saturation).

A Hausdorff polynormed space is *metrizable* (that is, its topology may be given by a metric) $\Longleftrightarrow$ it has an, at most,

---

[2] This is the accepted name instead of the more correct but dissonant name (even for a mathematical textbook) "polyseminormed spaces". [The translator has agreed to use this instead of "locally convex spaces" at the author's particular request.]

countable defining system of seminorms. "$\Leftarrow$" is proved by producing the metric

$$d(x, y) = \sum_{n=1}^{\infty} \frac{\|x - y\|_n}{2^n(1 + \|x - y\|_n)}$$

where $\|\cdot\|_n$ are the seminorms of the defining system. It is not difficult to prove "$\Rightarrow$" also (see, for example, [90, pp.28 and 48]).

If E/E_0 is a quotient space of the seminormed space $(E; \|\cdot\|)$, then it may be furnished with the so-called *quotient seminorm* $\|\dot{x}\|^\wedge := \inf\{\|x\| : x \in \dot{x}\}$. It is easy to verify that the latter is a norm (and we then call it the "*quotient norm*") $\Leftrightarrow$ E_0 is closed in E.

Now let E_0 be a subspace of the polynormed space E. Then E_0 and E/E_0 may also be considered as polynormed spaces: the first — with respect to the restrictions of the original seminorms, and the second - with respect to the system of quotient seminorms taken from all the seminorms in the saturation of the original system.[3] It is easy to show that *E/E_0 is Hausdorff $\Leftrightarrow$ E_0 is closed in E.*

Finally, the *Cartesian product* of the family $(E_\mu; \mu \in \Xi)$ of polynormed spaces is also a polynormed space. Namely, if $\{\|\cdot\|_\nu; \nu \in \Lambda_\mu\}$ is the system of seminorms in E_μ, then one introduces a system of seminorms $\|\cdot\|_\nu^\mu$ in $\bigtimes_\mu E_\mu$ according to the rule $\|\overline{x}\|_\nu^\mu = \|x_\mu\|_\nu$, where x_μ is the μ-th coordinate of the "row" $\overline{x}$ (or, if convenient, one can introduce the saturation of this system). It is easy to verify that the polynormed space topology of $\bigtimes_\mu E_\mu$ coincides with the topology of the Cartesian product of the corresponding topological spaces (see §0.3°).

A *Cauchy net* in a seminormed space is defined in exactly the same way as in a normed space. In a polynormed space $(E; \|\cdot\|_\nu; \nu \in \Lambda)$ a net is said to be a *Cauchy net* if, for each $\nu \in \Lambda$, it is a Cauchy net in the seminormed space $(E; \|\cdot\|_\nu)$. A polynormed space is said to be *complete* if every Cauchy net converges. (If it is metrizable, then for this it is sufficient that all Cauchy sequences converge; in general such cases are rare; see, for example, [94, p.480].) Recall that if we have completeness, then a normed space is said to be a *Banach* space, a pre-Hilbert space is said to be a *Hilbert* space and a metrizable polynormed space is said to be a *Fréchet space.* A subspace of a complete space is complete $\Leftrightarrow$ it is closed. A quotient space

---

[3] The preliminary transfer to the saturation is necessary: the system of quotient seminorms of a non-saturated (original) system is, generally speaking, not equivalent to the system of quotient seminorms of its saturation.

of a Banach space with respect to a closed subspace is itself Banach; the same is also true for Fréchet spaces.

Every polynormed space E is a dense subspace of a larger complete polynormed space, called the *completion of E*. For a general proof of the existence of the completion and its uniqueness (in a well-defined sense) for Hausdorff spaces see, for example, [65, Ch.VI].

* * *

A few words about examples. We assume that the reader is acquainted with the standard examples of Banach spaces and, in particular, Hilbert spaces (all the same, many of them will, however, be repeated in Chapters I and II — as Banach algebras). With regard to more general polynormed spaces, it is appropriate to give a few instructive examples now. We shall begin with the following class which plays much the same role among polynormed spaces as the class of discrete spaces plays among topological spaces.

Definition (0.1.5). A linear space E furnished with the system of all seminorms (without exception) that exist on it is called the *space with strongest seminorm system* (constructed on E).

It is very useful to remember these strongest spaces and to "try out" concepts of the general theory on them. Here is a first illustration (there will be a lot of them).

Exercise (0.1.6). Show that a sequence of elements in a space with strongest seminorm system converges $\Longleftrightarrow$ it is totally contained by a certain finite-dimensional subspace, and there converges in the sense of the standard convergence in the finite-dimensional space (coordinate-wise, in any base).

Now let us describe one of the main methods of giving a "polynormed structure" to a function space.

Let Ω be a topological space, E a linear space consisting of (not necessarily all) continuous functions on Ω. For any $f \in E$ and any compact subset $K \subset \Omega$ we put $\|f\|_K := \|f|_K\|_0$ (see Example 1.2). The system of seminorms $\|\cdot\|_K$, taken over all compact subsets, and the topology which it gives on E, is said to be *compact-open*. Obviously convergence in this topology is uniform convergence on all compact subsets.

Example (0.1.7). The space $\mathbb{C}^M$, where M is an arbitrary set, is polynormed with respect to the system of seminorms $\|f\|_K := \|f|_K\|_0$, where K runs over all finite subsets in M. Obviously this system is a particular case of a compact-open topology (when we suppose M to have the discrete topology) and at the same time it is a particular case of the standard saturated system in $\mathop{\times}\limits_{\mu} E_\mu$ (when $E_\mu = \mathbb{C}$ for all μ). Convergence in $\mathbb{C}^M$ is pointwise (simple) convergence.

Example (0.1.8). If U is a domain in $\mathbb{C}^n$, then the set $\mathcal{O}(U)$ of holomorphic or analytic functions in U (see, for example, [37]) is a polynormed space with pointwise operations and the compact-open system of seminorms. This is a Fréchet space; metrizability is implied by the existence of a countable covering of U by compact sets and the assertion that it is complete is precisely, when $n = 1$, the classic Weierstrass theorem.

§1.2. Basic principles of operator theory

The reader should be well acquainted with the concepts of a bounded operator and an operator norm in the context of normed spaces and also with the basic fact that continuity of an operator is equivalent to its being bounded. All this can be transferred word for word from normed spaces to seminormed spaces, only now, of course, it is necessary to talk about the "*operator seminorm*"; the latter is denoted by $\|\cdot\|_0$ [4] or simply $\|\cdot\|$. The *multiplicative inequality* $\|\varphi \circ \psi\|_0 \le \|\varphi\|_0 \|\psi\|_0$ also holds in the "seminormed" case.

When going from seminormed spaces to polynormed spaces the criterion for an operator to be continuous has the following natural generalization:

Proposition (0.1.9). *Let* $(E; \|\cdot\|_\mu, \mu \in \Lambda)$ *and* $(F; \|\cdot\|_\nu, \nu \in \Lambda')$ *be polynormed spaces, where the first system of seminorms is saturated. Then the operator* $\varphi : E \longrightarrow F$ *is continuous* $\Longleftrightarrow$ *for any* $\nu_0 \in \Lambda'$ *there exists* $\mu_0 \in \Lambda$ *such that* φ *is bounded as an operator between the seminormed spaces* $(E; \|\cdot\|_{\mu_0})$ *and* $(F; \|\cdot\|_{\nu_0})$.

In particular, the functional $f : E \longrightarrow \mathbb{C}$ *is continuous* $\Longleftrightarrow$ *there exists* $\mu_0 \in \Lambda$ *such that* f *is bounded as a functional on* $(E; \|\cdot\|_{\mu_0})$. ▸■

The property of an operator being open (see §0.3°) can also be formulated without difficulty in terms of seminorms:

Proposition (0.1.10). (I) *An operator* $\varphi : E \longrightarrow F$ *between seminormed spaces is open* $\Longleftrightarrow$ *there exists* $C > 0$ *(called the reversibility constant) such that for any* $y \in F$ *there exists* $x \in E$ *with* $\varphi(x) = y$ *and* $\|x\|_E \le C\|y\|_F$.

(II) *If* $(E; \|\cdot\|_\mu, \mu \in \Lambda)$ *and* $(F; \|\cdot\|_\nu, \nu \in \Lambda')$ *are polynormed spaces where the second system of seminorms is saturated, then the operator* $\varphi : E \longrightarrow F$ *is open* $\Longleftrightarrow$ *for any* $\mu_0 \in \Lambda$ *there exists*

---

[4] As also is the uniform norm in §1; but no ambiguity arises here.

$\nu_0 \in \Lambda'$ such that φ is open as an operator between $(E; \|\cdot\|_{\mu_0})$ and $(F; \|\cdot\|_{\nu_0})$.

A continuous operator between polynormed spaces is called a *topological isomorphism* (as for normed spaces) if it has a continuous inverse operator; this is equivalent to its being at the same time topologically injective and topologically surjective. In the case of seminormed spaces an operator which preserves the seminorm ($\|\varphi(x)\| \equiv \|x\|$) is called an *isometric operator* and if, in addition, it is bijective, it is called an *isometric isomorphism. Topologically isomorphic* polynormed spaces (*isometrically isomorphic* seminormed spaces, respectively) are those between which there exists a topological (isometric) isomorphism.

Any operator induced by a continuous operator between polynormed spaces (see §0.2°) is itself, obviously, continuous. Moreover (a useful observation!), *if $\varphi : E \longrightarrow F$ is topologically surjective, then it generates a topological isomorphism $\psi : E/\mathrm{Ker}\,\varphi \longrightarrow F$.*

If E and F are polynormed spaces, then the subspace of $\mathcal{L}(E, F)$ consisting of the continuous operators, is denoted by $\mathcal{B}(E, F)$. We shall write $\mathcal{B}(E)$ instead of $\mathcal{B}(E, E)$ and E^* instead of $\mathcal{B}(E, \mathbb{C})$; the latter is called the *dual* of E. We note that in the case of a space E with the strongest seminorm system (see Definition 1.5) $\mathcal{B}(E, F) = \mathcal{L}(E, F)$ for any F, which also means $E^* = E^\#$.

When E and F are seminormed spaces we consider, similarly, $\mathcal{B}(E, F)$ with the operator seminorm; the resulting topology is said to be *uniform*. Moreover $\mathcal{B}(E, F)$ is Hausdorff and/or complete $\Longleftrightarrow$ F is Hausdorff and/or complete; in particular, E^* is always a Banach space in which the norm is often denoted by $\|\cdot\|_E^*$ or $\|\cdot\|^*$. Banach spaces which are dual to a seminormed space (and hence also to a Banach space — show it!) and all spaces isometrically isomorphic to these are called *dual spaces*.

Given an arbitrary polynormed space E we can introduce a system of seminorms in E^* by taking all possible $\|\cdot\|_x$; $x \in E$ as in §1.1. This system, and also its corresponding topology, is called the *weak* topology*.[5] To facilitate future reference we cite the following

Proposition (0.1.11). *The space E^* is Hausdorff and for each point f in E^* the sets $U_{f,x,\varepsilon} = \{g \in E^* ; |g(x) - f(x)| < \varepsilon\}$, where $x \in E$, $\varepsilon > 0$, constitute a subbase for the neighbourhoods of f.* ▶■

---

[5] The star in this traditional terminology emphasizes the fact that our structures are given in the dual space but are based on the initial space.

Thus, we can introduce the uniform topology in $\mathcal{B}(E, F)$ if both spaces are seminormed and we can introduce the weak* topology if $F = \mathbb{C}$. Since dual spaces are "in the intersection" they admit both topologies. Moreover a "weak* open" set is of course "uniformly open" but the converse is only true for finite-dimensional spaces (explain why). There is a deeper interrelation between these two structures that is expressed by a theorem which turns out to be very useful for us.

Banach-Alaoglu Theorem. *Let E be a seminormed space. Then the unit ball in $(E^*; \|\cdot\|^*)$ is compact with respect to the weak* topology.*

For the proof see, for example, [67. 3.15]. For comparison let us recall the well-known theorem of Riesz: *in any normed space the unit ball is compact in the norm topology $\Leftrightarrow$ this space is finite-dimensional.*

* * *

Our next aim is to recall the three classic theorems all connected with the name of Banach and often called the three basic principles or "three whales",[6] of linear functional analysis. We set out these theorems and also some of their alternative versions and corollaries in the generality we need for future reference.

To formulate the first result we need to consider the scalar field to be, not the complex numbers as usual, but the reals. Recall that a function $p : E \longrightarrow \mathbb{R}^{[+]}$ on a real linear space E is said to be *convex* if $p(x + y) \le p(x) + p(y)$ and $p(tx) = tp(x)$ when $x, y \in E$, $t \ge 0$.

Hahn-Banach Theorem ("extension principle"). *Let E be a real linear space, p a convex function on E, E_0 a subspace of E and let f_0 be a (real) functional, given on E_0, such that $f_0(x) \le p(x)$ for all $x \in E_0$. Then there exists a functional f on the whole of E that extends f_0 and is such that $f(x) \le p(x)$ for all $x \in E$.*

We need a corollary of this theorem (in fact the assertion is equivalent to it) which is essentially of a geometrical character. Recall that a set M in a linear space E (over $\mathbb{C}$ or $\mathbb{R}$) is said to be *absorbing* if, for every $x \in E$, $tx \in M$ for all $t \in \mathbb{R}$ of sufficiently small modulus. A point y of the set M in E is called a *linearly interior* point in M if $M - y := \{x - y : x \in M\}$ is an absorbing set.

Theorem (0.1.12). *Let E be a real linear space, let M be a convex subset, let E_0 be a subspace of E and let f_0 be a*

---

[6] [This refers to an ancient description of the world and its supports.]

(real) functional on E_0 *such that* $f_0(x) \geq 0$ *when* $x \in M \cap E_0$. *Further, let* M *have a linearly interior point* y *belonging to* E_0. *Then there exists a functional* f *on* E *extending* f_0 *and such that* $f(x) \geq 0$ *when* $x \in M$.

(Thus, if $M \cap E_0$ lies on one side of a given hyperplane in E_0, then, when the condition on y is satisfied, the whole of M lies "on the same side" of some hyperplane in E containing the given hyperplane in E_0.)

▶ If we assume that $f_0 \neq 0$ (the assertion would be trivial otherwise) and use the condition on y, we easily see that $\theta := f_0(y) > 0$; put $g_0 := -\theta^{-1} f_0$. Now consider the convex absorbing set $M_1 = M - y$ and the function $p : E \longrightarrow \mathbb{R}^{(+)} : x \longmapsto \inf\{t > 0 : t^{-1}x \in M_1\}$ ("the Minkowski gauge function of the set M_1"); it is easy to verify that p is convex and $g_0(x) \leq p(x)$ when $x \in E_0$. Then by the Hahn-Banach theorem, there exists a functional g on E extending g_0 and also satifying $g(x) \leq p(x)$. All that remains is to put $f := -\theta g$ and check the required properties. ■

We select also some corollaries of the Hahn-Banach theorem which are perhaps closer to a university course. From this point on we again formulate results only for complex linear spaces.

Corollary A. *Any bounded functional given on a subspace of a seminormed space may be extended to the whole space without changing the norm.*

In turn we deduce from Corollary A:

Corollary B. *If* E *is a seminormed space and* $x \in E$ *is such that* $\|x\| > 0$, *then there exists* $f \in E^*$ *such that* $f(x) = \|x\|$ *and* $\|f\|^* = 1$.

Corollary C. *If* E *is a polynormed space,* E_0 *is a closed subspace and* $x \in E \backslash E_0$, *then there exists* $f \in E^*$ *such that* $f|_{E_0} = 0$ *but* $f(x) > 0$. *In particular, if* E *is Hausdorff, then for any* $x \in E \backslash \{0\}$ *there exists* $f \in E^*$ *such that* $f(x) > 0$ (one says "there are sufficiently many continuous functionals on a Hausdorff polynormed space").

The following result is often called the principle of uniform boundedness.

Banach-Steinhaus Theorem. *Let* E *be a Banach space, and* F *a seminormed space and let* M *be a certain set in* $\mathcal{B}(E, F)$. *Then, if, for every* $x \in E$, $\sup\{\|\varphi(x)\| : \varphi \in M\} < \infty$, *then also* $\sup\{\|\varphi\|_0 : \varphi \in M\} < \infty$.

We note that this theorem, after appropriate rethinking, may be considerably generalized and, in particular, carried over to the case when E is a Fréchet space and F an arbitrary polynormed space; see, for example, [90, III, Theorem 4.2].

The third classic theorem ("the open mapping principle") may be formulated in various ways, for example, thus:

Banach Theorem. *Every continuous bijective operator between two Banach spaces is a topological isomorphism.*

It is not difficult to see that this theorem is equivalent to several other assertions, in particular, the following:

— *any injective continuous operator between Banach spaces which has a closed image is topologically injective.*

— *any surjective continuous operator between Banach spaces is open.*

When we refer later in the text to the Banach theorem we shall have in mind whichever of the three assertions we have just formulated is appropriate. In certain cases we need a further equivalent formulation of these theorems.

Closed Graph Theorem. *Let* $\varphi : E \longrightarrow F$ *be an operator between Banach spaces with the following property: if a sequence* $x_n \in E$ *tends to zero and* $\varphi(x_n)$ *tends to some* $y \in F$, *then* $y = 0$ (in other words, the graph of this operator, that is, the set of pairs $(x, \varphi(x))$; $x \in E$, is closed in $E \times F$). *Then* φ *is continuous.*

All four theorems can be carried across word for word from Banach spaces to Fréchet spaces.

We need also a theorem discovered much later than the three "principles" formulated above. However, it has already long proved itself one of the most powerful tools in modern analysis.

Let M be a set in a linear space. Recall that $x \in M$ is called an *extreme point* of this set if it lies in the interior of no line segment with end-points in M.

Krein-Milman Theorem. *Every compact convex set in a polynormed space coincides with the closed convex hull of its extreme points.*

§1.3. Certain classes of operators on Banach and Hilbert spaces

The reader should be acquainted with the concept of a compact, or completely continuous, operator between Banach spaces E and F, and with the fact that every finite-dimensional (= having a finite-dimensional image) continuous operator is compact. There is another class of very interesting operators which is intermediate between the classes of finite-dimensional and compact operators. Namely, an operator $\varphi : E \longrightarrow F$ is said to be *nuclear* if it is the sum of an absolutely convergent series of one-dimensional operators in $\mathcal{B}(E, F)$. The compact operators form a closed subspace in $\mathcal{B}(E, F)$ denoted by $\mathcal{K}(E, F)$, but the nuclear operators form a subspace denoted by $\mathcal{N}(E, F)$ which is not, in general, a closed subspace of $\mathcal{K}(E, F)$ (we write $\mathcal{K}(E)$ or $\mathcal{N}$(E) when $E = F$). However, $\mathcal{N}(E, F)$ is a Banach space with respect to another norm $\|\varphi\|_{\mathcal{N}} := \inf \sum_{n=1}^{\infty} \|\varphi_n\|_0$, where the lower bound is taken over all possible representations of φ as an absolutely convergent series of one-dimensional operators φ_n. Obviously $\|\cdot\|_{\mathcal{N}} \geq \|\cdot\|_0$. The operator $\varphi : E \longrightarrow F$ transmits the property of being nuclear (as also that of being compact, finite-dimensional or one-dimensional) on going over to the dual (or adjoint) operator $\varphi^* : F^* \longrightarrow E^*$.

* * *

It is well known that throughout the theory of operators the richest and most developed applications are to the theory of operators on Hilbert spaces. The basic reason is in the much bigger role played by the dual operator. Let us recall that Riesz' theorem on the general form of a continuous functional on a Hilbert space implies that for a continuous operator $S : H_1 \longrightarrow H_2$ between Hilbert spaces one can identify its dual $S^* : H_2^* \longrightarrow H_1^*$ with an operator from H_2 to H_1 (without the stars!) which is defined uniquely by $\langle Sx, y \rangle = \langle x, S^*y \rangle$ and denoted by the same symbol. In particular, making this identification, $\mathcal{B}(H)$ turns out to be closed under the operation $S \longmapsto S^*$. This provides an important additional structure on $\mathcal{B}(H)$ (which is missing in $\mathcal{B}(E)$ if E is not Hilbert), to which we shall pay much attention later on.

The "star" operation is related to two other structures defined in $\mathcal{B}(H)$ — composition and the norm — by some extremely important identities.

Theorem (0.1.13). *For any* $S \in \mathcal{B}(H)$ *we have*

$$\|S^*S\| = \|S\|^2. \tag{1}$$

Consequently, if S *is self-adjoint* (= equal to its dual) *then* $\|S^2\| = \|S\|^2$.

▶It follows from $\|Sx\|^2 = \langle Sx, Sx \rangle = \langle x, S^*Sx \rangle \leq \|S^*Sx\| \|x\|$; where $x \in H$, that $\|S\|^2 \leq \|S^*S\|$. The opposite inequality arises from $\|S^*\| = \|S\|$ (which the reader should know) and the multiplicative inequality for norms (c.f. beginning of §1.2). ■

The identity we have just proved is called the *C^*-property* (for operators in H). Later on (Chapter IV.§7) we shall see that this property should properly be put in an abstract form leading to one of the most important ideas in this book — a so-called C^*-algebra.

By virtue of the fact that the norm in Hilbert spaces can be characterized by the parallelogram rule, an *operator* $U : H_1 \longrightarrow H_2$ between such spaces is *isometric* $\Longleftrightarrow$ $\langle Ux, Uy \rangle = \langle x, y \rangle$ for all $x, y \in H_1$. Hence an isometric isomorphism is precisely a bijective operator preserving the scalar product. Such an operator is said to be *unitary*; it is easy to show that its definition is equivalent to the condition $U^* = U^{-1}$.

Two operators $R : H_1 \longrightarrow H_1$ and $S : H_2 \longrightarrow H_2$ acting on Hilbert spaces are said to be *unitarily equivalent* if there exists a unitary operator U such that the diagram

$$\begin{array}{ccc} H_1 & \xrightarrow{\ R\ } & H_1 \\ {\scriptstyle U}\big\downarrow & & \big\downarrow{\scriptstyle U} \\ H_2 & \xrightarrow{\ S\ } & H_2 \end{array}$$

is commutative; in other words, $S = URU^{-1}$. The problem of classifying operators up to unitary equivalence is one of the main types of problem in the "Hilbert" theory of operators. It has been completely solved for the class of *normal* operators (= operators which commute with their duals); this class, of course, includes all self-adjoint and all unitary operators. We need only the part of the relevant result that consists of the following theorem.

Theorem (on the representation of self-adjoint operators). *Let S be a self-adjoint operator in an (arbitrary) Hilbert space. Then there exist a measure space (X, μ) and a bounded measurable function $f : X \longrightarrow \mathbb{R}$ such that S is unitarily equivalent to the operation of multiplication by f in $L^2(X, \mu)$* (that is, the operator $g \longmapsto fg$; where $g \in L^2(X, \mu)$).

We note that the complete classification of normal operators up to unitary equivalence may be deduced from the Hilbert spectral theorem which completely describes the structure of normal operators and is one of the main results of the whole theory of operators. Later on we shall show in what way the spectral theorem may be obtained by methods in the theory of Banach algebras (see Theorem IV.7.51(I) later).

We give another two, simpler, facts which prove useful later on.

Theorem ("criterion for operators to be positive"). *The following properties of an operator $S \in \mathcal{B}(H)$ are equivalent:*

(I) *S is self-adjoint and for any $\lambda \in \mathbb{C}\backslash\mathbb{R}^{[+]}$ the operator $S - \lambda 1$ is invertible* (= *there exists $R \in \mathcal{B}(H)$ such that $R(S - \lambda 1) = (S - \lambda 1)R = 1$*);

(II) *$\langle Sx, x\rangle \geq 0$ for any $x \in H$.*

For the proof see, for example, [67, Theorem 12.32].

Now let $\check{H}$ be a pre-Hilbert space and let H be its completion. The map $\mathcal{R} : \check{H} \times \check{H} \longrightarrow \mathbb{C}$ is called a *sesquilinear functional on* $\check{H}$ if $\mathcal{R}(x + y, z) = \mathcal{R}(x, z) + \mathcal{R}(y, z)$;

$\mathcal{R}(\lambda x, y) = \lambda\mathcal{R}(x, y)$, $\mathcal{R}(x, y + z) = \mathcal{R}(x, y) + \mathcal{R}(x, z)$ and $\mathcal{R}(x, \lambda y) = \bar{\lambda}\mathcal{R}(x, y)$ for any $x, y, z \in \check{H}$, $\lambda \in \mathbb{C}$.

Theorem. *Let* $\mathcal{R}$ *be a sesquilinear functional on* $\check{H}$ *which is bounded in the sense that there exists* $\|\mathcal{R}\| := \sup\{|\mathcal{R}(x, y)| : x, y \in \underline{S}_{\check{H}}\} < \infty$. *Then there exists a unique bounded operator* R *in* H *such that* $\langle Rx, y\rangle = \mathcal{R}(x, y)$ *for all* $x, y \in \check{H}$. *Moreover* $\|R\| = \|\mathcal{R}\|$.

(Obviously $\mathcal{R}$ has a unique continuous extension to a sesquilinear functional on the whole of H. The next step in the argument is based on Riesz' theorem and is essentially no different from the process of constructing the dual operator; see, for example, [67, Theorem 12.8].)

§1.4. Banach complements and examples of non-complemented subspaces

The material for this section goes beyond the university textbooks we have cited and we present it in greater detail.

Let E be a polynormed space, E_0 be a subspace and let E_1 be a linear complement of E_0; then, obviously, the operator $\kappa : E_0 \times E_1 \longrightarrow E : (x, y) \longmapsto x + y$ is bijective and continuous. Furthermore, if κ is a topological isomorphism then E_1 is called a *topological complement* to E_0 (in E); in the case of a Banach space E we shall, in accordance with tradition, call it a "Banach complement". A subspace of a polynormed space which has a topological complement, is said to be *complemented*. Obviously, a complemented subspace is always closed, as is also its topological complement.

Being complemented may be expressed in several different ways (as will be clear in §2, essentially in terms of category theory).

Proposition (0.1.14). *The following properties of the subspace* E_0 *of a polynormed space* E *are equivalent:*

(I) E_0 *is complemented;*

(II) *there exists a continuous operator* $\sigma : E \longrightarrow E_0$ *such that* $\sigma \circ \mathbf{in}$ *is the identity on* E_0, *where* $\mathbf{in} : E_0 \longrightarrow E$ *is the natural embedding;*

(III) *there exists a continuous operator* $\rho : E/E_0 \longrightarrow E$ *such that* $\mathbf{pr} \circ \rho$ *is the identity on* E/E_0, *where* $\mathbf{pr} : E \longrightarrow E/E_0$ *is the natural projection.*

▶ (I) ⇒ (II), (III). We consider the commutative diagram

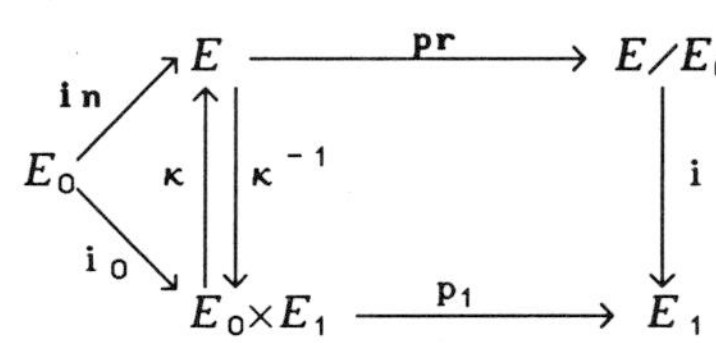

in which E_1 is a topological complement to E_0, $i_0 : x \longmapsto (x, 0)$, $p_1 : (x, y) \longmapsto y$ and the operator i is induced by the operator $p_1 \circ \kappa^{-1}$ (obviously i is bijective and continuous). We put $p_0 : E_0 \times E_1 \longrightarrow E_0 : (x, y) \longmapsto x$ and $\sigma := p_0 \circ \kappa^{-1}$, then σ is continuous and $\sigma \circ \mathbf{in} = p_0 \circ i_0$ is the identity on E_0. We also put $i_1 : E_1 \longrightarrow E_0 \times E_1 \ : y \longmapsto (0, y)$ and $\rho := \kappa \circ i_1 \circ i$, then ρ is continuous and $\mathbf{pr} \circ \rho = i^{-1} \circ (i \circ \mathbf{pr}) \circ \rho$ $= i^{-1} \circ p_1 \circ (\kappa^{-1} \circ \rho)$ $= i^{-1} \circ (p_1 \circ i_1) \circ i$ $= i^{-1} \circ i$ is the identity on E/E_0.

(II) $\Rightarrow$ (I). Obviously $\operatorname{Ker} \sigma$ is a linear complement to E_0 and the operator inverse to $\kappa : E_0 \times (\operatorname{Ker} \sigma) \longrightarrow E$ is given by $z \longmapsto (\sigma(z), z - \sigma(z))$. Therefore κ^{-1} is continuous because σ is.

(III) $\Rightarrow$ (I). Obviously $\operatorname{Im} \rho$ is a linear complement to E_0 and the operator inverse to $\kappa : E_0 \times (\operatorname{Im} \rho) \longrightarrow E$, is given by $z \longmapsto (z - \rho \circ \mathbf{pr}(z), \rho \circ \mathbf{pr}(z))$. Therefore κ^{-1} is continuous because $\mathbf{pr}$ and ρ are. ∎

Exercise (0.1.15). Show that E_0 being complemented is also equivalent to the existence of an operator $P \in \mathcal{B}(E)$ such that $P^2 = P$ and $\operatorname{Im} P = E_0$.

An important characteristic feature (sometimes reckoned among the so-called pathologies) of the geometry of Banach and polynormed spaces is the existence and even the abundance of non-complemented closed subspaces. Since this feature has a notable effect on several definitions and constructions later on, we feel that it is necessary to examine thoroughly at least one example of this kind.

We let c_b denote the Banach space of all bounded complex sequences furnished with the sup-norm (see 1.2) and let c_0 be the closed subspace consisting of sequences converging to zero. As a rule, we denote the terms of the sequence ξ by $(\xi_n;\ n \in \mathbb{N})$ and let $\mathbf{p}^k$ denote the sequence with 1 in the k-th place and zeros elsewhere.

For any $M \subseteq \mathbb{N}$ we put $c_b(M) := \{\xi \in c_b\ ;\ \xi_n = 0$ when $n \notin M\}$ and for $f \in c_b^*$ we put for brevity $f_M := f|_{c_b(M)}$. We note that if the subsets $M_k \subset \mathbb{N}$; $k = 1, 2, \cdots$ are mutually disjoint, then the definition of the norms in c_b and c_b^* obviously imply that $\sum_{k=1}^{\infty} \| f_{M_k} \|^*_{c_b(M_k)} \leq \| f \|^*$.

Theorem (0.1.16) (Phillips). *The subspace* c_0 *is not complemented in* c_b.

▸ We assume the contrary; then by Proposition 1.14(II) there exists a continuous operator $\sigma : c_b \longrightarrow c_0$ such that $\sigma(\xi) = \xi$ for all $\xi \in c_0$. We denote by $f^{(k)}$ the functional on c_b formed by taking, for each ξ, the k-th term of the sequence $\sigma(\xi) \in c_0$; it is obviously continuous. We note that $f^{(k)}(\mathbf{p}^l)$ is one when $k = l$ and zero when $k \neq l$.

By induction we construct natural numbers $n_0 < n_1 < \cdots$ and infinite subsets $M^{(0)} \supset M^{(1)} \supset \cdots$ embedded in $\mathbb{N}$. We put $n_0 := 1$ and $M^{(0)} := \mathbb{N}$; let $n_0, \cdots, n_k$ and $M^{(0)}, \cdots, M^{(k)}$ be already constructed. We consider an arbitrary partitioning of $M^{(k)}$ into non-intersecting infinite sets $\{M_m^{(k+1)};\ m \in \mathbb{N}\}$, and the associated functionals $f^{(n_k)}$ and $\{f^{(n_k)}_{M_m^{(k+1)}};\ m \in \mathbb{N}\}$. Since the series made up from the norms of the latter functionals converges (see above), then there exists m such that $\| f^{(n_k)}_{M_m^{(k+1)}} \|^* < \frac{1}{2}$ and $n_0, \cdots, n_k \notin M_m^{(k+1)}$. We put $M^{(k+1)}$ equal to $M_m^{(k+1)}$ and we take n_{k+1} to be any number in $M^{(k+1)}$ greater than n_k.

We now let $\zeta^{(k)}$; $k = 0, 1, \cdots$ be the sequence with one at places numbered $n_k, n_{k+1}, \cdots$ and zero at the remaining places and put $\zeta := \zeta^{(0)}$. Obviously $\zeta = \mathbf{p}^{n_0} + \cdots + \mathbf{p}^{n_k} + \zeta^{(k+1)}$, $\zeta^{(k+1)} \in c_b(M^{(k+1)})$ and $\| \zeta^{(k+1)} \| = 1$ (for all k). Further, $f^{(n_k)}(\mathbf{p}^{n_0}) = \cdots = f^{(n_k)}(\mathbf{p}^{n_{k-1}}) = 0$ and $f^{(n_k)}(\mathbf{p}^{n_k}) = 1$; hence by the choice of $M^{(k+1)}$, for any k, $| f^{(n_k)}(\zeta) | = | 1 + f^{(n_k)}(\zeta^{(k+1)}) | > \frac{1}{2}$. But in fact $f^{(n_k)}(\zeta)$ is the n_k-th term of the sequence $\sigma(\zeta) \in c_0$ and therefore these numbers tend to zero when $k \longrightarrow \infty$. A contradiction. ∎

See [70, §17.7] for some generalizations of Phillip's theorem.

The existence of non-complemented closed subspaces was discovered by Banach and Mazur, essentially by showing that all subspaces in $C[0, 1]$, isometrically isomorphic to l_1 or $L^1[0, 1]$, are non-complemented. This and other examples are mentioned in [70, p.374] and [208, p.256], where the reader will find the necessary references. Another important example pointed out by Newman and Rudin is the subspace of $C(\mathbb{T})$ consisting of the boundary values of analytic functions (our $\mathcal{A}(\mathbb{T})$ later on; see Chapter III §1.4). The proof uses the theory of Fourier series and is not at all the same as the Cantor diagonal method, which is what in fact was used in Theorem 1.16; see [23, p.149–155] and [67, p.125–130].

In conclusion we give a beautiful and conclusive result which says that the feature we are considering crops up everywhere.

Assertion (0.1.17). *Let E be a Fréchet space, in which every closed subspace is complemented. Then E is topologically isomorphic to either* (I) *a Hilbert space or* (II) $\mathbb{C}^{\mathbb{N}}$ (see Example 1.7) *or* (III) $\mathbb{C}^{\mathbb{N}} \times H$, *where H is an infinite-dimensional*

Hilbert space. As a consequence, every Banach space with the above property is topologically isomorphic to a Hilbert space.

The "Banach" part of Assertion 1.17 is due to Lindenstrauss and Tzafriri [191]; the rest was obtained by combining with an earlier result of Grothendieck [26] (see also [208, §9]).

§2. A MINIMUM OF CATEGORY THEORY. CERTAIN CATEGORIES OF FUNCTIONAL ANALYSIS AND FUNCTORS CONNECTED WITH THEM

As certain basic concepts in the language of category theory have now penetrated (at last!) university functional analysis textbooks [41], we assume we are right in supposing that at least these rudiments are known. It is necessary to comment on the remaining essential information. All the material in this section is covered by textbooks such as [11], [51]; with the exception of our examples taken from linear functional analysis which are directly related to the subject of this book. We recommend the monographs [70], [88] to a reader wishing for more details on the categorical aspect of the theory of Banach spaces.

§2.1. Categories and certain types of morphisms. Examples

First of all we assume the concept of a *category* itself is known as well as everything included in its definition — *objects*, *morphisms* and so on and also the concept of the dual category. We let $\mathbf{h}_{\mathcal{K}}(X, Y)$ denote the set of morphisms between the objects X and Y in the category $\mathcal{K}$. We write the morphisms themselves, in the generally accepted form $\varphi : X \longrightarrow Y$ or $X \xrightarrow{\varphi} Y$ (as for maps), the composition of morphisms is written in the form $\varphi \circ \psi$ (again as for maps), the local identity on X is written as 1_X (like an identity map); the noted similarity of notation does not cause confusion. The category dual to $\mathcal{K}$ is denoted by $\mathcal{K}^{op}$. A *diagram* in $\mathcal{K}$ and its *commutativity* are defined word for word as in 0.1° for a diagram of sets and its commutativity — replacing the word "set" by "object" and the word "map" by "morphism".

Recall that in specific examples of categories the objects are nearly always (but not always) sets with some structure and morphisms are maps of sets, in some way compatible with this structure. This is true for very many "working" categories in our account, from these we pick out various categories of polynormed spaces. The narrowest (and of course the most important) of them is the category of Banach spaces $\mathcal{B}an$, while the widest is the category of all polynormed spaces $\mathcal{PN}$; among the intermediate ones we note the category of Fréchet space $\mathcal{F}r$ and the category of complete Hausdorff polynormed spaces $\mathcal{CHPN}$. (We have indicated the objects of these categories; their morphisms are all the continuous operators.) In addition we need the category of linear spaces (and operators) $\mathcal{L}in$, the category of compact topological spaces (and continuous maps) $\mathcal{CT}op$ and finally the category of all sets (and all maps) $\mathcal{S}ets$.

A category $\mathcal{K}_1$ is called a *subcategory* of the category $\mathcal{K}_2$ if the objects and morphisms in $\mathcal{K}_1$ belong to $\mathcal{K}_2$. The subcategory is said to be *complete* if $\mathbf{h}_{\mathcal{K}_1}(X, Y) = \mathbf{h}_{\mathcal{K}_2}(X, Y)$ for any of its objects.

A morphism φ in the category $\mathcal{K}$ is called a *monomorphism* if $\varphi \circ \psi_1 = \varphi \circ \psi_2$ in $\mathcal{K}$ always implies $\psi_1 = \psi_2$ ("φ can be cancelled on the left"). Analogously, a morphism, "which can be cancelled on the right" is called an *epimorphism.* Here is an exercise for the better understanding of these concepts.

Exercise (0.2.1). Describe in set-theoretic terms the monomorphisms and epimorphisms in all the specific categories indicated.

Answer. The monomorphisms are everywhere the injective maps. The epimorphisms are the maps with dense image in $\mathcal{B}an$, $\mathcal{F}r$, $\mathcal{CHPN}$ and $\mathcal{CT}op$ and the surjective maps in the simpler categories (including $\mathcal{PN}$).

A morphism $\varphi : X \longrightarrow Y$ is called a *retraction* if it has a right inverse morphism (in the given category), that is, ψ such that $\varphi \circ \psi = 1_Y$. A *coretraction* is defined similarly in terms of a *left inverse* morphism (ψ with $\psi \circ \varphi = 1_X$). X is called a *retract* of Y if there exists a retraction $\varphi : Y \longrightarrow X$ or, what is equivalent, a coretraction $\psi : X \longrightarrow Y$.

If φ has both a right and left inverse morphism, then these coincide (show this!), and their common value is called the *inverse* morphism to φ and is denoted by φ^{-1}. A morphism having an inverse is called an *isomorphism* and objects between which an isomorphism exists are *isomorphic.* (Isomorphic objects can be thought of as "effectively the same" if you like, being different aspects of the same object.) It is obvious that in our categories of polynormed spaces an isomorphism is precisely a topological isomorphism, in $\mathcal{L}in$ it is a linear isomorphism, in $\mathcal{CT}op$ it is a homeomorphism, and in $\mathcal{S}ets$ it is simply a bijection.

Proposition (0.2.2). (I) *Every coretraction is a monomorphism and every retraction an epimorphism.*

(II) *An isomorphism is precisely a coretraction which is an epimorphism, and a retraction which is a monomorphism.* ▶■

Exercise (0.2.3). Show that in $\mathcal{S}ets$ and $\mathcal{L}in$ the concepts of a monomorphism and a coretraction, an epimorphism and a retraction, respectively, coincide.

However, this is not so, as a rule, in the categories of analysis and topology.

Proposition (0.2.4). *Let* $\varphi : E \longrightarrow F$ *be a morphism in any of the categories of polynormed spaces under consideration* (let us say in $\mathcal{B}an$). *Then* (I) φ *is a coretraction* $\Longleftrightarrow$ *it is topologically injective and its image is complemented in* F; (II) φ *is a retraction* $\Longleftrightarrow$ *it is open and its kernel is complemented in* E.

▸ This is obviously implied by Proposition 1.14. ■

If we combine Proposition 2.4 and the results of §1.4 we see that, in the indicated categories, even a topologically injective operator is not necessarily a coretraction and nor is a topologically surjective operator a retraction.

Topologically injective operators and open operators themselves can also be given a category theoretic interpretation.

Exercise (0.2.5). Let φ be a morphism in any of the categories of polynormed spaces under consideration (or in $\mathcal{CT}op$). Prove that it is topologically injective (topologically surjective, respectively) $\Longleftrightarrow$ $\varphi = \rho \circ \sigma$ (where ρ is a monomorphism and σ an epimorphism) implies that σ (or ρ, respectively) is an isomorphism.

Remark (0.2.6). Very many different sorts of problems in topology, algebra and functional analysis can be reduced to two types: is a given monomorphism (epimorphism, respectively) a coretraction (retraction)? (In practice we are usually talking about natural embeddings or projections.) General methods have now been devised for investigating such problems — the so-called homological methods. The simplest homological phenomenon, already observed in topology (perhaps all "homology" began from this) is that the circle ($\mathbb{T}$) is not a retract of the disc ($\overline{\mathbb{D}}$), while in algebra it is that $\mathbf{in} : 2\mathbb{Z} \longrightarrow \mathbb{Z}$ is not a coretraction. Such problems as these that arise in the theory of "the algebras of analysis" and their representations will be considered in Chapter VII.

§2.2. Functors and natural functor transformations; examples. Additive categories and functors

The next concept in category theory, after that of a category itself, is that of a *functor*, in the "covariant" or "contravariant" version, and we shall assume that this also is known; we recall only for conscience sake that a contravariant functor from $\mathcal{K}_1$ into $\mathcal{K}_2$ is just the same as a covariant functor from $\mathcal{K}_1$ into $\mathcal{K}_2^{op}$ or equally from $\mathcal{K}_1^{op}$ into $\mathcal{K}_2$. We cite a trivial but very useful proposition.[7]

Proposition (0.2.7). *Let* $F : \mathcal{K}_1 \longrightarrow \mathcal{K}_2$ *be a* (for definiteness, covariant) *functor. Then, if* φ *is a retraction, a coretraction or an isomorphism, then so is* $F(\varphi)$. ▸■

With each category $\mathcal{K}$ there are connected two families of standard functors from $\mathcal{K}$ into $\mathcal{S}ets$. Namely, we fix $X \in \mathcal{K}$ and associate the set $\mathbf{h}_{\mathcal{K}}(X, Y)$ with each $Y \in \mathcal{K}$ and the map $\mathbf{h}_{\mathcal{K}}(X, \varphi) : \mathbf{h}_{\mathcal{K}}(X, Y_1) \longrightarrow \mathbf{h}_{\mathcal{K}}(X, Y_2)$: $\psi \longmapsto \varphi \circ \psi$ with each morphism $\varphi : Y_1 \longrightarrow Y_2$ in $\mathcal{K}$. This is the construction of the covariant functor from $\mathcal{K}$ into $\mathcal{S}ets$, that is denoted by $\mathbf{h}_{\mathcal{K}}(X, ?)$ and called the *morphism functor*. Similarly, given $X \in \mathcal{K}$ there is the *contravariant morphism functor* $\mathbf{h}_{\mathcal{K}}(?, X)$, which may be defined as $\mathbf{h}_{\mathcal{K}^{op}}(X, ?)$ (give a "direct" definition!).

If $\mathcal{K}$ is one of the categories of polynormed spaces or $\mathcal{L}in$, then the construction described gives, not simply sets and maps but, linear spaces and operators; therefore the morphism functors we get — now it is better to say "*operator functors*" — are obviously from $\mathcal{K}$ into $\mathcal{L}in$. If, however, we are talking specifically about $\mathcal{K} = \mathcal{B}an$, then the operator norm in $\mathbf{h}_{\mathcal{K}}(E, F) = \mathcal{B}(E, F)$ enables us to talk about the operator functors $\mathbf{h}_{\mathcal{K}}(E, ?)$ and $\mathbf{h}_{\mathcal{K}}(?, E)$ with values in the same category $\mathcal{B}an$. We notice that the contravariant functor $\mathbf{h}_{\mathcal{B}an}(?, \mathbb{C})$, sometimes also denoted by $(*) : \mathcal{B}an \longrightarrow \mathcal{B}an$ and called the *star functor*, is none other than the functor of taking the dual spaces and dual operators.

Finally, we recall a third fundamental categorical concept. Let F and G be functors from $\mathcal{K}_1$ into $\mathcal{K}_2$, for definiteness, covariant. A *natural transformation* between these functors is a collection of morphisms $\alpha = \{\alpha_X : F(X) \longrightarrow G(X);\ X \in \mathcal{K}_1\}$ in $\mathcal{K}_2$, such that for any $\varphi : X \longrightarrow Y$ in $\mathcal{K}_1$ the diagram

---

[7] A few years ago the author went, out of curiosity, to some lectures on supermanifolds. One day the lecturer had to establish that a particular operator was bijective. He gave a long and tedious proof of the required property "with bare hands", then, without finishing the argument, suddenly woke us up by exclaiming: "Why am I doing this? Isn't my operator induced by a diffeomorphism of supermanifolds and so is the result of applying a functor to an isomorphism in a category!" And everything instantly became clear.

$$\begin{array}{ccc} F(X) & \xrightarrow{F(\varphi)} & F(Y) \\ \downarrow{\scriptstyle \alpha_X} & & \downarrow{\scriptstyle \alpha_Y} \\ G(X) & \xrightarrow{G(\varphi)} & G(Y) \end{array}$$

is commutative. A natural transformation is called a *natural equivalence* (and the corresponding functors are *naturally equivalent* or isomorphic) if α_X is an isomorphism for every $X \in \mathcal{K}_1$.

Here is a classic example, from the point of view of analysis.

Example (0.2.8). We let α_E denote the canonical isometric operator from $E \in \mathcal{B}an$ into E^{**} that to each $x \in E$ associates the functional on E^* taking f into $f(x)$. Then all the α_E; $E \in \mathcal{B}an$ form a natural transformation between $\mathbf{1}_{\mathcal{B}an}$ (the identity functor in $\mathcal{B}an$) and the functor $(**) = (*) \circ (*)$ of taking the second dual. This construction, when considered on the complete subcategory of reflexive spaces in $\mathcal{B}an$, gives an example of a natural functor equivalence.

Exercise (0.2.9). Construct a natural equivalence between the functors $\mathbf{1}_{\mathcal{B}an}$ and $\mathbf{h}(\mathbb{C}, ?) : \mathcal{B}an \longrightarrow \mathcal{B}an$.

A functor from a category $\mathcal{K}$ into $\mathcal{S}ets$ (for definiteness covariant) is said to be *representable* if it is naturally equivalent to a functor $\mathbf{h}_{\mathcal{K}}(X, ?)$ for some $X \in \mathcal{K}$ (the "representative object" of our functor). An enormous number of problems from various branches of mathematics are actually problems about whether a given functor is representable, or if so, what is its representative object. We refer the reader to [11, I, §4] for comprehensive examples; see also 3.11, 3.27 and 3.48 later on.

As to which categories may be assumed to be "effectively indistinguishable", there are various approaches; we describe the most "rigid". Categories are said to be *isomorphic* if there exists a covariant functor between them that can be realized by bijections of both the objects and the morphisms.

Example (0.2.10). As any operator between spaces with strongest seminorm system is continuous (this follows easily from Proposition 1.9), $\mathcal{L}in$ is obviously isomorphic to the complete subcategory of $\mathcal{PN}$ (and of $\mathcal{CHPN}$), consisting of these "strongest" spaces. In future we shall identify these two categories.

For another definition, less rigid than that of isomorphism — the so-called equivalence of categories — see, for example [41, p.11] or [6, pp.3–4].

* * *

An object $\odot$ of the category $\mathcal{K}$ is called an *initial object* (*final object*, respectively) if, for any $X \in \mathcal{K}$, there exists precisely one morphism from $\odot$ into X (from X into $\odot$). The category is said to be *pointed* if it has an object which is simultaneously initial and final; such an object is called the *zero object* and is denoted by $0_{\mathcal{K}}$ or simply 0. In this case, for any $X, Y \in \mathcal{K}$, there exists a unique morphism $0_{X,Y}$ determined by the commutative diagram

$$\begin{array}{ccc} X & \xrightarrow{\quad 0_{X,Y} \quad} & Y \\ & \searrow \quad \nearrow & \\ & 0 & \end{array}$$

which is called the *zero morphism*.

A pointed category $\mathcal{K}$ is said to be *pre-additive* if, for any $X, Y \in \mathcal{K}$, the set $\mathbf{h}_{\mathcal{K}}(X, Y)$ is furnished with the structure of an Abelian group so that the composition of morphisms $\varphi \circ \psi$ is additive with respect to both arguments (hence, in particular, $0_{X,Y}$ is the zero element in the group $\mathbf{h}_{\mathcal{K}}(X, Y)$). The diagram $X \underset{\sigma}{\overset{i}{\rightleftarrows}} Y \underset{\rho}{\overset{j}{\rightleftarrows}} Z$ in a pre-additive category is called the *direct sum diagram* if $\sigma \circ i = 1_X$, $j \circ \rho = 1_Z$ and $i \circ \sigma + \rho \circ j = 1_Y$; in such a diagram Y is called the *direct sum* of X and Z, and the latter are the *direct summands* in Y. Finally, a pre-additive category in which any two objects have a direct sum is called an *additive category*.[8]

Of course, $\mathcal{L}in$ and all the categories of polynormed spaces considered above are additive: the Cartesian product plays the role of the direct sum. We recommend the reader to indicate all the morphisms in the corresponding diagram.

Every direct summand is, of course, a retract. For the specific categories just mentioned the converse is also true:[9] if $\sigma : F \longrightarrow E$ is a retraction, then, as is easily verified, we have a direct sum diagram $E \underset{\sigma}{\overset{i}{\rightleftarrows}} F \underset{\mathrm{in}}{\overset{j}{\rightleftarrows}} \operatorname{Ker}\sigma$ in which i is an

---

[8] Another approach leading to an additive category is indicated in [6, I §3].

[9] A similar assertion is true for all additive categories in which the morphisms have so-called (category) kernels [11, Ch.5 §3].

(arbitrary) right inverse morphism for σ and j takes $x \in F$ into $x - i \circ \sigma(x) \in \operatorname{Ker} \sigma$.

A functor $F : \mathcal{K}_1 \longrightarrow K_2$ between pre-additive categories is said to be *additive* if $F(\varphi + \psi) = F(\varphi) + F(\psi)$ for any $X, Y \in \mathcal{K}_1$ and morphisms $\varphi, \psi : X \longrightarrow Y$. We can take as examples the functors $\mathbf{h}_{\mathcal{K}}(X, ?)$ and $\mathbf{h}_{\mathcal{K}}(?, X)$ in §2.1 in particular, $(*) : \mathcal{B}an \longrightarrow \mathcal{B}an$.

The following obvious proposition turns out to be very useful.

Proposition (0.2.11). *Let $F : \mathcal{K}_1 \longrightarrow \mathcal{K}_2$ be an additive functor, and let (D) be a direct sum diagram in $\mathcal{K}_1$. Then $F(D)$, that is, the result of applying F to the objects and morphisms from (D), is a direct sum diagram in $\mathcal{K}_2$.* ▸■

§3. TENSOR PRODUCTS

Tensor products originally arose long ago in abstract algebra [12] and have from the time of the memoirs by Schatten [92] and Grothendieck [26] been firmly rooted in functional analysis; moreover the presence of a topology in the spaces under consideration led to the appearance of various versions of this concept. The common significance behind the construction of a tensor product in polynormed spaces (at least for the two versions of it we shall define) is that it enables us to consider, not only continuous bilinear but also, continuous multilinear operators, as ordinary continuous operators — but only when they are given on spaces with a more complex structure than the originals.

We have to give a precise meaning to this fairly vague sentence. We begin by refining the concept of a "continuous bilinear operator" itself.

§3.1. Bilinear operators and their continuity types. The freezing isomorphisms

Let E, F and G be linear spaces. Recall that a map $\mathcal{R} : E \times F \longrightarrow G$ is called a *bi-operator* (= bilinear operator) if the maps $\mathcal{R}_y : E \longrightarrow G : x \longmapsto \mathcal{R}(x, y)$ and $\mathcal{R}_x : F \longrightarrow G : y \longmapsto \mathcal{R}(x, y)$, obtained by fixing either of the arguments, are operators. The set of bi-operators from $E \times F$ into G is denoted by $\mathcal{L}(E \times F, G)$; this is a linear space with pointwise operations.

Various interpretations of continuity are possible for bi-operators which connect polynormed spaces; we limit ourselves to the two most important ones.[10] Namely (with the same notation

---

[10] See, for example [90, III.5] for a lot of others, including so-called hypocontinuity.

as above) $\mathcal{R}$ is said to be *separately continuous* if all the operators $\{\mathcal{R}_y; y \in F\}$ and $\{\mathcal{R}_x; x \in E\}$ are continuous and *jointly continuous* if it is continuous as a map from the topological space $E \times F$ (see 0.3° and §1) into G.

Both concepts may be formulated simply in terms of seminorms. In the case of separate continuity we let the reader obtain the appropriate criteria starting from Proposition 1.9. With regard to joint continuity, it is easy to verify that for seminormed spaces E, F and G this property is equivalent to the *boundedness* of $\mathcal{R}$, that is the existence of a (finite) number $\sup\{\|\mathcal{R}(x, y)\| : \|x\| \le 1, \|y\| \le 1\}$, known as the *bi-operator seminorm* and denoted by $\|\mathcal{R}\|_0$ or $\|\mathcal{R}\|$. The following proposition holds in the general case.

Proposition (0.3.1). *Let* $(E; \|\cdot\|_\lambda, \lambda \in \Lambda_1)$, $(F; \|\cdot\|_\mu, \mu \in \Lambda_2)$ *and* $(G; \|\cdot\|_\nu, \nu \in \Lambda_3)$ *be polynormed spaces, where the first two systems are saturated. Then the bi-operator* $\mathcal{R} : E \times F \longrightarrow G$ *is jointly continuous* $\Leftrightarrow$ *for any* $\nu_0 \in \Lambda_3$ *there exists* $\lambda_0 \in \Lambda_1$ *and* $\mu_0 \in \Lambda_2$ *such that* $\mathcal{R}$, *when considered with respect to the seminormed spaces* $(E; \|\cdot\|_{\lambda_0})$, $(F; \|\cdot\|_{\mu_0})$, *and* $(G; \|\cdot\|_{\nu_0})$, *is bounded*. ▸■

The set of separately continuous bi-operators from $E \times F$ into G is denoted by $\mathcal{B}^s(E \times F, G)$ and the set of jointly continuous operators is denoted by $\mathcal{B}^j(E \times F, G)$. Both are obviously subspaces of $\mathcal{L}(E \times F, G)$ and moreover in the case where E and F both have the strongest seminorm system, the first of them coincides with the whole of $\mathcal{L}(E \times F, G)$. If E, F and G are seminormed spaces, then $\mathcal{B}^j(E \times F, G)$ is also a seminormed space with respect to the bi-operator seminorm; it is Hausdorff or complete $\Leftrightarrow$ G has the corresponding property.

Of course, "jointly" implies "separately" but the converse, generally speaking, is not true. We let c_{00} denote the space of finite (eventually zero) sequences and c_∞ $(= \mathbb{C}^{\mathbb{N}})$ denote the space of all sequences. As usual (cf. §1.4), the terms of the sequence ξ are denoted by $(\xi_n; n \in \mathbb{N})$.

Exercise (0.3.2). Show that

$$f : c_{00} \times c_{00} \longrightarrow \mathbb{C} : (\xi, \eta) \longrightarrow \sum_{n=1}^{\infty} n\xi_n\eta_n,$$

where c_{00} is considered with the uniform norm, is separately, but not jointly, continuous.

Exercise (0.3.3). Show that the "pairing" $f : c_{00} \times c_\infty \longrightarrow \mathbb{C} : (\xi, \eta) \longmapsto \sum_{n=1}^{\infty} n\xi_n\eta_n$ is separately continuous with respect to the strongest system of seminorms in c_{00} and any system containing the compact-open system of seminorms in c_∞, but (the more difficult part) it is not jointly continuous for any system of seminorms in c_{00} and c_∞.

Hint. Whatever the seminorms in c_{00} and c_∞, $\{f(\xi, \eta) : \|\xi\| \le 1, \quad \|\eta\| \le 1\} = \mathbb{C}$.

Nevertheless in the classes of spaces that are most important in our account, the two forms of continuity coincide.

Theorem (0.3.4). *Let E be a Banach space and let F and G be (arbitrary) seminormed spaces. Then any separately continuous bi-operator $\mathcal{R} : E \times F \longrightarrow G$ is bounded (= jointly continuous).*

▶ For each $y \in F$ we consider the operator $\mathcal{R}_y : E \longrightarrow G$ (see above) and put $M = \{\mathcal{R}_y : \|y\| \le 1\}$. Then, obviously, the separate continuity of $\mathcal{R}$ implies that M is a subset of $\mathcal{B}(E, G)$, which satisfies the conditions of the Banach–Steinhaus theorem (see §1.2). By this theorem, $\sup\{\|\mathcal{R}_y\|_0 : \|y\| \le 1\} < \infty$, and obviously this is precisely the number that figures in the definition of a bounded bi-operator. ■

Thus we are justified in talking about (simply) *continuous bi-operators* in the class of Banach spaces while implying either of the two types of continuity under discussion. In this case we use the notation $\mathcal{B}(E \times F, G)$ for continuous bi-operator spaces and put $(E \times F)^* := \mathcal{B}(E \times F, \mathbb{C})$.

Remark (0.3.5). The assertion of the theorem is also true under weaker assumptions, for example when E is a Fréchet space, F is metrizable and G is an arbitrary polynormed space [90, III.5].

* * *

Let us recall one of the fruitful, common methods in mathematics which is to consider, instead of a function of two variables, a function of one variable whose values are a function in the other. If we are given a bi-operator $\mathcal{R} : E \times F \longrightarrow G$, then such a method associates with this the operator $x \longmapsto \mathcal{R}_x$ from E into $\mathcal{L}(F, G)$. If we vary the bi-operator, we obviously get a linear isomorphism $\mathcal{L}(E \times F, G) \xrightarrow{\simeq} \mathcal{L}(E, \mathcal{L}(F, G))$. We shall call this isomorphism, and the other similar operator $\mathcal{L}(E \times F, G) \xrightarrow{\simeq} \mathcal{L}(F, \mathcal{L}(E, G))$, *freezing isomorphisms* (of the first or second argument, according to the sense).

In the case of polynormed spaces the question arises of how to characterize those operators in $\mathcal{L}(E, \mathcal{L}(F, G))$ that are separately or jointly continuous, "frozen" bi-operators. As for joint continuity, we can give an answer in the following special case.

Proposition (0.3.6). *Let E, F and G be seminormed spaces. Then the freezing isomorphism is an isometric isomorphism from*

$\mathcal{B}^j(E \times F, G)$ *onto* $\mathcal{B}(E, \mathcal{B}(F, G))$ *(or, according to the sense, onto* $\mathcal{B}(F, \mathcal{B}(E, G))$*, where all the spaces are considered to have the (bi)operator seminorm.* ▸■

In particular, in the class of Banach spaces, there are isometric isomorphisms

$$B(F, \mathcal{B}(E, G)) \xleftarrow{\simeq} B(E \times F, G) \xrightarrow{\simeq} B(E, \mathcal{B}(F, G)),$$

including the case $\mathcal{B}(F, E^*) \xleftarrow{\simeq} (E \times F)^* \xrightarrow{\simeq} \mathcal{B}(E, F^*)$; we shall call them *freezing isomorphisms.*

Separately continuous "freezing" operators can also be described fairly simply in the general case. Let F and $(G; \|\cdot\|_\nu, \nu \in \Lambda)$ be polynormed spaces. We consider in $\mathcal{B}(F, G)$ the so-called *strong operator system* of seminorms $\|\varphi\|_{x,\nu} := \|\varphi(x)\|_\nu$, taken over all $x \in F$ and $\nu \in \Lambda$. The corresponding topology in $\mathcal{B}(F, G)$ is called the *strong operator topology*; there is an obvious meaning for convergence in it, "convergence on every vector in F".

Exercise (0.3.7). Let E, F and G be polynormed spaces. Show that the freezing isomorphism (for definiteness, of the first argument) is a linear isomorphism from $\mathcal{B}^s(E \times F, G)$ onto $\mathcal{B}(E, \mathcal{B}(F, G))$, where $\mathcal{B}(F, G)$ is considered to have the strong operator system of seminorms.

§3.2. Universality property and the algebraic tensor product. Tensor product seminorm

Let E and F be polynormed spaces, and let $\mathfrak{M}$ be a certain (for the time being arbitrary) class of separately continuous bi-operators from $E \times F$ into polynormed spaces.

Definition (0.3.8). If Θ is a polynormed space and $\theta : E \times F \longrightarrow \Theta$ is a bi-operator, we say that the pair (Θ, θ) has the *universality property with respect to* $\mathfrak{M}$ if (I) $\theta \in \mathfrak{M}$ and (II) for any bi-operator $\mathcal{R} : E \times F \longrightarrow G$ in $\mathfrak{M}$ there is a unique continuous operator $R : \Theta \longrightarrow G$ (said to be *associated with* $\mathcal{R}$) such that the diagram

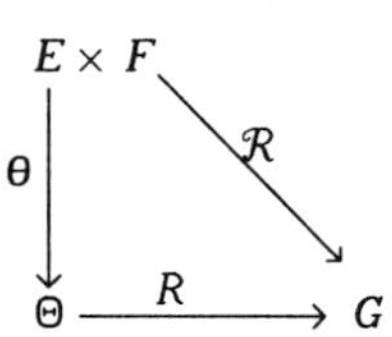

$(\mathcal{UP})$

is commutative.

We show that all pairs satisfying this definition are "essentially the same".

Theorem (0.3.9) (uniqueness). *Let* (Θ_k, θ_k); $k = 1, 2$, *be two pairs having the universality property with respect to the (same) class* $\mathfrak{M}$. *Then there is a topological isomorphism* $i : \Theta_1 \longrightarrow \Theta_2$ *such that the diagram*

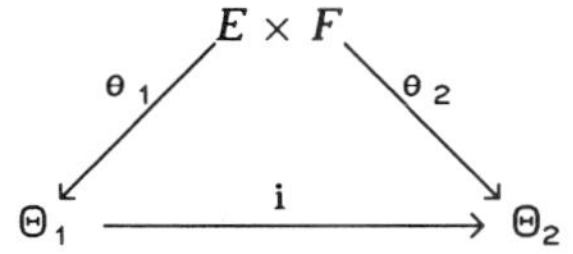

is commutative.

▶ If we put $\Theta := \Theta_1$, $G := \Theta_2$ and $\mathcal{R} := \theta_2$ in the diagram $(\mathcal{UP})$, we get an operator $i : \Theta_1 \longrightarrow \Theta_2$, whereas if we interchange the indices 1 and 2 we get an operator $j : \Theta_2 \longrightarrow \Theta_1$. If we put $\Theta = G = \Theta_1$ and $\theta = \mathcal{R} = \theta_1$ in $(\mathcal{UP})$ and use the uniqueness of the associated operator, we get that $j \circ i$ is the identity on Θ_1; similarly $i \circ j$ is the identity on Θ_2. ■

The question of whether there do exist pairs which have the universality property is fairly deep and its solution depends on how well one chooses the class $\mathfrak{M}$. We consider first the purely algebraic universality property (corresponding, if you like, to considering only spaces with the strongest seminorm system, cf. Example 2.10). This corresponds to the following definition.

Definition (0.3.10). Let E and F be linear spaces. The pair (Θ, θ), where Θ is a linear space and $\theta : E \times F \longrightarrow \Theta$ is a bi-operator, is called an *(algebraic) tensor product of* E *and* F if it has the universality property with respect to all bi-operators from $E \times F$ into linear spaces.

Exercise (0.3.11). Give the category theory interpretation of the space Θ as a representative object (see §2.2) of a certain functor $F : \mathcal{L}in \longrightarrow \mathcal{S}ets$.

Hint. $F(G) = \mathcal{L}(E \times F, G)$.

The existence of an algebraic tensor product is proved by producing some explicit construction or other. We shall deal with this in the following way. [11]

Let $E \circ F$ be the space of formal linear combinations (over $\mathbb{C}$) of the elements of $E \times F$. If we let $x \circ y := 1(x, y)$ be the elements of the natural basis for $E \circ F$, we can consider the set M of elements of any of the following kinds:

---

[11] In some textbooks one of these constructions is taken as the initial definition of a tensor product. For another explicit construction see, for example, [81, II §1.1].

$$(x_1 + x_2) \circ y - x_1 \circ y - x_2 \circ y,$$

$$x \circ (y_1 + y_2) - x \circ y_1 - x \circ y_2,$$

$$(\lambda x) \circ y - \lambda(x \circ y), \quad x \circ (\lambda y) - \lambda(x \circ y);$$

$$x_1, x_2, x \in E, \quad y_1, y_2, y \in F, \quad \lambda \in \mathbb{C}.$$

We introduce the notation $E \otimes F$ for $E \circ F/\ell.h.(M)$, $x \otimes y$ for $x \circ y + \ell.h(M)$ and $\vartheta : E \times F \longrightarrow E \otimes F : (x, y) \longmapsto x \otimes y$. Elements of the form $x \otimes y$ are called *elementary tensors.* Obviously to each kind of element in M there corresponds a certain identity; for example, $(x_1 + x_2) \otimes y = x_1 \otimes y + x_2 \otimes y$ corresponds to the first kind and so on.

Theorem (0.3.12) (existence). *The pair* $(E \otimes F, \vartheta)$ *is a tensor product of the spaces* E *and* F.

▶ Let $\mathcal{R} : E \times F \longrightarrow G$ be a bi-operator. It defines uniquely the operator $R^0 : E \circ F \longrightarrow G$, that takes $x \circ y$ to $\mathcal{R}(x, y)$, while R^0, by the choice of M, induces (see §0.2°) $R : E \otimes F \longrightarrow G$; moreover the diagram $(\mathcal{UP})$ is obviously commutative when $\Theta := E \otimes F$ and $\theta := \vartheta$. Finally, in view of $E \otimes F = \ell.h.(\operatorname{Im} \vartheta)$, R is uniquely determined by the requirement that this diagram is commutative. ■

Proposition (0.3.13). *We can represent every* $u \in E \otimes F$, $u \neq 0$, *in the form* $u = \sum_{k=1}^{n} x_k \otimes y_k$, *where the vectors* x_k *are linearly independent in* E *and the* y_k *are linearly independent in* F. ▶■

The *tensor product* of functionals f on E and g on F is the functional $f \otimes g$ on $E \otimes F$ that is associated with the bifunctional $f \times g : (x, y) \longmapsto f(x)g(y)$; it is obviously uniquely defined by $(f \otimes g)(x \otimes y) = f(x)g(y)$.

Proposition (0.3.14). *For every* $u \in E \otimes F$, $u \neq 0$, *there exist* $f \in E^{\#}$ *and* $g \in F^{\#}$ *such that* $f \otimes g(u) \neq 0$.

▶ We represent u as in Proposition 3.13 and we take f with $f(x_1) \neq 0$ and $f(x_k) = 0$ for $k > 1$ and also g with $g(y_1) \neq 0$. The rest is clear. ■

Exercise (0.3.15). Show that, if $x, y \in E$, then $x \otimes y = y \otimes x$ in $E \otimes E$ $\Longleftrightarrow$ x and y are linearly dependent.

We urge the reader to examine carefully the following two exercises.

Exercise (0.3.16). Show that (I) $\mathbb{C}^m \otimes \mathbb{C}^n$ is isomorphic to $\mathbb{C}^{mn}$, and (II) $\mathbb{C}^M \otimes \mathbb{C}^N$ is isomorphic to the subspace of those functions in $\mathbb{C}^{M \times N}$ that can be represented as $\sum_{k=1}^{n} f_k(s)g_k(t)$; $f_k \in \mathbb{C}^M$, $g_k \in \mathbb{C}^N$.

* * *

Now let E and F be given a seminorm. For $u \in E \otimes F$ we put $\|u\|_p := \inf \sum_{k=1}^{n} \|x_k\| \|y_k\|$, where the lower bound is taken over all possible representations $u = \sum_{k=1}^{n} x_k \otimes y_k$; $x_k \in E$, $y_k \in F$. Obviously $\|\cdot\|_p$ is a seminorm; it is called the *projective tensor product of the given seminorms* or, if they are fixed, the *projective seminorm* in $E \otimes F$. The seminormed space $(E \otimes F;\ \|\cdot\|_p)$ will be denoted by $E \otimes_p F$.

Exercise (0.3.17). Show that the unit ball in $E \otimes_p F$ is the closed convex hull of the set $\{x \otimes y\ ;\ x \in \underline{S}_E,\ y \in \underline{S}_F\}$.

Proposition (0.3.18). *Let* $\mathcal{R} : E \times F \longrightarrow G$ *be a bounded bi-operator between seminormed spaces. Then the operator* $R_0 : E \otimes_p F$ *associated with it is also bounded and* $\|R_0\|_0 = \|\mathcal{R}\|_0$.

▶ Let $u = \sum_{k=1}^{n} x_k \otimes y_k$; then $R_0(u) = \sum_{k=1}^{n} \mathcal{R}(x_k, y_k)$ implies that $\|R_0(u)\| \le \|\mathcal{R}\|_0 \sum_{k=1}^{n} \|x_k\| \|y_k\|$; hence from the definition of $\|\cdot\|_p$, $\|R_0\|_0 \le \|\mathcal{R}\|_0$. Further, $\mathcal{R}(x, y) = R_0(x \otimes y)$ and it is obvious that $\|x \otimes y\|_p \le \|x\| \|y\|$ and these imply that

$$\|\mathcal{R}\|_0 \le \|R_0\|_0 \sup\{\|x \otimes y\|_p : x \in \underline{S}_E,\quad y \in \underline{S}_F\} \le \|R_0\|_0. \blacksquare$$

Corollary (0.3.19). *For* $f \in E^*$ *and* $g \in F^*$ *the norm of* $f \otimes g$, *as a functional on* $E \otimes_p F$, *is equal to* $\|f\|^* \|g\|^*$.

We now introduce in $E \otimes F$ another seminorm by putting $\|u\|_w := \sup |f \otimes g(u)|$, where the supremum is taken over all $f \in \underline{S}_{E^*}$ and $g \in \underline{S}_{F^*}$. This seminorm is called the weak (more rarely — injective) *tensor* product of the given seminorms.[12] The space $(E \otimes F;\ \|\cdot\|_w)$ is indicated by $E \otimes_w F$.

Proposition (0.3.20). $\|\cdot\|_w \le \|\cdot\|_p$.

▶ This is implied by

$$|f \otimes g(u)| \le \|f\|^* \|g\|^* \sum_{k=1}^{n} \|x_k\| \|y_k\|,$$

where $u = \sum_{k=1}^{n} x_k \otimes y_k$. ■

---

[12] In our account the seminorm $\|\cdot\|_w$ plays a purely subsidiary role. However, in other branches of analysis, for example in the theory of nuclear spaces or summing maps [26], [59] it stands on a par with $\|\cdot\|_p$.

Proposition (0.3.21). *For any* $x \in E$ *and* $y \in F$

$$\|x \otimes y\|_p = \|x \otimes y\|_w = \|x\| \|y\|.$$

▶ If $\|x\| > 0$ and $\|y\| > 0$, then, if we take for x a functional f furnished by Corollary B of the Hahn-Banach theorem (see §1.2) and a similar functional g for y, we see that $\|x \otimes y\|_w \geq |f \otimes g(x \otimes y)| = \|x\| \|y\|$. The rest is clear. ■

Proposition (0.3.22). *Let* $u \in E \otimes F$ *be represented as* $u = \sum_{k=1}^{n} x_k \otimes y_k$, *where the* x_k *are linearly independent and* $\|x_1\| > 0$ *and* $\|y_1\| > 0$. *Then* $\|u\|_w > 0$ *and, consequently,* $\|u\|_p > 0$.

▶ From Corollary B of the Hahn-Banach Theorem (§1.2) there exist $f \in E^*$ and $g \in F^*$ such that $f(x_1) > 0$, $f(x_2) = \cdots = f(x_n) = 0$ and $g(y_1) > 0$. Then

$$\|f\|^* \|g\|^* \|u\|_w \geq |(f \otimes g)u| = |\sum_{k=1}^{n} f(x_k) g(y_k)| = |f(x_1) g(y_1)| > 0. \ ■$$

By combining Propositions 3.22 and 3.13 we immediately get the following.

Corollary (0.3.23). $E \otimes_p F$ *is a normed space* $\Leftrightarrow$ $E \otimes_w F$ *is a normed space* $\Leftrightarrow$ E *and* F *are normed spaces.*

§3.3. The Banach tensor product. Bounds for certain norms. Tensor multiplication by $L^1(\cdot)$ and other examples

Turning to the tensor product of polynormed spaces, we begin with Banach spaces and examine this most important case "in close-up". Everywhere in this section E and F are Banach spaces.

Definition (0.3.24). A pair (Θ, θ), where Θ is a Banach space and $\theta : E \times F \longrightarrow \Theta$ is a bi-operator, is called the *projective tensor product* of E and F if the following properties are satisfied:

(I) this pair has the universality property with respect to the class of all continuous bi-operators from $E \times F$ into Banach spaces;

(II) if $\mathcal{R}$ is a bi-operator from this class, and R is the operator associated with it, then $\|R\|_0 = \|\mathcal{R}\|_0$.

Later on we shall, as a rule, omit the word "projective" as we shall not consider other forms of Banach tensor product, except in the cases specially mentioned. In addition, when we speak of the properties of the tensor product, we shall often mean the properties of the space.

The definition immediately implies the following proposition.

Proposition (0.3.25). *If* Θ *is a tensor product of* E *and* F, *then* $\mathcal{B}(\Theta, G) \simeq \mathcal{B}(E \times F, G)$ *holds for any* $G \in \mathcal{B}an$, *and in* *particular* $\Theta^* \simeq (E \times F)^*$, *where the sign* $\simeq$ *denotes an isometric isomorphism.* ▶■

By Theorem 3.9 the tensor product of Banach spaces is unique in the sense indicated there. However, in this case we have, in addition, the following theorem.

Theorem (0.3.26) (uniqueness). *If* (Θ_k, θ_k), $k = 1, 2$, *are two tensor products, then the operator* i, *appearing in Theorem* 3.9, *is an isometric isomorphism.*

▶ If E and F are non-trivial, then there automatically exist non-zero continuous bi-operators on $E \times F$ (for instance, of the form $f \times g$). Hence, according to Definition 3.24(I), for any tensor product (Θ, θ), $\Theta \neq 0$. Consequently, as $\mathbf{1}_\Theta$ is associated with θ, then, by Definition 3.24(II), $\|\theta\| = \|\mathbf{1}_\Theta\| = 1$. Therefore, by the same Definition 3.24(II), $\|i\| = \|\theta_2\| = 1$ and $\|i^{-1}\| = \|\theta_1\| = 1$. The rest is clear. ■

The Banach tensor product, like the algebraic tensor product (see Exercise 3.11), furnishes a specific example of a representable functor. However, the appropriate category here turns out not to be $\mathcal{B}an$ but its subcategory $\mathcal{B}an_1$ with the same objects but only with operators of norm ≤ 1 as morphisms.

Exercise (0.3.27). Show that the tensor product of the Banach spaces E and F is the representative object of some functor $\Phi : \mathcal{B}an_1 \longrightarrow \mathcal{S}ets$.

Hint. $\Phi(G)$ is the unit ball of $\mathcal{B}(E \times F, G)$.

We now construct a Banach tensor product in an explicit form. We let $E \hat{\otimes} F$ denote the completion of the normed (see Corollary 3.23) space $E \otimes_p F$ and we set $\hat{\vartheta} : E \times F \longrightarrow E \hat{\otimes} F : (x, y) \longmapsto x \otimes y$.

Theorem (0.3.28) (existence). *The pair* $(E \hat{\otimes} F, \vartheta)$ *is a projective tensor product of the Banach spaces* E *and* F.

▶ Let $\mathcal{R} : E \times F \longrightarrow G$ be a continuous bi-operator into a Banach space, and let $R : E \otimes_p F \longrightarrow G$ be the "algebraic" operator associated with it (in the sense of Definition 3.10); then, according to Proposition 3.18, $\|R_0\|_0 = \|\mathcal{R}\|_0$. If we extend R_0 by continuity, we get $R : E \hat{\otimes} F \longrightarrow G$, which obviously is the unique operator satisfying both requirements of Definition 3.24. ■

The norm in $E \hat{\otimes} F$ is also denoted by $\|\cdot\|_p$ (or simply by $\|\cdot\|$).

Theorem (0.3.29). $\|u\|_p = \inf \sum_{k=1}^{\infty} \|x_k\| \|y_k\|$ *for* $u \in E \hat{\otimes} F$, *where the lower bound is taken over all possible representations of* u *as an absolutely convergent series* $\sum_{k=1}^{\infty} x_k \otimes y_k$.

▶Obviously it is sufficient, given any $\varepsilon > 0$, to produce an absolutely convergent series with $u = \sum_{k=1}^{\infty} x_k \otimes y_k$ and $\sum_{k=1}^{\infty} \|x_k\| \|y_k\| < \|u\|_p + \varepsilon$. Since $E \otimes_p F$ is dense in $E \hat{\otimes} F$, u can be represented as $\sum_{k=1}^{\infty} u_n$, where $u_n \in E \otimes F$ and $\sum_{k=1}^{\infty} \|u_n\|_p < \|u\|_p + \frac{\varepsilon}{2}$. Further, from the definition of $\|\cdot\|_p$, u_n can be represented as $\sum_{i=1}^{m_n} x_{n_i} \otimes y_{n_i}$, where $\sum_{i=1}^{m_n} \|x_{n_i}\| \|y_{n_i}\| < \|u_n\|_p + \frac{\varepsilon}{2 \cdot 2^n}$. All that remains is to consider the series $\sum_{n=1}^{\infty} \left(\sum_{i=1}^{m_n} x_{n_i} \otimes y_{n_i} \right)$. ■

The question of calculating the precise norm of some element or other in $E \hat{\otimes} F$ often turns out to be fairly difficult. However, for certain classes of elements there are methods for estimating their norms — both from above and from below. Here is a useful assertion of such a kind.

Proposition (0.3.30). *Let* $u \in E \hat{\otimes} F$ *be represented as* $\sum_{k=1}^{n} x_k \otimes y_k$, *where* $\left\| \sum_{k=1}^{n} \xi^k x_k \right\| \le C_1$ *and* $\left\| \sum_{k=1}^{n} \xi^k y_k \right\| \le C_2$ *for any (complex) n-th root* ξ *of* 1. *Then* $\|u\|_p \le C_1 C_2$.

▶ Some fairly simple calculations ("collecting similar terms") show that u can be represented also as

$$\frac{1}{n} \sum_{k=1}^{n} \left[\left(\sum_{l=1}^{n} \zeta^{lk} x_l \right) \otimes \left(\sum_{m=1}^{n} \zeta^{-mk} y_m \right) \right],$$

where ζ is a primitive n-th root of 1. Hence, by the definition of a projective norm

$$\|u\| \le \frac{1}{n} \sum_{k=1}^{n} \left\| \sum_{l=1}^{n} \zeta^{lk} x_l \right\| \cdot \left\| \sum_{m=1}^{n} \zeta^{-mk} y_m \right\| \le \frac{1}{n} n C_1 C_1 = C_1 C_2 . \ ■$$

Here is one specific situation in which the bound one obtains can be applied.

Corollary (0.3.31). *Let* Ω_1 *and* Ω_2 *be compact spaces and let* u *be an element in* $C(\Omega_1) \hat{\otimes} C(\Omega_2)$ *of the form* $\sum_{k=1}^{n} x_k \otimes y_k$. *Then*

$$\|u\|_p \le \left\| \sum_{k=1}^{n} |x_k| \right\|_0 \left\| \sum_{k=1}^{n} |y_k| \right\|_0 .$$

In particular, if $x_k x_l = y_k y_l = 0$; $k \ne l$, *and* $\|x_k\|_0, \|y_k\|_0 \le 1$, $1 \le k \le n$, *then, also,* $\|u\|_p \le 1$.

Exercise (0.3.32). Show that (I) the norm of $u = \sum_{k=1}^{n} \mathbf{p}^k \otimes \mathbf{p}^k$ in $c_0 \hat{\otimes} c_0$ or $c_b \hat{\otimes} c_b$ (for the notation see §1.4) is equal to 1; (II) the norm of $u = \sum_{k=1}^{n} P_k \otimes P_k$ in $\mathcal{B}(H) \hat{\otimes} \mathcal{B}(H)$ is equal to 1, where P_k are orthogonal projections in the Hilbert space H onto pairwise orthogonal subspaces.

Hint. An upper bound is implied by Proposition 3.30. For the other way round it is necessary to describe a functional f with $\|f\|^* = 1$ and $f \otimes f(u) = 1$.

With regard to methods for finding lower bounds for norms, we give the following assertion without proof.

Assertion (0.3.33). Given $u \in E \hat{\otimes} F$ suppose that there exist $f_k \in E^*$, $g_k \in F^*$, $1 \le k \le n$, such that $f_k \otimes g_l(u)$ is equal to one for $k \le l$ and zero for $k > l$ (in other words, the numbers $f_k \otimes g_l(u)$ form a triangular $n \times n$-matrix). Then $\|u\|_p \ge \frac{1}{8\pi} \ln 2n - 1$.[13]

Exercise (0.3.34). Let Ω be a compact space, and let $x_k \in C(\Omega)$; $1 \le k \le n$, be such that $\|x_k\|_0 = 1$ and $x_k x_l = 0$ for $k \ne l$. Show that $\sum_{k \le l} x_k \otimes x_l \in C(\Omega) \hat{\otimes} C(\Omega)$ satisfies the conditions of the preceding assertion.

* * *

Let us turn to examples and make some remarks of a general character. From the point of view of analysis, tensor products are very useful in that, for many of the classical function spaces, going over from scalar functions to vector functions of the "same kind" can be described in terms of tensor products (a particular case is in going from functions of one variable to the "same" functions in several variables).

In this respect one of the most convenient spaces to tensor by is the Banach space $L^1(X, \mu)$, which should be well known to the reader, where X is a set with measure μ. Given a Banach space E we let $L^1_E(X, \mu)$ denote the Banach space of the E-valued functions (more precisely, equivalence classes "modulo μ-measure zero") that are integrable with respect to μ on X with norm $\|\bar{f}\| = \int_X \|\bar{f}(t)\| \, d\mu$ (cf. [27, Ch.III]).

Theorem (0.3.35) (Grothendieck). *To within isometric isomorphism* $L^1(X, \mu) \hat{\otimes} E = L^1_E(X, \mu)$.

---

[13] The given bound is due to Johnson [133]; very similar results were obtained in [157], [269]; see also [81, Theorem II.2.48].

▶ We consider the bi-operator $\mathcal{R} : L^1(X, \mu) \times E \longrightarrow L^1_E(X, \mu) : (f, x) \longmapsto f(t)x, \quad t \in X$ of norm 1 and the operator R associated with it; it is sufficient to establish that R maps isometrically a certain dense subspace M of $L^1(X, \mu) \hat{\otimes} E$ onto a dense subspace of $L^1_E(X, \mu)$. We let L_0 denote the subspace of simple functions (taking a finite number of values) in $L^1(X, \mu)$ and we put $M := \ell.h.\{f \otimes x : f \in L_0, x \in E\}$; then M and its image (the set of simple vector functions) are dense in the corresponding spaces. But any $u \in M$ can be represented as $\sum_{k=1}^{n} \chi_k \otimes y_k$, where $y_k \in E$ and the χ_k are characteristic functions of disjoint subsets of X. Therefore $\|R(u)\| = \sum_{k=1}^{n} \|\chi_k\| \|y_k\| \geq \|u\|$, while, since $\|R\| = \|\mathcal{R}\| = 1$, we have $\|R(u)\| = \|u\|$ for all $u \in M$. The rest is clear. ■

Corollary (0.3.36). *If X_k is a space with a measure μ_k; $k = 1, 2$, then, to within isometric isomorphism, $L^1(X_1, \mu_1) \hat{\otimes} L^1(X_2, \mu_2) = L^1(X_1 \times X_2, \mu_1 \times \mu_2)$. In particular, $l_1 \hat{\otimes} l_1 = l_1^2$, where $l_1^2 := L^1(\mathbb{N} \times \mathbb{N})$ is the space of convergent double series.*

The (projective) tensor products of the various Banach spaces of continuous functions, generally speaking, do not admit such a clear description. In this connection we introduce an important map.

Let E and F be two Banach spaces consisting of continuous functions on the compact spaces Ω_1 and Ω_2 respectively, with norms not less than the uniform norm. We consider the bi-operator

$$\tilde{\gamma} : E \times F \longrightarrow C(\Omega_1 \times \Omega_2) : (x, y) \longrightarrow x(s)y(t); \quad s \in \Omega_1, \quad t \in \Omega_2,$$

and let γ denote the operator of norm ≤ 1 associated with it; a particular case is $\gamma : C(\Omega_1) \hat{\otimes} C(\Omega_2) \longrightarrow C(\Omega_1 \times \Omega_2)$. It turns out that the latter, although also injective, is not surjective (for infinite Ω_1 and Ω_2); $\operatorname{Im} \gamma$ is a proper dense subspace of $C(\Omega_1 \times \Omega_2)$, sometimes called the *Varopoulos space*. For more details about this see [109], [39] and also [81, II §2.2].

Exercise (0.3.37). Assuming that Assertion 3.33 is known, deduce from this that the operator under discussion is not surjective.

Hint. Use Exercise 3.34.

We formulate another important example as an assertion which plays a significant role in the theory of operators.

Assertion (0.3.38) (Schatten-von Neumann). Let H be a Hilbert space. Then there exists an isometric isomorphism between $H \hat{\otimes} H^*$ and $\mathcal{N}(H)$ (see §1.3), associated with the bi-operator

$\mathcal{R} : (x, f) \longmapsto S$, where S is the one-dimensional operator $y \longmapsto \langle y, f \rangle x$.

Remark (0.3.39). It is not difficult to prove that, for an arbitrary Banach space E, the analogously defined operator $R : E \hat{\otimes} E^* \longrightarrow \mathcal{N}(E)$ is, as before, surjective, however, the question of whether it is injective is nothing like so simple. When it is injective (and it is then an isometric isomorphism) we say that E has the *approximation property*; maybe the reader knows another definition of this famous property better; that every compact operator from an arbitrary Banach space into E can be approximated (with respect to the operator norm) by finite-dimensional operators. The approximation property arises in many questions that are outwardly unrelated; for example, $\gamma : C(\Omega_1) \hat{\otimes} C(\Omega_2) \longrightarrow C(\Omega_1 \times \Omega_2)$, considered above, is injective precisely because the spaces $C(\Omega)$ have this property (see also Remarks 3.42 and 3.46). The question of whether every Banach space has the approximation property remained open for many years until Enflo constructed the first counterexample [299] in 1972; afterwards it turned out that $\mathcal{B}(H)$: $\dim H = \infty$ also has this property (Szankowski [284]). Further details about the approximation property, its equivalent formulations and applications (including the problem of defining the trace of a nuclear operator) are contained in [88], [81].

§3.4. The tensor product of operators and the functor "$\hat{\otimes}$". A remark on the weak tensor product

Like the operation "$\mathbf{h}_{\mathcal{B}an}(?, ?)$", considered in §2.2, the operation "$? \hat{\otimes} ?$" can be defined not only for the objects in $\mathcal{B}an$, but also for the morphisms in this category.

Proposition (0.3.40). *For any pair of continuous operators $\varphi : E_1 \longrightarrow E_2$ and $\psi : F_1 \longrightarrow F_2$ acting between Banach spaces, there exists a unique continuous operator $\varphi \hat{\otimes} \psi : E_1 \hat{\otimes} F_1 \longrightarrow E_2 \hat{\otimes} F_2$ such that $(\varphi \hat{\otimes} \psi)(x \otimes y) = \varphi(x) \otimes \psi(y)$ for all $x \in E_1$, $y \in F_1$. Moreover $\|\varphi \hat{\otimes} \psi\| = \|\varphi\| \, \|\psi\|$.*

▸ We put $\mathcal{R} : E_1 \times F_1 \longrightarrow E_2 \hat{\otimes} F_2 : (x, y) \longmapsto \varphi(x) \otimes \psi(y)$; obviously this is a bi-operator with norm $\|\varphi\| \, \|\psi\|$. We let $\varphi \hat{\otimes} \psi$ denote the associated operator. The rest is clear. ∎

The operator $\varphi \hat{\otimes} \psi$ we have constructed is called the *(projective) tensor product of the operators* φ *and* ψ. In the particular case of functionals, that is, when $E_2 = F_2 = \mathbb{C}$, $\varphi \hat{\otimes} \psi$ maps $E_1 \hat{\otimes} F_1$ into $\mathbb{C} \hat{\otimes} \mathbb{C} \simeq \mathbb{C}$, and thus it is itself a functional; as is easily seen, it extends by continuity the functional $\varphi \otimes \psi$ (on $E_1 \otimes F_1$) defined in §3.2.

Exercise (0.3.41). Show that the tensor product of two surjective operators (in $\mathcal{B}an$) is itself surjective.

Remark (0.3.42). However, the tensor product of two injective, even topologically injective, operators is not necessarily injective. This is connected with the fact that certain Banach spaces do not have the approximation property; see Remark 3.39.

If we fix $E \in \mathcal{B}an$, we can associate to each $F \in \mathcal{B}an$ the space $E \hat{\otimes} F$ and to each continuous operator $\varphi : F_1 \longrightarrow F_2$ the operator $1_E \hat{\otimes} \varphi : E \hat{\otimes} F_1 \longrightarrow E \hat{\otimes} F_2$. We obviously get a covariant functor from $\mathcal{B}an$ into $\mathcal{B}an$, denoted by $E \hat{\otimes} ?$ and called the tensoring functor (by E on the left). Analogously, by fixing F we get the covariant functor $? \hat{\otimes} F : \mathcal{B}an \longrightarrow \mathcal{B}an$ ("tensoring by F on the right"). Both functors, as is easily seen, are additive; consequently (see Proposition 2.11), for any $E, F, G \in \mathcal{B}an$, to within a topological isomorphism, $E \hat{\otimes} (F \oplus G) = (E \hat{\otimes} F) \oplus (E \hat{\otimes} G)$ and $(E \oplus F) \hat{\otimes} G = (E \hat{\otimes} G) \oplus (F \hat{\otimes} G)$ (anyhow, both equations can be verified immediately without difficulty).

Several other important identities hold for the operation "$\hat{\otimes}$" besides those we have just considered.

Theorem (0.3.43). *For any $E, F, G \in \mathcal{B}an$ the following isometric isomorphisms of Banach spaces exist and are uniquely defined:*

(I) $\mathbb{C} \hat{\otimes} E \xrightarrow{\simeq} E : \lambda \otimes x \longmapsto \lambda x$;

(II) $E \hat{\otimes} F \xrightarrow{\simeq} F \hat{\otimes} E : x \otimes y \longmapsto y \otimes x$ (*"commutativity"*);

(III) $E \hat{\otimes} (F \hat{\otimes} G) \xrightarrow{\simeq} (E \hat{\otimes} F) \hat{\otimes} G : x \otimes (y \otimes z) \longmapsto (x \otimes y) \otimes z$ (*"associativity"*);

(IV) $\mathcal{B}(E \hat{\otimes} F, G) \xrightarrow{\simeq} \mathcal{B}(E, \mathcal{B}(F, G)) : \varphi \longmapsto \psi$, *where* $[\psi(x)]y = \varphi(x \otimes y)$ (*"dual associativity"*).

▸ The maps defined on elementary tensors as indicated in (I) - (III) extend uniquely, by linearity, to isometric isomorphisms between dense subspaces in the relevant spaces; this implies the first three assertions. With regard to the fourth, Proposition 3.25 and the freezing isomorphism (see the text after Proposition 3.6) imply that both the spaces under consideration are isometrically isomorphic to $\mathcal{B}(E \times F, G)$. ■

Remark (0.3.44). The reader who is acquainted with the concept of a functor in several arguments (see, for example, [52, II.3]) is recommended to verify that in each of the isometric isomorphisms indicated there is concealed an assertion about the natural equivalence of corresponding "multiple" functors.

Special attention is drawn to the relation between "$\mathbf{h}_{\mathcal{B}an}$" and "$\hat{\otimes}$" expressed by the law of dual associativity. Here is one of its many specific applications.

Exercise (0.3.45). Assuming that the Schatten – von Neumann Theorem (3.38) is known, construct the isometric isomorphism $\mathcal{N}(H)^* \xrightarrow{\simeq} \mathcal{B}(H)$.

The associativity of the tensor product enables a unique meaning to be given to the expression $E_1 \hat{\otimes} \cdots \hat{\otimes} E_n$, $E_k \in \mathcal{B}an$, $1 \leq k \leq n$. This space, coupled with the corresponding n-linear operator, also has the universality property for continuous n-linear operators from $E_1 \times \cdots \times E_n$ into Banach spaces. We leave the details to the reader.

Of course all the constructions given above and the formulae in this section also hold, with obvious modifications, for the purely algebraic tensor product. The details of the corresponding parallel account are also left to the reader.

Remark (0.3.46). We emphasize that although $\varphi \hat{\otimes} \psi$ and $\varphi \otimes \psi$ are similar, they generally denote different things; one is an operator, that is, an element of $\mathcal{B}(E_1 \hat{\otimes} F_1, E_2 \hat{\otimes} F_2)$, and the other is an elementary tensor in $\mathcal{B}(E_1, E_2) \hat{\otimes} \mathcal{B}(F_1, F_2)$. Due to the universality property, these two spaces are connected by the canonical operator

$$\mathbf{ott} : \mathcal{B}(E_1, E_2) \hat{\otimes} \mathcal{B}(F_1, F_2) \longrightarrow \mathcal{B}(E_1 \hat{\otimes} F_1, E_2 \hat{\otimes} F_2) : \varphi \otimes \psi \longmapsto \varphi \hat{\otimes} \psi,$$

which is often called the *operator of tensor twisting*. In the case of "pure" linear spaces such an operator is obviously injective and this enables one space to be identified with a part of the other. However, in the "Banach case" **ott** may not only reduce the norm by any amount, but may even have a non-trivial kernel.

Again, an important role in this question (cf. Remark 3.42) is played by the approximation property: for example, if E^* or F^* has this property, then $\mathbf{ott} : E^* \hat{\otimes} F^* \longrightarrow (E \hat{\otimes} F)^*$ is automatically injective. See [81, II §2.3 – 4] for some details.

* * *

The next most important version of the concept of a tensor product of Banach spaces, after the projective version, is the *weak* tensor product; this is what we call the completion $E \check{\otimes} F$ of the normed space $E \otimes_w F$ (see §3.2) together with the bi-operator $\check{\vartheta} : E \times F \longrightarrow E \check{\otimes} F : (x, y) \longmapsto x \otimes y$. Much has been said about it, for example, in [26] and [88]. (Our definition by an explicit construction is connected with the fact that the class of bi-operators for which this pair has the universality property — they are called integral operators — is not at all easy to describe.) The operation "$\check{\otimes}$", as well as "$\hat{\otimes}$", gives a functor. For certain spaces the weak tensor product is more convenient than the projective one: for example, to within isometric isomorphism $C(\Omega_1) \check{\otimes} C(\Omega_2) = C(\Omega_1 \times \Omega_2)$ (cf. Theorem 3.35) and $H \check{\otimes} H^* = \mathcal{K}(H)$ (cf. Assertion 3.38). The last result is closely connected with the fact that $\mathcal{K}(H)^* = \mathcal{N}(H)$ and so (see Exercise 3.45) $\mathcal{K}(H)^{**} = \mathcal{B}(H)$. (Both duals play an outstanding role in the theory of operator algebras [31], [75], [58].)

For another version of the tensor product specifically for Hilbert spaces, and defined only within that class, see, for example, [75, IV.1] and also [81].

§3.5. The inductive and projective tensor products of polynormed spaces. Remark on nuclear spaces

The reader will remember that, when talking about continuous bi-operators between polynormed spaces, which are more general than Banach spaces, we really ought to explain which type of continuity we have in mind. Therefore, in defining a "polynormed" tensor product by analogy with Definition 3.24, we get at least two different versions of this concept.

Definition (0.3.47). Let E and F be polynormed spaces. The pair (Θ, θ), where Θ is a polynormed space and $\theta : E \times F \longrightarrow \Theta$ is a bi-operator, is called an *inductive* (*projective*, respectively) *tensor product* of E and F if it has the universality property with respect to the class of all separately (jointly, respectively) continuous bi-operators from $E \times F$ into all possible polynormed spaces.[14]

If we consider only complete Hausdorff polynormed spaces instead of arbitrary polynormed spaces we get another two, similarly defined, forms of the tensor product: the *complete Hausdorff inductive* and the *complete Hausdorff projective* tensor product.

Exercise (0.3.48) (cf. Exercises 3.11 and 3.27). Show that in each of these four definitions the space Θ is the representative object of a certain functor.

Obviously, for each of the tensor products we have introduced there is a uniqueness theorem which is a particular case of Theorem 3.9. We pass on to questions of existence.

Warning. In order to avoid clumsy notation we shall, to the end of this section, consider only saturated systems of seminorms.

---

[14] Of course the requirement in Definition 3.24 about the norms of the bi-operator and of the associated operator coinciding now loses its meaning.

So, let $\{\|\cdot\|_\mu;\ \mu \in \Lambda_1\}$, be a system of seminorms in E, and $\{\|\cdot\|_\nu;\ \nu \in \Lambda_2\}$ be a system in F. First of all, we consider the system of seminorms in $E \otimes F$ consisting of all possible $\|\cdot\|'$ with the following property: for any $x_0 \in E$ and $y_0 \in F$ the maps $F \longrightarrow \mathbb{R} : y \longmapsto \|x_0 \otimes y\|'$ and $E \longrightarrow \mathbb{R} : x \longmapsto \|x \otimes y_0\|'$ are continuous. The corresponding polynormed space is denoted by $E \otimes_i F$. Further, given the seminorms $\|\cdot\|_\mu$ and $\|\cdot\|_\nu$, we let $\|\cdot\|_{p(\mu,\nu)}$ denote their projective tensor product and $\|\cdot\|_{w(\mu,\nu)}$ denote their weak tensor product. We can then define the polynormed space $(E \otimes F;\ \|\cdot\|_{p(\mu,\nu)}, (\mu, \nu) \in \Lambda_1 \times \Lambda_2)$ which is denoted by $E \otimes_p F$ and $(E \otimes F;\ \|\cdot\|_{w(\mu,\nu)}, (\mu, \nu) \in \Lambda_1 \times \Lambda_2)$ which is denoted by $E \otimes_w F$. It is obvious that the topology of neither space depends on the choice of the defining system of seminorms in E and F.

Proposition (0.3.49). *The topology in $E \otimes_i F$ is not weaker than that in $E \otimes_p F$ and the topology in the latter is not weaker than that in $E \otimes_w F$.*

▶ The first assertion is implied by the fact that the system of seminorms in $E \otimes_i F$ obviously contains all the $\|\cdot\|_{p(\mu,\nu)}$. The second is implied by the inequality $\|\cdot\|_{w(\mu,\nu)} \leq \|\cdot\|_{p(\mu,\nu)}$ (Proposition 3.20). ■

Proposition (0.3.50). *$E \otimes_i F$ is Hausdorff $\Longleftrightarrow$ $E \otimes_p F$ is Hausdorff $\Longleftrightarrow$ $E \otimes_w F$ is Hausdorff $\Longleftrightarrow$ both E and F are Hausdorff.*

▶ This is implied by Propositions 3.22 and 3.13 in conjunction with the criteria for a space to be Hausdorff in terms of seminorms (Proposition 1.4). ■

Recall the bi-operator $\vartheta : E \times F \longrightarrow E \otimes F$ that occurs in the explicit construction of the algebraic tensor product.

Obviously ϑ, when considered to have values in $E \otimes_i F$, is separately continuous, whereas when it is considered to have values in $E \otimes_p F$ it is jointly continuous.

Theorem (0.3.51) (existence). *The pair $(E \otimes_i F, \vartheta)$ $((E \otimes_p F, \vartheta)$, respectively) is an inductive (projective, respectively) tensor product of E and F.*

▶ According to Theorem 3.12, for any polynormed space $(G;\ \|\cdot\|_\lambda,\ \lambda \in \Lambda_3)$, and any bi-operator $\mathcal{R} : E \times F \longrightarrow G$, there is a uniquely defined operator $R : E \otimes F \longrightarrow G$ such that $R \circ \vartheta = \mathcal{R}$. Therefore our problem is to prove that the separate (joint, respectively) continuity of $\mathcal{R}$ implies the continuity of R as an operator from $E \otimes_i F$ (from $E \otimes_p F$, respectively) into G.

If $\mathcal{R}$ is separately continuous, then for each $\lambda_0 \in \Lambda_3$, $x_0 \in E$ and $y_0 \in F$, the maps $E \longrightarrow \mathbb{R}$, given by $x \longmapsto \|\mathcal{R}(x, y_0)\|_{\lambda_0} = \|R(x \otimes y_0)\|_{\lambda_0}$ and $F \longrightarrow \mathbb{R}$, given by $y \longmapsto \|\mathcal{R}(x_0, y)\|_{\lambda_0} = \|R(x_0 \otimes y)\|_{\lambda_0}$ are both continuous. Thus

$\|\cdot\|'_{\lambda_0} := \|R(\cdot)\|_{\lambda_0}$ is one of the seminorms in $E \otimes_i F$. Hence $R: E \otimes_i F \longrightarrow G$ is bounded as an operator from $(E \otimes F;\ \|\cdot\|'_{\lambda_0})$ into $(G;\ \|\cdot\|_{\lambda_0})$, and so is continuous (Proposition 1.9).

If, however, $\mathcal{R}$ is jointly continuous, then for each $\lambda_0 \in \Lambda_3$ there is a certain $\mu_0 \in \Lambda_1$ and $\nu_0 \in \Lambda_2$ as described in Proposition 3.1. But then Proposition 3.18 implies that R is bounded as an operator from $(E \otimes F;\ \|\cdot\|_{p(\mu_0,\nu_0)})$ into $(G;\ \|\cdot\|_{\lambda_0})$. Hence, by Proposition 1.9, $R: E \otimes_p F \longrightarrow G$ is continuous. ■

Example (0.3.52). If E and F are spaces with strongest seminorm system, then obviously so is $E \otimes_i F$. Therefore (see Example 2.10) the algebraic tensor product may be considered as a special case of the inductive tensor product.

An analogous existence theorem is true also for complete Hausdorff inductive and projective tensor products; it is only necessary to pass from the spaces $E \otimes_i F$ and $E \otimes_p F$ (Hausdorff by Proposition 3.50) to their completions, which are usually denoted by $E \bar{\otimes} F$ and $E \hat{\otimes} F$, respectively. We leave the details to the reader; see also [81, II §4.1].

Proposition (0.3.53). *If E and F are metrizable, then so is $E \otimes_p F$. As a consequence, if E and F are Fréchet spaces, so is $E \hat{\otimes} F$.* ▸■

Just as in the "Banach" case, each of the selected tensor products may be defined not only for spaces but for operators also. We let the reader give corresponding analogues of the concepts and facts of the preceding section.

It was noted earlier (Theorem 3.35) that the $L^1(X, \mu)$ are, among Banach spaces, "nice" to tensor by. In passing to more general polynormed spaces the scope of such examples broadens; in particular, for the tensor product "$\hat{\otimes}$", it includes practically all the spaces of test functions and generalized functions that are used in analysis. There is a theorem of a general nature in that direction due to Grothendieck [26, II Theorem 13]; we limit ourselves to one specific example which we suggest in the form of an exercise.

Let U be a domain in $\mathbb{C}^n$ and let E be a polynormed space. A function $\overline{w}: U \longrightarrow E$ is said to be *holomorphic* if for any $f \in E^*$ the (complex-valued) function $f \circ \overline{w}: U \longrightarrow \mathbb{C}$ is holomorphic. Holomorphic vector valued functions are discussed for example, in [84, III §2], [67, pp.78 - 81] or (with emphasis on functions of several variables) in [16, II]. We require only the simplest case where $U = \mathbb{D}$ and E is a Banach space. Then the function $\overline{w}: \mathbb{D} \longrightarrow E$ is holomorphic $\Longleftrightarrow$ it can be represented as a series $\overline{w}(z) = \sum_{k=0}^{\infty} x_k z^k$; $x_k \in E$, which is absolutely convergent for all $z \in \mathbb{D}$. The space $\mathcal{O}(\mathbb{D}, E)$ of holomorphic E-valued functions on $\mathbb{D}$ is a Fréchet space with respect to pointwise operations and the system of norms $\|\bar{w}\|_n := \max\{\|\bar{w}(z)\| : |z| \le 1 - \frac{1}{n}\}$. It does, however, have another defining system of norms, namely $\|\cdot\|'_n$; $\left\| \sum_{k=0}^{\infty} x_k z^k \right\|'_n := \sum_{k=0}^{\infty} \|x_k\| \left(1 - \frac{1}{n}\right)^k$ ($n \in \mathbb{N}$ everywhere).

Exercise (0.3.54). Let E be a Banach space. Show that there exists a topological isomorphism between $\mathcal{O}(\mathbb{D}) \hat{\otimes} E$ and $\mathcal{O}(\mathbb{D}, E)$, that takes $w \otimes x$ into the vector function $\bar{v}(z) := w(z)x$.

Hint. The operator R, associated with $\mathcal{R} : (w, x) \longmapsto \bar{v}$, maps the dense subspace $L := \{u = \sum_{k=1}^{m} z^k \otimes x_k\}$ of $\mathcal{O}(\mathbb{D}) \hat{\otimes} E$ onto a dense subspace in $\mathcal{O}(\mathbb{D}, E)$. For $u \in L$, it is not difficult to verify that $\|u\|_{p(n)} \le \|R(u)\|'_n$ where $\|\cdot\|_{p(n)}$ is the projective tensor product of the norm $\|w\|_n = \max\{|w(z)| : |z| \le 1 - \frac{1}{n}\}$ in $\mathcal{O}(\mathbb{D})$ by the norm in E.

Actually this fact is true for arbitrary domains in $\mathbb{C}^n$ and also for arbitrary complete Hausdorff polynormed spaces; it remains valid on replacing $\mathcal{O}(U)$ by other "classical" polynormed spaces of test functions. In particular, the tensor product of the spaces of all infinitely smooth functions on manifolds $\mathcal{M}_1$ and $\mathcal{M}_2$ is the space of all infinitely smooth functions on $M_1 \times \mathcal{M}_2$; a similar fact is true for holomorphic functions, rapidly decreasing infinitely smooth functions and so on.

We refer the reader to [26] and to the excellent book [76] for details; see also [90] and [81].

* * *

There is also a unique natural analogue of the weak tensor product mentioned in §3.3 on passing from Banach spaces to general polynormed spaces. Namely, given polynormed spaces E and F, their *weak tensor product* is what we call — the reader will certainly have already guessed — the space $E \otimes_w F$ introduced above, or, when considering the "complete Hausdorff" version, the completion of this space, denoted by $E \check{\otimes} F$ (for both of these, we take the bi-operator $(x, y) \longmapsto x \otimes y$).

Together with the projective tensor product, the weak tensor product features in the definition of one of the most remarkable classes of spaces, discovered by Grothendieck: a polynormed Hausdorff space E is said to be *nuclear* if for any polynormed Hausdorff space F the identity operator $E \otimes_p F \longrightarrow E \otimes_w F$ is a topological isomorphism. (This is only one of the many definitions of nuclear spaces, which characterize them from various points of view.) An overwhelming majority of the spaces of test and generalized functions considered in analysis — spaces of holomorphic functions, smooth functions, distributions with compact supports and so on — are nuclear spaces; $\mathbb{C}^M$ is also nuclear for any M. On the other hand a Banach space is nuclear $\Longleftrightarrow$ it is finite-dimensional. (However, as was shown quite recently by Pisier [209], there exist infinite-dimensional Banach spaces E for which $1: E \otimes_p E \longrightarrow E \otimes_w E$ is a topological isomorphism).

The basic monographs [26], [76], already referred to, are devoted to nuclear spaces and their application to analysis; see also [59], [90].

§4. COMPLEXES AND THEIR HOMOLOGY

The material set out here will be used (with one exception in Ch.II §2.2) only in the last chapter of the book. However, the scope of its applications, going far beyond the bounds of our subject, forms a very significant part of modern mathematics. Therefore, even if our reader wishes to limit himself only to the initial chapters (which are devoted to the "classic" problems of the theory) we would, all the same, recommend him not to omit this section; it should prove useful at some time.

§4.1. Complexes, homotopy and splittability

Let

$$\cdots X_{n-1} \xleftarrow{d_{n-1}} X_n \xleftarrow{d_n} X_{n+1} \longleftarrow \cdots$$

$(\cdots X^{n-1} \xrightarrow{\delta^{n-1}} X^n \xrightarrow{\delta^n} X^{n+1} \longrightarrow \cdots,$ respectively) be a sequence[15] of objects and morphisms in a certain pointed category. It is called a *chain* (*cochain*, respectively) *complex* if the composition of any pair of neighbouring morphisms is the zero morphism. The difference between chain and cochain complexes is purely conventional (does it really matter in which direction the arrows go?); however, in practice, as we shall see, it is convenient to work at times with both concepts.

When we are talking about a complex, the morphisms d_n (or δ^n) are called *differentials*. We shall often write (co)chain complexes in the form $\{X, d\}$ $(\{X, \delta\})$, implying X_n (X^n) and d_n (δ^n).

A sequence and, in particular, a complex of the form $0 \longleftarrow X \longleftarrow Y \longleftarrow Z \longleftarrow 0$ is called a *short sequence* or *complex*.

---

[15] We consider this to be either infinite in both directions, or to end on one or both sides with the zero object ("zero"); in this case we identify it with the sequence that is infinite on both sides and is obtained by extending the zeros to the left and/or to the right.

A *morphism between the complexes* $\{X, d\}$ and $\{Y, d'\}$ is a family of morphisms $\underline{\varphi} : \{\varphi_n : X_n \longrightarrow Y_n\}$ such that the diagram

$$\begin{array}{ccccccc} \cdots \longleftarrow & X_n & \xleftarrow{d_n} & X_{n+1} & \longleftarrow \cdots \\ & \Big\downarrow{\scriptstyle \varphi_n} & & \Big\downarrow{\scriptstyle \varphi_{n+1}} & \\ \cdots \longleftarrow & Y_n & \xleftarrow{d'_n} & Y_{n+1} & \longleftarrow \cdots \end{array}$$

is commutative; morphisms of cochain complexes are defined similarly.

The (co)chain complexes in a category $\mathcal{K}$ and their morphisms form a pointed category in themselves, denoted by $\underline{\mathcal{K}}$ $(\overline{\mathcal{K}})$, and this is also additive if $\mathcal{K}$ is additive. Every co(contra)variant functor $F : \mathcal{K}_1 \longrightarrow \mathcal{K}_2$ between two pointed categories evidently generates a co(contra)variant functor from $\underline{\mathcal{K}_1}$ into $\underline{\mathcal{K}_2}$ $(\overline{\mathcal{K}_2})$, denoted by $\underline{F}$ $(\overline{F})$.

Example (0.4.1) The functor $(*) : \mathcal{B}an \longrightarrow \mathcal{B}an$ (see §2.2) generates $\overline{(*)} : \underline{\mathcal{B}an} \longrightarrow \overline{\mathcal{B}an}$. Obviously, given a complex $\mathcal{X} \in \mathcal{B}an$, the complex $\overline{(*)}(\mathcal{X})$, called the *dual* of $\mathcal{X}$ and denoted also by $\mathcal{X}^*$, consists of the duals of the spaces and operators that constitute $\mathcal{X}$.

Now let $\mathcal{K}$ be additive. Then one can distinguish, from among its complexes, the class of those that, in a well-defined sense, are constructed in the best way. As a preliminary we introduce one very important concept. We consider morphisms $\underline{\varphi}, \underline{\psi} : \mathcal{X} = \{X, d\} \longrightarrow \mathcal{Y} = \{Y, d'\}$ between two (for definiteness, chain) complexes from $\mathcal{K}$.

Definition (0.4.2). A family of morphisms $s = \{s_n : X_n \longrightarrow Y_{n+1}, n \in \mathbb{Z}\}$ is called a *homotopy* between $\underline{\varphi}$ and $\underline{\psi}$ if $d'_n s_n + s_{n-1} d_{n-1} = \varphi_n - \psi_n$ for each n. Two morphisms of complexes are called *homotopic* if there is a homotopy between them.

(For the historical roots of the homotopy concept, which are geometrical, see, for example [51, pp.38 - 39].)

Definition (0.4.3). A morphism of complexes $\underline{\varphi} : \mathcal{X} \longrightarrow \mathcal{Y}$ is called a *homotopy equivalence* if there exists $\underline{\psi} : \mathcal{Y} \longrightarrow \mathcal{X}$ such that $\underline{\psi} \circ \underline{\varphi}$ is homotopic to the identity morphism on $\mathcal{X}$ and $\underline{\varphi} \circ \underline{\psi}$ is homotopic to the identity morphism on $\mathcal{Y}$. The two complexes are called *homotopically equivalent* if there is a homotopy equivalence between them.

Definition (0.4.4). A complex $\mathcal{X}$, for which the identity morphism on $\mathcal{X}$ is homotopic to the zero morphism, is said to *split* or to be *contractible*. The homotopy between the identity and the zero morphism is called a *contracting homotopy* for $\mathcal{X}$.

Thus a split complex $\mathcal{X} = \{X, d\}$ is one that can be augmented to get a diagram

$$\cdots \underset{}{\overset{d_{n-2}}{\rightleftarrows}} X_{n-1} \underset{s_{n-1}}{\overset{d_{n-1}}{\rightleftarrows}} X_n \underset{s_n}{\overset{d_n}{\rightleftarrows}} X_{n+1} \rightleftarrows \cdots \tag{1}$$

where $d_n s_n + s_{n-1} d_{n-1} = 1_{X_n}$ for all n. Here, $s = \{s_n : X_n \longrightarrow X_{n+1}\}$ is its contracting homotopy.

Proposition (0.4.5). *Any additive functor between two additive categories preserves the homotopy relation between morphisms of complexes. As a consequence, it preserves the relation of homotopy equivalence between complexes and takes split complexes to split complexes.* ▶■

Split complexes are the "best" kind of complex; the reader will be gradually convinced of this on observing their behaviour from various points of view. First of all we note that we are essentially already acquainted with short split complexes; these are complexes that can be augmented to get a direct sum diagram (bordered by zeros).

We shall now show that, for complexes of polynormed spaces, splittability has a fairly intuitive geometrical meaning.

Proposition (0.4.6). *A complex* $\mathcal{X}$ *in* $\mathcal{PN}$ *(or in any of its subcategories indicated in* §2.1) *splits* $\Leftrightarrow$ *it is isomorphic to a complex of the form*

$$\cdots \; Y_n \oplus Y_{n+1} \overset{d'_n}{\longleftarrow} Y_{n+1} \oplus Y_{n+2} \longleftarrow \cdots \tag{$\mathcal{Y}$}$$

where d'_n *coincides with the identity operator on* Y_{n+1} *and with the zero operator on* Y_{n+2}; $n \in \mathbb{Z}$. *In particular, a short complex in these same categories splits* $\Leftrightarrow$ *it is isomorphic to a complex of the form*

$$0 \longleftarrow X \overset{j}{\longleftarrow} X \oplus Z \overset{i}{\longleftarrow} Z \longleftarrow 0,$$

in which $i : z \longmapsto (0, z)$ *and* $j : (x, z) \longmapsto x$.

▶ "$\Leftarrow$" is obvious; we prove "$\Rightarrow$". Let the diagram (1) above depict our complex and its contracting homotopy. We take $\mathcal{Y}$ to be the complex for which $Y_n = \operatorname{Ker} d_{n-2}$. Then the morphism $\underline{\varphi} ; \mathcal{X} \longrightarrow \mathcal{Y}$ formed from $\varphi_n : X_n \longrightarrow Y_n \oplus Y_{n+1} : x \longmapsto (d_{n-1}x, d_n s_n x)$ has an inverse morphism, namely $\underline{\psi}$, where $\psi_n : Y_n \oplus Y_{n+1} \longrightarrow X_n : (y, z) \longmapsto s_{n-1}y + z$. In fact, $\psi_n \varphi_n(x) = s_{n-1}d_{n-1}x + {} + d_n s_n x = x$, and

$$\varphi_n\psi_n(y, z) = (d_{n-1}s_{n-1}y + d_{n-1}z,\ d_ns_ns_{n-1}y + d_ns_nz) =$$
$$= (y - s_{n-2}d_{n-2}y + 0,\ (1 - s_{n-1}d_{n-1})s_{n-1}y + z - s_{n-1}d_{n-1}z) =$$
$$= (y,\ s_{n-1}s_{n-2}d_{n-2}y + z) = (y, z).$$

Thus $\underline{\varphi}$ is an isomorphism. ■

Other equivalent definitions of splittability will be given in the next section.

§4.2. Exact sequences and their interpretation in the simplest cases. Comparison of exactness and splittability

Unlike the case for complexes, it is convenient for us to consider exact sequences only for those categories whose objects are linear spaces (maybe, with some additional structure), for which the zero space is the zero object and in which every morphism is an operator.[16] Such categories include all the categories of polynormed spaces considered and also a lot of others we shall meet later on.

Definition (0.4.7). A sequence

$$\cdots \longleftarrow X_{n-1} \xleftarrow{\varphi_{n-1}} X_n \xleftarrow{\varphi_n} X_{n+1} \longleftarrow \cdots$$

of linear spaces and operators is said to be *exact at the term* X_n if the kernel of the operator φ_{n-1} coincides with the image of φ_n and *exact* if it is exact at all its terms.

Since complexes, in the class of categories under consideration, are obviously characterized by the condition $\operatorname{Im}\varphi_n \subseteq \operatorname{Ker}\varphi_{n-1}$, $n \in \mathbb{Z}$, an exact sequence in any of them is automatically a complex; we shall often say "exact complex" instead of "exact sequence".

We recall that it is commonplace that exact sequences turn out to be a very convenient and compact language for expressing many facts in algebra, topology and analysis: by proving that a sequence is exact we obtain important information about its objects and morphisms. Here are several very simple examples.

Proposition (0.4.8). *Let* $\mathcal{K} = \mathcal{L}in$ *or* $\mathcal{PN}$ *and let* $\varphi : E \longrightarrow F$ *be a morphism in* $\mathcal{K}$. *Then* φ *is injective, surjective, bijective, respectively* $\Leftrightarrow$ *there exists an exact sequence in* $\mathcal{K}$ *of the form* $\cdots \xleftarrow{\psi} F \xleftarrow{\varphi} E \longleftarrow 0$, $0 \longleftarrow F \xleftarrow{\varphi} E \xleftarrow{\psi} \cdots$,

[16] In fact there is also the concept of an exact sequence (and also of a short exact sequence) [296] in general category theory, but it is superfluous.

$0 \longleftarrow F \overset{\varphi}{\longleftarrow} E \longleftarrow 0$, *respectively. If, however,* $\mathcal{K} = \mathcal{B}an$ *or* $\mathcal{F}r$, *then these assertions still hold if, on the left of the sign "⟺", we replace "injective" by "topologically injective", "surjective" by "topologically surjective" (= "open") and "bijective" by "a topological isomorphism".*

▶ All the assertions can be verified immediately for $\mathcal{L}in$ and $\mathcal{PN}$; we only note that in forming the first exact sequence we can put $\psi = \mathbf{pr} : F \longrightarrow F/\mathrm{Im}\,\varphi$, and in forming the second we can put $\psi = \mathbf{in} : \mathrm{Ker}\,\varphi \longrightarrow E$. In exactly the same way, but involving the Banach theorem in an obvious way, the assertions can be verified for $\mathcal{B}an$ and $\mathcal{F}r$. ■

The interpretation of the exactness of short sequences deserves special attention.

Proposition (0.4.9). *A short sequence* $\underline{\mathcal{S}}$ *in* $\mathcal{K} = \mathcal{L}in$ *or* $\mathcal{PN}$ *is exact* ⟺ *we have a commutative diagram in* $\mathcal{K}$

$$\begin{array}{ccccccccc} 0 & \longleftarrow & G & \overset{j}{\longleftarrow} & F & \overset{i}{\longleftarrow} & E & \longleftarrow & 0 \\ & & \uparrow & & \| & & \downarrow & & \\ 0 & \longleftarrow & F/\mathrm{Im}\ i & \longleftarrow & F & \longleftarrow & \mathrm{Im}\ i & \longleftarrow & 0 \end{array} \qquad (\underline{\mathcal{S}})$$

in which the vertical arrows represent continuous bijective operators. ▶■

Definition (0.4.10). A short exact sequence $\underline{\mathcal{S}}$ in which i is topologically injective and j is topologically surjective, is called *topologically exact.*

Proposition (0.4.11). *Any short exact sequence in* $\mathcal{B}an$ *or* $\mathcal{F}r$ *is topologically exact.*

▶ This is obviously another equivalent formulation of the Banach theorem. ■

Proposition (0.4.12). *If* $\underline{\mathcal{S}}$ *is a topologically exact sequence in* $\mathcal{PN}$ *(for example, any short exact sequence in* Ban *or* $\mathcal{F}r$*),*[17] *then the operators depicted by vertical arrows in the diagram of Proposition* 4.9 *are necessarily topological isomorphisms.* ▶■

Thus in the above situation the short exact sequence can be interpreted as a sequence of "subobject ⟶ object ⟶ quotient-object".

---

[17] Or even in $\mathcal{L}in$; cf. Example 2.10.

* * *

For the categories described at the beginning of the section there exists a useful method of reducing "long" complexes, in particular, exact sequences, to short ones. Namely, together with the given complex, consisting of linear spaces and operators, we consider the commutative diagram

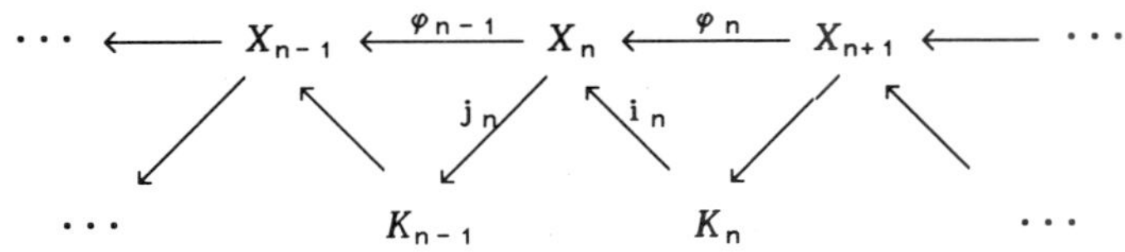

$(\mathcal{X})$

in which $K_n := \operatorname{Ker} \varphi_{n-1}$ for all n, and i_n is the natural embedding; this then defines j_n uniquely. Obviously, each of the sequences

$$0 \longleftarrow K_{n-1} \xleftarrow{j_n} X_n \xleftarrow{i_n} K_n \longleftarrow 0 \qquad (\underline{\mathcal{S}}_n)$$

is a complex and to be given $\mathcal{X}$ is equivalent to being given the (infinite or finite) collection of $\underline{\mathcal{S}}_n$. These $\underline{\mathcal{S}}_n$ are called the *components* of $\mathcal{X}$ and $\mathcal{X}$ is the *Yoneda product* of the $\underline{\mathcal{S}}_n$.

Proposition (0.4.13). *$\mathcal{X}$ is exact $\Leftrightarrow$ each of its component complexes is exact.* ▶■

The reader will see that the exactness property, like the splittability property introduced above, characterizes "good" complexes. How are these two properties connected? First of all, for any of the categories considered here, we have the following theorem.

Theorem (0.4.14). *Every split complex is exact.*

▶ If $s = \{s_n\}$ is a contracting homotopy for $\mathcal{X} = \{X, d\}$, then $x \in \operatorname{Ker} d_{n-1}$ implies that $x = d_n s_n x + s_{n-1} d_{n-1} x = d_n(s_n x) \in \operatorname{Im} d_n$. ■

Now let $\mathcal{K}$ denote any of the categories of polynormed spaces introduced in §2. It is possible, for these categories, to give a precise characterization of those properties that are "needed for exact complexes to split". We begin with short complexes.

Theorem (0.4.15). *The following properties of the complex* $0 \longleftarrow G \xleftarrow{j} F \xleftarrow{i} E \longleftarrow 0$ $(\underline{\mathcal{S}})$ *in* $\mathcal{K}$ *are equivalent:*

(I) $\underline{\mathcal{S}}$ *splits;*

(II) $\underline{\mathcal{S}}$ *is exact,* j *is a retraction in* $\mathcal{K}$ *and* i *is topologically injective;*

(III) $\underline{\mathcal{S}}$ *is exact,* i *is a coretraction in* $\mathcal{K}$ *and* j *is open.*

(We note straight away that Proposition 4.11 shows that, when $\mathcal{K} = \mathcal{B}an$ or $\mathcal{F}r$, the conditions on i in (II) and on j in (III) are automatically satisfied.)

▶ (I) ⇒ (II), (III). This is obviously implied by Theorem 4.14 and the form of a contracting homotopy for $\underline{\mathcal{S}}$.

(II) ⇒ (I). Let ρ be a right inverse to j in $\mathcal{K}$. Then $x - \rho jx \in \operatorname{Ker} j = \operatorname{Im} i$ is satisfied for any $x \in F$ and therefore, by the condition on i, the operator $\sigma : F \longrightarrow E : x \longmapsto (i|^{\operatorname{Im} i})^{-1}(x - \rho jx)$ is continuous. As is easily verified, ρ and σ furnish a contracting homotopy for $\underline{\mathcal{S}}$ in $\mathcal{K}$.

We leave the reader to deduce (I) from (III) by similar arguments. ■

It is evident from this theorem that already in $\mathcal{B}an$ (and hence in all larger categories) the splittability property is essentially stronger than the exactness property. In fact, by Proposition 1.14 and Theorem 1.16, the operator **in** in the exact complex $0 \longrightarrow c_0 \xrightarrow{\mathbf{in}} c_b \xrightarrow{\mathbf{pr}} c_b/c_0 \longrightarrow 0$ is not a coretraction (and **pr** is not a retraction) and therefore by Theorem 4.15 this complex does not split.

Passing from short complexes to "long" complexes, we note that if we have a complex in any of the categories under consideration its component short complexes also belong to the same category. Moreover, the following obviously holds.

Proposition (0.4.16). *A complex in* $\mathcal{K}$ *splits* ⟺ *all its component complexes split.* ▶■

Theorem (0.4.17). *For an (arbitrary) complex* $\mathcal{X}$ *in* $\mathcal{K}$ *to split it is necessary, and when* $\mathcal{K} = \mathcal{B}an$ *(or* $\mathcal{F}r$*) it is sufficient, that* $\mathcal{X}$ *be exact and the kernel of every differential in* $\mathcal{X}$ *be complemented.*

▶ Necessity. The complex $\mathcal{Y}$ in Proposition 4.6 has automatically both the required properties; therefore so has the complex $\mathcal{X}$ that is isomorphic to it.

Sufficiency (when $\mathcal{K} = \mathcal{B}an$ or $\mathcal{F}r$). By Proposition 4.16 our problem is to prove that each of the component short and, by Proposition 4.13, exact complexes of $\mathcal{X}$ splits. Let $0 \longleftarrow G \xleftarrow{j} F \xleftarrow{i} E \longleftarrow 0$ be any one of these complexes; then, according to Propositions 4.9 and 4.12, it is sufficient to prove that the complex $0 \longleftarrow F/\operatorname{Im} i \xleftarrow{\mathbf{pr}} F \xleftarrow{\mathbf{in}} \operatorname{Im} i \longleftarrow 0$ splits. But $\operatorname{Im} i$ is precisely the kernel of a certain differential in $\mathcal{X}$, and therefore it is complemented in F. Hence, in view of

Proposition 1.14, this last complex has properties (II) and (III) in Theorem 4.15. The rest is clear. ■

We note that the situation in $\mathscr{Lin}$ is considerably simpler:

Proposition (0.4.18). *A complex in* $\mathscr{Lin}$ *is exact* $\Leftrightarrow$ *it splits.*

▸ As is easily seen, every surjective morphism in $\mathscr{Lin}$ is a retraction (and an injective morphism is a coretraction); therefore if we consider $\mathscr{Lin}$ as a subcategory of $\mathscr{PN}$ and use Theorem 4.15 we get directly that every short exact complex in $\mathscr{Lin}$ splits. All that remains, on passing from short complexes to an arbitrary complex, is to apply Propositions 4.13 and 4.16 in an obvious way. ■

* * *

The principal difference between the properties of exactness and splittability of complexes is that exactness may be lost after applying an additive functor (cf. Proposition 4.5). The ubiquitous problem of whether a given functor preserves exactness turns out to be closely connected with the problem mentioned in Remark 2.6 and it is also related to the subject of "homology theory". This problem will also be considered in this book for certain functors that are important in the theory of "the algebras of analysis". For the moment we present as an exercise one specific positive result which has important applications ([30] [81, 0.5.2 and VII.1.3]).

Exercise (0.4.19). Show that a complex $\mathcal{X}$ in the category $\mathscr{Ban}$ is exact $\Leftrightarrow$ its dual complex $\mathcal{X}^*$ is exact.

Hint. Using the Banach and Hahn-Banach theorems establish that (I) if $\mathcal{X}$ is exact at X_{n-1} and at X_n, then $\mathcal{X}^*$ is exact at $\mathcal{X}^*_n$; (II) if $\mathcal{X}^*$ is exact at X^*_{n+1}, then $\operatorname{Im} d_n$ is closed in X_n and finally (III) if $\mathcal{X}^*$ is exact at X^*_n, then $\operatorname{Im} d_n$ is dense in $\operatorname{Ker} d_{n-1}$. The arguments are facilitated if one passes from "long" complexes to the component short complexes.

However, the functor $h_{\mathscr{Ban}}(?, E)$, where E is other than $\mathbb{C}$, need not preserve exactness except at certain terms:

Exercise (0.4.20). Let $0 \longleftarrow E_1 \overset{\sigma}{\longleftarrow} E_2 \longleftarrow E_3 \longleftarrow \cdots$ be an exact sequence in $\mathscr{Ban}$, and let F be another Banach space. Show that $0 \longrightarrow \mathscr{B}(E_1, F) \longrightarrow \mathscr{B}(E_2, F) \longrightarrow \mathscr{B}(E_3, F) \longrightarrow \cdots$ is exact at the first two terms (not counting zero), but it is cetainly not automatically exact at the third term if $F = \operatorname{Ker} \sigma$ and $\operatorname{Ker} \sigma$ is not complemented in E_2.

The tensor product functor also need not preserve exactness and again this is implied by the existence of non-complemented subspaces.

Exercise (0.4.21) (Grothendieck). Let F be a reflexive non-complemented closed subspace of the Banach space E, and suppose that $0 \longrightarrow F \xrightarrow{\mathbf{in}} E \longrightarrow \dots$ is a certain exact sequence. Show that the operator $\varphi := \mathbf{in} \hat{\otimes} \mathbf{1}_{F*}$ is certainly not topologically injective and, as a consequence, the sequence $0 \longrightarrow F \hat{\otimes} F^* \xrightarrow{\varphi} E \hat{\otimes} F^* \longrightarrow \dots$ is not exact at least at one of the terms shown.

Hint. Using dual associativity, identify φ^* with $\mathbf{h}_{\mathcal{B}an}(\mathbf{in}, F)$ which is automatically not surjective. Next apply the Hahn-Banach theorem.

§4.3. Homology of a complex. "Basic lemma of homological algebra"

In this section $\mathcal{K}$ again denotes any of the standard categories of polynormed spaces, including $\mathcal{L}in$. (In fact, the concepts and facts stated here can be extended in a natural way to a far wider class of categories, including the category of modules with which we shall become acquainted later on.)

We fix an integer n and take a chain complex $\mathcal{X} = \{X, d\}$ in $\mathcal{K}$. Since $d_{n-1}d_n = 0$, there is a subspace $\operatorname{Im} d_n$ of $\operatorname{Ker} d_{n-1}$ which, in general, is not closed. We put $H_n(\mathcal{X}) := \operatorname{Ker} d_{n-1}/\operatorname{Im} d_n$; although not necessarily Hausdorff, this is obviously a polynormed space, in particular, a complete seminormed space when $\mathcal{K} = \mathcal{B}an$.

Definition (0.4.22). $H_n(\mathcal{X})$ is called the *n-dimensional homology* and the family $\{H_n(\mathcal{X});\ n \in \mathbb{Z}\}$ is called the *homology* of the complex $\mathcal{X}$.

Let $\underline{\varphi} = \{\varphi_n\}$ be a morphism from $\mathcal{X} = \{X, d\}$ into $\mathcal{Y} = \{Y, d'\}$. Then obviously $\varphi_n(\operatorname{Ker} d_{n-1}) \subseteq \operatorname{Ker} d'_{n-1}$ and $\varphi_n(\operatorname{Im} d_n) \subseteq \operatorname{Im} d'_n$ for every n. Therefore $(\varphi_n|_{\operatorname{Ker} d_{n-1}})|^{\operatorname{Ker} d'_{n-1}}$ induces (in the sense of §0.2) a continuous operator $H_n(\underline{\varphi}) : H_n(\mathcal{X}) \longrightarrow H_n(\mathcal{Y})$, which for fixed n will often be abbreviated to φ_*.

The correspondence $\mathcal{X} \longmapsto H_n(\mathcal{X})$; $\underline{\varphi} \longmapsto H_n(\underline{\varphi})$, denoted by $H_n : \underline{\mathcal{K}} \longrightarrow \mathcal{PN}$, is obviously a covariant functor; it is called the *n-dimensional homology functor*. The *cohomology of a cochain complex* and the *n-dimensional cohomology functor* $H^n : \overline{\mathcal{K}} \longrightarrow \mathcal{PN}$ are defined analogously. In the case of Banach structures it is appropriate, on the basis of what has been noted above, to talk about the functors $H_n : \mathcal{B}an \longrightarrow \langle\mathcal{B}an\rangle$ and $H^n : \mathcal{B}an \longrightarrow \langle\mathcal{B}an\rangle$, where $\langle\mathcal{B}an\rangle$ is the category of complete seminormed spaces and continuous operators.

Of course, for any $\mathcal{K}$ we are justified in assuming that both functors, H_n and H^n, take values (simply) in $\mathcal{L}in$; and for many, although not all, applications this is sufficient.

Of course, $H_n(\mathcal{X})$ being equal to zero for all n means that the complex is exact (= it is an exact sequence). It can be said that the functor H_n "measures the degree of non-exactness of the complex at its n-th term".

When calculating the n-dimensional (co)homology functor the following two facts are very useful; we limit ourselves to formulating them for H_n.

Theorem (0.4.23). *If two morphisms between complexes* $\underline{\varphi}, \underline{\psi} : \mathcal{X} \longrightarrow \mathcal{Y}$ *are homotopic, then the operators* $H_n(\underline{\varphi}), H_n(\underline{\psi}) : H_n(\mathcal{X}) \longrightarrow H_n(\mathcal{Y})$ *coincide for all* $n \in \mathbb{Z}$. ▶■

Corollary (0.4.24). *If* $\mathcal{X}$ *and* $\mathcal{Y}$ *are homotopically equivalent, then there is a topological isomorphism between* $H_n(\mathcal{X})$ *and* $H_n(\mathcal{Y})$ *for every* $n \in \mathbb{Z}$.

The following theorem plays a fundamental role in most of the questions connected with the functors under discussion. In a purely algebraic context (as a rule, that of modules over rings) it is often called the "basic lemma of homological algebra".

A short sequence

$$0 \longleftarrow \mathcal{X}'' \xleftarrow{\underline{\varphi}} \mathcal{X} \xleftarrow{\underline{\psi}} \mathcal{X}' \longleftarrow 0 \qquad (\underline{\mathcal{S}})$$

of complexes from $\mathcal{K}$ is said to be *exact* if in the diagram below, where $\underline{\mathcal{S}}$ is written out in expanded form, every column is exact.

$$\begin{array}{ccccccc}
 & & 0 & & 0 & & \\
 & & \downarrow & & \downarrow & & \\
\cdots \longleftarrow & & X'_{n-1} & \xleftarrow{d'_{n-1}} & X'_n & \longleftarrow \cdots & \\
 & \text{II} & \downarrow \psi_{n-1} & \text{IV} & \downarrow \psi_n & & \\
\cdots \longleftarrow & & X_{n-1} & \xleftarrow{d_{n-1}} & X_n & \longleftarrow \cdots & \qquad (\mathcal{D}) \\
 & & \downarrow \varphi_{n-1} & \text{I} & \downarrow \varphi_n & \text{III} & \\
\cdots \longleftarrow & & X''_{n-1} & \xleftarrow{d''_{n-1}} & X''_n & \longleftarrow \cdots & \\
 & & \downarrow & & \downarrow & & \\
 & & 0 & & 0 & &
\end{array}$$

Theorem (0.4.25). *Let* $\underline{\mathcal{S}}$ *be a short exact sequence in* $\underline{\mathcal{K}}$. *Then operators* $\xi_n : H_n(\mathcal{X}'') \longrightarrow H_{n-1}(\mathcal{X}')$; $n \in \mathbb{Z}$, *exist such that the sequence*

$$\cdots \leftarrow H_{n-1}(\mathcal{X}') \xleftarrow{\xi_n} H_n(\mathcal{X}'') \xleftarrow{H_n(\underline{\varphi})} H_n(\mathcal{X}) \xleftarrow{H_n(\underline{\psi})} H_n(\mathcal{X}') \xleftarrow{\xi_{n+1}} \cdots \tag{$H(\mathcal{S})$}$$

is exact. If, in addition, $\mathcal{K} = \mathcal{B}an$ *or* $\mathcal{F}r$, *then* ξ_n *is continuous (in other words,* $H(\underline{\mathcal{S}}) \in \underline{\mathcal{PN}}$).

▶ For brevity we put $\hat{X}_n := \varphi_n^{-1}(\operatorname{Ker} d''_{n-1})$, $\check{X}_n := \varphi_n^{-1}(\operatorname{Im} d''_n)$ and $\tilde{\psi}_n := \psi_n|^{\operatorname{Im}\psi_n}$; these last operators, in view of the columns being exact, have inverses.

We take $x \in \hat{X}_n$; square I being commutative implies that $d_{n-1}x \in \operatorname{Ker} \varphi_{n-1} = \operatorname{Im} \psi_{n-1}$, and therefore $\tilde{\psi}_{n-1}^{-1} \circ d_{n-1}x \in X'_{n-1}$ is defined; square II being commutative implies that $d'_{n-2}\tilde{\psi}_{n-1}^{-1} \circ d_{n-1}x = 0$. This means that the operator $\eta : \hat{X}_n \longrightarrow \operatorname{Ker} d'_{n-2} : x \longmapsto \tilde{\psi}_{n-1}^{-1} \circ d_{n-1}x$ is defined.

Now let $x \in \check{X}_n$. Since the columns are exact and since square III is commutative, this implies that $\varphi_n x = \varphi_n d_n x'$ for some $x' \in X_{n+1}$ and hence $x - d_n x' \in \operatorname{Ker} \varphi_n = \operatorname{Im} \psi_n$. But then square IV being commutative implies that $\eta(x)$, being equal to $\eta(x - d_n x')$, is $d'_{n-1}\tilde{\psi}_n^{-1}(x - d_n x') \in \operatorname{Im} d'_{n-1}$.

We have shown that $\eta(\check{X}_n) \subseteq \operatorname{Im} d'_{n-1}$. Consequently η generates an operator $\xi_n : \hat{X}_n/\check{X}_n \longrightarrow \operatorname{Ker} d'_{n-2}/\operatorname{Im} d'_{n-1} = H_{n-1}(\mathcal{X}')$. But the first quotient space obviously coincides, to within a topological isomorphism, with $H_n(\mathcal{X}'')$; in this way we have constructed the desired ξ_n. We leave the reader to carry out the elementary arguments proving that the sequence $H(\underline{\mathcal{S}})$ is exact.

Finally, we note that for $\mathcal{K} = \mathcal{B}an$ or $\mathcal{F}r$ the Banach theorem implies that $\tilde{\psi}_{n-1}^{-1}$ is continuous; hence so also is η and with it also ξ_n. ■

$H(\underline{\mathcal{S}})$ is called the *long exact homology sequence* associated with $\underline{\mathcal{S}}$ and the ξ_n are called the *connecting operators.* Theorem 4.25 is carried across in an obvious way to cochain complexes (the connecting operators in the latter case are denoted by ξ^n); we recommend the reader to formulate carefully the corresponding assertions.

We give (taking into account Proposition 4.8) one of the standard corollaries of this theorem which describes a situation that is met fairly frequently in practice.

Corollary (0.4.26). *If one of the three complexes in* $\underline{\mathcal{S}}$ *is exact, then all the non-trivial operators in* $H(\underline{\mathcal{S}})$ *are linear isomorphisms.*

It should be noted that these isomorphisms are, in general, between non-Hausdorff spaces, and need not be topological. All the same it has been successfully proved in certain cases that they are.

Exercise (0.4.27). Let $\underline{\mathcal{S}}$ consist of complexes in $\mathcal{B}an$ or $\mathcal{F}r$ and let one of its edge terms be exact. Show that all the non-trivial operators in $H(\underline{\mathcal{S}})$ are topological isomorphisms.

Hint. If, let us say, $\mathcal{X}''$ is exact, then $\psi_* : H_n(\mathcal{X}') \longrightarrow H_n(\mathcal{X})$ is bijective; therefore the operator

$$\tilde{\psi} : \operatorname{Ker} d'_{n-1} \oplus X_{n+1} \longrightarrow \operatorname{Ker} d_{n-1} : (x', x) \longmapsto \psi_n(x') + d_n x$$

is surjective and hence open. But $\tilde{\psi}$ occurs in the commutative diagram

$$\begin{array}{ccccc} \operatorname{Ker} d'_{n-1} \oplus X_{n+1} & \xrightarrow{j} & \operatorname{Ker} d'_{n-1} & \xrightarrow{\sigma} & H_n(\mathcal{X}') \\ \downarrow{\scriptstyle \tilde{\psi}} & & & & \downarrow{\scriptstyle \psi^*} \\ \operatorname{Ker} d_{n-1} & & \xrightarrow{\sigma'} & & H_n(\mathcal{X}) \end{array}$$

in which j is the projection onto a direct summand and σ and σ' are natural projections; all these operators are also continuous and open. This implies that ψ_* is open.

See [81, VI], [213] for applications of the result in this Exercise.

Chapter I
Initial concepts and first results

The subject of this book is a kind of synthesis. In the well-known textbook by Rickart it is said jokingly, but not without justification, that the head of the theory is in algebra and the feet in analysis [64, vi]. To characterize the present chapter such a comparison could be extended. Here our subject is presented as a new-born baby and so it has a disproportionately large head and small feet: in fact the reader will see that at the beginning of the account algebra is somewhat dominant compared with analysis.

In the second chapter — as is customary with the growth of a body — the proportion begins to even out and there appear the first significant results based on an equally balanced interaction between the two "basic" disciplines — theorems on spectra and on the structure of polynormed skew fields. This, in its turn, will lay the foundation for the central results of the general theory contained in Chapter IV.

§1. ABSTRACT ALGEBRAS

§1.1. Complex associative algebras and examples of these. Homomorphisms, characters and representations

Definition (I.1.1). Let A be a linear space. A bi-operator[1] $m : A \times A \longrightarrow A$ is called a *multiplication bi-operator* or simply a *multiplication* in A if, writing ab instead of $m(a, b)$, it satisfies the identity $(ab)c = a(bc)$; $a, b, c \in A$. The space itself when furnished with a multiplication (in other words, the pair (A, m) consisting of a linear space and a multiplication given in it) is called a *complex associative algebra* or more briefly an *algebra*.

---

[1] This is just a bilinear operator; cf. Ch.0 §3.1.

The element ab is called the *product* of the elements a and b and the identity indicated above is called the *associativity identity.*

Accepting the usual freedom of speech, we shall in future, as a rule, talk about the "algebra A" assuming that the multiplication is fixed.

Here are several standard examples. The simplest algebra is, of course, that which consists of just the zero (it is denoted by $\mathbb{O}$), and, among algebras with non-zero multiplication, $\mathbb{C}$. We may assume that the next in order of complexity is $\mathbb{C}^n$, $n \in \mathbb{N}$, with coordinate-wise mutliplication.

Further, among the linear spaces consisting of complex-valued functions, many are algebras with respect to pointwise multiplication. We select, in particular, the algebra c_∞ of all sequences (= functions on the natural numbers) and the algebra c_{00} of all finite (eventually zero) sequences. More generally, there are at least two algebras associated with each set M (any kind of set); one is $\mathbb{C}^M$, consisting of all functions on M, and another is $\mathbb{C}^M_{00}$, consisting of those functions on M that are zero except at a finite number of points (so in this notation $c_\infty = \mathbb{C}^{\mathbb{N}}$ and $c_{00} = \mathbb{C}^{\mathbb{N}}_{00}$). Finally, if Ω is a topological space, then the set of all continuous functions on Ω is also an algebra, denoted, as usual, by $C(\Omega)$.

Let us recall two other algebras $\mathcal{P}_n$ and $\mathcal{R}_n$. The first consists of the polynomials, and the second consists of the rational functions of n variables with complex coefficients (the multiplication and other operations should be well known to the reader). In future we shall write $\mathcal{P}$ instead of $\mathcal{P}_1$ and $\mathcal{R}$ instead of $\mathcal{R}_1$.

Definition (I.1.2). We say that the elements a and b of an algebra A *commute* if $ab = ba$. An algebra in which each pair of elements commute is said to be *commutative.*

Of course, all the examples of algebras introduced above are commutative. The simplest non-commutative algebra is the algebra of $n \times n$ dimensional matrices, $n > 1$, with complex elements, considered with the standard matrix operations; it is denoted by $\mathcal{M}_n$. The next example is the algebra $\mathcal{L}(E)$ of all operators on a linear space E with composition as the multiplication; Theorem 1.15, later, will show that this is "more than just an example". With the same multiplication, the subspace $\mathcal{F}(E)$ of $\mathcal{L}(E)$, consisting of the finite-dimensional operators is also an algebra. Both these algebras, obviously, are non-commutative, except when $\dim E = 1$.

A further important example can be constructed, starting from a given group G. We let $\mathbb{C}(G)$ denote the space of formal linear combinations of elements of G with complex coefficients; if we identify $g \in G$ with $1 \cdot g \in \mathbb{C}(G)$, we are justified in assuming G to be an algebraic basis in $\mathbb{C}(G)$. We furnish $\mathbb{C}(G)$ with the multiplication $\left(\sum_{k=1}^{m} \lambda_k g_k\right)\left(\sum_{l=1}^{n} \mu_l g_l\right) := \sum_{k,l} \lambda_k \mu_l g_k g_l$, which is uniquely determined by the fact that, on G, it coincides with the group multiplication. The algebra thus constructed is called the *group algebra* of the group G; obviously, it is commutative $\Leftrightarrow$ G is commutative.

A subspace B of an algebra A with multiplication m is called a *subalgebra* of A if, whenever it contains two elements, it also contains their product; of course, B itself is then an algebra with respect to the multiplication $m|_{B\times B}$. Thus, $C(\Omega)$ is a subalgebra of $\mathbb{C}^{\Omega}$ and $\mathcal{F}(E)$ is a subalgebra of $\mathcal{L}(E)$.

For every algebra A the set $\{a \in A : ab = ba$ for all $b \in A\}$ is obviously a subalgebra; it is called *the centre* of A.

For any subset M in the algebra A there exists a smallest subalgebra among all the subalgebras in A containing M; this is obviously $\ell.h.\{a_1 \cdots a_n : a_1, \cdots, a_n \in M,\ n \in \mathbb{N}\}$. When this subalgebra is the whole of A, M is called a *set of generators* and its elements are called the *generators* of the algebra A. If there exists a finite set of generators in the algebra A, it is said to be *finitely-generated*; $\mathcal{P}_n$ is an example but $\mathcal{R}_n$ is not (check it!).

Exercise (I.1.3). Find a minimal set of generators for the algebra $\mathcal{M}_n$.

Hint. It will contain altogether just two elements!

Given an algebra A, its underlying space, when furnished with a new multiplication $a \circ b$, equal to the "previous" ba, is obviously also an algebra, called the *opposite* to A and denoted by A^{op}. Of course, A is commutative $\Leftrightarrow$ A, as an algebra, coincides with A^{op}.

We shall call the subspace $\ell.h.\{ab : a, b \in M\}$, denoted by $M^{[2]}$ the *algebraic square* of the subset M in A.

Whenever we introduce a new structure — the structure of an algebra in this case — it is necessary to distinguish the class of maps with which it is compatible.

Definition (I.1.4). A map $\kappa : A \longrightarrow B$ between two algebras is called a *homomorphism* (of algebras) if it is a linear operator and, in addition, $\kappa(ab) = \kappa(a)\kappa(b)$, for any $a, b \in A$.

Algebras and their homomorphisms form a category in which the law of composition is the composition of homomorphisms as maps (show that it is pointed but not additive). When we speak of an "*algebra isomorphism*" we always have in mind an isomorphism in this category, often denoted later on by the sign $\simeq$; obviously this is precisely a bijective homomorphism. Here is an instructive problem about homomorphisms.

Exercise (I.1.5). Show that

(I) every isomorphism $\kappa : \mathcal{L}(E) \longrightarrow \mathcal{L}(F)$, where $\dim E = \dim F < \infty$, is given by $\kappa(S) = U^{-1}SU$, $S \in \mathcal{L}(E)$, in terms of some linear isomorphism $U : E \longrightarrow F$;

(II) every isomorphism $\kappa : \mathcal{L}(E) \longrightarrow \mathcal{M}_n$, where $\dim E = n$, is given by a certain basis in E and associates an operator with its matrix in terms of this basis;

(III) every isomorphism $\kappa : \mathcal{M}_n \longrightarrow \mathcal{M}_n$ is given by a non-singular matrix a so that $\kappa(b) = aba^{-1}$; $b \in \mathcal{M}_n$, and moreover in all the cases (I) - (III) every non-zero homomorphism between the indicated algebras is an isomorphism.

Hint. If S is a one-dimensional projection in $\mathcal{L}(E)$, then under conditions (I) $\kappa(S)$ is a one-dimensional projection in $\mathcal{L}(F)$. The required U is well-defined by $U(Tx) = \kappa(T)y$, where x and y are fixed non-zero vectors from $\operatorname{Im} S$ and $\operatorname{Im} \kappa(S)$ respectively, and T is any operator in $\mathcal{L}(E)$.

It is well known that when considering the majority of categories encountered in practice, valuable (and if you are lucky, comprehensive) information on the structure of the objects is furnished by studying their morphisms into certain standard — the simplest or at least "well-understood" — objects. Here is an important class of such morphisms for the category of algebras.

Definition (I.1.6). A homomorphism (any homomorphism) from A into $\mathbb{C}$ is called a *character.*

Exercise (I.1.7). Describe all the characters of the algebra $\mathcal{P}_n$.

We shall see later (Chapter IV) that characters are especially effective as a means of investigating commutative algebras. With regard to algebras which are "nowhere near commutative", it is, generally speaking, of little use to investigate their characters, as is illustrated by the following.

Exercise (I.1.8). Show that the algebras $\mathcal{M}_n$; $n > 1$, do not have any characters, except the zero character.

The following wider class of homomorphisms plays a fundamental role in studying non-commutative algebras.

Definition (I.1.9). Every homomorphism from A into $\mathcal{L}(E)$ is called a *representation* of the algebra A on the linear space E.

Since it is obvious that $\mathcal{L}(\mathbb{C}) \simeq \mathbb{C}$, every character may be considered as a particular case of a representation. Another important example occurs when $E = A$: if we associate to each $a \in A$ the operator $m_a : A \longrightarrow A : b \longmapsto ab$, we get a homomorphism from A into $\mathcal{L}(A)$ (check it!) called the *regular representation* of this algebra. Indeed in general, we shall as a rule use the widespread notation T for a representation and write T_a in place of $T(a)$.

§1.2. The identity and adjoining an identity, inverse elements. Polynomials and rational functions of an element of the algebra

Definition (I.1.10). An element e_A of the algebra A is called an *identity* of this algebra if $e_A a = a = ae_A$ for all $a \in A$.

If it is clear which algebra we are talking about, we shall write e instead of e_A.

If only the left or only the right of the two equalities in $e'a = a = ae'$; $a \in A$, is satisfied, in general, for some $e' \in A$, then e' is called a *left* (or *right*, respectively) *identity* in our algebra.

Proposition (I.1.11). *If there exists in A a left and a right identity, then they coincide and are the (ordinary) identity in A. In particular, there cannot be more than one identity in an algebra.*

▶ If e' and e'' are the corresponding left and right identities, then each of them is equal to $e'e''$. ■

Definition (I.1.12). An algebra $A \neq \mathbb{O}$, that has an identity, is said to be *unital*. A homomorphism $\kappa : A \longrightarrow B$ between two unital algebras is said to be *unital* if $\kappa(e_A) = e_B$.

We note the following obvious proposition.

Proposition (I.1.13). *Every non-zero character of a unital algebra is unital.* ▶■

Of course, the polynomial algebra $\mathcal{P}$ (in the variable x) can be included among the unital algebras. It is distinguished by the following remarkable property.

Proposition (I.1.14). *For any unital algebra A and any $a \in A$ there exists a unique unital homomorphism $\kappa : \mathcal{P} \longrightarrow A$ such that $\kappa(x) = a$. This homomorphism is determined by $\kappa(p) := p(a)$, where if $p = \lambda_0 + \lambda_1 x + \cdots + \lambda_n x^n$, $\lambda_0, \cdots, \lambda_n \in \mathbb{C}$, then the symbol $p(a)$ denotes the element $\lambda_0 e + \lambda_1 a + \cdots + \lambda_n a^n \in A$.*[2] ▶■

Elements of the form $p(a) \in A$; $p \in P$ are called *polynomials* in a. Obviously the set of all polynomials in a is the smallest subalgebra in A containing e and a.

It is not difficult to see that the property indicated in Proposition 1.14 uniquely characterizes the algebra $\mathcal{P}$ up to isomorphism. We suggest the reader formulates the appropriate assertion carefully and proves both this and its analogue for the algebra $\mathcal{P}_n$.

---

[2] As usual, in the notation for multiplication of operations we write a^n instead of $a \cdots a$ (n times).

Later on "life will teach us" that it is substantially easier to work with unital algebras than with others. In this connection methods for augmenting an arbitrary algebra to give one that is automatically unital deserve attention. The most economic of these methods[3] is the following. Let A be the original algebra. We put $A_+ := A \oplus \mathbb{C}$ and we introduce in A_+, using the notation $a + \lambda e_+$ instead of (a, λ), the multiplication $(a + \lambda e_+)(b + \mu e_+) := (ab + \mu a + \lambda b) + \lambda\mu e_+$; $a, b \in A_+$, $\lambda, \mu \in \mathbb{C}$ (all the required properties can easily be verified). Obviously A_+ is a unital algebra with identity $0 + 1e_+ = (0, 1)$, in future denoted simply by e_+ (or even e if there is no danger of confusion). Moreover A may be identified, up to an isomorphism, with the subalgebra of A_+ consisting of the elements $a + 0e_+ = (a, 0)$, $a \in A$, in future denoted simply by a.

The algebra A_+ thus constructed is called the *unitization* of the algebra A. Given a homomorphism $\kappa : A \longrightarrow B$ between arbitrary algebras we let $\kappa_+ : A_+ \longrightarrow B_+$ denote the homomorphism that coincides with κ on A and takes one adjoined identity into the other; this κ_+ is called the *unitization of the homomorphism* κ. (Thus unitization is a functor sending the category of algebras into its "unital" subcategory. We leave the reader to give a precise meaning to this sentence.)

Here is the first application of the given construction.

Theorem (I.1.15). *For any algebra A there exists an injective representation (in other words, any algebra is isomorphic to a certain subalgebra of $\mathcal{L}(E)$ for a certain E).*

▶ For any $a \in A$ we consider the operator $m_{a+} : b \longmapsto ab$ acting in A_+; then $m_{a+}(e_+) = a$, and therefore $a \neq 0$ implies $m_{a+} \neq 0$. Hence it is sufficient to take as the required representation the restriction to A of the regular representation of the algebra A_+. ■

We emphasize that unitization was defined for all algebras, including those having identities; the "former" identity in A simply fails to be an identity any longer in A_+. (We recommend considering how this happens in the simplest example $A := \mathbb{C}$ when $A_+ \simeq \mathbb{C}^2$.) In a lot of cases, instead of such a "universal" unitization $A \longmapsto A_+$ it is more reasonable to adjoin an identity only to those algebras that have not got one. In such cases we shall let $A_\times$ denote A itself if it is unital and the algebra

---

[3] In certain situations another method is very useful which, in contrast, gives us a very extensive and, in a definite sense, maximal unital algebra containing the original algebra. For these algebras of double centralizers, and also for multipliers, see for example [129], [317], [58, 3.12]

A_+ otherwise. We shall call the algebra $A_\times$ the *conditional unitization* of the algebra A.

* * *

An element a of a unital algebra A is said to be *invertible on the left* if there exists a certain $a_l^{-1} \in A$, called a *left inverse* of a, such that $a_l^{-1}a = e$. An element a that is *invertible on the right* and a *right inverse* a_r^{-1} are defined analogously. An element a that is invertible on the left and on the right, is said to be (simply) *invertible*; in this case $(a_l^{-1}a)a_r^{-1} = a_l^{-1}(aa_r^{-1})$ immediately implies that any left inverse of a coincides with any right inverse of a and thus we have a single element denoted by a^{-1} and such that $a^{-1}a = aa^{-1} = e$; it is called the *inverse* of a. The set of all invertible elements in A is denoted by $\operatorname{Inv} A$; obviously this is a group with respect to the multiplication inherited from A.

Exercise (I.1.16). Show that there exist algebras with elements that are invertible on the left (or on the right) but which are not invertible.

Hint. They can be found among the $\mathcal{L}(E)$ with infinite-dimensional E.

In the next three propositions A is a unital algebra.

Proposition (I.1.17). *Let* $a_1, \cdots, a_n \in A$. *Then for* $a_1 \cdots a_n$ *to be invertible, it is sufficient, and if these elements commute it is also necessary, for* $a_1, \cdots, a_n$ *themselves all to be invertible.* ▸■

The preceding exercise shows, of course, that it is impossible to disregard the condition that they commute in the "necessary" part of Proposition 1.17.

Proposition (I.1.18). *If* $a \in A$, $b \in \operatorname{Inv} A$, *then* $ab = ba \iff ab^{-1} = b^{-1}a$. ▸■

Proposition (I.1.19). *For* $a, b \in A$ *the element* $e - ab$ *is invertible on the left (invertible on the right, invertible, respectively)* $\iff$ $e - ba$ *is invertible on the left (invertible on the right, invertible, respectively).*

▶ If c is a left (right, respectively) inverse to $e - ab$, then, as is easily verified,[4] $e + bca$ is a left (right) inverse to $e - ba$. ■

In our account the name "skew field" is adopted for those unital algebras in which each non-zero element is invertible[5] while "field" is taken to mean a commutative field; such as, let us say, $\mathbb{C}$ and $\mathcal{R}_n$.

To finish the algebraic introduction let us define yet another useful concept. Let a be an element of a unital algebra A, and let $r \in \mathcal{R}$ be a rational function which can be written as $r = p/q$; $p, q \in \mathcal{P}$, where $q(a)$ is invertible in A. (Proposition 1.17 easily implies that this condition is equivalent to $q(a)$ being invertible for the standard representation of r as a quotient of co-prime polynomials.) In this situation we let $r(a)$ denote the element $p(a)(q(a))^{-1} = (q(a))^{-1}p(a) \in A$ (equality is implied by Proposition 1.18) and call the elements of this kind *rational functions* of a. Obviously $r(a)$ does not depend on the choice of $p, q \in \mathcal{P}$; $q(a) \in \operatorname{Inv} A$ such that $r = p/q$.

Remark (I.1.20). Having discovered what a polynomial, and then a rational function, of an element of an abstract algebra are, the reader may be curious to know: are there similar definitions for more general functions of a complex variable? Can one, let us say, under certain conditions, talk about the sine of an element of algebra or about its logarithm? This question — we warn in advance — turns out to be far from an idle one; "not only it is possible, but it is also essential". However, to define this important idea properly — the so-called functional calculus (see Chap.II §2 and Chap.IV §7.2) — a purely algebraic structure in A is not enough. A topology is required.

---

[4] The reader will certainly want to ask: how can one guess that the inverse to $e - ba$ must be found just in this form? This can be motivated, for example, by matrix arithmetic. Let a and b be two $n \times n$ matrices (any n) with sufficiently small coefficients. Then the matrix $(1 - ba)^{-1}$ can obviously be represented as the sum of the series
$$1 + ba + (ba)^2 + \cdots + (ba)^n + \cdots = 1 + b[1 + ab + (ab)^2 + \cdots]a.$$
The sum of the series in square brackets is just the matrix $(1 - ab)^{-1}$.

[5] That is, those which are skew fields in the sense of ring theory. We emphasize that skew fields which are not complex associative algebras (the classical example is the skew field of quaternions) will not be considered in this book.

§2. BANACH ALGEBRAS AND OTHER POLYNORMED ALGEBRAS; PRELIMINARY INFORMATION AND EXAMPLES

§2.1. Definition of the basic structures. Automatic continuity of the characters of Banach algebras

Let us turn to the definition of those algebraic-topological structures that are the basic objects in our monograph.

Definition (I.2.1). A Banach space furnished with a multiplication bi-operator (= an algebra) is called a *Banach algebra* if this bi-operator is continuous.

As we want to give an analogous definition for general polynormed spaces we first ought to distinguish which basic type of continuity is the one we are taking for the bi-operator (cf. Chap.0 §3.1).

Definition (I.2.2). A polynormed space A furnished with a multiplication bi-operator m is called a *polynormed algebra* if m is separately continuous (in other words, for any $a \in A$ the regular representation operator $m_a : b \longmapsto ab$ and the analogous operator $m_a^{op} : b \longmapsto ba$ are continuous).

In addition to the classes of Banach and (all) polynormed algebras we shall consider a lot of intermediate classes, including *(semi)normed algebras*[6] *and Fréchet algebras.* The meaning of such terms is obvious.

Recall that the construction of the projective tensor product enables us to consider, instead of continuous bi-operators between Banach spaces, continuous operators associated with them (Chap.0 §3.3). This enables us to give an equivalent definition of a Banach algebra as a Banach space A furnished with a continuous operator $\underline{m} : A \hat{\otimes} A \longrightarrow A$ such that the diagram

$$\begin{array}{ccc} A \hat{\otimes} A \hat{\otimes} A & \xrightarrow{\underline{m} \hat{\otimes} 1} & A \hat{\otimes} A \\ {\scriptstyle 1 \hat{\otimes} \underline{m}} \downarrow & & \downarrow {\scriptstyle \underline{m}} \\ A \hat{\otimes} A & \xrightarrow{\underline{m}} & A \end{array} \tag{1}$$

is commutative; we let the reader verify that this condition is equivalent to the multiplication bi-operator being associative.

---

[6] Obvious but, alas, non-standard. Thus, by a "normed algebra" we shall mean a normed space furnished with a separately continuous multiplication. However, many authors use the term "normed algebra" to mean an algebra equipped with a multiplicative norm.

Similarly, a polynormed algebra may be defined as a polynormed space A furnished with a continuous operator $\underline{m}: A \otimes_i A \longrightarrow A$ such that the diagram obtained from (1) by replacing "$\hat{\otimes}$" by "$\otimes_i$" is commutative. Of course, the original, purely algebraic concept has an equivalent definition formulated in a similar way in terms of the algebraic tensor product "$\otimes$" (we did not mention it immediately in §1.1 to avoid tiresome repetition).

Such a "tensor" approach to algebras has its virtues; see Definition 2.26 later and also [251, §1], even within the framework of pure algebra [51, Ch.VI].

One additional assumption that notably facilitates the study of polynormed algebras, is the following.

Definition (I.2.3). Let A be an algebra furnished with a (semi)norm $\|\cdot\|$. This (semi)norm is said to be *multiplicative* (or sometimes *submultiplicative*) if $\|ab\| \leq \|a\|\|b\|$ for all $a, b \in A$. If A is unital, then the (semi)norm is also said to be *unital* whenever $\|e\| = 1$.

Definition (I.2.4). A polynormed space furnished with a multiplication is called a *multiplicatively polynormed algebra* or briefly, a *multinormed algebra* if it has a defining system (see Chap.0 §1.1) of multiplicative seminorms[7] and it is called an *Arens-Michael algebra*[8] if it is, in addition, Hausdorff and complete.

Obviously every multinormed algebra is a polynormed algebra (since the first class is narrower than the second, we shall discuss this later). On the other hand, every Banach algebra is an Arens-Michael algebra. This is shown by the following.

Proposition (I.2.5). *If* $(A; \|\cdot\|)$ *is a Banach algebra, then it has a multiplicative norm equivalent to the original one, and if A has an identity this norm may be assumed to be unital.*

▸ In view of Theorem 0.3.4 the multiplication bi-operator is bounded; this means that $\|ab\| \leq C\|a\|\|b\|$; $a, b \in A$ for some $C > 0$. But then the norm $\|a\|' := C\|a\|$ is obviously multiplicative. If, however, A is also unital, then the norm $\|a\|'' := \sup\{\|ab\|' : \|b\|' \leq 1\}$ has the necessary properties. ■

It is natural to consider maps that are simultaneously linear continuous operators and algebra homomorphisms — briefly — *continuous homomorphisms*, to be those maps that are compatible with the structures introduced above. If we take some class or

---

[7] In another terminology multinormed algebras are called locally multiplicatively convex.

[8] In honour of Arens [99] and Michael [50] who initiated this research.

other of polynormed algebras as objects and all continuous homomorphisms acting between these algebras as morphisms, we get a set of categories, among which the narrowest (but at the same time, the most important in our account) turns out to be the *category of Banach algebras* and the widest is the *category of all polynormed algebras*.

An isomorphism in any of these categories is called a *topological isomorphism*;[9] obviously this is an algebraic isomorphism (that is a pure algebra isomorphism) that is simultaneously a homeomorphism of topological spaces. The Banach theorem (Chap.0 §1.2) immediately implies that *every continuous algebraic isomorphism between Banach algebras is a topological isomorphism*; the same, of course, applies to the wider category of Fréchet algebras. A topological isomorphism of a polynormed algebra onto itself is called an *automorphism*.

In the class of seminormed algebras, in particular, Banach algebras, we can distinguish among the homomorphisms those that are isometric operators; they are called *isometric homomorphisms*. A bijective isometric homomorphism is called an *isometric isomorphism*; obviously this is a special case of a topological isomorphism.

If M is a subset of the polynormed algebra A, then the closure of its algebraic square $M^{[2]}$ (see §1.1) will be called the *topological square* of this set and denoted by M^2.

* * *

But now we interrupt for a moment the list of definitions we need to tell the reader about the first substantial result on these structures which gives some idea of one of the distinctive features of this theory. We are talking about the fact that in certain well-defined situations the continuity of maps between polynormed algebras, primarily Banach algebras, may be deduced from their purely algebraic properties. Here is an example (the oldest) of such a "theorem on automatic continuity".

Theorem (I.2.6). *Let A be a Banach algebra. Then all its characters are continuous and form a bounded set in the Banach space A^*; if the norm in A is also multiplicative, then they are contained in the unit ball of this space.*

▶ Let χ be a character on our algebra. In view of Proposition 2.5, assuming that the norm is multiplicative, it is sufficient to show that there does not exist $a \in A$ such that $\|a\| < 1$ and $\chi(a) = 1$. Let us assume otherwise; then, because

---

[9] The same name is used for an isomorphism in all categories of polynormed spaces (cf. Chap.0 §1.2); therefore, as there is a danger of confusion, we shall always specify what structures we are talking about.

$\|a^n\| \le \|a\|^n$ and because A is complete, the series $\sum_{n=1}^{\infty} a^n$ converges to a certain $b \in A$. The continuity of multiplication implies that $a + ab = a + \lim_{n\to\infty} a \sum_{k=1}^{n} a^k = b$. But then $\chi(a + ab) = \chi(b)$, that is $1 + \chi(b) = \chi(b)$, which, of course, cannot be so. ■

Other results of a similar kind will be given later; see IV.5.18, IV.5.25 and IV.7.62 and VI.2.10, VI.2.62, VI.2.65 and VI.2.66. We note that the book [72] is devoted to problems of "automatic continuity" as is also the survey [142] and a considerable part of the collection [62].

The reader may possibly be interested in a question which at present still remains open.

Question. Is every character of a Fréchet algebra (let it even be "simultaneously" an Arens-Michael algebra) continuous?

This question is discussed in detail in the recently published monograph [86].

* * *

Let A be a polynormed algebra, and let $\overline{A}$ be the completion of its underlying polynormed space. As A is dense in $\overline{A}$, there can exist in the latter at most one separately continuous multiplication extending the original multiplication (given in A). We shall consider the question of when such a multiplication does in fact exist, first of all for normed algebras.

Proposition (I.2.7). *Let* $(A; \|\cdot\|)$ *be a normed algebra. Then* $\overline{A}$ *is a Banach algebra with respect to a certain multiplication that coincides, on* A, *with the original multiplication* $\Leftrightarrow$ *this original multiplication is jointly continuous. Moreover, if the norm in* A *is multiplicative, then so also is the norm in* $\overline{A}$.

▸ "⇒". Since $\overline{A}$ is a Banach algebra, its multiplication is automatically jointly continuous; consequently so is multiplication in any subalgebra, including A itself.

"⇐". The joint continuity of the multiplication means that $\|a'b'\| \le C\|a'\|\|b'\|$ for a certain $C > 0$ for all $a', b' \in A$. We take $a, b \in \overline{A}$ and consider sequences $a_n, b_n \in A$ such that $a = \lim_{n\to\infty} a_n$ and $b = \lim_{n\to\infty} b_n$. Then

$$\|a_n b_n - a_m b_m\| \le C\|a_n - a_m\|\|b_n\| + C\|a_m\|\|b_n - b_m\|$$

obviously implies that $a_n b_n$ is a Cauchy sequence in $\overline{A}$. We denote its limit by ab and consider the bi-operator $(a, b) \longmapsto ab$; obviously this is a multiplication in $\overline{A}$

extending the original and satisfying $\|ab\| \le C\|a\|\|b\|$. In particular, if we put $C = 1$, we get the last assertion. ■

An analogous argument apart from some obvious modifications (instead of sequences it is now necessary to consider nets and so on) gives us the following.

Proposition (I.2.8). *Let A be a multinormed Hausdorff algebra. Then $\overline{A}$ is also a multinormed algebra with respect to the unique multiplication that coincides, on A, with the original multiplication.* ▶■

We shall call the Banach algebra (or, according to the sense, the Arens-Michael algebra) $\overline{A}$ with the multiplication indicated in Proposition 2.7 or 2.8 the *completion* of the given algebra A (of corresponding type).

§2.2. **Some preliminary examples**

"Grau, teurer Freund, ist alle Theorie,
Und grün des Lebens goldner Baum"

teaches Goethe's Mephistopheles. Can such a categorical statement by the Prince of Darkness that all theory is dull also refer to mathematics? To speak seriously, then certainly in mathematics any theory that is poor in interesting examples is dull; it is just such examples that can be likened to that "green tree of life" mentioned subsequently in "Faust".

The examples within the scope of the present book are chiefly specific classes of algebras connected with various branches of analysis and topology. The reader will become acquainted with the basic stock of them in Chap.III which is specially devoted to them. However, it is expedient to give a certain number of examples now, which are urgently needed so as to be able, as soon as possible, to "try out", as one might say, the general definitions, constructions or facts on them; it is only then that these can be really understood.

Example (I.2.9) Any polynormed space (in particular any Banach space) is a polynormed (Banach) algebra with respect to the zero multiplication bi-operator. This example is "not as stupid as it looks": it is useful to recall it when examining a whole range of later concepts.

Example (I.2.10) ("the main function algebra"). If Ω is a compact space, then the algebra $C(\Omega)$ (see §1.1) is a Banach algebra with respect to the uniform norm. More generally, if Ω is a locally compact space, then the subalgebra $C_0(\Omega)$ of $C(\Omega)$ is a Banach algebra with respect to the same norm, where $C_0(\Omega)$ consists of the functions that vanish at infinity (that is the functions f having the property that, for any $\varepsilon > 0$, there exists a compact set $K \subset \Omega$ such that $|f(s)| < \varepsilon$ for all

$s \notin K$).[10] In particular, $C_0(\mathbb{N})$ is the algebra of sequences that converge to zero; it is more often denoted by c_0. Obviously $C_0(\Omega)$ is unital $\Leftrightarrow$ Ω is a compact space and then of course $C_0(\Omega) = C(\Omega)$.

Example (I.2.11) (the two main operator algebras). Let E be a Banach space. Then the space $\mathcal{B}(E)$ of all bounded operators and the space $\mathcal{K}(E)$ of all compact operators in E are Banach algebras with composition taken as the multiplication and with the operator norm. $\mathcal{B}(E)$ is always unital (1_E is the identity) and $\mathcal{K}(E)$, as follows from Riesz' theorem, is not unital when E is infinite-dimensional.

We warn the reader that among all the specific algebras featuring in this section two, $C_0(\Omega)$ and $\mathcal{B}(H)$, where H is a Hilbert space should be treated "with great respect": they turn out to be the image domains of two canonical homomophisms which play a fundamental role in the general theory (see Chap.IV §2.2 and §5.3).

Example (I.2.12) ("a typical algebra of harmonic analysis"). The Banach space $l_1(\mathbb{Z})$ consisting of the sequences $a = \{a_n \in \mathbb{C};\ n \in \mathbb{Z}\}$ such that $\|a\| = \sum_{n=-\infty}^{\infty} |a_n| < \infty$, with coordinate-wise linear operations, is a Banach algebra with respect to the so-called convolution multiplication,[11] denoted by $*$ and defined by

$$(a * b)_n = \sum_{k=-\infty}^{\infty} a_k b_{n-k}; \qquad a = \{a_k\}, \quad b = \{b_k\}.$$

Example (I.2.13). We consider the (specific) Hilbert space $L^2[0, 1]$ and let $\mathcal{B}$ denote the subalgebra of $\mathcal{B}(L^2[0, 1])$, consisting of the operators of pointwise multiplication by continuous functions, that is operators of the form $S_a : g \longmapsto ag$, where $a \in C[0, 1]$. Obviously this is a Banach algebra with respect to the operator norm, and moreover the map $a \longmapsto S_a$ is a well-defined isometric isomorphism between $C[0,1]$ and $\mathcal{B}$.

Example (I.2.14). We consider the subalgebra in $C(\mathbb{T})$ ($\mathbb{T}$, as usual, is the unit circle in the complex plane) consisting of those functions f that can be represented as an absolutely

---

[10] If Ω is not locally compact, then there may turn out to be too few such functions; we recommend the reader to consider the case when Ω is homeomorphic to an infinite-dimensional Banach space.

[11] One can also introduce another (coordinatewise) multiplication in this space, but this is obviously less interesting.

convergent series $f(z) = \sum_{n=-\infty}^{\infty} c_n z^n$. It is easy to verify, that this is a Banach algebra with respect to the norm $\|f\| = \sum_{n=-\infty}^{\infty} |c_n|$; this is the so-called *Wiener algebra*, denoted by W. We note that the map $j ; l_1(\mathbb{Z}) \longrightarrow W$ that takes the sequence $a = \{a_n\}$ to the function f; $f(z) = \sum_{n=-\infty}^{\infty} a_n z^n$; $z \in \mathbb{T}$, is an isometric isomorphism between the two Banach algebras.

The last two examples show, in particular, that function algebras with pointwise multiplication may sometimes appear in a "disguised" form ($C[0,1]$ in the disguise of $\mathcal{B}$, W in the disguise of $l_1(\mathbb{Z})$). We shall find out later which general facts and constructions lie behind these observations (Chap.IV §§3.2 and 7.1).

The next example is, evidently, the most important algebra of complex analysis; it goes beyond the framework of Banach algebras.

Example (I.2.15). The space $\mathcal{O}(U)$ consisting of functions that are holomorphic (analytic)[12] in the domain $U \subseteq \mathbb{C}^n$, furnished with pointwise multiplication and with the seminorms giving the compact-open topology (see Example 0.1.8), is obviously a polynormed algebra belonging to the class of Arens-Michael algebras and also Fréchet algebras.

Example (I.2.16). The algebra $\mathbb{C}^M$, where M is an arbitrary set (see §1.1) is also, obviously, an Arens-Michael algebra with respect to the seminorms given by the topology of pointwise convergence (see Example 0.1.7). Moreover $\mathbb{C}^M$ is a Fréchet algebra $\Longleftrightarrow$ M is at most countable; therefore c_∞ is a Fréchet algebra.

* * *

We give a few more concepts and examples connected with (non-Banach) polynormed algebras.

Example (I.2.17). Any ("pure") algebra may be considered as a polynormed algebra with the strongest seminorm system (see Definition 0.1.5 and Proposition 0.1.9). Moreover, it is obvious that the category of algebras may be identified with the corresponding complete subcategory in the category of all (or all complete) polynormed algebras. In particular, if we speak of $\mathcal{P}_n$ or $\mathcal{F}(E)$ we are always considering them as polynormed algebras of the class indicated.

---

[12] In this book the words "holomorphic" and "analytic" are always used as synonyms when referring to a function of one or several complex variables.

Definition (I.2.18). A polynormed algebra A is said to be *jointly polynormed* if its multiplication $m : A \times A \longrightarrow A$ is a jointly continuous bi-operator.

The characterization of jointly continuous bi-operators in terms of seminorms (Proposition 0.3.1) immediately implies the following proposition.

Proposition (I.2.19). *A polynormed algebra A with a saturated system of seminorms $\|\cdot\|_\nu$; $\nu \in \Lambda$, is jointly polynormed $\Longleftrightarrow$ for any $\nu \in \Lambda$ there exists $\mu \in \Lambda$ and $C > 0$ such that $\|ab\|_\nu \leq C\|a\|_\mu\|b\|_\mu$ for any $a, b \in A$.* ▸■

Example (I.2.20). It is easy to see that the algebra c_{00} (see §1.1), furnished with the norm $\|\xi\| = \sum_{n=1}^{\infty} \frac{1}{n} |\xi_n|$, is a normed algebra but not a jointly normed algebra.

Exercise (I.2.21). Let E be a linear space with a countable algebraic basis (let us say, $\mathcal{P}$ or c_{00}). Show that $\mathcal{F}(E)$, although it is polynormed (in the sense of Example 2.17), is not a jointly polynormed algebra.

Hint. By considering the multiplication of one-dimensional operators, reduce the problem to a question of the type of continuity of the bifunctional in Exercise 0.3.3.

Thus, there are fewer jointly polynormed algebras than "simply" (separately) polynormed algebras. On the other hand, there are more than there are multinormed algebras: this is implied by Proposition 2.19 and the result of the following exercise.

Exercise (I.2.22) (Arens). Show that the space $L^\omega[0, 1] = \bigcap_p L^p[0, 1]$, furnished with pointwise multiplication and the system of norms $\|f\|_p = (\int |f(t)|^p dt)^{1/p}$ ($1 \leq p < \infty$ always) is a jointly polynormed but not multinormed algebra.

Finally, we generalize Propositions 2.7 and 2.8 to the classes of algebras that are closed under the operation of completion.

Exercise (I.2.23). Let A be a Hausdorff polynormed algebra, and let $\overline{A}$ be its completion as a polynormed space. Show that for there to exist a separately continuous multiplication in $\overline{A}$ extending the original it is sufficient, and if A is metrizable then it is also necessary, that the multiplication given in A be jointly continuous.

§2.3. Some constructions on polynormed algebras (tensor product, unitization etc). Approximate identity

Every *subalgebra of a polynormed algebra* is itself, of course, a polynormed algebra with respect to the restrictions of the given system of seminorms. Moreover, if the original algebra is complete (as a polynormed space), then a subalgebra of it is complete $\Leftrightarrow$ the subalgebra is closed. In particular, a closed subalgebra of a Banach algebra is itself a Banach algebra.

The Cartesian product of any family of polynormed algebras may be considered to be a polynormed algebra with respect to the obvious coordinate-wise operations and the system of seminorms indicated in Chap.0 §1.1. We note that in the case of a finite family of normed (Banach, respectively) algebras A_k; $1 \le k \le n$ the algebra $A = \bigtimes_k A_k$ is itself, up to topological isomorphism, normed (or Banach, respectively) with respect to either of the equivalent norms $\|\cdot\|_1$ or $\|\cdot\|_2$, where $\|\bar{x}\|_1 = \max_k \|x_k\|$ and $\|\bar{x}\|_2 = \sum_k \|x_k\|$, $\bar{x} = \{x_1, \cdots, x_n\}$. Moreover, if the norms in the original algebras are multiplicative, then so also are both the above norms in A.

Example (I.2.24). The Cartesian product of a family of copies of the complex plane, indexed by a certain set M, is none other than $\mathbb{C}^M$ from Example 2.16.

The construction of the tensor product may also be extended from spaces to algebras, at least if we take a finite number of them. In its definition we essentially use the alternative approach to the concept of an algebra discussed in §2.1. We basically concentrate on Banach algebras.

Let A and B be Banach algebras and let $\underline{m}_A : A \hat{\otimes} A \longrightarrow A$ and $\underline{m}_B : B \hat{\otimes} B \longrightarrow B$ be the associated multiplication operators. We let $\omega : A \hat{\otimes} B \longrightarrow B \hat{\otimes} A$ be the isometric isomorphism of Banach spaces indicated in 0.3.43(II); then, because the tensor product is associative (0.3.43(III)), the diagram

$$(A \hat{\otimes} B) \hat{\otimes} (A \hat{\otimes} B) = A \hat{\otimes} (B \hat{\otimes} A) \hat{\otimes} B \xrightarrow{1 \hat{\otimes} \omega \hat{\otimes} 1} A \hat{\otimes} (A \hat{\otimes} B) \hat{\otimes} B =$$

$$= (A \hat{\otimes} A) \hat{\otimes} (B \hat{\otimes} B) \xrightarrow{\underline{m}_A \hat{\otimes} \underline{m}_B} A \hat{\otimes} B.$$

makes sense. We let $\underline{m}$ denote the composition mapping and we put

$$m : (A \hat{\otimes} B) \times (A \hat{\otimes} B) \longrightarrow A \hat{\otimes} B : (u, v) \longmapsto \underline{m}(u \otimes v)$$

(in other words, we take the bi-operator associated with $\underline{m}$).

Proposition (I.2.25). *The bi-operator* m *is a multiplication turning* $A \hat{\otimes} B$ *into a Banach algebra. It is uniquely defined by the fact that (in the usual multiplicative notation)* $(a_1 \otimes b_1)(a_2 \otimes b_2) = a_1a_2 \otimes b_1b_2$ *for any* $a_1, a_2 \in A$, $b_1, b_2 \in B$. *Moreover, if the norms in* A *and* B *are multiplicative, then so also is the norm in* $A \hat{\otimes} B$.

▶ It is obvious that the indicated equation is, in fact, satisfied and this together with the fact that the linear hull of the elementary tensors is dense in $A \hat{\otimes} B$ immediately implies that the multiplication m is associative as well as unique. The assertion about the norms follows easily from 0.3.29. ■

Definition (I.2.26). The Banach algebra $A \hat{\otimes} B$ with the multiplication introduced above is called the *(projective) tensor product of the Banach algebras* A *and* B.

Of course, this concept has an algebraic prototype — in the construction of the tensor product of pure algebras the algebraic tensor product "$\otimes$" is used (instead of "$\hat{\otimes}$"). In exactly the same way, using the versions "$\otimes_i$" and "$\bar{\otimes}$" of the inductive tensor product, we introduce the *inductive tensor product* and the *complete Hausdorff inductive tensor product of polynormed algebras*, respectively. We leave the details of the arguments and statements to the reader.[13] We note that a tensor product of pure algebras may be considered as a particular case of the tensor products "$\otimes_i$" and "$\bar{\otimes}$" of polynormed algebras (cf. Example 2.17 and Chap.0 §3.5).

The operation of taking the tensor product of Banach algebras has obvious commutativity and associativity properties (cf. 0.3.43(II) and (III)). Associativity means that we need not worry about the order of the brackets in the expression $A_1 \hat{\otimes} \cdots \hat{\otimes} A_n$ and in this way we can define the *tensor product of any finite number of Banach algebras*. Of course this is also true for other types of tensor product.

We note that none of the tensor products introduced takes the class of commutative algebras out of itself. The same is also true for unital algebras: $e_A \otimes e_B$ is the identity in the tensor product.

Proposition (I.2.27). *Let* A *and* B *be unital Banach algebras. Then given any two characters* χ_1 *on* A *and* χ_2 *on* B *the*

---

[13] The projective tensor product "$\otimes_p$" and the complete Hausdorff projective tensor product "$\hat{\otimes}$" can be used with jointly polynormed algebras (see Chap.0 §3.5). The class of Fréchet algebras is closed under "$\hat{\otimes}$" as also is the class of Arens-Michael algebras.

functional $\chi_1 \hat{\otimes} \chi_2$ (*see Chap.*0 §3.4) *is a character on* $A \hat{\otimes} B$ *and moreover all the characters of the latter algebra are of this form.*

▶ If we are given a character χ on $A \hat{\otimes} B$ (it is automatically continuous by Theorem 2.6), then we put $\chi_1 : A \longrightarrow \mathbb{C} : a \longmapsto \chi(a \otimes e_B)$ and $\chi_2 : B \longrightarrow \mathbb{C} : b \longmapsto \chi(e_A \otimes b)$; it is obvious that χ_1 and χ_2 are characters and $(a \otimes b) = (a \otimes e_B)(e_A \otimes b)$ implies that $\chi(a \otimes b) = (\chi_1 \hat{\otimes} \chi_2)(a \otimes b)$. All the rest is implied by the fact that the linear hull of elementary tensors is dense in $A \hat{\otimes} B$. ■

Exercise (I.2.28). Show that the condition that the algebras in Proposition 2.27 are unital may be discarded.

Hint. Show that if $\chi \neq 0$, there exist $a \in A$ and $b \in B$ such that $\chi(a \otimes b) = \chi(a^2 \otimes b) = \chi(a \otimes b^2) = 1$ and deduce that $\chi(a \otimes b') = \chi(a^2 \otimes b')$ and $\chi(a' \otimes b) = \chi(a' \otimes b^2)$ for all $a' \in A$ and $b' \in B$. We can now put $\chi_1 : a \longmapsto \chi(a' \otimes b)$ and $\chi_2 : b' \longmapsto \chi(a \otimes b')$ and use $\chi(aa' \otimes bb') = \chi(a' \otimes b')$ to prove that $\chi = \chi_1 \hat{\otimes} \chi_2$. Let us collect together some simple but still useful facts.

Proposition (I.2.29). (I) *For any* $E, F \in \mathcal{Ban}$ *the operator* **ott** : $\mathcal{B}(E) \hat{\otimes} \mathcal{B}(F) \longrightarrow \mathcal{B}(E \hat{\otimes} F)$ (see Remark 0.3.46) *is a Banach algebra homomorphism with norm* 1; *if either* E *or* F *is also finite-dimensional, then it is a topological isomorphism*;

(II) *for any algebra* A *the algebra* $A \otimes \mathcal{M}_n$ *is isomorphic to the matrix algebra with elements from* A (*with the usual matrix mutliplication*). ▶■

An obvious corollary of either of the two assertions in Proposition 2.29 is the fact that $\mathcal{L}(E) \otimes \mathcal{L}(F) \simeq \mathcal{L}(E \otimes F)$ when either E or F is finite-dimensional. We suggest also showing that, up to an algebra isomorphism, $\mathcal{P}_m \otimes \mathcal{P}_n = \mathcal{P}_{m+n}$ $(m, n \in \mathbb{N})$ and $\mathbb{C}(G) \otimes \mathbb{C}(\Gamma) = \mathbb{C}(G \times \Gamma)$, where G and Γ are arbitrary groups.

Some examples of the algebras of analysis based on the tensor product construction will be given later in Chap.III (see Definition III.2.18 and also just after Assertions III.1.20 and III.3.16). Besides these it is also worth mentioning the so-called *Varopoulos algebra* $C(\Omega_1) \hat{\otimes} C(\Omega_2)$ which can be identified, using the operator γ considered in Chap.0 §3.3, with a certain dense subalgebra of $C(\Omega_1 \times \Omega_2)$; it plays an important role in harmonic analysis [39], [109]. Another interesting example is indicated in [105].

Our next aim is to carry across to Banach and polynormed algebras the construction in §1.2 of adjoining an identity. Let $(A; \|\cdot\|)$ be a seminormed algebra. We let $\|\cdot\|_+$ denote the seminorm in its unitization A_+ defined by $\|a + \lambda e\|_+ := \|a\| + |\lambda|$, $a \in A$, $\lambda \in \mathbb{C}$. Obviously, if $(A; \|\cdot\|')$

is a Banach algebra (or more generally $(A; \|\cdot\|^{\nu}; \nu \in \Lambda)$ is a polynormed algebra), then its unitization A_+ is also a Banach (polynormed) algebra with respect to the norm $\|\cdot\|'_+$ (the system of seminorms $\{\|\cdot\|^{\nu}_+; \nu \in \Lambda\}$, respectively). In future, when we talk about the unitization we shall mean the (semi)norms constructed in the way described — unless some other is explicity stated.

Example (I.2.30). Let Ω be a locally compact space. Then $C_0(\Omega)_+$ coincides, up to topological[14] isomorphism, with $C(\Omega_+)$, where Ω_+ is the one point compactification of Ω (see Chap.0 §0.3°). This isomorphism sends the element $f + \lambda e$; $f \in C_0(\Omega)$, $\lambda \in \mathbb{C}$, to the function $g : \Omega_+ \longrightarrow \mathbb{C}$ such that $g(s) = f(s) + \lambda$ where $s \in \Omega$ and $g(\infty) = \lambda$. (In other words, we identify $f + \lambda e$ with the function $f(s) + \lambda$ on Ω and to this we associate its unique continuous extension to Ω_+.)

* * *

Actually, the adjoining of an identity does not by any means appear so natural for all algebras as it does for the last example; furthermore this operation sometimes leads to the loss of some useful properties (the property of being an ideal in some larger algebra, say; see later §3). At the same time for a lot of non-unital algebras there exist well-defined nets which successfully fill the role of an identity in certain crucial questions.

Definition (I.2.31). Let A be a polynormed algebra. A net $(e_{\nu}; \nu \in \Lambda)$, in A is called an *approximate identity* (abbreviated to a.i.) of this algebra if $\lim\limits_{\nu} e_{\nu}a = a = \lim\limits_{\nu} ae_{\nu}$ for each $a \in A$, and it is called a *bounded approximate identity* (abbreviated to b.a.i.) if in addition the elements e_{ν} form a bounded set in A.

Thus, a b.a.i. in the Banach algebra $(A; \|\cdot\|)$ is an a.i. $(e_{\nu}; \nu \in \Lambda)$, such that $\|e_{\nu}\| \leq C$ for all ν, where C is a constant. Furthermore, if $C = 1$, then our net is called a *metric a.i.*

If only the left-hand of the double equation for the limits in Definition 2.31 holds in general, then we talk about a *left a.i.* and if only the right-hand holds then we talk about a *right a.i.*; a "one-sided" b.a.i. is defined analogously.

Here are a few illustrations clarifying the concepts introduced.

Exercise (I.2.32). Let Ω be a locally compact space, and let Λ be its collection of compact subsets, ordered by inclusion

---

[14] But not isometric — check it!

("precedes" means "$\subseteq$"). For $K \in \Lambda$ choose e_K to be an arbitrary function in $C_0(\Omega)$ that is equal to one on K and satisfies $0 \le e_K \le 1$ everywhere on Ω. Show that $(e_K;\ K \in \Lambda)$, is a metric a.i. in $C_0(\Omega)$.

Example (I.2.33). Let H be a separable Hilbert space, let p_n; $n = 1, 2, \cdots$, be the orthogonal projection onto the linear hull of the first n vectors of a fixed orthonormal basis. Then it is not difficult to show (for example, by identifying H with l_2 and using the known criterion for sets in l_2 to be precompact) that $(p_n;\ n \in \mathbb{N})$ is a metric a.i. in $\mathcal{K}(H)$. In general, if H is not separable, one can take as a metric a.i. in $\mathcal{K}(H)$ the net consisting of all orthogonal projections onto finite-dimensional subspaces in H; we leave the details to the reader.

Later on the two results of Exercises 2.32 and 2.33 turn out to be particular cases of a fairly general theorem IV.7.79.

Exercise (I.2.34). Show that there is an a.i. but no b.a.i. in l_1 (the space of absolutely convergent series) considered as a Banach algebra with respect to coordinate-wise multiplication.

Exercise (I.2.35). Let E be a linear space, and let Λ be its collection of finite-dimensional subspaces ordered by inclusion, let $p_F \in \mathcal{L}(E)$ be an (arbitrarily chosen) projection onto F for $F \in \Lambda$. Show that $(p_F;\ F \in \Lambda)$ is a left but not a right a.i. in the algebra $\mathcal{F}(E)$ (the latter is the hardest part of the problem!).

We pass from examples to some facts of a general character.

Proposition (I.2.36). *Let a polynormed algebra A have an a.i. with no limit (as a net). Then A is not unital.* ▸

If $(e_\nu;\ \nu \in \Lambda)$, is an a.i. and e is an "ordinary" identity in A, then the net $e_\nu = e_\nu e$ will converge to e. The rest is clear. ■

Exercise (I.2.37) (Dixon). Suppose there exists both a left and a right b.a.i. in a Banach algebra. Show that there also exists a ("two-sided") b.a.i.

Hint. Consider the family $e'_\mu + e''_\nu - e'_\mu e''_\nu$, where e'_μ and e''_ν are the corresponding "one-sided" b.a.i.

Finally, we give an important assertion without proof.

Assertion (I.2.38) ("Cohen's factorization theorem"). If a Banach algebra has a b.a.i., then any element of it can be represented as the product of two others.

It should be noted that in the majority of propositions it is not the specific form of the b.a.i. that is important but only the fact that it exists; the last assertion, in particular, bears witness to this. For more details of the various types of a.i. and their interrelation and tests for their existence see [9, §11]; Cohen's theorem is proved there.

§2.4. Topological properties of the group of inverse elements

Let A be a unital polynormed algebra, and let $\operatorname{Inv} A$ be the group of its invertible elements (see §1.2). We show here that, in the case of a Banach algebra, this group automatically has some nice properties, from a topological point of view; as for more general algebras, then the presence of such properties depends on the behaviour of the group $\operatorname{Inv} A$ "near the identity".

The map that sends an element to its inverse will be denoted by $\mathbf{inv} : \operatorname{Inv} A \longrightarrow \operatorname{Inv} A$. We also put $aM = \{ab : b \in M\}$ and $Ma = \{ba : b \in M\}$ for $a \in A$ and $M \subseteq A$.

Proposition (I.2.39). *Let A be a unital Banach algebra. Then*

(I) *e is an interior point of $\operatorname{Inv} A$. Moreover, if the norm in A is multiplicative, then for every $a \in A$ with $\|a\| < 1$ the element $e - a$ is invertible and its inverse can be represented as a series $(e-a)^{-1} = e + \sum_{n=1}^{\infty} a^n$ (the "Neumann series");*

(II) *the map* **inv** *is continuous at e.*

▸ In view of Proposition 2.5 we first assume that the norm is multiplicative.

(I) We put $b_n = e + \sum_{k=1}^{n} a^k$; $n \in \mathbb{N}$. Since $\|a^k\| \le \|a\|^k$, when $\|a\| < 1$ the sequence b_n is a Cauchy sequence and, since A is complete, it converges to a certain $b \in A$. The obvious equation $(e-a)b_n = b_n(e-a) = e - a^{n+1}$ and multiplication being continuous imply that $(e-a)b = b(e-a) = e$.

(II) The form of the Neumann series implies that, when $\|a\| < 1$,

$$\|e - (e-a)^{-1}\| \le \sum_{n=1}^{\infty} \|a^n\| \le \sum_{n=1}^{\infty} \|a\|^n = \frac{\|a\|}{1 - \|a\|}.$$

Hence

$$\lim_{b \to e} \mathbf{inv}(b) = \lim_{a \to 0} (e-a)^{-1} = e = \mathbf{inv}(e). \blacksquare$$

In the next three propositions A is an arbitrary unital polynormed algebra.

Proposition (I.2.40). *If $a \in \operatorname{Inv} A$, and $U \subseteq A$ is open (in A), then aU and Ua are also open.*

▶ aU, Ua respectively, is the preimage of U under the continuous operator $m_{a^{-1}}$, $m^{op}_{a^{-1}}$ respectively (see Definition 2.2). ■

Proposition (I.2.41). *Suppose that* e *is an interior point in* $\operatorname{Inv} A$. *Then* $\operatorname{Inv} A$ *is open in* A.

▶ Let $a \in \operatorname{Inv} A$ and let U be a neighbourhood of the identity lying in $\operatorname{Inv} A$. Then, according to Proposition 2.40, aU (and also Ua) is a neighbourhood of the point a, which is easily seen to lie (cf. Proposition 1.17), in $\operatorname{Inv} A$. ■

Proposition (I.2.42). *Let the map* **inv** *be continuous at* e. *Then it is continuous (everywhere) on* $\operatorname{Inv} A$.

▶ Let $a \in \operatorname{Inv} A$, and let U be a neighbourhood of the point a^{-1}. Then, according to Proposition 2.40, aU is a neighbourhood of the identity and, in view of the hypothesis, there is another neighbourhood V of e such that $\mathbf{inv}(V) = \{b^{-1} : b \in V\}$ lies in aU. Consequently, the set $\mathbf{inv}(Va)$, which coincides, as is easily seen, with $a^{-1}(\mathbf{inv}\,(V))$, lies in U. Since Va, again according to Proposition 2.40, is a neighbourhood of the point a, we have verified that **inv** is continuous at this point. ■

We combine Propositions 2.39, 2.41 and 2.42.

Theorem (I.2.43). *Let* A *be a unital Banach algebra. Then* $\operatorname{Inv} A$ *is an open subset of it and the map* **inv** *is continuous. As a consequence,* $\operatorname{Inv} A$ *is a metrizable topological group.* ▶■

Among the papers devoted to studying the topological group $\operatorname{Inv} A$, we single out the important papers of Arens [100] and Royden [229]; see also [254] and the references contained there.

For which (non-Banach) polynormed algebras are the conditions of Proposition 2.41 valid and for which are those of Propositon 2.42 true? It turns out that, roughly speaking, the first kind of algebra (for an example of such an algebra see later in Chap.III §1.3) is met far less frequently than the second. As an illustration we have the following.

Exercise (I.2.44). Verify that the algebra $\mathcal{O}(U)$, where U is, say, the unit disc in $\mathbb{C}$, and the algebra $\mathbb{C}^M$ with infinite M satisfy the conditions of Proposition 2.42 but not Proposition 2.41.

More will be said later on about algebras in which the map **inv** is continuous; see Chap. II §1 and Chap.V §1.

§2.5. Topologies on the set of characters

As the reader knows, the dual spaces (formed by the linear functionals) play a big part in studying polynormed spaces. In a similar way the topological spaces formed by the characters are very important in the study of polynormed algebras, especially commutative algebras.

Let A be a (for the time being arbitrary) polynormed algebra. We let $X\alpha\rho\, A$ be the set of all its non-trivial continuous characters and let $(X\alpha\rho\, A)_+$ be the set of all its continuous characters. For fixed A we shall use the abbreviated notation $X\alpha\rho$ and $X\alpha\rho_+$. We use the notation χ_∞ for the trivial (zero) character; thus $X\alpha\rho_+ = X\alpha\rho \cup \{\chi_\infty\}$.

Since a character is a special case of a functional, then $X\alpha\rho_+$ and with it $X\alpha\rho$ is a subset of A^*, the dual of A. Therefore any topology in A^* induces a topology on both these subsets. We limit ourselves to one — for our purposes the most important — of the possible topologies: the one that corresponds to the weak* topology in A^* (see Chap.0 §1.2). We shall also call this the weak* topology and, in future, when talking about $X\alpha\rho$ and $X\alpha\rho_+$ as topological spaces we shall mean just this topology.

0.1.1 immediately implies the following proposition.

Proposition (I.2.45). (I) *Both spaces* $X\alpha\rho$ *and* $X\alpha\rho_+$ *are Hausdorff.*

(II) *For each* $\chi \in X\alpha\rho$ $(X\alpha\rho_+$ *respectively) the sets of the form* $U_{\chi,a,\varepsilon} = \{\chi' \in X\alpha\rho\ (\chi' \in X\alpha\rho_+) : |\chi'(a) - \chi(a)| < \varepsilon\}$, *taken over all* $a \in A$ *and* $\varepsilon > 0$, *form a neighbourhood subbasis for this point.* ▶■

Here is a first illustration.

Exercise (I.2.46). Show that for the polynormed algebra $\mathcal{P}_n$ (see Example 2.17) the map $\chi \longmapsto (\chi(x_1), \cdots, \chi(x_n))$, where $x_1, \cdots, x_n$ are the independent variables, is a homeomorphism between $X\alpha\rho\, \mathcal{P}_n$ and $\mathbb{C}^n$ (considered with the usual topology).

We turn to considering the topological properties of some subsets in A^* consisting of characters.

Proposition (I.2.47). *For any* $a \in A$ *and* $t \geq 0$ *the set* $W_{a,t} = \{\chi \in X\alpha\rho_+ : |\chi(a)| \geq t\}$ *is closed in* A^*.

▶ Let $f \in A^*$ be a (weak*) limit of a certain net $(\chi_\nu \in W_{a,t};\ \nu \in \Lambda)$. Then, firstly, for any $b, c \in A$,

$$f(bc) = \lim_\nu \chi_\nu(bc) = \lim_\nu \chi_\nu(b)\chi_\nu(c) =$$

$$= (\lim_\nu \chi_\nu(b)) \times (\lim_\nu \chi_\nu(c)) = f(b)f(c)$$

so that $f \in X\alpha\rho_+$. Secondly, $f(a) = \lim_\nu \chi_\nu(a)$; and by the definition of $W_{a,t}$ this implies that $|f(a)| \geq t$. ∎

If we put $t = 0$, we get the following.

Corollary (I.2.48). *$X\alpha\rho_+$ is closed in A^*.*

In the case of unital A we put $a = e$ and $t = 1$ and we get the following corollary.

Corollary (I.2.49). *If A is unital, then $X\alpha\rho$ is closed in A^* (and thus χ_∞ is an isolated point in $X\alpha\rho_+$).*

Let us indicate a situation when the topological spaces under consideration are particularly nice.

Theorem (I.2.50). *Let A be a Banach algebra. Then*

(I) *the space $X\alpha\rho$ is locally compact and $X\alpha\rho_+$ is compact and coincides, up to homeomorphism, with the one-point compactification of $X\alpha\rho$;*

(II) *if, in addition, A is unital, then $X\alpha\rho$ is also compact.*

▶ We take, using Proposition 2.5, a multiplicative norm $\|\cdot\|'$ in A which is equivalent to the original norm and let B^* denote the unit ball in A^*, the dual of $(A; \|\cdot\|')$. Then, according to Theorem 2.6, $X\alpha\rho_+ \subseteq B^*$. But by the Banach-Alaoglu theorem (see Chap.0 §1.2) B^* is compact in the weak* topology. Therefore Corollary 2.48 implies that $X\alpha\rho_+$ is a closed subset of a compact (Hausdorff) space and therefore is itself compact. But then $X\alpha\rho = X\alpha\rho_+ \setminus \{\chi_\infty\}$ obviously implies the remaining part of (I). Finally (I) and Corollary 2.49 imply (II). ∎

We now establish an important connection between the characters on A and its unitization A_+. Obviously any character χ of the first algebra may be uniquely extended to a non-trivial (= unital) character χ_+ of the second algebra; this latter is defined by $\chi_+(a + \lambda e_+) = \chi(a) + \lambda$; $a \in A$, $\lambda \in \mathbb{C}$. Moreover, if χ is continuous, then so also is χ_+.

Proposition (I.2.51). *The map $\alpha : (X\alpha\rho\, A)_+ \longrightarrow X\alpha\rho\, (A_+) : \chi \longmapsto \chi_+$ is a homeomorphism.*

▶ First of all α is bijective: its inverse is the map sending a character on A_+ to its restriction to A. Further, we put $U'_{\chi,a,\varepsilon} = \{\chi' \in (X\alpha\rho\, A)_+ : |\chi'(a) - \chi(a)| < \varepsilon\}$, where $\chi \in (X\alpha\rho\, A)_+$, $a \in A$ and $\varepsilon > 0$, and we put $U''_{\chi_1,b,\delta} = \{\chi'' \in X\alpha\rho\, (A_+) : |\chi''(b) - \chi_1(b)| < \varepsilon\}$, where $\chi_1 \in X\alpha\rho\, (A_+)$, $b \in A_+$ and $\delta > 0$. By Proposition 2.45(II) this gives neighbourhood subbases for points in both of the spaces being compared. Therefore the obvious relations $\alpha^{-1}(U''_{\chi_+,b,\delta}) = U'_{\chi,c,\delta}$, where $b \in A_+$, is of the form

$b = c + \lambda e_+$; $c \in A$, $\lambda \in \mathbb{C}$, and $\alpha(U'_{\chi,a,\varepsilon}) = U''_{\chi_+,a,\varepsilon}$, where $a \in A \subset A_+$, imply that α and α^{-1} are continuous. ■

If we combine Proposition 2.51 and Theorem 2.50(I) we get the following corollary.

Corollary (I.2.52). *Let A be a Banach algebra. Then the space $X\alpha\rho\,(A_+)$ coincides, up to homeomorphism, with the one-point compactification of the space $X\alpha\rho\,A$.*

In conclusion we show that the operation of taking the tensor product of Banach algebras corresponds to the operation of taking the Cartesian product of their topological character spaces.

Proposition (I.2.53). *Let A and B be unital Banach algebras. Then the map $\omega : X\alpha\rho\,A \times X\alpha\rho\,B \longrightarrow X\alpha\rho\,(A \hat{\otimes} B)$: $(\chi_1, \chi_2) \longmapsto \chi_1 \hat{\otimes} \chi_2$* (see Proposition 2.27) *is a homeomorphism.*

▸ According to Proposition 2.27 ω is bijective. Therefore, since the spaces being compared are compact (Theorem 2.50(II)), by Alexandrov's theorem (Chap.0 §0.3°), it is sufficient to verify that the map ω^{-1} is continuous.

Let a certain net $(\chi_1^\nu \hat{\otimes} \chi_2^\nu;\ \nu \in \Lambda)$, converge to the character $\chi_1 \hat{\otimes} \chi_2$ in $X\alpha\rho\,(A \hat{\otimes} B)$; this means that $\lim_\nu \chi_1^\nu \hat{\otimes} \chi_2^\nu(u) = \chi_1 \hat{\otimes} \chi_2(u)$ for any $u \in A \hat{\otimes} B$. But then for any $a \in A$

$$\lim_\nu \chi_1^\nu(a) = \lim_\nu(\chi_1^\nu \hat{\otimes} \chi_2^\nu)(a \otimes e_B) = (\chi_1 \hat{\otimes} \chi_2)(a \otimes e_B) = \chi_1(a),$$

that is, $\lim_\nu \chi_1^\nu = \chi_1$ in $X\alpha\rho\,A$; similarly $\lim_\nu \chi_2^\nu = \chi_2$ in $X\alpha\rho\,B$.

Hence $\lim_\nu \omega^{-1}(\chi_1^\nu \hat{\otimes} \chi_2^\nu) = \lim_\nu (\chi_1^\nu, \chi_2^\nu) = (\chi_1, \chi_2) = \omega^{-1}(\chi_1 \hat{\otimes} \chi_2)$ in $X\alpha\rho\,A \times X\alpha\rho\,B$. The rest is clear. ■

§3. IDEALS

§3.1. Ideals and examples of ideals

The following "threefold" concept is one of the most important in this book — perhaps second in importance after the concept of an algebra itself.

Definition (I.3.1). A subspace I of an algebra A is called a *left* (*right*, respectively) *ideal* if $ax \in I$ ($xa \in I$, respectively) for all $x \in I$ and $a \in A$. A subspace, that is simultaneously a left and a right ideal of an algebra is called a *two-sided ideal* or briefly a *bi-ideal*.

Thus, I is a right ideal in A $\Leftrightarrow$ it is a left ideal in A^{op} and "vice versa". The term "*ideal*" (without epithets) will be used in future as the standard name for all three types. In the case of a commutative algebra, where all three concepts coincide, these epithets are, of course, superfluous.

The left ideals in A, obviously, may be also defined as subspaces that are invariant under all the operators m_a; $a \in A$, of the regular representation of our algebra.

There is in each algebra at least one ideal; the algebra itself. Naturally, we are interested in ideals that are distinct from the whole algebra; they are called *proper ideals*. In particular, when $A \neq \mathbb{O}$, $\{0\}$ is a proper ideal.

Proposition (I.3.2). *An ideal of a unital algebra is proper* $\Leftrightarrow$ *it does not contain the identity.* ▸■

Corollary (I.3.3). *A left ideal (right ideal, respectively) of a unital algebra is proper* $\Leftrightarrow$ *it contains no element that is invertible on the left (right).*

For any $a \in A$ there exists a smallest left ideal in A containing a: this is obviously $\{ba + \lambda a : b \in A, \lambda \in \mathbb{C}\}$ (coinciding, in the case of a unital algebra, with $Aa := \{ba : b \in A\}$. Such a left ideal is called the *principal ideal generated by* a.

Exercise (I.3.4). Let M be an arbitrary subset of A. Prove that there exists a smallest right ideal containing M and describe its elements.

In algebras furnished with a topology the most interesting ideals are those that are closed. For polynormed algebras the definitions imply that the closure of every ideal — left, right or two-sided — is again an ideal of the same type. The following proposition is more interesting.

Proposition (I.3.5). *The closure of a proper ideal in a unital Banach algebra is again a proper ideal.*

▸ If we combine Corollary 3.3 and Proposition 2.39(I) we see that the identity is not a point in the closure of a proper ideal. ■

* * *

Before continuing the general theory we shall "try out" the concept of an ideal on examples. At the moment, in talking about the ideals of specific algebras, we are only sorting out what lies on the surface; the deeper facts are encountered in Chap. III.

Example (I.3.6). Let A be a certain function algebra on the set M with pointwise multiplication (= a certain subalgebra of

$\mathbb{C}^M$), and let Δ be a subset of M. We put $I_\Delta := \{f \in A : f|_M = 0\}$; then obviously I_Δ is an ideal in A.

Example (I.3.7). Consider the Banach algebra $C(\Omega)$, where Ω is a compact space. Then, as a special case of Example 3.6, every subset Δ of Ω furnishes an ideal I_Δ in $C(\Omega)$. All these ideals are closed, and moreover, for any $\Delta_1, \Delta_2 \subseteq \Omega$, $I_{\Delta_1} = I_{\Delta_2}$ $\Leftrightarrow$ the closure of Δ_1 and Δ_2 in Ω coincide (check it!).

Example (I.3.8). We take, in the same algebra $C(\Omega)$, the set I_t^0 of functions that are equal to zero in some neighbourhood (each having its own neighbourhood) of a fixed point $t \in \Omega$. Then I_t^0 is also an ideal, and moreover it is distinct from those just described: it is, generally speaking, not closed, and its closure is easily seen to be $I_{\{t\}}$.

Example (I.3.9). Let U be a domain in $\mathbb{C}$, let $z_1, \cdots, z_n$ be a collection of points of U, and let $m_1, \cdots, m_n$ be a collection of natural numbers (the "multiplicities of these points"). Then the set of functions that are holomorphic in U and have a zero of multiplicity $\geq m_k$ at z_k; $1 \leq k \leq n$, is a closed ideal in the Fréchet algebra $\mathcal{O}(U)$.

Exercise (I.3.10). Show that in the algebra $\mathcal{P}$ every ideal is a principal ideal. Deduce from this that an ideal is determined, like the ideals described above in $\mathcal{O}(\mathbb{C})$, by a collection of points of $\mathbb{C}$ with multiplicities.[15]

Hint. Use the Euclidean algorithm for the division of polynomials with remainders.

Example (I.3.11). Let E be a linear space, and let F be a subspace of it. Then $I := \{\varphi : \operatorname{Ker}\varphi \supseteq F\}$ is a left ideal and $J := \{\varphi : \operatorname{Im}\varphi \subseteq F\}$ is a right ideal in $\mathcal{L}(E)$. When E is a Banach space, closed left and right ideals in $\mathcal{B}(E)$ may be constructed in exactly the same way from a closed F. The set of finite-dimensional operators in E, that is $\mathcal{F}(E)$, is a bi-ideal in $\mathcal{L}(E)$, while, when E is a Banach space, the set of compact operators, that is $\mathcal{K}(E)$, is a closed bi-ideal in $\mathcal{B}(E)$ (although $\mathcal{K}(E) \cap \mathcal{F}(E)$ is, generally speaking, not closed).

Exercise (I.3.12). Show that when E is finite-dimensional all the ideals of $\mathcal{L}(E)$ are included among the left and right ideals indicated above, and as a consequence $\mathcal{L}(E)$ has no proper bi-ideals except $\{0\}$.

---

[15] There are other ideals, besides these, in $C(\mathbb{C})$ itself (check it!).

Exercise (I.3.13). Let E have, as a linear space, a countable basis ($E = \mathcal{P}$ or c_{00}, say). Show that in this case $\mathcal{F}(E)$ is the only proper bi-ideal in $\mathcal{L}(E)$, besides $\{0\}$.

Example (I.3.14). In the Banach algebra $l_1(\mathbb{Z})$ we consider the set of sequences with zero sum. Obviously this is a closed ideal of codimension 1.

Example (I.3.15). Let M be an infinite set. Then $\mathbb{C}^M_{00}$ is a dense ideal in the Arens-Michael algebra $\mathbb{C}^M$; and therefore it is not closed.

§3.2. Maximal and modular ideals

We return to algebras of a general kind. In order to avoid tedious repetition a few of the following definitions and results will be formulated only for left ideals; their "right" and "bi-" analogues are obvious.

Definition (I.3.16). A proper left ideal is said to be *maximal* if it is not contained in any other proper left ideal of the given algebra. A non-zero left ideal is said to be *minimal* if it does not contain any other left ideal except zero.

Example (I.3.17). Every algebra A is a maximal ideal (of any of the three types) in its unitization A_+. If A itself is also unital, then $\ell.h.\{e_+ - e_A\}$ is a minimal ideal (again of any type) in A_+.

Running on ahead, we note that, in spite of the similarity of the definitions, maximal ideals play a far more significant role than minimal ideals in the theory of polynormed algebras (and, in particular, Banach algebras). In many respects this is explained by the following fact.

Theorem (I.3.18) (Krull's lemma). *Every proper left ideal of a unital algebra is contained in a maximal left ideal.*

▶ Let us perform the ritual dance around Zorn's lemma.[16] Let Λ be the set of left ideals of our algebra that contain the given ideal, ordered by inclusion, and let Λ' be any linearly ordered subset of Λ. We put $J := \bigcup\{I : I \in \Lambda'\}$; obviously this is also a left ideal. Further, Proposition 3.2 implies that $e \notin I$ for any $I \in \Lambda'$ (this is why an identity is necessary!); hence $e \notin J$ and therefore J is a proper ideal. Thus J is an upper bound for Λ' in Λ.

We have shown that each linearly ordered subset in Λ is bounded above; consequently, by Zorn's lemma, there is a maximal

[16] Curiously enough, Krull's lemma (in the classical version it is really for commutative rings) is equivalent to Zorn's lemma and therefore also to the Axiom of Choice [273].

element in Λ. The latter is obviously a maximal left ideal containing the given ideal. ■

For comparison we note that there are no minimal ideals in many unital algebras: such as, for example $\mathcal{P}$ (cf. Exercise 3.10).

Theorem (I.3.19). *The following properties of an element a in a unital algebra A are equivalent:*

(I) *a is invertible on the left;*

(II) *a does not belong to any proper left ideal in A;*

(III) *a does not belong to any maximal left ideal in A.*

▶ If a is not invertible on the left, then the principal ideal it generates does not contain the identity; therefore (II) implies (I). Further, by Theorem 3.18 (III) implies (II) and by Corollary 3.3 (I) implies (III). ■

Theorem (I.3.20). *The following properties of a unital algebra A are equivalent:*

(I) *A is a skew field;*

(II) *A has no proper left ideal, except zero;*

(III) *A has no maximal left ideal, except zero;*

(IV) *A has no proper right ideal, except zero;*

(V) *A has no maximal right ideal, except zero.*

▶ In the implications (I) ⇒ (III) ⇒ (II) ⇒ (I) the first is implied by Theorem 3.19 ((I) ⇒ (III)) and the second by Theorem 3.18; we deal with the third. By Theorem 3.19 ((II) ⇒ (I)) any $a \in A \setminus \{0\}$ is invertible on the left. But then for such an a the element a_l^{-1} is invertible both on the left and on the right; consequently a_l^{-1} is (simply) invertible and hence a, being its right inverse, is also its (two-sided) inverse. Thus a is invertible and so A is a skew field.

To complete the proof it is sufficient to establish (I) ⇒ (V) ⇒ (IV) ⇒ (I). These are the obvious "right" versions of the previous implications. ■

When applied to Banach algebras, Proposition 3.5 immediately implies the following.

Theorem (I.3.21). *A maximal ideal of a unital Banach algebra is always closed.* ▶■

In some familiar polynormed algebras non-closed maximal ideals are by no means rare: this is so, for example, for the maximal ideals containing c_{00} in the "very respectable" (Fréchet and Arens-Michael) algebra c_∞.

Exercise (I.3.22). Describe the maximal ideals in Examples 3.7, 3.8, and 3.11 and Exercise 3.10.

How essential is the assumption in Theorem 3.18 that there is an identity? It is impossible to simply omit it.

Exercise (I.3.23). Show that it is impossible to extend the ideal c_{00} in c_0 to a maximal ideal.

Hint. If the sequence $\{\xi_n\}$ does not belong to a proper ideal $I \supset c_{00}$, then there is some $J \supsetneqq I$ not containing $\{\sqrt{|\xi_n|}\}$.

Nevertheless in non-unital algebras there also exists an important class of ideals that can be extended to maximal ideals.

Definition (I.3.24). A left ideal I of an (arbitrary) algebra A is said to be *modular* if there exists $u \in A$ such that $a - au \in I$ for any $a \in A$. This u is called a *right modular identity* for I.

Of course, every ideal of a unital algebra is modular; on the contrary none of the ideals in an algebra with zero multiplication (see Proposition 2.7) is modular.

Proposition (I.3.25). *Every modular proper left ideal of an (arbitrary) algebra is contained in a maximal modular left ideal.*

▶ If u is a right modular identity for I, then obviously $u \notin J$ for any proper ideal $J \supseteq I$. Therefore we can repeat the proof of Theorem 3.18 where e is replaced by u. ■

Another "geometric" approach to the concept of a modular ideal is also useful. Here two lemmas are necessary, the first of them is a simple fact from linear algebra.

Lemma (I.3.26). *Let E be a linear space, let E' be a subspace of codimension* 1, *and let F be another subspace such that $E' \cap F \neq F$. Then* $\operatorname{codim}_F E' \cap F = 1$. ▶■

Lemma (I.3.27). *Let I be a left ideal in A, and let J be a left ideal in A_+, where $I = A \cap J$ and $I \neq J$. Then*

(I) *I is proper (in A) $\Longleftrightarrow$ J is proper (in A_+);*

(II) *I is maximal (in A) $\Longleftrightarrow$ J is maximal (in A_+).*

▶ (I) The previous lemma (with $E = A_+$, $E' = A$ and $F = J$) implies that $\operatorname{codim}_J I = 1$, hence $I = A \Longleftrightarrow J = A_+$.

(II) "$\Rightarrow$". In view of (I), "$\Rightarrow$", J is a proper ideal. If J_0 is a proper ideal containing J, then in view of (I), "$\Leftarrow$", $A \cap J_0$ is a proper ideal containing I and hence equal to I. But then, by Lemma 3.26 (with $F = J_0$) $\operatorname{codim}_{J_0} I = 1$; this, together with $\operatorname{codim}_J I = 1$, as already shown, gives $J_0 = J$.

(III) "$\Leftarrow$". In view of (I), "$\Leftarrow$", I is a proper ideal; let I_0 be a proper ideal in A containing it. Then the ideal $J_0 = I_0 + J$, since it is obviously proper, coincides with J. Hence $I_0 = J_0 \cap A = I$. ■

Theorem (I.3.28). *Let I be a left ideal in an (arbitrary) algebra A. Then I is modular $\Longleftrightarrow$ it has the form $I = A \cap J$, where J is a left ideal in A_+ distinct from I. Moreover I*

is a proper, or maximal respectively, ideal in $A \iff J$ *is a proper, or maximal respectively, ideal in* A_+.

▶ Let u be a left modular identity for I. We note that $e_+ - u \notin A$ and we put $J = \{x + \lambda(e_+ - u) : x \in I, \ \lambda \in \mathbb{C}\}$. Then it is easy to verify that J is a left ideal in A_+ with $I = J \cap A$ and $J \neq I$.

Conversely, let $I = A \cap J$ with $J \neq I$; then we have $x_0 \in J \setminus A$. Now $\operatorname{codim}_{A_+} A = 1$ implies that $e_+ = u + \lambda x_0$ for some $u \in A$ and $\lambda \in \mathbb{C}$. Therefore, for any $a \in A$, $a - au = a - a(e_+ - \lambda x_0) = \lambda a x_0$ belongs to A as well as to J. This obviously means that u is a right modular identity for I.

The second assertion follows from Lemma 3.27 that has already been proved. ■

We can now somewhat strengthen Theorem 3.21.

Theorem (I.3.29). *Every maximal modular ideal* I *of an (arbitrary) Banach algebra* A *is closed.*

▶ According to the previous theorem, $I = A \cap J$, where J, as well as A, is a maximal ideal in the Banach algebra A_+. By Theorem 3.21 this means that I is the intersection of two closed sets. ■

§3.3. **Quotient algebras**

Let A be an algebra and I a subspace; we consider the quotient space A/I. We would like to define a multiplication in A/I "through representatives", that is by the formula

$$(a + I)(b + I) = ab + I, \tag{1}$$

in the same way as one induces linear operators. When can this be done?

Proposition (I.3.30). *The following properties of* I *are equivalent*:

(I) *formula* (1) *gives a well-defined binary operation in* A/I;

(II) I *is a bi-ideal in* A.

If these hold, then the operation above is a multiplication.

▶ Property (I) means precisely that $(a + x)(b + y) + I = ab + I$ for any $a, b \in A$ and $x, y \in I$, (hence, in particular, $(a + 0)(0 + y) + I = a \cdot 0 + I$ for all $a \in A$ and $y \in I$), that is $ay \in I$. Therefore (I) implies that I is a left ideal and as a similar argument shows, that it is a right ideal and hence satisfies (II). The converse deduction and also the verification of the properties of the multiplication are obvious. ■

In this way we are justified in giving the following definition.

Definition (I.3.31). Let I be a bi-ideal in A. The space A/I, furnished with the multiplication given by formula (I), is called the *quotient algebra* of A with respect to I.

In the operation under discussion the natural projection $\mathbf{pr} : A \longrightarrow A/I$ (cf. Chap.0 §0.2°) is obviously an algebra homomorphism. Moreover if A has an identity, then $e_A + I$ is an identity in A/I and $\mathbf{pr}$ is a unital homomorphism.

Here is the first illustration.

Exercise (I.3.32). Show that any quotient algebra of the algebra $\mathcal{P}$ coincides, up to isomorphism, with the algebra of the polynomials in a certain matrix in Jordan normal form.

Hint. Use the result of Exercise 3.10.

In the next proposition I is a bi-ideal in A and $\mathbf{pr} : A \longrightarrow A/I$ is the natural projection.

Proposition (I.3.33). *The map* $\widetilde{\mathbf{pr}} : J \longmapsto \dot{J} = \{\mathbf{pr}(a) : a \in J\}$, *defined on the set of left ideals in* A *that contain* I, *is a bijection onto the set of all left ideals in* A/I. *Moreover* J *is a maximal or modular ideal in* $A \iff \dot{J}$ *is a maximal or a modular ideal in* A/I.

▶ Obviously, all the $\dot{J}$ are left ideals in A/I and so $\widetilde{\mathbf{pr}}$ is in fact a map between the indicated sets. Further it is easy to see that the map $\dot{J} \longmapsto J = \{a : \mathbf{pr}(a) \in \dot{J}\}$, defined on the set of left ideals in A/I, is the inverse to $\widetilde{\mathbf{pr}}$. The assertion about maximal ideals is implied by the fact that $\widetilde{\mathbf{pr}}$ and $\widetilde{\mathbf{pr}}^{-1}$ preserve the inclusion relation between ideals. Finally, it can be immediately verified that u is a left modular identity for $J \iff \dot{u} = \mathbf{pr}(u)$ is a left modular identity for $\dot{J}$; this implies the assertion about modular ideals. ■

Of course, the result remains valid if left ideals are replaced by right or two-sided ideals. The next application plays a special role later on.

Theorem (I.3.34). *Let* I *be an ideal in a commutative algebra* A. *Then it is a maximal modular ideal* $\iff$ A/I *is a field.*

▶ By Proposition 3.33 the first condition is equivalent to the zero ideal in A/I being a maximal modular ideal. Since the modular identity for a zero ideal is, obviously, the (ordinary) identity of the algebra, this last condition means precisely that A/I is unital and contains no proper ideals, except the zero ideal. All that is left is to apply Theorem 3.20. ■

Exercise (I.3.35). Let I be a bi-ideal in A. Show that A/I is unital $\iff$ I is simultaneously a modular left ideal and a modular right ideal.

If I is a bi-ideal in A, then obviously it is also a bi-ideal in A_+, and moreover A/I turns out to be a subalgebra of $(A/I)_+$ as well as of A_+/I. There should be no confusion here.

Proposition (I.3.36). *There is a (necessarily unital) isomorphism between $(A/I)_+$ and A_+/I leaving A/I fixed.*

* * *

Now let A be a polynormed algebra and I be a bi-ideal (for the moment arbitrary). In this situation A/I is not only an algebra but also a polynormed space and when we talk about it we shall always have in mind the system of quotient seminorms corresponding to a saturated system of seminorms (see Chap.0 §1.1).

Proposition (I.3.37). *A/I is a polynormed algebra and $\mathbf{pr} : A \longrightarrow A/I$ is a continuous homomorphism.*

▶ For any $\dot{a} \in A/I$, $\dot{a} = a + I$, $a \in A$, the operators $m_{\dot{a}}$ and $m_{\dot{a}}^{op}$ (see Definition 2.2) are induced by the operators m_a and m_a^{op} respectively and consequently they are either both continuous or neither is continuous. ■

We shall show that there are certain special classes of polynormed algebras which, when they contain an algebra also contain all its quotient algebras or, at least all its quotient algebras with respect to closed bi-ideals.

Proposition (I.3.38). *Let I be a bi-ideal in the algebra A, let $\|\cdot\|$ be a seminorm in A, and let $\|\cdot\|'$ be its quotient seminorm in A/I. Then if $\|\cdot\|$ is multiplicative, so also is $\|\cdot\|'$.*

▶ For any $\dot{a}, \dot{b} \in A/I$

$$\|\dot{a}\dot{b}\|' = \inf\{\|c\| : c \in \dot{a}\dot{b}\} \leq \inf\{\|ab\| : a \in \dot{a},\ b \in \dot{b}\} \leq$$

$$\leq \inf\{\|a\|\,\|b\| : a \in \dot{a},\ b \in \dot{b}\} = \|\dot{a}\|'\,\|\dot{b}\|'. \ ■$$

Corollary (I.3.39). *If A is a multiplicatively polynormed algebra, then a quotient algebra with respect to any bi-ideal is also a multiplicatively polynormed algebra.*

Proposition (I.3.40). (I) *The quotient algebra of a Banach algebra with respect to any closed bi-ideal is a Banach algebra.*
(II) *The quotient algebra of a commutative Banach algebra with respect to any maximal modular ideal is a Banach field.*

▶ The first assertion follows immediately from the analogous property for quotient spaces of Banach spaces (mentioned in Chap.0

§1.1). If we combine it and Theorem 3.29 and Theorem 3.34, we get the second assertion. ■

Later on we shall be able to say something far greater about the quotient algebras indicated in Proposition 3.40(II) (see Theorem II.1.17).

Of course, a result similar to Proposition 3.40(I) (but not to 3.40(II)!) holds for more general Fréchet algebras.

Exercise (I.3.41). Let Ω be a compact space, and let Δ be a closed subset. Show that the quotient algebra $C(\Omega)/I_\Delta$ (see Example 3.7) is isometrically isomorphic to $C(\Delta)$.

Exercise (I.3.42). Let A be a jointly polynormed algebra (see Definition 2.18). Show that any quotient algebra of it is jointly polynormed.

We give a few simple but useful results concerning quotient algebras (or obtained with their help). In order to avoid duplication we shall formulate them for polynormed algebras of varying generality, leaving the reader to establish for each its purely algebraic prototype.

Proposition (I.3.43). *A subset I of a Banach (polynormed, respectively) algebra A is a closed bi-ideal $\Leftrightarrow$ it is the kernel of a continuous homomorphism from A into a certain Banach (Hausdorff polynormed, respectively) algebra.*

▶ "$\Leftarrow$" is obvious; "$\Rightarrow$" follows immediately from a consideration of the homomorphism $\mathbf{pr}: A \longrightarrow A/I$. ■

Proposition (I.3.44). *Let I and J be bi-ideals in the polynormed algebras A and B respectively, and let $\kappa: A \longrightarrow B$ be a continuous homomorphism mapping I into J. Then the operator $\dot\kappa: A/I \longrightarrow B/J$, induced by κ* (see Chap.0 §0.2°) *is also a continuous homomorphism between polynormed algebras. Moreover $\dot\kappa$ is injective $\Leftrightarrow$ $\kappa^{-1}(J) = I$.* ▶■

This implies a fundamental fact, called by tradition the "*homomorphism theorem*".

Theorem (I.3.45). *Let $\kappa: A \longrightarrow B$ be a surjective continuous homomorphism between Banach algebras (or, more generally, Fréchet algebras). Then there exists a unique topological isomorphism between Banach algebras (Fréchet algebras) $\dot\kappa$ such that the diagram below is commutative.*

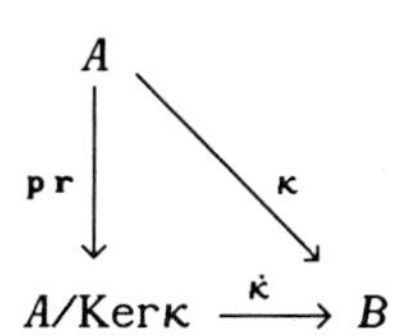

▶ We consider $\dot{\kappa}$ from the preceding proposition in the case when $I = \operatorname{Ker}\kappa$ and $J = \{0\}$; obviously this is a continuous algebra isomorphism. All we have to do then is to use the Banach theorem. ■

Another observation concerns the topological spaces of characters introduced in §2.5. Let A be a polynormed algebra. We let $I_\cap$ denote the intersection of the kernels of all the characters in $X\alpha\rho\, A$. Obviously $I_\cap$ is a closed bi-ideal in A; according to Proposition 3.44 applied to $I = I_\cap$, $B = \mathbb{C}$ and $J = \{0\}$, every continuous character χ on A generates a continuous character $\dot{\chi}$ on the polynormed algebra $A/I_\cap$; moreover $\chi \neq 0$ implies $\dot{\chi} \neq 0$.

Proposition (I.3.46). *$\beta : X\alpha\rho\, A \longrightarrow X\alpha\rho\, A/I_\cap : \chi \longmapsto \dot{\chi}$ is a homeomorphism between the character spaces.*

▶ First of all β has an inverse: it is easy to show that β^{-1} is the map taking $\dot{\chi}$ to $\chi = \dot{\chi}\cdot \mathbf{pr}$, where $\mathbf{pr} : A \longrightarrow A/I_\cap$ is the natural projection. Further, it can be immediately verified that β maps the set $U_{\chi,a,\varepsilon}$ from the standard neighbourhood subbasis for the point $\chi \in X\alpha\rho\, A$ (see Propositiion 2.45(II)) onto the set $U_{\dot{\chi},a+I_\cap,\varepsilon}$ from the standard neighbourhood subbasis for the point $\dot{\chi} \in X\alpha\rho\, A/I_\cap$. This implies in an obvious way, taking into account that β has already been proved to be bijective, that β is a homeomorphism. ■

§3.4. The radical and semisimplicity

There is among the ideals of the algebra one special ideal — called the radical; it consists of elements whose algebraic properties are very nearly those of the zero. In the general theory this ideal appears to be "rather a nuisance": the fact that it could be non-trivial, as a rule, prevents results being given in a more complete form.[17] Fortunately this fact is not really "fatal", since the majority of algebras met in analysis have a trivial radical.

Let us proceed with the precise definitions.

Definition (I.3.47). An element r of the algebra A is said to be *quasi-null* if for any a in the algebra $A_\times$ (see §1.2) the element $e - ar$ is invertible on the left.

---

[17] We shall see later on that the radical prevents algebras from being function algebras (Chap.IV §2) or operator algebras (Chap.IV §5), prevents irreducible representations from separating the elements (Chap.VI §2.2), homomorphisms from being "automatically continuous" (Chap.VI §2.3) and so on.

Example (I.3.48). Every *nilpotent* element r (that is, $r^n = 0$ for some power n) of a commutative algebra A is quasi-null. In fact, for any $a \in A$, $e - ar$ has an inverse, namely $e + ar + a^2r^2 + \cdots + a^{n-1}r^{n-1}$.

Exercise (I.3.49). Show that no *idempotent* element (equal to its square) of an arbitrary algebra can be quasi-null, unless it is zero.

Exercise (I.3.50). Show that no non-zero element (even if it is nilpotent!) of the matrix algebra $\mathcal{M}_n$ is quasi-null.

Definition (I.3.51). The set of quasi-null elements of the algebra A is called its *radical* or, more fully, its *Jacobson radical*, and is denoted by $\operatorname{Rad} A$ (and often simply by R).

The definition of the radical has a lot of equivalent formulations. For the first group of these it is convenient to introduce a symbolic notation. We denote the original Definition 3.47 itself ("as a piece of text") by $\{[ar]$; the definitions obtained from this by replacing the words "invertible on the left" by "invertible on the right" or (simply) "invertible" are denoted by $[ar]\}$ and $[ar]$ respectively, and finally, the definitions obtained from these three by replacing ar by ra are denoted by $\{[ra], [ra]\}$ and $[ra]$ respectively. For example, the definition $[ra]\}$ gives us the set of those $r \in A$ for which $e - ra$ is invertible on the right for any $a \in A_\times$, that is, precisely the set $\operatorname{Rad} A^{\mathrm{op}}$, and so on.

Proposition (I.3.52). *All six definitions are equivalent (in other words, each of them gives us the same set* $\operatorname{Rad} A$). *In particular,* $\operatorname{Rad} A = \operatorname{Rad} A^{\mathrm{op}}$.

▶ It is sufficient to establish the chain of logical equivalences

$$\begin{array}{ccccc} \{[ar] & \underset{1}{\Longleftrightarrow} & [ar] & & [ar]\} \\ 3\Big\updownarrow & & 4\Big\updownarrow & & 5\Big\updownarrow \\ \{[ra] & & [ra] & \underset{2}{\Longleftrightarrow} & [ra]\} \end{array}$$

First let r satisfy $\{[ar]$, and let a be an arbitrary element in $A_\times$. Then $b(e - ar) = e$ for a certain $b \in A_\times$ which is therefore invertible on the right. Hence, if we take $(-ba)$ instead of a in the definition $\{[ar]$, we get that $b = e + bar$ is invertible also on the left and it hence is an inverse for $e - ar$. We have proved that $\{[ar]$ implies $[ar]$, that is, since the converse is obvious, we have established the equivalence "$\underset{1}{\Longleftrightarrow}$".

We now note that the definition $[ra]\}$ is none other than $\{[ar]$ for the algebra A^{op}; hence, by what has already been proved, it is equivalent to the definition [*ar*] for A^{op}, that is the definition [*ra*] for the original algebra. This establishes the equivalence "$\underset{2}{\Longleftrightarrow}$".

The remaining equivalences denoted by the vertical logical arrows follow immediately from Proposition 1.19. ∎

Theorem (I.3.53). *For any algebra A its radical coincides with the intersection*

(I) *of all maximal left ideals in $A_\times$;*

(Ia) *of all maximal modular left ideals in A or with A itself if there are no such ideals;*

(II) *of all maximal right ideals in $A_\times$;*

(IIa) *of all maximal modular right ideals in A, or with A itself if there are no such ideals.*

▶ We limit ourselves to proving (I) and (II) assuming that A is unital (that is, $A_\times = A$). All the rest easily reduces to the "unital" case with the help of Theorem 3.28; we leave the details of the corresponding argument to the reader.

Assertion (I) may be rephrased thus: for any $s \in A$, $s \notin \operatorname{Rad} A \iff$ there exists a maximal left ideal I in A such that $s \notin I$. If the left-hand statement is satisfied, then $e - as$ is not invertible on the left for a certain $a \in A$; this means, in view of Theorem 3.19(III) that $e - as \in I$ for a certain maximal left ideal I. Hence as (which lies in $e + I$) does not belong to I and so neither does s. Conversely, if I is a maximal left ideal such that $s \notin I$, then the left ideal $\{as + x : a \in A, x \in I\}$, which strictly contains I, necessarily coincides with A. Hence there exist $a \in A$ and $x \in I$ such that $as + x = e$ and hence $e - as$ (equal to $x \in I$) is not invertible on the left. We have verified that $s \notin \operatorname{Rad} A$.

So the intersection of the maximal left ideals in A coincides with $\operatorname{Rad} A$ and hence the intersection of maximal right ideals coincides with $\operatorname{Rad} A^{op}$. All we have to do now is to use the preceding proposition. ∎

If we combine Theorems 3.53 and 3.28, we immediately get:

Corollary (I.3.54). $\operatorname{Rad} A = \operatorname{Rad} A_+$ *for any A (even if it is unital).*

Exercise (I.3.55). Prove this fact, starting immediately from the definition of a quasi-null element.

(There exists another approach to the concept of the radical which is important in the theory of representations; we leave it until Chap.IV §2.)

The theorem we have proved, taking into account Theorem 3.29, implies:

Corollary (I.3.56). *The radical of an algebra is a bi-ideal, which is closed if we are talking about a Banach algebra.*

(The question of the closedness of the radical in more general polynormed algebras is discussed later in Chap.V §3.)

Example (I.3.57). Let A be the algebra of $n \times n$ matrices that have zero below the main diagonal. Then, obviously, Rad A is the set of matrices that have zeros on the main diagonal also.

Example (I.3.58). Let $A = C(\Omega)/I_t^0$, where I_t^0 is the ideal described in Example 3.8. Then it is easy to verify that Rad A is the set of cosets (or, as we say, of germs) of those $f \in C(\Omega)$ for which $f(t) = 0$.

This example shows that the quasi-null elements are not necessarily nilpotent (cf. Example 3.48). However, they are always nilpotent in finite-dimensional radicals.

Proposition (I.3.59). *Let the radical in the algebra A have finite dimension n. Then $r_1 \cdots r_{n+1} = 0$ for any $r_1, \cdots, r_{n+1} \in$ Rad A.*

▶ Since the system of elements $r_1, r_1r_2, \cdots r_1 \cdots r_{n+1} \in$ Rad A is linearly dependent, then $r_1 \cdots r_m = \lambda_{m+1}(r_1 \cdots r_{m+1}) + \cdots + \lambda_{n+1}(r_1 \cdots r_{n+1})$ for some $m \leq n+1$ and $\lambda_{m+1}, \cdots, \lambda_{n+1} \in \mathbb{C}$. Then for

$$r = \lambda_{m+1}r_{m+1} + \cdots + \lambda_{n+1}(r_{m+1} \cdots r_{n+1})$$

$r_1 \cdots r_m(e_{A_X} - r) = 0$ is satisfied. As $e_{A_X} - r$ is invertible, this means that $r_1 \cdots r_m$ must be zero and so $r_1 \cdots r_{n+1}$ is also zero. ∎

* * *

With the help of the concept of the radical we can pick out two classes of algebras which in many respects have opposite properties.

Definition (I.3.60). An algebra A is said to be *semisimple* if Rad $A = \{0\}$ and *radical* if Rad $A = A$.[18]

We recommend the reader to verify that the specific algebras featuring as examples in §1.1 and §2.2 are semisimple — excluding

---

[18] In the literature of abstract algebra where the algebras under consideration are almost always unital, the radical algebras correspond to the so-called local algebras — algebras with a unique maximal ideal. A classical example is the algebra of formal power series (see, for example [47, Chap.VI §3]).

the algebra with zero multiplication (Example 2.9) which is radical. Here is another radical algebra which is popular in analysis.

Exercise (I.3.61). In the Banach space $L^1[0,1]$ let us consider the bi-operator $(f,g) \longmapsto f \mathbin{\hat{*}} g : f \mathbin{\hat{*}} g(s) = \int_0^s f(t)g(s-t)dt$ (similar to convolution). Show that this is a multiplication that converts $L^1[0,1]$ into a commutative radical Banach algebra.

Hint. Associativity and commutativity (the latter with the help of the substitution $t \longmapsto s-t$) follow from Fubini's theorem and radicality follows from the existence of a dense set of nilpotent elements (what do they look like?).

A fairly extensive literature is devoted to the algebra $L^1[0,1]$ and its variations; see for example [142, §7] and also the review [62] and the references contained in it.

On the other hand, in function algebras with the usual — pointwise — multiplication such phenomena are impossible:

Proposition (I.3.62). *For any algebra* A *and any homomorphism* $\kappa : A \longrightarrow \mathbb{C}^M$, *where* M *is an arbitrary set,* $\operatorname{Rad} A \subseteq \operatorname{Ker} \kappa$. *As a consequence every subalgebra of* $\mathbb{C}^M$ *is semisimple.*

▶ For any $t \in M$, it is obvious that $I^t = \{a \in A : [\kappa(a)](t) = 0\}$ is a bi-ideal in A of codimension ≤ 1. Hence either $I^t = A$ or I^t is a maximal modular left (or right) ideal in A (with any $u \in A$ such that $[\kappa(u)](t) = 1$ as a modular identity). Hence, by Theorem 3.53(Ia), $r \in \operatorname{Rad} A$ implies that $r \in I^t$ for all $t \in M$, that is, $\kappa(r) \equiv 0$. ■

We shall now show that every algebra is, in a well-defined sense, a combination of a semisimple algebra and a radical algebra.

Theorem (I.3.63). *For any algebra* A

(I) $R = \operatorname{Rad} A$ *is a radical algebra;*

(II) $A/\operatorname{Rad} A$ *is a semisimple algebra.*

▶ Let us agree that the identity in $A_\times$ be denoted by e and the identity in $R_\times$ by e'.

(I) Let $r \in R$; our problem is to show that for any $a \in R_\times$, $e' - ar$ is invertible on the left in $R_\times$. Since $ar \in R \subseteq A_\times$, then $e - ar$ has a left inverse b in $A_\times$, and, further, $b(e - ar) = e$ implies that $b = e + s$, where $s \in R$. But then $(e+s)(e-ar) = e$ in $A_\times$ immediately implies that $(e'+s)(e'-ar) = e'$ in $R_\times$.

(II) Let $\dot{a} = a + R \in A/R$. If $\dot{a} \neq 0$, then $a \notin R$, and, by Theorem 3.53(Ia), $a \notin J$ for a certain maximal modular left ideal J in A. But then $\dot{\tilde{J}} = \tilde{\mathbf{pr}}(J)$, where $\tilde{\mathbf{pr}}$ is the map in Proposition 3.33 where we suppose $I = R$, is a maximal modular left ideal in A/R and moreover it is obvious that $\dot{a} \notin \dot{J}$. By the same Theorem 3.53(Ia), this means that $\dot{a} \neq 0$ implies $\dot{a} \notin \mathrm{Rad}(A/R)$. ■

To what extent do the above semisimple and radical "derivatives" of the given algebra determine its structure? The results on this subject form the so-called extension theory that we touch on in Chap.VII §2.2. A specimen is the following classical (known since the beginning of the century) "finite-dimensional" fact.

Assertion (I.3.64) ("Wedderburn's theorem").[19] *Every finite-dimensional algebra decomposes, as a linear space, into the direct sum of a certain (necessarily semisimple) subalgebra and the radical.*

See, for example, [78, Vol.1, p.471-2] for the proof. Later on we shall show (by homological methods) in what way this assertion may be reduced to the question of constructing finite-dimensional semisimple algebras (see further VII.3.39).

In the classic period when abstract algebra was being established the term "semisimple" was used for a considerably narrower class of algebras. The following modern definition reminds us of this.

Definition (I.3.65). A unital algebra A is said to be *classically semisimple* if for each maximal left ideal I there exists a (necessarily minimal) left ideal J such that $A = I \oplus J$.

The beautiful and in many respects complete theory of classically semisimple algebras belongs essentially to ring theory [29], [46], [78] and goes beyonds the scope of this book. Nevertheless it is difficult to resist the temptation to quote some of its results.

Exercise (I.3.66). Prove that a classically semisimple algebra coincides with the linear hull of a finite number of principal minimal left ideals which are generated by idempotents.

Hint. The linear hull of all minimal ideals cannot be a proper ideal. An idempotent in the minimal ideal J complementing I (see Definition 3.65) is the projection of e along I.

---

[19] Wedderburn proved his theorem for a wide class of scalar fields. For the particular case we have given, Molin and E. Cartan essentially already knew of the result.

Exercise (I.3.67). Prove that a classically semisimple algebra is semisimple and moreover (this is the more difficult part of the exercise!) for finite-dimensional algebras the two concepts coincide.

Hint. For any ideal J which is complementary to a maximal ideal, $\text{Rad}\, A \cdot J = \{0\}$. In a finite-dimensional semisimple algebra one can find a finite number of maximal ideals with zero intersection.

When he has solved both exercises the reader has in fact carried out part of the proof of the following fundamental result. We let $\mathcal{M}_n(A)$ denote the algebra of all $n \times n$ matrices with coefficients in A.

Assertion (I.3.68 (the "Wedderburn-Artin theorem").[20] *An algebra is classically semisimple* $\Leftrightarrow$ *it coincides, up to isomorphism, with the Cartesian product of a finite number of algebras of the form* $\mathcal{M}_n(A)$, *where* A *is a skew field* (the n and A for these algebras are, in general, not the same).

As a corollary (taking into account that $\mathbb{C}$ is algebraically closed and using Exercise 3.67) every finite-dimensional[21] classically semisimple algebra is isomorphic to the Cartesian product of complete matrix algebras with complex coefficients. └

* * *

Remark (I.3.69). Since there are semisimple algebras — surely, the reader has thought — there ought to be simple algebras. In fact, there is such a concept: an algebra is said to be *simple* if it has no proper bi-ideals except zero. (Theorem 3.20 thus implies that the unital commutative simple algebras are just fields.) *Every finite-dimensional simple algebra is isomorphic to* $\mathcal{M}_n$ *for a certain* n (see, for example, [91, Chap.II).

A polynormed algebra is said to be *topologically simple* if it has no closed proper ideals, except, perhaps, zero. An old open question — almost of the same age as the theory of Banach algebras itself — consists of the following: do there exist topologically simple commutative Banach algebras apart from $\mathbb{C}$ with the usual multiplication and $\mathbb{C}$ with the zero multiplication? (Such an algebra is necessarily radical — check it!). Important partial results on this problem are contained in [303].

---

[20] We give one of the many equivalent formulations of this theorem; cf. [46 §3.4 Proposition 6].

[21] In the next chapter (see II.1.20) we shall see that also every Banach algebra .....

Remark (I.3.70). In Definition 3.60, the classes of semisimple and radical algebras appear with equal prominence. However, in the theory of "the algebras of analysis", which is the subject of our book, immeasurably more is known about the first than about the second — this applies to general theory as well as interesting examples. It should, however, be noted that in recent years considerable strides have been made in the study of radical Banach algebras; see, in particular the paper of Esterle on the hierarchy of types of these algebras [302] and other papers from the collection [62].

Chapter II
Around the spectrum

The Latin word spectrum means something like "apparition", "spirit" or "ghost". But whereas the ghosts of kings and dukes, as a rule, inhabit ancient half-abandoned castles, the habitat of operator ghosts and, more generally, that of the elements of abstract algebras, is the complex plane, to which they were confined by Hilbert in his classic definition. In both cases the ghosts, if asked properly, can tell us many interesting things. Hamlet, the Prince of Denmark, in his day found this out in conversations with his father's ghost. We hope that our reader will be no less impressed by the theorems on the representation of skew fields with continuous inverse (Theorem 1.17) and on the holomorphic functional calculus of a Banach algebra element (Theorem 2.15).

§1. SPECTRA AND THE CHARACTERIZATION OF CERTAIN POLYNORMED SKEW FIELDS

§1.1. Definition of the spectrum and its simplest properties

Definition (II.1.1). Let a be an element of the algebra A. A point $\lambda \in \mathbb{C}$ is said to be *regular* if the element $a - \lambda e$ of the algebra $A_{\times}$ is invertible, and to be a *singular* point otherwise. The set of all singular points of the element a is called its *spectrum* and is denoted by $\mathrm{Sp}_A(a)$ or simply $\mathrm{Sp}(a)$. Finally the map $\mathcal{R} := \mathcal{R}(\lambda) : \mathbb{C}\backslash\mathrm{Sp}(a) \longrightarrow A_{\times} : \lambda \longmapsto (\lambda e - a)^{-1}$ is called the *resolvent* of the element a.

(This definition, in particular, implies that the spectrum of an element in a non-unital algebra automatically contains 0.)

Example (II.1.2). The spectrum of an element of the algebra $\mathbb{C}^M$, that is, of a function on M, is the set of its values.

Example (II.1.3). The spectrum of a non-constant element of the polynomial algebra $\mathcal{P}_n$ is the whole of $\mathbb{C}$. In contrast, the spectrum of a non-constant element of the set of rational functions $\mathcal{R}_n$ is the empty set.

Let us consider what can happen to spectra after certain algebraic constructions.

Lemma (II.1.4). *Let A be unital, $a \in A$ and $\lambda \in \mathbb{C}\setminus\{0\}$. Then $a - \lambda e$ is invertible in A $\Longleftrightarrow$ $a - \lambda e_+$ is invertible in A_+.*

▶ If b is the inverse of $a - \lambda e_A$ in A, then it is easy to verify that $b - \lambda^{-1}(e_+ - e_A)$ is the inverse of $a - \lambda e_+$ in A_+. If c is the inverse of $a - \lambda e_+$ in A_+, then $e_A c$ is the inverse of $a - \lambda e_A$ in A. ■

This lemma immediately implies the following.

Proposition (II.1.5). *If A is an arbitrary algebra and $a \in A$, then $\mathrm{Sp}_{A_+}(a) = \mathrm{Sp}_A(a) \cup \{0\}$.* ▶■

Proposition (II.1.6). *Let $\kappa : A \longrightarrow B$ be an algebra homomorphism. Then for any $a \in A$ we have $(\mathrm{Sp}_B(\kappa(a)) \cup \{0\}) \subseteq (\mathrm{Sp}_A(a) \cup \{0\})$. If, however, A, B and κ are also unital, then $\mathrm{Sp}_B(\kappa(a)) \subseteq \mathrm{Sp}_A(a)$.*

▶ By passing from κ to κ_+ and using Proposition 1.5, we can obviously reduce the general case to the "unital" case; therefore it is sufficient to prove the second assertion. Let $\lambda \in \mathbb{C}\setminus\mathrm{Sp}_A(a)$, and let $b \in A$ be the inverse of $a - \lambda e_A$. Then $\kappa(a - \lambda e_A) = \kappa(a) - \lambda e_B$ implies that $\kappa(b) \in B$ is the inverse of $\kappa(a) - \lambda e_B$ and hence $\lambda \in \mathbb{C}\setminus\mathrm{Sp}_B(\kappa(a))$. ■

Corollary (II.1.7). *Let A be a subalgebra of B. Then for any $a \in A$ we have $(\mathrm{Sp}_B(a) \cup \{0\}) \subseteq (\mathrm{Sp}_A(a) \cup \{0\})$; and if both algebras have a common identity, then $\mathrm{Sp}_B(a) \subseteq \mathrm{Sp}_A(a)$.*

If we go over to an algebra containing the given one, the spectrum of an element may in fact get smaller; the simplest example is the pair $\mathcal{P} \subset \mathcal{R}$ (see Example 1.3). For analysts the examples $\mathcal{A}(\mathbb{T}) \subset C(\mathbb{T})$ or $C_b(\mathbb{R}) \subset C(\mathbb{R})$ are more typical, where $\mathcal{A}(\mathbb{T})$ consists of those continuous functions that are boundary values of functions holomorphic in the open unit disc and $C_b(\mathbb{R})$ consists of the continuous bounded functions on the line (give relevant examples!). We return to both these algebras in Chap. III.

Proposition (II.1.8). *$\mathrm{Sp}(ab) \cup \{0\} = \mathrm{Sp}(ba) \cup \{0\}$ for any $a, b \in A$; and if A is unital and a is invertible, then $\mathrm{Sp}(ab) = \mathrm{Sp}(ba)$.*

▶ In the general case Proposition I.1.19 implies that for any complex number $\lambda \neq 0$ the element $ab - \lambda e = -\lambda(e - (\lambda^{-1}a)b)$ is invertible in $A_\times$ $\Longleftrightarrow$ $ba - \lambda e = -\lambda(e - b(\lambda^{-1}a))$ is invertible in $A_\times$. If furthermore $A = A_\times$ and $a \in \mathrm{Inv}\, A$, then

$0 \notin \mathrm{Sp}(ab) \Leftrightarrow ab$ is invertible $\Leftrightarrow a^{-1}(ab)a = ba$ is invertible $\Leftrightarrow 0 \notin \mathrm{Sp}(ba)$. The rest is clear. ■

In the next two propositions we assume that A is unital and we treat elements of the algebras $\mathcal{P}$ and $\mathcal{R}$ as functions of a complex variable.

Proposition (II.1.9). *For any* $a \in A$ *and any polynomial* $p \in \mathcal{P}$ *we have* $\mathrm{Sp}(p(a)) = \{p(\lambda) : \lambda \in \mathrm{Sp}(a)\}$.

▶ We take $\mu \in \mathbb{C}$ and consider the polynomial $p(\lambda) - \mu$; according to a basic corollary of the "fundamental theorem of algebra" $p(\lambda) - \mu$ can be written as $\lambda_0(\lambda - \lambda_1) \cdots (\lambda - \lambda_n)$ for certain $\lambda_0, \cdots, \lambda_n \in \mathbb{C}$ and moreover we are justified in assuming that $\lambda_0 \neq 0$. But then it is easy to verify that the element $p(a) - \mu e$ in A is equal to $\lambda_0(a - \lambda_1 e) \cdots (a - \lambda_n e)$ with the same $\lambda_0, \cdots, \lambda_n$. Hence, by Proposition I.1.17 $p(a) - \mu e$ is invertible $\Leftrightarrow a - \lambda_k e$ is invertible for all $k = 1, \cdots, n \Leftrightarrow \lambda_k \notin \mathrm{Sp}(a)$ for all $k = 1, \cdots, n \Leftrightarrow \lambda_0(\lambda - \lambda_1) \cdots (\lambda - \lambda_n) \neq 0$ for any $\lambda \in \mathrm{Sp}(a) \Leftrightarrow p(\lambda) \neq \mu$ for any $\lambda \in \mathrm{Sp}(a)$. The rest is clear. ■

The fact that we have established is a particular case of the following.

Theorem (II.1.10). *For any* $a \in A$ *and* $r \in \mathcal{R}$, *the element* $r(a)$ *exists* (see the end of Chap.I §1) $\Leftrightarrow$ *the function* $r(\lambda)$ *is defined for all* $\lambda \in \mathrm{Sp}(a)$. *In this case* $\mathrm{Sp}(r(a)) = \{r(\lambda) : \lambda \in \mathrm{Sp}(a)\}$.

▶ Let $r = p/q$, where p and q are coprime polynomials. Then, by Proposition 1.9, $r(a)$ exists $\Leftrightarrow q(a)$ is invertible $\Leftrightarrow 0 \notin \mathrm{Sp}(q(a)) \Leftrightarrow q(\lambda) \neq 0$ for all $\lambda \in \mathrm{Sp}(a) \Leftrightarrow r(\lambda)$ is defined for all $\lambda \in \mathrm{Sp}(a)$. Further, if these properties hold, then Propositions I.1.18 and I.1.17 imply that for $\mu \in \mathbb{C}$ the element $r(a) - \mu e = (p(a) - \mu q(a))(q(a))^{-1}$ is invertible $\Leftrightarrow p(a) - \mu q(a)$ is invertible $\Leftrightarrow 0 \notin \mathrm{Sp}(p(a) - \mu q(a))$. The latter, by Proposition 1.9 is equivalent to the fact that $p(\lambda) - \mu q(\lambda) \neq 0$ for all $\lambda \in \mathrm{Sp}(a)$, that is, to the fact that $r(\lambda) \neq \mu$ for all $\lambda \in \mathrm{Sp}(a)$. The rest is clear. ■

§1.2. Conditions for the spectrum to be compact and non-empty. The Gel'fand-Mazur Theorem

Examples 1.2 and 1.3 show that any subset of $\mathbb{C}$, including the empty set could be the spectrum of an element in a "pure" algebra. Certain kinds of topological conditions change the picture completely.

Theorem (II.1.11). *Let* A *be a polynormed unital algebra such that* e *is an interior point of* $\mathrm{Inv}\, A$. *Then* $\mathrm{Sp}(a)$ *is compact* (= closed + bounded) *for any* $a \in A$.

▶ The condition implies that there is some $\varepsilon > 0$ such that $e - \mu a \in \operatorname{Inv} A$ for all $\mu \in \mathbb{C}$ with $|\mu| < \varepsilon$. But then $a - \lambda e = -\lambda(e - \lambda^{-1}a)$ is also invertible whenever $|\lambda| > 1/\varepsilon$. This means that $\operatorname{Sp}(a)$ is bounded.

Further, according to Proposition I.2.41, any point in $\operatorname{Inv} A$ is an interior point; in particular, $a - \lambda_0 e$ is an interior point of $\operatorname{Inv} A$, for any $\lambda_0 \in \mathbb{C}\backslash\operatorname{Sp}(a)$. But then for some $\delta = \delta(\lambda_0) > 0$ the element $a - \lambda e$ is invertible whenever $|\lambda - \lambda_0| < \delta$. This means that any point in $\mathbb{C}\backslash\operatorname{Sp}(a)$ is an interior point and therefore $\operatorname{Sp}(a)$ is closed. ■

The question of when the spectrum of an element is automatically non-empty turns out to be deeper and requires different topological conditions on the algebra under consideration. We recall the map $\mathbf{inv} : \operatorname{Inv} A \longrightarrow \operatorname{Inv} A$ $: a \longmapsto a^{-1}$ considered in Chap. I §2.4. It was there, essentially, that the discussion of the following class of algebras was begun.

Definition (II.1.12). A polynormed unital algebra is called an *algebra with continuous inverse* if **inv** is continuous at e (or equivalently, by Proposition I.2.42, if it is continuous everywhere on $\operatorname{Inv} A$).

It should be pointed out that this definition does not assume that the set $\operatorname{Inv} A$ is "abundant" (open, say). For example, when $A = \mathcal{P}$ (see Example I.2.17) the required property is satisfied for the trival reason that $\operatorname{Inv} \mathcal{P}$ consists of the non-zero constants. Of course, in this case, $\operatorname{Inv} A$ has no interior points.

We now have to consider the behaviour of the resolvent (see Definition 1.1).

Proposition (II.1.13) ("Hilbert identity"). *Let a be an element of the unital algebra A (for the moment, arbitrary), and let $\lambda, \mu \in \mathbb{C}\backslash\operatorname{Sp}(a)$. Then $\mathcal{R}(\lambda) - \mathcal{R}(\mu) = (\mu - \lambda)\mathcal{R}(\mu)\mathcal{R}(\lambda)$.*

▶ If we multiply both sides of the equation by $\mu e - a$ on the left and by $\lambda e - a$ on the right, then we get the same element $(\mu - \lambda)e$. ■

We now come to the point in our account when classical complex analysis comes into play. The next proposition enables us to enlist its help in looking at a number of questions.

Proposition (II.1.14). *Let A be an algebra with continuous inverse (for example, a Banach algebra), let a be an element of A and let U be an open set of $\mathbb{C}$ lying in $\mathbb{C}\backslash\operatorname{Sp}(a)$. Then for any continuous functional $f : A \longrightarrow \mathbb{C}$ the function $w_f : U \longrightarrow \mathbb{C} : \lambda \longmapsto f(\mathcal{R}(\lambda))$ is holomorphic* (= analytic).

▶ We fix $\lambda_0 \in U$; then the Hilbert identity implies that for all $\lambda \in U$

$$\frac{w_f(\lambda) - w_f(\lambda_0)}{\lambda - \lambda_0} = -f(\mathcal{R}(\lambda)\mathcal{R}(\lambda_0)). \tag{1}$$

Further, by assumption (cf. Definitions 1.1 and 1.12) $\mathcal{R}(\lambda) \longrightarrow \mathcal{R}(\lambda_0)$ when $\lambda \longrightarrow \lambda_0$. Since the multiplication in A is separately continuous and f is also continuous, this means that when $\lambda \longrightarrow \lambda_0$ the right-hand side of (1) has a limit, namely $-f(\mathcal{R}(\lambda_0)^2)$. We have verified that the function w_f is $\mathbb{C}$-differentiable, that is, by definition it is analytic (holomorphic) "in the sense of Riemann". ■

Theorem (II.1.15). *Let* A *be a Hausdorff polynormed algebra with continuous inverse. Then the spectrum of any element is non-empty.*

▶ Otherwise, suppose that $a \in A$ is such that $\mathrm{Sp}(a) = \varnothing$. If we take $f \in A^*$ arbitrarily and put $U = \mathbb{C}$ in Proposition 1.14, we see that $w_f(\lambda) := f(\mathcal{R}(\lambda))$ is an entire analytic function. Further, the fact that **inv** is continuous at e implies that $\lim_{\lambda\to\infty} \mathcal{R}(\lambda) = \lim_{\lambda\to\infty}(\lambda^{-1}(e - \lambda^{-1}a)^{-1}) = 0$, and hence in particular $\lim_{\lambda\to\infty} w_f(\lambda) = 0$. Consequently the function w_f is bounded and, by the classical Liouville theorem, it is constant.

So $f(\mathcal{R}(\lambda) - \mathcal{R}(\mu)) = w_f(\lambda) - w_f(\mu) = 0$ for any $f \in A^*$ and $\lambda, \mu \in \mathbb{C}$. Since A is Hausdorff, a standard corollary of the Hahn-Banach theorem (see Chap.0 §1.2) gives $\mathcal{R}(\lambda) = \mathcal{R}(\mu)$. Thus $\mathcal{R} : \mathbb{C} \longrightarrow A$ is a constant vector function, which by its already known behaviour when $\lambda \longrightarrow \infty$ means that $\mathcal{R}(\lambda) \equiv 0$. This gives us a contradiction to the fact that $\mathcal{R}(\lambda)$ is invertible for any λ and therefore is non-zero. ■

Recall that Theorem I.2.43 ensures that both the theorems in this section, Theorems 1.11 and 1.15, are satisfied for Banach algebras. Therefore, taking into account that unitization does not take us out of the class of Banach algebras (Chap.I §2.3), we have the following.

Corollary (II.1.16). *Let* A *be an* (arbitrary) *Banach algebra. Then the spectrum of any element of* A *is a non-empty compact subset in* $\mathbb{C}$.

The following theorem forms the basis for the majority of the subsequent results of the general theory.

Theorem (II.1.17) (Gel'fand-Mazur). *Let* A *be a Hausdorff polynormed skew field with continuous inverse (for example – a Banach skew field). Then* $A = \{\lambda e_A : \lambda \in \mathbb{C}\}$ *and thus* A *is topologically isomorphic to the field of complex numbers.*

▶ By Theorem 1.15, for any $a \in A$ there is at least one $\lambda \in \mathbb{C}$ such that $a - \lambda e$ is not invertible. Since A is a skew field, this means that $a - \lambda e = 0$. The rest is clear. ■

* * *

Remark (II.1.18). As the reader can verify, all the particular examples of polynormed algebras we have met up to now are algebras with continuous inverse (see the end of Chap.I §2.4). In fact algebras with "discontinuous inverse" are not found to be very abundant but they do exist all the same. In particular, Theorem 1.17 implies that any skew field, other than $\mathbb{C}$, is automatically such an example if only one can give it the structure of a Hausdorff polynormed algebra. The appropriate construction for the field $\mathcal{R}$ (of rational functions) is set out in [7, p.277]. There are some simpler examples:

Exercise (II.1.19). Show that the algebra $L^{\omega}[0, 1]$ in Exercise I.2.22 is not an algebra with continuous inverse.

Remark (II.1.20). One of the consequences of the Gel'fand-Mazur theorem is a "Banach" version of the Wedderburn-Artin structure theorem formulated earlier (Assertion I.3.68). If we combine the two theorems, we get the following assertion: every classically semisimple Banach algebra is isomorphic to the Cartesian product of a number of complete matrix algebras with complex coefficients (and as a consequence it is finite-dimensional).

In concluding this section we recall one typical peculiarity of Banach algebras connected with properties of the spectra of their elements.

Proposition (II.1.21). *In a unital Banach algebra A there is no pair $a, b \in A$ such that $ab - ba = e$.*

▸ Suppose that on the contrary $ab = ba + e$ for some a and b. Then, obviously, $\mathrm{Sp}(ab) = \{\lambda + 1 : \lambda \in \mathrm{Sp}(ba)\}$. Further the relation between a and b remains valid if a is replaced by $a + \mu e$ with any $\mu \in \mathbb{C}$; therefore, choosing μ with a modulus sufficiently large and using the boundedness of the spectrum (Corollary 1.16), we are justified in assuming that a is invertible. But then Proposition 1.8 implies that $\mathrm{Sp}(ab)$ coincides with $\mathrm{Sp}(ba)$ and by what has been said above, it is a subset of $\mathbb{C}$ which is invariant under translation by $1 \in \mathbb{C}$. Since such a set must be either empty or unbounded, we have reached a contradiction again with Corollary 1.16. ■

For another proof of this same fact without using spectra see [79, p.333].

Remark (II.1.22). The fact we have just established has an "application" (in the Pickwickian sense) to quantum mechanics and throws some light on why it is, at least to the eyes of a pure mathematician, such a complicated science. You see, Proposition 1.21 applied to $A = \mathcal{B}(H)$ implies that there is no pair of bounded operators S and T on a Hilbert space that would satisfy the "commutativity relation" $ST - TS = -i\hbar 1_H$, where $\hbar$ is the Plank constant. But it is precisely a pair like this which must figure in the familiar basic mathematical models of quantum mechanics (giving, in particular, the mathematical representation of the Heisenberg Uncertainty Principle). So we have to seek the pairs we need among the unbounded operators which are a little more difficult to work with than bounded ones (cf. the epigraph to Chap. VIII in [63]).

§1.3. The spectral radius and Banach algebra radicals

Within this section A is a Banach algebra with norm $\|\cdot\|$. The properties of spectra, described in Corollary 1.16, enable us to give the following definition.

Definition (II.1.23). The number $\max\{|\lambda| : \lambda \in \mathrm{Sp}(a)\}$ denoted by $\underline{r}(a)$ is called the *spectral radius* of the element $a \in A$.

Thus the definition of the spectral radius does not depend on the norm; nevertheless — and this constitutes the main result of this section — it does have an explicit expression in terms of the norm.

Proposition (II.1.24). *If* $\|\cdot\|$ *is a multiplicative norm, then* $\underline{r}(a) \le \|a\|$.

▸ By Proposition I.2.39(I) for any $C > \|a\|$ the element $e - C^{-1}a$ is invertible in $A_\times$; consequently so is $-C(e - C^{-1}a) = a - Ce$. The rest is clear. ■

Lemma (II.1.25). *For any* $a \in A$ *both* $\overline{\lim\limits_{n\to\infty}}\sqrt[n]{\|a^n\|}$ *and* $\underline{\underline{\lim}}_{n\to\infty}\sqrt[n]{\|a^n\|}$ *exist and do not change when the norm is replaced by an equivalent norm.*

▸ Let $\|\cdot\|_1$ be a norm equivalent to the original one; then $C_1\|a^n\|_1 \le \|a^n\| \le C_2\|a^n\|_1$ for some $C_1, C_2 > 0$ and any $n \in \mathbb{N}$ and hence $\sqrt[n]{C_1}\sqrt[n]{\|a^n\|_1} \le \sqrt[n]{\|a^n\|} \le \sqrt[n]{C_2}\sqrt[n]{\|a^n\|_1}$. As $\lim\limits_{n\to\infty}\sqrt[n]{C_1} = \lim\limits_{n\to\infty}\sqrt[n]{C_2} = 1$, this means that the sequences $\sqrt[n]{\|a^n\|}$ and $\sqrt[n]{\|a^n\|_1}$ have the same upper and the same lower limit. We use Proposition I.2.5 and the fact that $\sqrt[n]{\|a^n\|_1} \le \sqrt[n]{\|a\|_1^n} = \|a\|_1$ for any n when the norm $\|\cdot\|_1$ is multiplicative to show that these limits are finite. ■

Lemma (II.1.26). $\underline{r}(a) \le \underline{\underline{\lim}}_{n\to\infty}\sqrt[n]{\|a^n\|}$ *for any* $a \in A$; *if however the norm in* A *is multiplicative, then in addition* $\underline{r}(a) \le \inf\limits_n \sqrt[n]{\|a^n\|}$.

▶ By passing from A to $A_\times$ and using Proposition I.2.5 together with Lemma 1.25, we are justified in assuming that $(A; \|\cdot\|)$ is a multiplicatively normed unital algebra. Proposition 1.9 in particular implies that $\lambda^n \in \mathrm{Sp}(a^n)$ for any $\lambda \in \mathrm{Sp}(a)$ and $n \in \mathbb{N}$; in conjunction with Proposition 1.24 this gives $|\lambda^n| \leq \|a^n\|$ and $|\lambda| \leq \sqrt[n]{\|a^n\|}$. This implies the second assertion and this, in turn, implies the first. ■

The next lemma requires more powerful tools from analysis.

Lemma (II.1.27). $\overline{\lim\limits_{n\to\infty}} \sqrt[n]{\|a^n\|} \leq \underline{r}(a)$ *for any* $a \in A$.

▶ Again we are justified in assuming that A is unital and $\|\cdot\|$ is multiplicative.

We put $U = \{\lambda \in \mathbb{C} : |\lambda| > \underline{r}(a)\}$; this set belongs automatically to the domain of definition of the resolvent of a. Moreover, if $|\lambda| > \|a\|$, then by Proposition 1.24, $\lambda \in U$ and then Proposition I.2.39(I) implies that

$$\mathcal{R}(\lambda) = (\lambda e - a)^{-1} = \lambda^{-1}(e - \lambda^{-1}a)^{-1} = \sum_{k=0}^{\infty} \lambda^{-(k+1)}a^k.$$

We fix (temporarily) some $f \in A^*$. According to Proposition 1.14 the function $w_f(\lambda) = f(\mathcal{R}(\lambda))$ is holomorphic in U; at the same time the form of the above series implies that, in a certain neighbourhood of infinity, w_f has the Laurent series

$$w_f(\lambda) = \sum_{k=0}^{\infty} \frac{f(a^k)}{\lambda^{k+1}}.$$

But then, by well-known properties of analytic functions, this same series represents the function $w_f(\lambda)$ in the whole of U as well. Hence, in particular, $\lim\limits_{n\to\infty} \lambda^{-(n+1)}f(a^n) = 0$ for any $\lambda \in U$.

We now fix $\lambda \in U$ and "free" f. The equation we have obtained implies that for any $f \in A^*$ the sequence $\langle \alpha_n, f\rangle$, where α_n is the functional on A^* that can be canonically identified with $\lambda^{-(n+1)}a^n$ (see Chap.0 §2.2), is bounded. But then the Banach-Steinhaus theorem (Chap.0 §1.2) implies that $\|\lambda^{-(n+1)}a^n\| < C$ for a certain $C > 0$ and all n and hence $\sqrt[n]{\|a^n\|} \leq |\lambda| \sqrt[n]{C|\lambda|}$.

If we take upper limits we get $\overline{\lim\limits_{n\to\infty}} \sqrt[n]{\|a^n\|} \leq |\lambda|$. It only remains to recall that λ was an arbitrary number with $|\lambda| > \underline{r}(a)$. ■

Taken together these lemmas also give the promised formula for the spectral radius.

Theorem (II.1.28). *Let a be an element of a Banach algebra. Then* $\lim\limits_{n\to\infty} \sqrt[n]{\|a^n\|}$ *exists and is equal to* $\underline{r}(a)$. *Furthermore, if the norm* $\|\cdot\|$ *is multiplicative, then both numbers coincide with*

$\inf \sqrt[n]{\|a^n\|}$. ▶■

We note that the crucial thing in this theorem is that the above limit coincides with the spectral radius; just the fact that the limit exists may be proved by comparatively elementary means.

Exercise (II.1.29). Show that, for any element a of a normed algebra with jointly continuous multiplication, $\lim_{n\to\infty} \sqrt[n]{\|a^n\|}$ exists and, if the norm is multiplicative, it coincides with $\inf_n \sqrt[n]{\|a^n\|}$.

Hint. Check that for any sequence $x_n > 0$ such that $x_{m+n} \leq x_m x_n$; $m, n \in \mathbb{N}$, $\lim_{n\to\infty} \sqrt[n]{x_n}$ exists and $= \inf_n \sqrt[n]{x_n}$.

Remark (II.1.30). Theorem 1.28 and, at the same time, the theorem saying that the spectrum of an element of a Banach algebra is non-empty, may also be obtained without applying the apparatus of the theory of analytic functions; see [219] and also [9, §5].

Theorem 1.28 enables us, in particular, to look afresh at the concept of the radical in the case of Banach algebras. We say that an element s of a Banach algebra A is *topologically nilpotent* if $\lim_{n\to\infty} \sqrt[n]{\|s^n\|} = 0$ (or, what is the same thing, if the sequence $(\|s^n\|;\ n \in N)$, decreases faster than any exponential $e^{-\alpha n}$; $\alpha > 0$).

Proposition (II.1.31). *An element $s \in A$ is topologically nilpotent $\Leftrightarrow$ $s - \lambda e$ is invertible in $A_\times$ for all $\lambda \in \mathbb{C}\setminus\{0\}$ $\Leftrightarrow$ $e - \mu s$ is invertible in $A_\times$ for all $\mu \in \mathbb{C}$.*

▶ The equivalence of the first and second properties follows from Theorem 1.28 and the definition of the spectral radius and the equivalence of the second and third follows from the fact that $s - \lambda e$ is a non-zero multiple of $e - \lambda^{-1}s$ when $\lambda \neq 0$. ■

Proposition (II.1.32). *Let I be a left or a right ideal of A all of whose elements are topologically nilpotent. Then $I \subseteq \operatorname{Rad} A$.*

▶ Let $s \in I$, $a \in A_\times$. If I is a left ideal of A (and hence also a left ideal of $A_\times$), then as is topologically nilpotent and, according to Theorem 1.31, $e - as$ is invertible. If however I is a right ideal, then an analogous argument shows that $e - sa$ is invertible. Therefore, by Proposition I.3.52, in both cases $s \in \operatorname{Rad} A$. ■

Theorem (II.1.33). *The radical of a Banach algebra is its largest left (or right, or two-sided, respectively) ideal consisting of topologically nilpotent elements.*

▶ By Proposition 1.31 the radical itself is a bi-ideal consisting of topologically nilpotent elements. It only remains to use the previous proposition. ■

The situation is much simpler in the commutative case.

Theorem (II.1.34). *The radical of a commutative Banach algebra is precisely the set of its topologically nilpotent elements.*

▶ Let A be our algebra, and let s be a topologically nilpotent element; by the previous theorem it is sufficient to show that $s \in \text{Rad}\, A$. We take any $a \in A_{\times}$; since it commutes with s, $(as)^n = a^n s^n$, hence $\sqrt[n]{\|(as)^n\|} \leq \sqrt[n]{\|a\|^n \|s^n\|} = \|a\| \sqrt[n]{\|s^n\|}$ (again, using Proposition I.2.5 and Lemma 1.25, we can assume that the norm is multiplicative). Therefore the fact that s is topologically nilpotent implies that as is topologically nilpotent and hence, according to Proposition 1.31, that $e - as$ is invertible. The rest is clear. ■

Corollary (II.1.35). *If the set of topologically nilpotent elements of a commutative Banach algebra contains a pair of elements, it contains their sum.*

Exercise (II.1.36). Prove this fact without using Theorem 1.34.

Hint. By using the binomial expansion of $(s+t)^n$, where s and t are our elements, one can obtain $\|(s+t)^n\| \leq 2^n \alpha_n$, where α_n are the Taylor coefficients in the power series $(1 + \sum_{n=1}^{\infty} \|s^n\| z^n)(1 + \sum_{n=1}^{\infty} \|t^n\| z^n)$; $z \in \mathbb{C}$.

Another two equivalent definitions of the radical of a commutative Banach algebra will be given later; see IV.1.4 and IV.2.11(II).

A classic example of a (generally speaking not nilpotent) topologically nilpotent element — in the algebra $\mathcal{B}(L^2[0, 1])$ — is the Volterra operator $Vf(x) = \int_0^x K(x, y)f(y)dy$ with square integrable kernel $K(x, y)$; see, for example, [41, p.222]. The following is a simpler example.

Exercise (II.1.37). Let H be a Hilbert space with orthonormal basis e_n; $n = 1, 2, \cdots$. Show that the operator that takes e_n to $\frac{1}{n+1} e_{n+1}$ is a topologically nilpotent element in $\mathcal{B}(H)$.

We also recommend the reader to consider the algebra $L^1[0, 1]$ in Exercise I.3.61 and to prove "from scratch" (omitting the result of the exercise) that all the elements of this algebra are topologically nilpotent.

§2. HOLOMORPHIC FUNCTIONS OF ELEMENTS IN A BANACH ALGEBRA

§2.1. One-dimensional holomorphic functional calculus. An exercise and its solution in the class of Banach algebras

At the end of Chap.I §1 we gave a fairly vague promise to make sense, in due course, of expressions like $w(a)$, where w is a function of a complex variable (sin or log, say) and a, instead of being a complex number, is an element of a certain algebra. In this section we concentrate on holomorphic functions and explain how and under what conditions it is possible to "apply such a function to elements of a Banach algebra". The precise statement of the problem is based on the following definition.

Let A be a unital polynormed algebra, let a be a fixed element of A, let U be a domain in $\mathbb{C}$ and let $\mathcal{O}(U)$ be the Fréchet algebra of holomorphic functions in U (see Example 0.1.8).

Definition (II.2.1). (Any) continuous[1] unital homomorphism $\Gamma o\lambda : \mathcal{O}(U) \longrightarrow A$, that takes the function $w(z) = z$ to a is called a *holomorphic functional calculus* (or just *holomorphic calculus*) for $a \in A$ in U.

Example (II.2.2). Let f be a continuous function on a compact space Ω, let U be a domain containing its set of values. Then, obviously, there exists in U a holomorphic calculus for the element f of the Banach algebra $A = C(\Omega)$ that takes $w \in \mathcal{O}(U)$ into the composition $w \circ f$.

An important special case of Definition 2.1 occurs when a is taken to be a continuous operator S in a Banach space E. Then a holomorphic calculus for S in U, where S is considered as an element of the Banach algebra $\mathcal{B}(E)$, is called an *(operator) holomorphic calculus for S in U.*

If a holomorphic calculus is given, then, for $w \in \mathcal{O}(U)$, we put $w(a) := \Gamma o\lambda(w)$ and talk about the "value of this function at a" (in this way, if S is an operator, it determines an operator $w(S)$ acting on the same Banach space).

We let $\mathcal{R}(U)$ denote the set of rational functions defined everywhere on U (= with poles outside U); obviously this is a subalgebra of $\mathcal{O}(U)$ as well as of $\mathcal{R}$. However, the expression

---

[1] Some A have algebraic properties that "automatically" imply that $\Gamma o\lambda$ is continuous (cf. Chap.I §2.1): for example, this is so for commutative Banach algebras with a finite-dimensional radical, including semisimple algebras (Dales [141]). However, in particular, if the radical is large, this theorem is no longer valid (see [141]). For further results and references on this topic see [142, §8].

"a rational function of a" (cf. Chap.I §1.2) does not raise any ambiguity:

Proposition (II.2.3). *Let $\Gamma o\lambda$ be a holomorphic calculus for a in U, and let $r \in \mathcal{R}(U)$. Then $\Gamma o\lambda(r)$ is none other than $r(a)$ in the sense of Chap.I §1.2. In particular, $\Gamma o\lambda(p) = p(a)$ for any polynomial p.*

▶ Every $r \in \mathcal{R}(U) \subset \mathcal{O}(U)$ can be represented as pq^{-1}, where $p, q \in \mathcal{P} \subset \mathcal{O}(U)$. Then by the algebraic properties of $\Gamma o\lambda$, $\Gamma o\lambda(r) = \Gamma o\lambda(p)(\Gamma o\lambda(q))^{-1}$. But if $p(z) = \sum_{k=0}^{n} \lambda_k z^k$, then $\Gamma o\lambda(p) = \sum_{k=0}^{n} \lambda_k(\Gamma o\lambda(z))^k = \sum_{k=0}^{n} \lambda_k a^k = p(a)$ and in exactly the same way $\Gamma o\lambda(q) = q(a)$. The rest is clear. ■

We now recall the well-known fact in complex analysis (see for example [77, pp.16–17]) that any function which is holomorphic in U, can be uniformly approximated on any compact set $K \subset U$ by rational functions with poles outside U, in other words, that the subalgebra $\mathcal{R}(U)$ is dense in $\mathcal{O}(U)$. By combining this fact with Proposition 2.3 and the continuity condition in Definition 2.1 we get the following result.

Theorem (II.2.4) (uniqueness). *If A is Hausdorff, then for given U and $a \in A$ there exists at most one holomorphic calculus.* ▶■

Remark (II.2.5). In this way both possible meanings of the symbol $w(f)$ in Example 2.2 — as a holomorphic function of an element of an algebra and as the usual composition — necessarily coincide. Later on (IV.2.18) we shall see that the same is true too for many other algebras which consist of functions. This explains, in particular, that for such A the composition of the function $f \in A$ with a holomorphic function — where it makes sense of course — does not take us outside A. (We point out that this is not at all obvious for an algebra such as, say, the Wiener algebra.)

Let us return to general algebras. We go on to the main question in the subject under discussion: when does a holomorphic calculus really exist? A simple observation shows that, generally speaking, it is possible there may not be any.

Proposition (II.2.6). *For a holomorphic calculus for a to exist in U it is necessary that $Sp_A(a)$ lies in U.*

▶ If $\lambda \in \mathbb{C}\backslash U$, then $w(z) = z - \lambda$ is invertible in $\mathcal{O}(U)$; therefore $a - \lambda e$, being the image of this under a unital homomorphism, is invertible in A. The rest is clear. ■

It turns out — and this is the fundamental result of the section — that in the case of Banach algebras this necessary condition becomes the sole criterion. We first show that the condition $U = \mathbb{C}$ is automatically sufficient.

Theorem (II.2.7). *Let A be a unital Banach algebra. Then for any $a \in A$ there exists a holomorphic calculus for a in the whole of $\mathbb{C}$ which sends the function $w(z) = \sum_{n=0}^{\infty} c_n z^n$ to the element $\sum_{n=0}^{\infty} c_n a^n \in A$* (here and everywhere later on, we put $a^0 = e$).

▶ If we assume that the norm in A is multiplicative (Proposition I.2.5), we see that, since $\sum_{n=0}^{\infty} c_n z^n$ is absolutely convergent when $z = \|a\|$, then the series $\sum_{n=0}^{\infty} \|c_n a^n\|$ is majorized by a convergent series. Hence $\sum_{n=0}^{\infty} c_n a^n$ is absolutely convergent in A; we denote its sum by $w(a)$ and put $\Gamma o\lambda : \mathcal{O}(\mathbb{C}) \longrightarrow A : w \longmapsto w(a)$. The absolute convergence of the last series and also the form of the Taylor coefficients of the sum and the product of holomorphic functions imply that $\Gamma o\lambda$ is an algebra homomorphism; furthermore it is obviously unital and $\Gamma o\lambda(z) = a$.

Further, our reader, if he has at any time done "complex analysis", will remember that the Taylor coefficients of an entire function w satisfy $|c_n| \le R^{-n}M$ for any $R > 0$, where $M = \max\{|w(z)| : |z| = R\}$. Hence $\|w(a)\| \le \sum_{n=0}^{\infty} |c_n| \|a\|^n \le$ $\le (R - \|a\|)^{-1}RM$ for any $R > \|a\|$ and this means that for any closed disc K with centre at 0 and radius greater than $\|a\|$ there exists $C > 0$ such that $\|w(a)\| \le C\|w\|_K$. By Proposition 0.1.9 and the definition of the norms in $\mathcal{O}(\mathbb{C})$ (see Example 0.1.8) this shows that the operator $\Gamma o\lambda$ is continuous. ■

Exercise (II.2.8). Show that the theorem is true not only for Banach algebras but also for Arens-Michael algebras.

Remark (II.2.9). In this exercise we have essentially reached the limit of possible generalizations: it is known that a Fréchet algebra in which "one can take entire functions" of any element is necessarily an Arens-Michael algebra (Mitiagin, Rolewicz, Żelazko [197]).

Warning. From now on we shall assume everywhere in this section that A is a unital Banach algebra.

Definition (II.2.10). For $w(z) = e^z \in \mathcal{O}(\mathbb{C})$ the element $w(a)$; $a \in A$ is called the *exponential* of the element a and is denoted by e^a (or $\exp a$).

Theorem 2.7 implies that e^a is the sum of the series $\sum_{n=0}^{\infty} \frac{1}{n!} a^n$.

Proposition (II.2.11). *e^a is invertible for any $a \in A$ and its inverse is e^{-a}.*

▶ The form of the Taylor expansion of $w(z) = e^{-z}$ implies that $w(a) = \sum_{n=0}^{\infty} \frac{1}{n!} (-a)^n = e^{-a}$. On the other hand, if $\Gamma o \lambda$ is an "entire" calculus for a, then $w(a) = \Gamma o \lambda(e^{-z}) = = \Gamma o \lambda((e^z)^{-1}) = (\Gamma o \lambda(e^z))^{-1} = (e^a)^{-1}$. ■

Exercise (II.2.12). Show (by generalizing Proposition 2.11) that $e^{a+b} = e^a e^b$ if $a, b \in A$ commute.

... When expressing our good intentions in Remark I.1.20 we referred, to be more convincing, to the sine and the logarithm. We now know, after Theorem 2.7, what $\sin a$; $a \in A$ is, but we do not yet know what $\log a$ is; $\log z$ cannot be defined on the whole of $\mathbb{C}$. How does one proceed in such a situation?

The technique of vector integration along paths in $\mathbb{C}$ comes to our aid. The appropriate concept of the integral combines two concepts with which the reader is certainly acquainted: the integral of a continuous vector function over an interval (from a course in functional analysis) and the integral of a complex function along a path (from "complex analysis"). Namely let I be an interval in $\mathbb{R}$ and let $\gamma : I \longrightarrow \mathbb{C}$ be a path (here and in future we assume that all paths are piecewise smooth: these are quite sufficient for what we want). Further, let E be a Banach space and let $\overline{u}$ be a continuous E-valued function given on the "curve" $\tilde{\gamma} := \operatorname{Im} \gamma$. Then the element $\int_I \overline{u}(\gamma(t))\gamma'(t)dt \in E$ is written $\int_\gamma \overline{u}(\lambda)d\lambda$ and called the *integral of the vector function $\overline{u}$ along the path γ*. A definition of the integral of a vector function along a composite path is based on the same idea as the integral of a complex function: for $\gamma = \{\gamma_1, \cdots, \gamma_n\}$ and $\overline{u} : \bigcup_{k=1}^{n} \tilde{\gamma}_k \longrightarrow E$ we put $\int_\gamma \overline{u}(\lambda)d\lambda := \sum_{k=1}^{n} \int_{\gamma_k} \overline{u}(\lambda)d\lambda$. This definition immediately implies that, for any functional $f \in E^*$, the number $f(\int_\gamma \overline{u}(\lambda)d\lambda)$, where γ is a simple or a composite path, is the ordinary (scalar) integral $\int_\gamma f(\overline{u}(\lambda))d\lambda$.

Let γ be a closed path (generally speaking, composite), and let $z \in \mathbb{C}$. We say that z *lies inside* γ if $\int_\gamma (\lambda - z)^{-1} d\lambda = 2\pi i$, and that z *lies outside* γ if

$\int_\gamma (\lambda - z)^{-1} d\lambda = 0$. (Speaking in graphical terms,[2] in the first case the given path goes round our point just once "in total" in a positive direction whereas in the second case, again "in total", it does not go round at all.) If M is a subset in $\mathbb{C}$, then we say that M *lies inside* (*outside*) γ if each point in M has the relevant property.

In the next few lemmas a is a fixed element of a Banach algebra A, γ is a (simple or composite) closed path with values in $\mathbb{C}\backslash \mathrm{Sp}_A(a)$. Recall (Proposition 1.14) that the resolvent $\mathcal{R} : \lambda \longmapsto (\lambda e - a)^{-1}$ is holomorphic on $\mathbb{C}\backslash \mathrm{Sp}_A(a)$; hence, in particular, $\int_\gamma v(\lambda)\mathcal{R}(\lambda)d\lambda$ automatically exists as an element of A for any continuous $v : \tilde{\gamma} \longrightarrow \mathbb{C}$.

Lemma (II.2.13). *Let* $\mathrm{Sp}_A(a)$ *lie inside* γ. *Then for any* $n = 0, 1, 2, \cdots$ $(2\pi i)^{-1}\int_\gamma \lambda^n(\lambda e - a)^{-1} d\lambda = a^n$.

▶ As the spectrum is bounded (Theorem 1.11) $\mathrm{Sp}_A(a)$ lies inside the path $\gamma_R : [0, 1] \longrightarrow \mathbb{C} : t \longmapsto Re^{2\pi i t}$ for sufficiently large $R > 0$; in particular $\tilde{\gamma}_R \subset \mathbb{C}\backslash \mathrm{Sp}_A(a)$. "Cauchy's integral theorem for a composite contour" implies that $\int_\gamma w(\lambda)d\lambda = \int_{\gamma_R} w(\lambda)d\lambda$ for any $w \in \mathcal{O}(\mathbb{C}\backslash \mathrm{Sp}_A(a))$; in particular (by Proposition 1.14), $\int_\gamma f(\lambda^n\mathcal{R}(\lambda))d\lambda = \int_{\gamma_R} f(\lambda^n\mathcal{R}(\lambda))d\lambda$ for all $f \in A^*$ and n.

But, according to Proposition I.2.39(I) we can write $(\lambda e - a)^{-1} = \lambda^{-1}(e - \lambda^{-1}a)^{-1} = \sum_{k=0}^{\infty} \lambda^{-(k+1)}a^k$ when $|\lambda|$ is sufficiently large, and therefore $f(\lambda^n\mathcal{R}(\lambda)) = \sum_{k=0}^{\infty} f(a^k)\lambda^{n-k-1}$; moreover, the inequality $|f(a^k)| \leq \|f\|\|a\|^k$ (again we consider the norm to be multiplicative) implies that this series of functions converges uniformly for $\lambda \in \tilde{\gamma}_R$, where $R > \|a\|$. Hence, for such an R, by the known rules for integration in complex analysis $\int_{\gamma_R} f(\lambda^n\mathcal{R}(\lambda))d\lambda = \sum_{k=0}^{\infty} f(a^k)\int_{\gamma_R} \lambda^{n-k-1}d\lambda = 2\pi i f(a^n)$.

---

[2] And very loosely... Unfortunately, such an incantation is often offered as a precise mathematical definition. As a result, to a poor student "everything is clear", while a good student is worried by the feeling that he has been swindled.

We have shown that $f((2\pi i)^{-1}\int_\gamma \lambda^n(\lambda e - a)^{-1}d\lambda) = f(a^n)$ for any $f \in A^*$. All that remains is to apply the Hahn-Banach theorem in the form of Corollary B in Chap.0 §1.2. ∎

Lemma (II.2.14). *Let* $\mathrm{Sp}_A(a)$ *lie inside* γ *and let the point* z *lie outside* γ. *Then for* $m = 1, 2, \cdots$

$$(2\pi i)^{-1}\int_\gamma (\lambda - z)^{-m}(\lambda e - a)^{-1}d\lambda = (a - ze)^{-m}.$$

▶ If b is any element of A then Lemma 2.13 (the case $n = 0$) and the fact that the operator $m_b : A \longrightarrow A : c \longmapsto bc$ is continuous imply that $b = (2\pi i)^{-1}\int_\gamma b(\lambda e - a)^{-1}d\lambda$. Hence, if we put $b = (a - ze)^{-m}$, we see that the difference between the two sides of the equation is

$$(2\pi i)^{-1}\int_\gamma [(\lambda - z)^{-m}e - (a - ze)^{-m}](\lambda e - a)^{-1}d\lambda =$$

$$= (2\pi i)^{-1}\int_\gamma (\lambda - z)^{-m}(a - ze)^{-m}[(a - ze)^m - (\lambda - z)^m e](\lambda e - a)^{-1}d\lambda =$$

$$= (2\pi i)^{-1}\int_\gamma (\lambda - z)^{-m}(a - ze)^{-m}\Big(\sum_{k=1}^{m} (\lambda - z)^{k-1}(a - ze)^{m-k}\Big)$$

$$(a - \lambda e)(\lambda e - a)^{-1}d\lambda =$$

$$= -(2\pi i)^{-1}\sum_{k=1}^{m}\left[\left(\int_\gamma (\lambda - z)^{k-m-1}d\lambda\right)(a - ze)^{-k}\right].$$

But, by the Cauchy Residue theorem, for all $k = 1, \cdots, m - 1$, the integral in the last expression is zero; since z lies outside γ, this also holds when $k = m$. The rest is clear. ∎

Theorem (II.2.15) (Gel'fand). *Let* A *be a unital Banach algebra, let* a *be an element in* A *and let* U *be a domain in* $\mathbb{C}$ *containing* $\mathrm{Sp}_A(a)$. *Then a holomorphic calculus for* a *exists in* U *and it is given by* $\Gamma\circ\lambda(w) := (2\pi i)^{-1}\int_\gamma w(\lambda)(\lambda e - a)^{-1}d\lambda$, *where* γ *is any closed path in* U (simple or composite) *such that* $\mathrm{Sp}_A(a)$ *lies inside* γ *and* $\mathbb{C}\backslash U$ *lies outside* γ.

▶ Let γ satisfy the condition; then for every $w \in \mathcal{O}(U)$ the above integral, denoted later by $w(a)$, is well defined and the map $\Gamma\circ\lambda : \mathcal{O}(U) \longrightarrow A : w \longmapsto w(a)$ is obviously a continuous operator. In addition, Lemma 2.13 (when $n = 0$ and $n = 1$)

implies that $\Gamma o\lambda(w) = e$ when $w(z) \equiv 1$ and $\Gamma o\lambda(w) = a$ when $w(z) = z$.

Further, it is known from elementary algebra that any rational function $r = r(\lambda)$ over $\mathbb{C}$ decomposes into "partial fractions", that is, it can be represented as a sum of a polynomial and a linear combination of functions of the form $(\lambda - z)^{-m}$, $z \in \mathbb{C}$, $m = 1, 2, \cdots$ Of course, where $r \in \mathcal{R}(U)$, none of the z belongs to U and therefore they lie outside γ. But then the two lemmas imply that, for any $r \in \mathcal{R}(U)$, $\Gamma o\lambda(r)$ is $r(a)$ in the sense of Chap.I §1.2. This has the obvious implication that the restriction of $\Gamma o\lambda$ to $\mathcal{R}(U)$ is a homomorphism and this, since $\mathcal{R}(U)$ is dense in $\mathcal{O}(U)$ and $\Gamma o\lambda$ is continuous, means that the latter is itself a homomorphism. ■

As a first application we get a new approach to the concept of a spectrum:

Proposition (II.2.16). *The spectrum of an element of a Banach algebra (in particular, of an operator) coincides with the intersection of all those domains in* $\mathbb{C}$ *in which there exists a holomorphic calculus for this element.*

▶ This is implied by Theorem 2.15 and Proposition 2.6. ■

Of course the path γ in Theorem 2.15 that features in the explicit construction of $\Gamma o\lambda(w)$ can be taken in any neighbourhood of the spectrum that lies in U. This, by the uniqueness theorem, implies the following proposition.

Proposition (II.2.17). *Let* $a \in A$ *and let* $\Gamma o\lambda_k : \mathcal{O}(U_k) \longrightarrow A$; *be holomorphic calculi for* a *in the domains* U_k; $k = 1, 2$, $w \in \mathcal{O}(U_1 \cup U_2)$. *Then* $\Gamma o\lambda_1(w|_{U_1}) = \Gamma o\lambda_2(w|_{U_2})$. ▶■

Thus we are justified in using the notation $w(a)$ without being concerned about indicating precisely the domain in which the holomorphic function w is determined — provided this domain contains $\mathrm{Sp}_A(a)$.

An important property of the holomorphic calculus for applications is that it commutes with the action of characters.

Theorem (II.2.18). *Suppose, under the assumptions of Theorem 2.15, that* $w \in \mathcal{O}(U)$ *and* $\chi \in X\alpha\rho A$. *Then* $\chi(w(a)) = w(\chi(a))$.

▶ First of all the fact that $\chi(a - \chi(a)e) = 0$ obviously implies that $\chi(a) \in \mathrm{Sp}_A(a)$ and therefore we have proved that the right-hand side of the equation makes sense. Further, let γ be a closed path in U such that $\mathrm{Sp}_A(a)$ and hence also the point $\chi(a) \in \mathbb{C}$ lie inside γ. Then, since χ is continuous, the (usual) Cauchy integral formula and Theorem 2.15 imply that

$$\chi(w(a)) = \chi\left((2\pi i)^{-1}\int_\gamma w(\lambda)(\lambda e - a)^{-1}d\lambda\right) =$$
$$= (2\pi i)^{-1}\int_\gamma w(\lambda)(\lambda - \chi(a))^{-1}d\lambda = w(\chi(a)). \blacksquare$$

Exercise (II.2.19) (generalizing the construction in Theorem 2.7). Let U be an open disc with centre z_0 and let $w \in \mathcal{O}(U)$. Show that, if $a \in A$ with $\mathrm{Sp}_A(a) \subset U$, then $w(a) = \sum_{n=0}^{\infty} c_n(a - z_0e)^n$, where c_n are the coefficients of the Taylor expansion of w in U. Formulate and prove a similar result for an annulus (in terms of a Laurent series).

Example (II.2.20). Consider the function $\sqrt[n]{z}$ in $\mathbb{C}^{(+)}$, taking $z = |z|e^{i\varphi}$; $-\pi < \varphi < \pi$, into $\sqrt[n]{|z|}e^{i\varphi/n}$ (the "principal branch of the n-th root"). Theorem 2.15 allows us to define $\sqrt[n]{a}$ as $w(a)$ for any $a \in A$ with $\mathrm{Sp}_A(a) \subset \mathbb{C}^{(+)}$, where $w(z) = \sqrt[n]{z}$.

Example (II.2.21) (the long promised logarithm). For the same $\mathbb{C}^{(+)}$ we consider the "principal branch of the logarithm" given by $\log z := \ln|z| + i\varphi$ ($\varphi = \arg z$ is the same as in Example 2.20) and, for any $a \in A$ with $\mathrm{Sp}_A(a) \subset \mathbb{C}^{(+)}$, define the element $\log a$ as $w(a)$ where $w(z) = \log z$.

Here is a comparatively simple sample of applications of the holomorphic calculus.

Proposition (II.2.22). *Suppose* $\mathrm{Sp}_A(a)$ *does not intersect some ray in* $\mathbb{C}$ *emanating from zero. Then*

(I) *for any* $n \in \mathbb{N}$ *there exists* $b \in A$ *such that* $b^n = a$; *in particular* $(\sqrt[n]{a})^n = a$ *is satisfied whenever* $\mathrm{Sp}_A(a) \subset \mathbb{C}^{(+)}$;

(II) *there exists* $b \in A$ *such that* $e^b = a$; *in particular* $e^{\log a} = a$ *is satisfied whenever* $\mathrm{Sp}_A(a) \subset \mathbb{C}^{(+)}$.

▶ If we replace a by an element proportional to it, we are justified in assuming that $\mathrm{Sp}_A(a) \subset \mathbb{C}^{(+)}$. Let $\Gamma o\lambda$ be a holomorphic calculus for a in $\mathbb{C}^{(+)}$.

(I) $(\sqrt[n]{a})^n = (\Gamma o\lambda(\sqrt[n]{z}))^n = \Gamma o\lambda((\sqrt[n]{z})^n) = \Gamma o\lambda(z) = a.$

(II) We put $p_n(z) = \sum_{k=0}^{n} \frac{1}{k!} z^k$ and $b_n = \sum_{k=0}^{n} \frac{1}{k!}(\log a)^k$. Then from the algebraic properties of $\Gamma o\lambda$, we have $\Gamma o\lambda(w_n) = b_n$ for $w_n(z) = p_n(\log z)$, and therefore because $w_n(z)$ converges to z in $\mathcal{O}(\mathbb{C}^{(+)})$ and $\Gamma o\lambda$ is continuous, $\lim_{n\to\infty} b_n = a$. On the other hand $p_n(z)$ converges to e^z in $\mathcal{O}(\mathbb{C})$ and since the holomorphic calculus for $\log a$ in $\mathbb{C}$ is continuous this implies that $\lim_{n\to\infty} b_n = \lim_{n\to\infty} p_n(\log a) = e^{\log a}$. ■

The next theorem generalizes the purely algebraic Theorem 1.10 to Banach algebras.

Theorem (II.2.23) ("Spectral Mapping Theorem"). *Under the conditions of Theorem* 2.15 $\mathrm{Sp}_A(w(a)) = \{w(\lambda) : \lambda \in \mathrm{Sp}_A(a)\}$ *for any* $w \in \mathcal{O}(U)$.

▶ We take $\mu \in \mathbb{C}$. First let $\mu = w(\lambda_0)$ for a certain $\lambda_0 \in \mathrm{Sp}_A(a)$. Then there is $v \in \mathcal{O}(U)$ such that $w(z) - \mu = (z - \lambda_0)v(z)$; hence $w(a) - \mu e = \Gamma o\lambda(w - \mu) = \Gamma o\lambda(z - \lambda_0)\Gamma o\lambda(v) = (a - \lambda_0 e)v(a)$. But $a - \lambda_0 e$ is not invertible and $\mathrm{Im}\,\Gamma o\lambda$ is a commutative algebra; therefore Proposition I.1.17 implies that $w(a) - \mu e$ is also not invertible and therefore $\mu \in \mathrm{Sp}_A(w(a))$.

Conversely, suppose $w(\lambda) - \mu \neq 0$ on $\mathrm{Sp}_A(a)$. Then, as a consequence, there is some domain V such that $\mathrm{Sp}_A(a) \subset V \subseteq U$ and the function $w_1(z) - \mu$, where $w_1 := w|_V$, is invertible in $\mathcal{O}(V)$. Hence the element $\Gamma o\lambda_1(w_1 - \mu) = \Gamma o\lambda_1(w_1) - \mu e$ is invertible in A, where $\Gamma o\lambda_1$ is a holomorphic calculus for a in V. But by Proposition 2.17, $\Gamma o\lambda_1(w_1) = \Gamma o\lambda(w) = w(a)$. The rest is clear. ■

Exercise (II.2.24) (generalizing the equation $(\sqrt[n]{a})^n = e^{\log a} = a$ in Proposition 2.22). Let $a \in A$, $U \supset \mathrm{Sp}_A(a)$, $w_1 \in \mathcal{O}(U)$, $V \supset \mathrm{Sp}_A(w_1(a))$, $w_2 \in \mathcal{O}(V)$. Show that the element $w_2 \circ w_1(a)$ (the composition $w_2 \circ w_1$ "taken at a") is equal to $w_2(w_1(a))$ (w_2 "taken at $w_1(a)$").

Hint. If we take the paths γ_1 in U and γ_2 in V such that $\mathrm{Sp}_A(a)$ lies inside γ_1 while $\mathrm{Sp}_A(w_1(a))$ and $w_1(\mathrm{Im}\,\gamma_1)$ lie inside γ_2, then we can put $w_2(w_1(\lambda)) = (2\pi i)^{-1}\int_{\gamma_2} w_2(\zeta)(\zeta - w_1(\lambda))^{-1}d\zeta$ in the expression $(2\pi i)^{-1}\int_{\gamma_1} w_2(w_1(\lambda))(\lambda e - a)^{-1}d\lambda$ and then interchange the order of integration.

§2.2. Some remarks on multi-dimensional holomorphic calculus: searches for joint spectra, Shilov and Taylor spectra, the basic Taylor theorem (sketch)

Can the concepts and facts in the previous section be extended to holomorphic functions of several complex variables? Because of the size of our book we can only touch on these problems in a very general way although interest in them has increased sharply in recent years. Here, as is usual when going from one-dimensional complex analysis to multi-dimensional complex analysis — one only needs to compare the first volume of the textbook by Shabat [89] with the second — new complications of a fundamental nature arise.

A formal statement of the question causes no problems: for a given domain $U \subseteq \mathbb{C}^n$ and a system $\alpha = (a_1, \cdots, a_n)$ of (necessarily commuting) elements of a unital polynormed algebra A can one construct a continuous unital homomorphism $\Gamma o\lambda : \mathcal{O}(U) \longrightarrow A$ such that $\Gamma o\lambda(w_k) = a_k$; $1 \leq k \leq n$ is satisfied for $w_k(z_1, \cdots, z_n) = z_k$? Such a $\Gamma o\lambda$, by analogy with Definition 2.1, is called a *holomorphic calculus for the system* α *in* U. Accordingly for a system $\mathcal{S} = (S_1, \cdots, S_n)$ of commuting continuous operators in the Banach space E we talk of a *holomorphic calculus for* $\mathcal{S}$ *in* U by putting $A = \mathcal{B}(E)$.

However, even in trying to imagine what the "multi-dimensional" analogues of the previous results might look like, primarily Theorem 2.15, a natural question arises: how can one define a "joint spectrum" of the system α that would be the set in $\mathbb{C}^n$ which would play the same role, in relation to α, as $\mathrm{Sp}_A(a)$ plays, in relation to a, in §2.1?

This question, which has a long history, has turned out to be very difficult and, at the moment, a solution in the class of general Banach algebras has not yet been found — if it exists, in general. But the initial research was very promising: in 1952 Shilov [291] succeeded in giving a definition, without raising any doubt that it was the "correct" one, in the commutative case:

Definition (II.2.25). Let A be a unital commutative Banach algebra, and let $\alpha = (a_1, \cdots, a_n)$ be a system of elements in A. The *joint Shilov spectrum* of this system (denoted by $\mathrm{Sp}_A(\alpha)$) is the set of those $\lambda = (\lambda_1, \cdots, \lambda_n) \in \mathbb{C}^n$ for which the set $\{\sum_{k=1}^{n} b_k(a - \lambda_k e) : b_1, \cdots, b_n \in A\}$ is a proper ideal in A (= does not contain e).

After a few years, and after a series of misadventures (a long-hidden mistake in the original argument was found; see [148] about this) a difficult theorem that is the deciding argument in favour of the given definition, was finally proved.

Assertion (II.2.26) (the "Shilov-Arens-Calderon-Waelbroeck theorem"; see for example [17, III §4]). *Let* A *and* α *be the same as in Definition* 2.25 *and let* U *be a domain in* $\mathbb{C}^n$ *containing* $\mathrm{Sp}_A(\alpha)$. *Then there exists in* U *a holomorphic calculus for* α *and it is unique.*

The Shilov spectrum and the "four mathematicians' theorem" have important applications in the theory of commutative Banach algebras and closely related areas (see IV.2.19 later on about one of them). Subsequently, it turned out that Definition 2.25 can be carried across word for word, without losing its attractive properties, to the case when A is not necessarily commutative but α belongs to its centre. In this case, as Taylor showed (using "his own" spectrum, about which more later) Assertion 2.26 remains as strong as before [249].

At the moment it is not clear[3] how to define $\mathrm{Sp}_A(\alpha)$ "correctly" without requiring that α lie in the centre of the algebra. Meanwhile, since the centre of $\mathcal{B}(E)$ is $\{\lambda 1 : \lambda \in \mathbb{C}\}$ (check it!), without a solution to this question the above approach can do nothing for systems of operators; it is hardly necessary to convince the reader of the importance of this case. One can of course, given $\mathcal{S}$, consider, instead of the whole of $\mathcal{B}(E)$, a closed subalgebra B with $\mathcal{S} \subseteq \mathrm{Centre}\, B$ and then take $\mathrm{Sp}_B(\mathcal{S})$. But such a manoeuvre for by-passing the problem has not brought the desired result; it is not clear in advance which subalgebras to choose as B and, what is even worse, each of the possible claimants results in a set that is unacceptable through lack of certain natural requirements. For more details about this see [249] and [239].

Nevertheless the problem has been completely solved for operators: a definition of a joint spectrum has been suggested which does all that is needed and, furthermore, does not depend on the choice of an extraneous object — the algebra B ambient to the system. This was done by Taylor ([249], in 1970) and is based on a fundamentally different, "homological" idea: we consider the points of the joint spectrum of a system $\mathcal{S}$ to be those $\lambda \in \mathbb{C}^n$ for which exactness fails for a certain complex constructed from $\mathcal{S}$ and λ (see Definition 0.4.7). Let us give the precise definitions.

Let E be a Banach space and let $\mathcal{S} = (S_1, \cdots, S_n)$ be a system of commuting continuous operators acting on E. For $m = 0, 1, \cdots, n$ we let $\underset{m}{\wedge}\mathbb{C}^n$ denote the m-th exterior power of $\mathbb{C}^n$, that is the $\binom{n}{m}$-dimensional space with a basis whose elements are enumerated by the multi-indices $(i_1, \cdots, i_m)$; $1 \le i_1 \le \cdots \le i_m \le n$ and denoted by $e_{i_1} \wedge \cdots \wedge e_{i_m}$. Further we put $E_m := E \otimes (\underset{m}{\wedge}\mathbb{C}^n)$ and for $m = 0, \cdots, n$ we introduce the operator $d_m : E_{m+1} \longrightarrow E_m$ uniquely determined by the formula [4]

$$d_m(x \otimes e_{i_1} \wedge \cdots \wedge e_{i_{m+1}}) = \sum_{l=1}^{m+1} (-1)^l S_{i_l}(x) \otimes e_{i_1} \wedge \cdots \wedge \hat{e}_{i_l} \wedge \cdots \wedge e_{i_{m+1}} .$$

---

[3] Possibly Harte [265], [266] is justified in defining $\mathrm{Sp}_A(\alpha)$, even without this condition, as the set of those $\lambda \in \mathbb{C}^n$ for which either the left or the right ideal generated by $a_1 - \lambda_1 e, \cdots, a_n - \lambda_n e$ is a proper ideal. However we do not know whether the analogue of Assertion 2.26 is true for Harte's spectrum. See [255], [315] for the relations between the Harte spectrum and the Taylor spectrum in the case $A = \mathcal{B}(E)$.

[4] The symbol "$\wedge$" denotes the omission of the factor of the exterior product over which it is placed.

It is easy to verify that $d_{m-1}d_m = 0$ for all m, in other words the sequence $0 \leftarrow E_0 \xleftarrow{d_0} E_1 \leftarrow \cdots \xleftarrow{d_{n-1}} E_n \leftarrow 0$ $(\mathcal{K}(E, \mathcal{S}))$ is a complex. It is called the *Koszul complex of the system* $\mathcal{S}$.

Definition (II.2.27). *The (joint) Taylor spectrum of the system* $\mathcal{S}$ is the set of all those $\lambda = (\lambda_1, \cdots, \lambda_n)$ for which the sequence $\mathcal{K}(E, \mathcal{S} - \lambda)$, where $\mathcal{S} - \lambda = (S_1 - \lambda_1 1, \cdots, S_n - \lambda_n 1)$, is not exact.

The Taylor spectrum of the system $\mathcal{S}$ is denoted by $\operatorname{Sp}\mathcal{S}$ or $\operatorname{Sp}(S_1, \cdots, S_n)$.

Obviously, by considering exactness at the extreme terms (cf. Proposition 0.4.8) we get two conditions:

$$\operatorname{Im}(S_1 - \lambda_1 1) + \cdots + \operatorname{Im}(S_n - \lambda_n 1) = E \qquad \text{and} \qquad \bigcap_{k=1}^{n} \operatorname{Ker}(S_k - \lambda_k 1) = \{0\}$$

which are necessary for $\lambda \notin \operatorname{Sp}\mathcal{S}$. (This, by the Banach theorem, implies that, when $n = 1$, the Taylor spectrum reduces to the usual spectrum of an operator.) It is also not difficult to decypher the exactness conditions at the remaining terms; we recommend the reader to do this for the term E_1 when $n = 2$.

Here are a few simple exercises clarifying the concept we have introduced.

Exercise (II.2.28). Show that (I), when $\dim E < \infty$, $\operatorname{Sp}\mathcal{S}$ is the set of those λ such that $S_k x = \lambda_k x$ for some $x \neq 0$ and all $k = 1, \cdots, n$; (II) any compact set $K \subset \mathbb{C}^n$ is the Taylor spectrum of the system of operators in $C(K)$ defined as multiplication by $z_1, \cdots, z_n$; (III) if $\mathcal{S} = (S, S^2, \cdots, S^n)$, then $\operatorname{Sp}\mathcal{S} = \{\lambda = (\lambda_1, \cdots, \lambda_1^n) : \lambda_1 \in \operatorname{Sp}(S)\}$.

The Taylor spectrum satisfies in practice all that could be required of an "ideal" concept of the joint spectrum. Like the classical spectrum for a single operator it is non-empty and compact. It has the "projection property": if $S_1, \cdots, S_{n+1} \in \mathcal{B}(E)$, commute, then $\operatorname{Sp}(S_1, \cdots, S_n)$ is the projection of $\operatorname{Sp}(S_1, \cdots, S_{n+1})$ onto the first n coordinates in $\mathbb{C}^{n+1}$. If A is a Banach algebra and $\alpha = (a_1, \cdots, a_n)$ lies in the centre of A and if $S_k := m_{a_k} : b \longmapsto a_k b$; $b \in A$; $k = 1, \cdots, n$, then $\operatorname{Sp}(S_1, \cdots, S_n)$ coincides with $\operatorname{Sp}_A(\alpha)$; this means that the Shilov spectrum may be considered as a special case of the Taylor spectrum. But the great merit of the Taylor spectrum is that the following fact is true.

Assertion (II.2.29) ("Taylor's theorem on a multi-operator holomorphic calculus"). *Let* $\mathcal{S} = (S_1, \cdots, S_n)$ *be a system of commuting operators in the Banach space* E, *and let* U *be a domain in* $\mathbb{C}^n$. *Then, whenever* $\operatorname{Sp}\mathcal{S} \subset U$, *a holomorphic calculus for* $\mathcal{S}$ *exists in* U. *If, however,* U *is, in addition, a domain of holomorphy* (see, for example, [18]), *then the condition* $\operatorname{Sp}\mathcal{S} \subset U$ *is necessary for the existence of a holomorphic calculus which in this case is unique.*

Remark (II.2.30). The requirement that U be a domain of holomorphy is essential for the necessity as well as for uniqueness in Assertion 2.29 and it is connected with the multi-dimensional phenomenon of "obligatory" analytic extension which does not occur in the one-dimensional case [18]. An example of the non-uniqueness of the holomorphic calculus is given in [213]; see also [81].

Taylor suggested, within the space of two years, two completely different proofs of his theorem, each of which entailed a whole flood of research. In the first of them [250] which may be considered as a remote (in truth really very remote) generalization of the arguments in Theorem 2.15, the required calculus is constructed with the help of a so-called abstract version of the Cauchy-Weil integral. The methods applied there turned out to be fruitful in many questions on spectral theory, including the study of spectral decompositions of families of operators. These aspects of the Taylor theory together with other questions on holomorphic multi-operator calculus are summed up in the monograph by Vasilescu [16].

The second of Taylor's proofs [252] (see also [271] and [81]) which, to the surprise of an analyst, no longer contains any integrals, is the result of deep rethinking over the essence of the problem from the position of module and homology theory. Here we shall only sketch the primary argument: it is necessary to think of the system of commuting operators, not as a subset of a certain algebra, but as giving a representation of the polynomial algebra $\mathcal{P}_n$ on E, which — and this is the problem — one must extend to a representation of the algebra $\mathcal{O}(U)$ which contains $\mathcal{P}_n$. The ideas and methods of the second proof were subsequently developed in papers by Putinar [213], [215], [319], [320], and this improved the algebraic apparatus that was used and gave, incidentally, an interpretation of the problem from the position of sheaf theory. Other connections with sheaf theory and also with K-theory are found in Levy [187], [188].

A specific holomorphic calculus was constructed in [250] that satisfied the "Spectral Mapping Theorem" (cf. Theorem 2.23) which asserts that, if the conditions of Assertion 2.29 are satisfied, then

$$\mathrm{Sp}(w_1(S_1, \cdots, S_n), \cdots, w_m(S_1, \cdots, S_n)) =$$

$$= \{(w_1(\lambda), \cdots, w_m(\lambda)) : \lambda \in \mathrm{Sp}(S_1, \cdots, S_n)\},$$

where $w_1, \cdots, w_m \in \mathcal{O}(U)$ (Theorem 4.8 in [250]). Later Putinar [318] proved the same equation for the calculus constructed in [252] and, what is just as important, established [319] that, among all the holomorphic calculi for $\mathcal{S}$ in U, only one has such a property; this immediately implies that the calculi in

[250] and [252] coincide. He also [213a] proved the "composition theorem" for these calculi thus generalizing Exercise 2.24.

In conclusion we note that Voiculescu [313] applied the Taylor theory to prove the existence of a common non-trivial invariant subspace for several commuting compact operators on a Banach space.

Chapter III
Basic stock of examples

Our collection of "on call" examples of algebras, given in Chap.I §2.2, is no longer enough if we want to move on. In this chapter we get to know some new specific algebras and we also study more thoroughly the examples we already know. The algebras considered here, in much the same way as our subject itself, belong to neighbouring branches of analysis — the theory of functions, operator theory, harmonic analysis; there are some questions in topology and in the mathematics of quantum mechanics that are also related to this same boundary layer. The application of the general theory of algebras of analysis to these disciplines is through just such algebras as these: to obtain information about a topological group it is necessary to involve its group algebra; to examine an "individual" operator, it is useful to include it in some operator algebra or other, and so on. On the other hand the material in this chapter is needed to get an informal grasp of the constructions and results of the general theory, to which we shall return in the next few chapters: there we shall have far more opportunity to show how all this works and what it turns out to be in specific situations.

Here before us is a new algebra; what do we want to know about it? A lot of questions, of course, will be of a specific character and will be expressed in terms appropriate to the way the algebra is defined and to the branch of analysis where it is used. But there will also be standard questions that are duplicated from one algebra to another; they form a sort of questionnaire which we shall distribute to our new acquaintances and ask them to fill in. What type of structure (from those defined in Chap.I §2.1) do you have? Have you an identity? Or if not, an approximate identity? What are your closed ideals, in particular your maximal ideals? Have you a radical? Finally (a very important question!) how are your characters constructed[1]

---

[1] If the algebra — and this is typical of non-commutative algebras — objects that it has only a few characters, then we shall ask it to give us some more details about its representations.

and (by the way) are they all continuous?

The reader will see that, when completing some box or other of the questionnaire, the algebras behave differently: some answer quickly and willingly and others with difficulty and as though reluctantly; it may be that an algebra has, up to now, flatly refused to reveal anything. (We can compare, for example, the state of the question about describing closed ideals for the three function algebras on the circle $C(\mathbb{T})$, $\mathcal{A}(\mathbb{T})$ and W; see §1.) In some cases we shall have to wait for an answer until the following chapters when the fact that interests us turns out to be a corollary (as a rule, a trivial one) of results in the general theory.

It should be noted that it is precisely the hard questions about specific algebras that have turned out to be the most fruitful for developing the neighbouring areas of mathematics by helping to solve internal problems that arise within those areas. (For example, one such problem is the "corona" problem in complex analysis, which we shall mention at the end of §1.)

§1. FUNCTION ALGEBRAS

§1.1. Posing certain types of question (whether ideals and characters are fixed, spectral synthesis, point ideals)

There are at least two polynormed algebras naturally associated with each completely regular (= Tikhonov) topological space Ω. These are: the algebra $C(\Omega)$ of all continuous functions on Ω with the compact-open system of seminorms (see Chap. 0 §1.1 and Chap.I §1.1) and the algebra of all bounded continuous functions on Ω with the uniform norm, denoted by $C_b(\Omega)$. Obviously $C(\Omega)$ is a multinormed algebra and $C_b(\Omega)$ is a Banach algebra. If Ω is locally compact, then the Banach algebra $C_0(\Omega)$ is also of interest (see Example I.2.10) as it is obviously a closed ideal of $C_b(\Omega)$. If Ω is compact all three algebras have exactly the same elements and topology.

For convenience later on we shall give the words "function algebra" the following precise meaning.[2]

Definition (III.1.1). Any subalgebra of $C(\Omega)$ that is a polynormed algebra with topology no weaker than the compact-open topology is called a *function algebra* (or, more fully, a *polynormed function algebra*) on Ω. Any subalgebra of $C_b(\Omega)$ that is a Banach algebra with respect to a certain norm $\|\cdot\| \geq \|\cdot\|_0$ is called a *Banach function algebra*.

---

[2] Different authors use these words in different ways but always in a narrower sense than "(any) algebra consisting of functions".

As a start to the discussion of the properties of function algebras we note that they are all, by Proposition I.3.62, semisimple.

Throughout this section, A is a certain function algebra on Ω. We put $\Omega_0 := \{t \in \Omega : \text{there exists } f \in A \text{ with } f(t) \neq 0\}$ (as a rule, in our examples $\Omega_0 = \Omega$), and for any $t \in \Omega_0$, we consider $\chi_t : A \longrightarrow \mathbb{C} : f \longmapsto f(t)$. Obviously χ_t is a continuous non-trivial character on A; therefore this defines a map $\alpha : \Omega_0 \longrightarrow \mathrm{X}\alpha\rho : t \longmapsto \chi_t$.

We consider, as well as $\mathrm{X}\alpha\rho = \mathrm{X}\alpha\rho\, A$, the set $\mathrm{X}\alpha$ of all (possibly discontinuous) non-trivial characters on A, the set Max of closed maximal modular ideals of A and the set Ma of all (possibly non-closed) maximal modular ideals in A. If we associate with each character χ its kernel we get injective maps $\omega : \mathrm{X}\alpha\rho \longrightarrow \mathrm{Max}$ and $\tilde{\omega} : \mathrm{X}\alpha \longrightarrow \mathrm{Ma}$ (describe the modular identity in $\operatorname{Ker}\chi$!).

We have the diagram

$$\begin{array}{ccccc} \Omega_0 & \xrightarrow{\alpha} & \mathrm{X}\alpha\rho & \xrightarrow{\omega} & \mathrm{Max} \\ & & {\scriptstyle \mathrm{in}_X}\downarrow & & \downarrow{\scriptstyle \mathrm{in}_M} \\ & & \mathrm{X}\alpha & \xrightarrow{\tilde{\omega}} & \mathrm{Ma} \end{array} \tag{1}$$

and a typical question about A is the following: when is one or other of the above maps bijective (= enables the corresponding sets to be identified)?

Running ahead, we warn the reader that, for the maps that form the square in the diagram, one should expect a negative answer only for algebras that are far from being Banach algebras. We already know that in the Banach case both vertical arrows are automatically bijections (Theorems I.2.6 and I.3.29) and the corresponding fact for the upper, and hence the lower, horizontal arrow of the square is Theorem IV.1.1 later. Later on (Proposition V.3.27), we shall also learn that ω (but not, generally speaking, $\tilde{\omega}$) remains bijective on passing to the wider class of multinormed algebras.

Although the maps round the square in the diagram could be considered for any polynormed algebra, α is connected directly with Definition 1.1. Recall that, given a family of functions on a certain set M, we say that it *separates the points of* M if, for any two distinct $t_1, t_2 \in M$, there is an f in our family with $f(t_1) \neq f(t_2)$. The following obviously holds.

Proposition (III.1.2). *α is injective $\Leftrightarrow$ the family $\{f|_{\Omega_0} : f \in A\}$ separates the points of Ω_0.* ▸■

We note that the above condition that α be injective is not burdensome: one can always assume that it is satisfied after a certain modification ("correction by gluing") of the original space Ω; for some examples of such a procedure see Exercise IV.3.3 later. A deeper question is "when is α surjective?".

Definition (III.1.3). A character (ideal, respectively) of a function algebra A is said to be *fixed* if it is of the form $\chi_t : f \longmapsto f(t)$ $(I_t = \operatorname{Ker} \chi_t = \{f \in A : f(t) = 0\}$, respectively) for some $t \in \Omega$.

In this terminology the question of whether all the maps in the diagram (or α and ω, α and $\mathbf{in}_X$, and finally, α alone, respectively) are surjective can be formulated as follows.

Question. Is every maximal modular ideal (every closed maximal modular ideal, every non-trivial character, and finally, every continuous non-trivial character, respectively) fixed?

Even if the answer to one of these questions is negative, it is by no means hopeless to attempt to describe $X\alpha\rho$ or the other sets in (1) in terms of Ω in some reasonable way; we shall see this quite often. On the other hand, if all the maximal ideals in A, or at least all the closed maximal ideals, are fixed, then we have the following supplementary question.

Question ("the problem of spectral synthesis").[3] Is every closed ideal of A of the form $I_\Delta = \{f \in A : f|_\Delta = 0\}$ for a certain subset Δ of Ω (in other words, is it the intersection of a certain family of fixed ideals)?

If this is so, then we say that *spectral synthesis holds in A*.

For any ideal I in A the set (necessarily closed) $\Delta(I) = \{t \in \Omega : f(t) = 0$ for all $f \in I\}$ is called its *zero set*. It is not difficult to see that the problem of spectral synthesis can be formulated as the question of whether every closed ideal I of A coincides with $I_{\Delta(I)}$.

A closed ideal with a one-point zero set is called a *point ideal*.[4] We pose another two typical questions about function algebras. Is every point ideal in A maximal (= fixed)? Is every closed ideal of A the intersection of a certain family of point ideals? (This question is sometimes called the *problem of weak spectral synthesis.*) Clearly, both these properties being satisfied is equivalent to having spectral synthesis (but neither property, as examples show, is implied by the other).

---

[3] The name is connected with the history of the question which was first posed for the classic algebras of harmonic analysis. For details see, for example, the commentary on Chapter 10 "Spectral synthesis" in [87].

[4] In the case where every maximal ideal of A is fixed, the zero set of an ideal is more commonly called its "hull" (see IV.1.22) and a point ideal is called a "primary" ideal.

§1.2. Algebras of continuous functions. Spectral synthesis in $C(\Omega)$

The first specific algebra we shall deal with on a compact space is $C(\cdot)$ and this turns out to be a model of good behaviour.

Theorem (III.1.4) (Gel'fand and Kolmogorov). *Every maximal ideal (and, as a consequence, every non-trivial character) of the algebra* $C(\Omega)$*, where* Ω *is a compact space, is fixed.*

▶ Suppose otherwise that there is a certain maximal ideal I such that, for each $t \in \Omega$ there exists $f_t \in I$ with $f_t(t) \neq 0$. Then, as f is continuous, $f_t \neq 0$ in a whole neighbourhood U_t of the point t. Using the fact that Ω is compact, we choose, from the open cover $\{U_t;\ t \in \Omega\}$, a finite subcover $U_k := U_{t_k}$; $1 \leq k \leq n$, and put $f_k := f_{t_k}$. Since I is an ideal, it contains not only the functions f_k but also $g = \sum_{k=1}^{n} \overline{f_k} f_k$, and this function is positive on the whole of Ω and, consequently, it is invertible in $C(\Omega)$. Hence, by Theorem I.3.19, I is not proper. Contradiction! ■

The following strengthens the Gel'fand–Kolmogorov theorem.

Theorem (III.1.5). *We have spectral synthesis in* $C(\Omega)$ *when* Ω *is a compact space.*

▶ Let I be a closed ideal in $C(\Omega)$ with zero set Δ; our problem is to show that $I = I_\Delta$. With this as our aim, we consider the ideal I_Δ^0 in $C(\Omega)$ consisting of functions that are zero in a whole neighbourhood (each in its own neighbourhood) of the set Δ. Since $I \subseteq I_\Delta$ and $I_\Delta^0 \subseteq I_\Delta$ and I is closed, it is sufficient to show that I_Δ^0 is dense in I_Δ and lies in I.

We take $f \in I_\Delta$ and $\varepsilon > 0$. We put $U_k = \{t \in \Omega : |f(t)| < k\varepsilon\}$ for $k = 1, 2$. According to Urysohn's theorem (Chap.0 §0.3°) there exists $g \in C(\Omega)$ with $0 \leq g \leq 1$, $g(s) = 0$ when $s \in U_1$ and $g(s) = 1$ when $s \notin U_2$. But then obviously $\|f - fg\|_0 < 2\varepsilon$, where since g belongs to I_Δ^0 so does fg. But f and ε were arbitrary, so this means that I_Δ^0 is dense in I_Δ.

We now let $h \in I_\Delta^0$ and let U be a neighbourhood of Δ such that $h|_U = 0$. We put $\Phi := \Omega \setminus U$ and consider the homomorphism $\tau : C(\Omega) \longrightarrow C(\Phi) : f \longmapsto f|_\Phi$. The Tietze–Urysohn theorem (Chap.0 §0.3°) implies that τ is surjective. But then it is obvious that $\tau(I)$ is an ideal in $C(\Phi)$ and moreover, since $\Delta \cap \Phi = \varnothing$, it follows that $\tau(I)$ does not lie in any fixed ideal and hence, by Theorem 1.4, it does not lie in any maximal ideal of the algebra $C(\Phi)$. Therefore $\tau(I)$ is not a proper ideal and so contains the identity in $C(\Phi)$, in other words, there exists a function $\tilde{e}$ in I with $\tilde{e}|_{\Phi} = 1$. Hence, since $h = h\tilde{e}$, $h \in I$. The rest is clear. ∎

If Ω is a locally compact space, then every ideal in $C_0(\Omega)$ may be considered as an ideal in the larger algebra $C(\Omega_+)$ (see Example 1.2.30). Therefore from the theorem we have proved we get the following generalization.

Corollary (III.1.6). *Spectral synthesis holds in* $C_0(\Omega)$ *for any locally compact space* Ω.

The two other algebras associated with non-compact spaces, $C(\Omega)$ and $C_b(\Omega)$, behave worse.

Example (III.1.7). Let Ω be a locally compact, non-compact space and let $C_{00}(\Omega)$ be the algebra of continuous functions on Ω with compact supports. Then any maximal ideal of $C(\Omega)$ or of $C_b(\Omega)$ containing $C_{00}(\Omega)$ is, of course, not fixed. In addition, $C_{00}(\Omega)$ is a dense subset of $C(\Omega)$ and therefore any maximal ideal in $C(\Omega)$ containing it is not closed.

Thus $C(\Omega)$ is "worse" than the Banach algebra $C_b(\Omega)$ in that its maximal ideals are not necessarily closed. But then in other respects it is "better".

Exercise (III.1.8). Prove that every continuous non-trivial character of the algebra $C(\Omega)$, where Ω is any completely regular space, is fixed.

Hint. If χ is not a fixed character then, for any compact set $K \subset \Omega$, $\operatorname{Ker}\chi$ contains a function f with $f|_K \equiv 1$ (cf. the proof of Theorem 1.4) and therefore $\operatorname{Ker}\chi$ is dense in $C(\Omega)$.

We ought to note that, in fact, if we use the general method of reducing a multinormed algebra to a system of Banach algebras, as described in Chapter V, it is possible to show, for any completely regular Ω, that spectral synthesis [7, p.275] holds in $C(\Omega)$.

In fact the situation, described in Example 1.7, holds for any completely regular space Ω: there are maximal ideals in $C(\Omega)$ and $C_b(\Omega)$ that are not fixed and moreover, for $C(\Omega)$ they are all non-closed. Nevertheless (all) maximal ideals of both algebras can be fully described: for $C(\Omega)$, as well as for $C_b(\Omega)$, they may be put in 1-1 correspondence with the points of a certain special compact space $\beta\Omega$, which contains Ω as a dense subset, and moreover so that the points of Ω correspond to the fixed ideals (see [7] and [115]). We shall discuss this compact space $\beta\Omega$, called the Stone-Čech compactification of Ω, in detail later on in Chapter IV §3.1; there it will be shown that all maximal ideals of the algebra $C_b(\Omega)$ can, loosely speaking, become fixed if we extend its functions by continuity from Ω to $\beta\Omega$.

...We leave $C_b(\Omega)$ and concentrate on the algebras $C(\Omega)$. What has been said about them up to now means that for any such algebra the upper horizontal arrows in diagram (1) are always bijections but that the right vertical arrow is a bijection only when Ω is compact. As for the left vertical arrow, it is a bijection far more often than the right vertical arrow.

Assertion (III.1.9). *The characters on $C(\Omega)$ are all continuous $\Longleftrightarrow$ Ω is homeomorphic to a closed subset of the Cartesian product of some family of copies of $\mathbb{R}$.*

The topological condition mentioned is automatically satisfied if Ω has the second axiom of countability or is separable and metrizable. A counterexample is provided by the space Ξ of countable ordinals with the order topology; moreover $C(\Xi)$ has a unique discontinuous character, associating to each function its limit at the "first uncountable ordinal" (it necessarily exists).

For a wide class of spaces the algebra $C(\Omega)$ is complete and so it is then an Arens-Michael algebra; for example, this is automatically so if Ω is locally compact or has the first axiom of countability. If, however, Ω is locally compact and is the union of a countable number of its compact subsets, then Ω is, in addition, a Fréchet algebra.

The reader can find a far more detailed account of the properties of the algebras $C(\Omega)$ in [7], where, in particular, criteria for completeness, metrizability and the automatic continuity of the characters of these algebras are fully examined.

§1.3. Algebras of smooth functions, the Wiener algebra, sequence algebras

If Ω has the structure of a smooth manifold (possibly with boundary), then one is naturally interested in the function algebras that "react" with this structure — that is, $C^n(\Omega)$, consisting of functions with continuous n-th derivative and $C^\infty(\Omega)$ consisting of infinitely differentiable functions. For such algebras it is not difficult to describe a system of seminorms giving the topology of uniform convergence on compact subsets contained in charts, for functions together with their partial derivatives — up to order n for $C^n(\Omega)$ and for any order for $C^\infty(\Omega)$. For example, for the algebra $C^\infty(\mathbb{R})$ which is important in the theory of generalized functions, one can put $\|f\|_{k,n} = \max\{|f^{(n)}(t)| : -k \le t \le k\}$, where $n = 0, 1, 2;$ $k = 1, 2, \cdots$. If Ω is compact, then the topology needed for $C^n(\Omega)$ can be given by a single norm; we suggest the reader describes such a norm for $C^2(\cdot)$ on a two-dimensional sphere, say.

Many properties of the algebras under discussion, primarily those for compact Ω, are easy to observe in the example

$\Omega = [0, 1]$ and we shall restrict ourselves to this because it makes the presentation simple.

We shall furnish $C^n[0, 1]$ with the norm $\|f\|_n := \sum_{k=0}^{n} \frac{1}{k!} \max\{|f^{(k)}(t)| : 0 \le t \le 1\}$ and $C^\infty[0, 1]$ with the system of norms $\|f\|_n$; $n = 0, 1, 2, \cdots$. Obviously $C^n[0, 1]$ is a Banach algebra and $C^\infty[0, 1]$ is an Arens-Michael algebra and a Fréchet algebra and moreover all the norms $\|\cdot\|_n$ are multiplicative.

The algebra $C^\infty[0, 1]$ provides an example of a non-Banach algebra A [5] with $\operatorname{Inv} A$ an open subset: $\{f ; \|1 - f\|_0 < 1\}$ is a neighbourhood of the identity consisting of invertible elements and therefore the conditions of Proposition I.2.41 are satisfied. Hence the same arguments as in Theorem I.3.21 show that maximal ideals are closed in this algebra as well as in the Banach algebras $C^n[0, 1]$. This same property also follows immediately for the algebras under discussion from the following proposition which brings out their similarity to the $C(\cdot)$ on compact spaces.

Proposition (III.1.10). *Every maximal ideal (and as a consequence every character) in* $C^n[0, 1]$; $n = 1, 2, \cdots,$ *is fixed.*

▶ The proof of Theorem 1.4 goes through with obvious modifications. ■

However, if we try to carry the theorem about spectral synthesis over from $C(\Omega)$ to algebras of smooth functions, we immediately run into a fiasco: their point ideals are not necessarily maximal.

Example (III.1.11). We fix some $t \in [0, 1]$. Then, for any $n = 1, 2, \cdots, \infty$ and any $k = 0, 1, 2, \cdots$ not exceeding n (in particular, any k, when $n = \infty$) we put $I_{\{t\}}^{(n,k)} := \{f \in C^n[0, 1] : f(t) = f'(t) = \cdots = f^{(k)}(t) = 0\}$ and $I_{\{t\}}^{(\infty,\infty)} := \{f \in C^\infty[0, 1] : f^{(l)}(t) = 0$ for all $l = 0, 1, 2, \cdots\}$. Obviously, all the above sets are distinct point ideals in the corresponding algebras and each has zero set $\{t\}$. Moreover, among those that are in $C^n[0, 1]$; $n = 1, 2, \cdots, \infty$, $I_{\{t\}}^{(n,0)}$ is maximal (and therefore, the biggest) while $I_{\{t\}}^{(n,n)}$ is the smallest.

Exercise (III.1.12). Prove that there are no other point ideals in $C^n[0, 1]$ with zero set $\{t\}$ except those described; $n = 1, 2, \cdots, \infty$.

---

[5] It is not difficult to show directly that this topology cannot be given by a single norm; see also Remark VII.2.11 later.

Hint. Each point ideal contains the ideal consisting of all functions that are zero in neighbourhoods of t and hence contains its closure $I_{\{t\}}^{(n,n)}$. Therefore by Proposition I.3.33 the exercise reduces to describing the ideals in $C^n[0,1]/I_{\{t\}}^{(n,n)}$.

We have the following result on the structure of arbitrary closed ideals.

Assertion (III.1.13). *Each closed ideal in* $C^n[0,1]$; $n \in \mathbb{N}$ *is the intersection of point ideals* (and so, although spectral synthesis does not hold for these algebras, we do have weak spectral synthesis).

The result we have formulated is a particular case of a deep theorem establishing a similar result for algebras of smooth functions in several variables that is due to Whitney. It was given by Malgrange in [199] together with other results about these algebras.

* * *

On the circle $\mathbb{T}$, there is, besides the algebras of continuous and smooth functions, another interesting Banach function algebra: this is the Wiener algebra W which we have already met (Example I.2.14). In §3 of this chapter we shall see that it is representative of a certain class of group algebras and exhibits many of their properties and as such it is advisable to say a few words about it now.

First of all, from the nature of its elements, W satisfies the relation $C^1(\mathbb{T}) \subset W \subset C(\mathbb{T})$; this is implied by the fact that the Fourier series of a smooth function is absolutely convergent and more refined observations ensure that both inclusions are strict (see, for example [39]).

The following property which makes W look more like $C(\mathbb{T})$ and $C^1(\mathbb{T})$ turns out to be very important for applications in harmonic analysis (see Theorem IV.1.9 later).

Theorem (III.1.14). *Every character on* W *is fixed (in other words,* $\mathrm{X\alpha\rho}\, W$, *as a set, can be identified with* $\mathbb{T}$*).*

▶ Let z_0 be the value of a character χ on $f(z) = z$; our problem is to show that $z_0 \in \mathbb{T}$ and $\chi(f) = f(z_0)$ for any $f \in W$.

Since $\|z\|_W = \|z^{-1}\|_W = 1$, Theorem I.2.6 implies that $|\chi(z)| \le 1$ and $|\chi(z^{-1})| = |\chi(z)|^{-1} \le 1$; hence $|\chi(z)| = 1$, that is, $z_0 \in \mathbb{T}$. Therefore for any $f \in W$, where $f(z) = \sum_{n=-\infty}^{\infty} c_n z^n$, the algebraic properties and the continuity of χ imply that

$$\chi(f) = \lim_{m\to\infty} \chi\left(\sum_{k=-m}^{m} c_k z^k \right) = \lim_{m\to\infty} \sum_{k=-m}^{m} c_k \chi(z^k) =$$

$$= \lim_{m\to\infty} \sum_{k=-m}^{m} c_k z_0^k = \sum_{n=-\infty}^{\infty} c_n z_0^n = f(z_0). \blacksquare$$

Later on (Corollary IV.1.6) we deduce from this theorem that all maximal ideals in W are also fixed. But now we give without proof a more refined result.

Assertion (III.1.15) (see [20, p.218]). *Every point ideal of the Wiener algebra is maximal* (note — as in $C(\mathbb{T})$ but not as in $C^1(\mathbb{T})$).

This fairly old result was born at a time when it was hoped there would be a positive solution to the problem of spectral synthesis for W. But this hope was not justified: Malliavin (1959) used fairly refined methods in harmonic analysis to show that there exist[6] closed ideals in W which are not of the form I_Δ; $\Delta \subseteq \mathbb{T}$. Subsequently the phenomenon discovered by Malliavin was clarified by Varopoulos who connected the question under discussion with properties of the algebra $C(\mathbb{T}) \hat{\otimes} C(\mathbb{T})$ (the "Varopoulos algebra", see Chap.I §2.3).

As for the problem of describing all closed ideals of the Wiener algebra, up to now this has not been solved and like the earlier one throws down a challenge to specialists in harmonic analysis.

The reader can find a detailed discussion of the properties of the algebra W that we have mentioned here, plus references, in the concise and masterly written book by Kahane [39].

* * *

The case when the domain of definition for the functions in an algebra is $\mathbb{N}$ with the discrete topology is of separate interest. The elements of the algebra are then sequences and we shall correspondingly talk about a *sequence algebra* (that could be Banach or polynormed, according to the meaning) instead of a "function algebra on $\mathbb{N}$". Moreover the compact-open topology, mentioned in Definition 1.1, becomes the topology of pointwise convergence.

The simplest examples are the Banach algebras c_0, c and c_b, consisting of all sequences converging to zero, all sequences having limits, and all bounded sequences respectively, considered with uniform norm and also the classic Banach spaces l_p; $1 \le p < \infty$ (with coordinatewise multiplication). Of course, c_0

---

[6] In fact Malliavin proved a more general theorem, which we shall say something about in §3.2. later.

is another notation for $C_0(\mathbb{N})$, c_b is another notation for $C_b(\mathbb{N})$ and c is topologically isomorphic to $(c_0)_+$ and $C(\mathbb{N}_+)$. The algebras c and c_b have identities, c_0 has a b.a.i., and l_p an a.i. but not a b.a.i. (check it!).

Of the non-Banach sequence algebras we first of all mention $c_\infty = C(\mathbb{N})$ (cf. Example I.2.16) consisting of all possible sequences (with the topology of pointwise convergence) and $c_{00} = C_{00}(\mathbb{N})$ (cf. Example 1.7) consisting of finite (eventually zero) sequences and considered with strongest seminorm system. The first algebra, of course, is unital and the second has an a.i. (what is it?).

Let us recall another algebra called the *algebra of slowly increasing sequences* and denoted by c_p. It consists of those $\xi = \{\xi_n \in \mathbb{C};\ n \in \mathbb{N}\}$ that have $|\xi_n| \le C|n|^k$ for some $C > 0$ and k (depending on ξ), and it is furnished with the system of seminorms $\|\xi\|_\alpha = \max_n |\xi_n \alpha_n|$, where $\alpha = \{\alpha_n\}$ runs over all sequences decreasing faster than any negative power of n.[7]

Exercise (III.1.16). Show that spectral synthesis holds in c_0, l_p, c_∞, c_{00} and c_p.

Exercise (III.1.17). Show that c_p is a jointly polynormed algebra but not an Arens-Michael algebra.

§1.4. Algebras of holomorphic functions. Remark on the interpretation of characters as points of the envelope of holomorphy

Among the algebras discussed in modern complex analysis the most important are algebras of holomorphic functions on complex analytic spaces. We restrict ourselves here to the algebras $\mathcal{O}(U)$; $U \subseteq \mathbb{C}^n$, introduced in Example I.2.15. We have already seen that these are Arens-Michael algebras and Fréchet algebras; let us consider some of their further properties.

Theorem (III.1.18). *When* $U \subseteq \mathbb{C}$ (= the one-dimensional case) *every continuous character on* $\mathcal{O}(U)$ *is fixed.*

▶ Let χ be the given character, λ_0 its value on $w(z) = z$. Since $(z-\lambda)^{-1} \in \mathcal{O}(U)$ for any $\lambda \notin U$, the number $\chi((z-\lambda)^{-1})$ exists and is the inverse of $\chi(z-\lambda) = \lambda_0 - \lambda$; hence $\lambda_0 \ne \lambda$, and therefore $\lambda_0 \in U$. Further, the algebraic properties of χ imply that $\chi(r) = r(\lambda_0)$ for any $r \in \mathcal{R}(U)$

---

[7] c_p crops up later at the end of §3.3, in another guise, as a certain algebra connected with the group $\mathbb{T}$ (the "distribution algebra").

(see Chap.II §2.1). All that remains is to use the facts that $\mathcal{R}(U)$ is dense in $\mathcal{O}(U)$ (see *ibid*) and χ is continuous.[8] ■

Passing to multi-dimensional domains we run into conflict with another situation:

Exercise (III.1.19). Let $U = \mathbb{C}^2 \setminus \{0\}$. We associate to each $w = w(z_1, z_2) \in \mathcal{O}(U)$ the number $(2\pi i)^{-1}\int_{|\zeta|=1} w(\zeta, 0)\zeta^{-1}d\zeta$. Show that this defines a continuous character on $\mathcal{O}(U)$ that is not fixed.

Hint. If we fix the first variable and compute the values of w by integrating round $\mathbb{T}$ with respect to the second variable in the Cauchy integral formula, we get that the function $w_1(z) = w(z, 0)$ is bounded in $\mathbb{D}\setminus\{0\}$ and hence we have a removable singularity at zero.

A reader, who is acquainted with the basics of multi-dimensional complex analysis, will know that $\mathbb{C}^2\setminus\{0\}$ is the simplest example of a domain, on which every holomorphic function has an extension to the same bigger domain — in the given case, the whole of $\mathbb{C}^2$. Therefore we can define the required character directly as the value of the extended holomorphic function at zero and do not need to write it as an integral.

It turns out that the question of whether there exists a character on $\mathcal{O}(U)$ that is not fixed has been completely solved, for an arbitrary U, in terms of simultaneous analytic extension. Moreover it is remarkable that all the characters of $\mathcal{O}(U)$ — fixed as well as not fixed — can be interpreted as points in a new extended domain of definition for the holomorphic functions in $\mathcal{O}(U)$ and furthermore this is the biggest possible such domain. However (as in the classic procedure for constructing the Riemann surface of an "individual" function), the resulting domain, generally speaking, no longer lies in $\mathbb{C}^n$.

We limit ourselves to giving a precise formulation of the relevant assertion; for a detailed proof see, for example, [18, I.G].

Let $(\tilde{U}, \tau)$ be a pair consisting of a topological space and a map from $\tilde{U}$ into $\mathbb{C}^n$. Such a pair is called a *complex manifold over* $\mathbb{C}^n$ (or, briefly, a manifold) if every $\tilde{z} \in \tilde{U}$ has a neighbourhood $V = V_{\tilde{z}}$ such that $\tau_V := (\tau|_V)|^{\tau(V)} : V \longrightarrow \tau(V)$ is a homeomorphism. A function $\tilde{w} : \tilde{U} \longrightarrow \mathbb{C}$ is said to be *holomorphic* if $w(z) := \tilde{w}(\tau_V^{-1}(z))$ is holomorphic (in the usual sense) on $\tau(V)$ for all such V.

---

[8] In fact, the assumption that χ is continuous is not necessary here. Any function f in $\mathcal{O}(U)$ satisfies $f(z) - f(\lambda_0) = (z - \lambda_0)g(z)$ for some $g \in \mathcal{O}(U)$ so that $\chi(f) - f(\lambda_0) = (\lambda_0 - \lambda_0)\chi(g) = 0$.

For every $\tilde{z}_0 \in \tilde{U}$ we let $d(\tilde{z}_0)$ denote the least upper bound of the radii of those polydiscs with centre at $\tau(\tilde{z}_0)$ that are homeomorphic images of neighbourhoods of the point $\tilde{z}_0$. Further, for $\tilde{z}_0$ and holomorphic $\tilde{w} : \tilde{U} \longrightarrow \mathbb{C}$ we let $r(\tilde{z}_0, \tilde{w})$ denote the radius of convergence of the expansion of the function $w(z) = \tilde{w}(\tau_V^{-1}(z))$ as a Taylor series (in powers of $z - \tau(\tilde{z}_0)$) near $\tau(\tilde{z}_0)$.

Now suppose that U is a domain in $\mathbb{C}^n$ and the manifold $(\tilde{U}, \tau)$ is such that $\tilde{U}$ contains U as an open subset. In this situation $(\tilde{U}, \tau)$ is called the *envelope of holomorphy* of the domain U if (I) each holomorphic function on U admits a unique extension to a holomorphic function on $\tilde{U}$ and (II) there exists a holomorphic function $\tilde{w}$ on $\tilde{U}$ such that $r(\tilde{z}, \tilde{w}) = d(\tilde{z})$ for all $\tilde{z} \in \tilde{U}$. Condition (II), loosely speaking, means that there is a function on $\tilde{U}$ which does not automatically extend to a holomorphic function on a larger manifold; in this way $\tilde{U}$ is the largest domain of definition for "its" holomorphic functions.

One can show that the envelope of holomorphy of a given domain is unique up to a certain natural identification (see, for example, [82, V.4]. We shall deal with the question of its existence.

Given U we take $\tilde{U}$ to be the topological space $\mathrm{X\alpha\rho}\, \mathcal{O}(U)$ (Chap.I §2.5) and put $\tau : \tilde{U} \longrightarrow \mathbb{C}^n : \chi \longmapsto (\chi(z_1), \cdots, \chi(z_n))$, where $z_k \in \mathcal{O}(U)$; $k = 1, \cdots, n$, are the coordinate functions. It is easily verified that $(\tilde{U}, \tau)$ is a complex manifold over $\mathbb{C}^n$, and moreover by identifying $z \in U$ with the fixed character χ_z we are justified in considering U to be open subset of $\tilde{U}$. The basic result is formulated in the following way.

Assertion (III.1.20). *Every domain* $U \subseteq \mathbb{C}^n$ *has an envelope of holomorphy and this is* $(\tilde{U} := \mathrm{X\alpha\rho}\, \mathcal{O}(U), \tau)$. *Moreover the correspondence* $\tilde{w} \longmapsto \tilde{w}|_U$ *is a topological isomorphism between the algebras* $\mathcal{O}(\tilde{U})$ *of holomorphic functions on* $\tilde{U}$, *with the compact-open topology, and* $\mathcal{O}(U)$. *As a consequence* $\mathrm{X\alpha\rho}\, \mathcal{O}(\tilde{U}) = \mathrm{X\alpha\rho}\, \mathcal{O}(U) = \tilde{U}$, *that is, all the continuous characters on* $\mathcal{O}(\tilde{U})$ *are fixed.*

Continuing our discussion of the algebras $\mathcal{O}(U)$; $U \subseteq \mathbb{C}^n$, we note that their characters are always continuous, however, the more general question about the automatic continuity of the characters of algebras of holomorphic functions on Stein spaces remains open [196], [261]. The closed maximal ideals of $\mathcal{O}(U)$, $U \subseteq \mathbb{C}$, as we shall show later on, are always fixed (see Corollary V.3.30). As for non-closed (and therefore, non-fixed) maximal ideals, they are found in abundance: as an example take any maximal ideal in $\mathcal{O}(\mathbb{C})$ containing $I := \{w : w|_{\mathbb{N}} \in c_{00}\}$.

Finally we recall that for any pair of domains $U \subseteq \mathbb{C}^m$ and $V \subseteq \mathbb{C}^n$ there exists a unique topological isomorphism between Fréchet algebras $i : \mathcal{O}(U) \mathbin{\overline{\otimes}} \mathcal{O}(V) \xrightarrow{\simeq} \mathcal{O}(U \times V)$, which takes an element of the form $w_1 \otimes w_2$ into the function $w(z, \zeta) := w_1(z)w_2(\zeta)$; $z \in U$, $\zeta \in V$. We suggest the reader proves this fact for the case $U = V = \mathbb{D}$ using the result of Exercise 0.3.54.

* * *

The algebras $\mathcal{O}(U)$, although giving exhaustive information about the structure of domains in $\mathbb{C}^n$ as complex manifolds, ignore the structure of their boundaries. The most important algebras that take this structure into account may be introduced in the following way.

Again let U be a domain in $\mathbb{C}^n$, from now on presumed to be bounded, and let $\overline{U}$ be its closure. We let $\mathcal{A}(\overline{U})$ be the algebra of those functions which are continuous on $\overline{U}$ and whose restrictions to U are holomorphic. If we furnish this with the uniform norm, we get a Banach algebra which is a closed subalgebra of $C(\overline{U})$. The simplest of these algebras is the so-called "*polydisc algebra*" $\mathcal{A}(\overline{\mathbb{D}}^n)$ (here $\mathbb{D}^n$ is the n-th Cartesian power of $\mathbb{D}$) and this, when $n = 1$, becomes the "*disc algebra*" $\mathcal{A}(\overline{\mathbb{D}})$. We limit ourselves to considering just these, mostly with $n = 1$.

According to the maximum principle, the map $w \longmapsto w|_{\mathbb{T}}$ is an isometric isomorphism of $\mathcal{A}(\overline{\mathbb{D}})$ onto a certain closed subalgebra of $C(\mathbb{T})$; $\mathbb{T} = \overline{\mathbb{D}}\backslash\mathbb{D}$; we denote this "aspect" of it by $\mathcal{A}(\mathbb{T})$. In exactly the same way, for any n, $\mathcal{A}(\overline{\mathbb{D}}^n)$ is isometrically isomorphic to $\mathcal{A}(\mathbb{T}^n)$ — the algebra of the restrictions of its functions to the torus $\mathbb{T}^n := \{z = (z_1, \cdots, z_n) : |z_k| = 1;$ $1 \le k \le n\}$. (We note that, when $n > 1$, the n-dimensional torus $\mathbb{T}^n$ is smaller than the $2n-1$-dimensional boundary of the polydisc.)

We can get at the algebras under consideration by other means.

Proposition (III.1.21). $\mathcal{A}(\overline{\mathbb{D}})$, $\mathcal{A}(\mathbb{T})$ *respectively, consists of precisely those functions that are the uniform limits of polynomials on* $\overline{\mathbb{D}}$, $\mathbb{T}$ *respectively.*

▶ Every $w \in \mathcal{A}(\overline{\mathbb{D}})$ is approximated uniformly by the functions $w_t(z) := w(tz)$; $0 < t < 1$, and each of the latter, being holomorphic in a larger disc than $\mathbb{D}$, is approximated on $\overline{\mathbb{D}}$ by the partial sums of its Taylor series.[9] The rest is clear. ■

Exercise (III.1.22). Show that $\mathcal{A}(\mathbb{T})$ consists of precisely those continuous functions w for which all the Fourier

---

[9] We hope that our reader does not remain happily convinced that the Taylor series of a given function must converge uniformly in $\mathcal{A}(\overline{\mathbb{D}})$ to that function.

coefficients of the corresponding 2π-periodic function $f(t) := w(e^{it})$; $t \in \mathbb{R}$, which have negative index, are zero.

Hint. Use Proposition 1.21 and the fact that every function in $C(\mathbb{T})$ is the uniform limit of its Féjer sums.

Proposition (III.1.23). *Every character of the algebra* $\mathcal{A}(\overline{\mathbb{D}})$ (but not $\mathcal{A}(\mathbb{T})$) *is fixed.*

▶ Let λ be the value of the character χ on $w(z) = z$; then the fact that $\|z\|_0 = 1$ and Theorem I.2.6 imply that $|\lambda| \leq 1$, that is $\lambda \in \overline{\mathbb{D}}$. Further, let w be any function from $\mathcal{A}(\overline{\mathbb{D}})$ and let p_n be a sequence of polynomials converging uniformly to w (see Proposition 1.21). Then, from the algebraic properties and the continuity of χ, we have $\chi(w) = \lim_{n\to\infty} \chi(p_n) = \lim_{n\to\infty} p_n(\lambda) = w(\lambda)$. The rest is clear. ■

We suggest the reader formulates and proves the multi-dimensional analogues of 1.21 - 1.23.

A few words about the ideals of the algebras under discussion. We shall show later that, as for W, all maximal ideals in $\mathcal{A}(\overline{\mathbb{D}})$ are fixed ($\mathcal{A}(\overline{\mathbb{D}}^n)$ has this same property for larger n). The question of describing arbitrary closed ideals in $\mathcal{A}(\overline{\mathbb{D}})$ is far more difficult, however Beurling (in an equivalent terminology) and Rudin have succeeded in finding the complete answer. The relevant theorem, in view of the variety of ideals it describes, has a fairly long formulation which we shall not give; see, for example, [23, Ch.VI] or [20, pp.235-239]. We only note that every closed ideal in the disc algebra is the closure of a principal ideal; at the same time it is not necessarily an intersection of point ideals. As to the question of the structure of the latter, this is simple and we shall give the answer.

Exercise (III.1.24). Show that any point ideal in $\mathcal{A}(\overline{\mathbb{D}})$ with zero set $\{z_0\} \subset \mathbb{D}$ is the principal ideal generated by the function $(z - z_0)^k$ for some $k \in \mathbb{N}$.

There are far more point ideals with zero set on the boundary of the disc.

Assertion (III.1.25). *Any point ideal with zero set* $\{z_0\} \subset \mathbb{T}$ *is, for a certain* t; $0 \leq t < \infty$, *the closure of the principal ideal generated by the function* $w_t(z) = (z - z_0)e^{t\left(\frac{z+z_0}{z-z_0}\right)}$. *As a consequence, there is a strictly decreasing continuum of point ideals with this zero set and they may be parametrized by the ray* $\mathbb{R}^{(+)}$.

The question of the structure of closed ideals in $\mathcal{A}(\overline{\mathbb{D}}^n)$ when $n > 1$, as far as we know, remains open. In comparison with the one-dimensional case, the situation is distinctly more complicated: for example, the closed ideals in $\mathcal{A}(\overline{\mathbb{D}})$ are no longer necessarily the closures of finitely generated ideals. See [66, Ch.IV] for this and some more recent results.

Let us recall another Banach algebra with uniform norm which is popular in complex analysis; it consists of all bounded holomorphic functions in $\mathbb{D}$ (irrespective of their behaviour near the boundary) and it is denoted by $H^\infty(\mathbb{D})$. If we study its properties we come up against — in contrast with $\mathcal{O}(\mathbb{D})$ and $\mathcal{A}(\overline{\mathbb{D}})$ — considerable difficulties at the level of trying to describe the characters: up to now this is unsolved and appears fairly hopeless. For a long time the more specific question of whether the disc $\mathbb{D}$, identified with the set of fixed characters on $H^\infty(\mathbb{D})$ is dense in $X\alpha\rho\, H^\infty(\mathbb{D})$, also defied solution. This problem, which can be translated without difficulty into the language of classical complex analysis (see, for example, [23, p.163]) was called the "corona problem". (The set $X\alpha\rho\, H^\infty(\mathbb{D})\backslash\mathbb{D}$ was called the corona — after the solar corona, with $\mathbb{D}$ being likened to the disc of the sun.) The corona problem which stimulated the appearance of much other work in complex analysis, was solved positively by Carleson [153] in 1962.

§2. OPERATOR ALGEBRAS

The theory of operators provides a large number of various algebras - Banach algebras and polynormed algebras; moreover a considerable proportion of them (although not all) are subalgebras of $\mathcal{B}(E)$, where E is a Banach space (see Example I.2.11). In 2.1 and 2.2 of this section the majority will be subalgebras of $\mathcal{B}(E)$ which are Banach algebras with respect to a certain norm $\|\cdot\| \geq \|\cdot\|_0$; when we say "*a Banach operator algebra on* E" we always have such an algebra in mind.

The case when our space is a Hilbert space is, of course, of greatest interest; we shall then always use H instead of E. The algebra $\mathcal{B}(H)$ has an important additional structure — the star operation $a \longmapsto a^*$ taking an operator to its dual (cf. Chap.0 §1.3). Accordingly, we distinguish those subalgebras that are closed under this operation; we shall call them *self-adjoint*. If these algebras are, in addition, closed in $\mathcal{B}(H)$ and therefore Banach algebras with respect to the operator norm, then they are called *realized* C^**-algebras* (why they are called this will become clear in Chap.IV §7).

Sometimes examples of algebras turn up which do not formally consist of operators but are also furnished with a natural star operation (or *involution*), denoted as before by an asterisk. We shall call both these algebras (we shall see that they are in fact operator algebras "in disguise") and "actual" self-adjoint algebras of operators, *-algebras.[10] A homomorphism $\kappa : A \longrightarrow B$ between two *-algebras is called a **-homomorphism* if $\kappa(a^*) = \kappa(a)^*$ for all $a \in A$; similar meanings are given to the words "**-isomorphism*" and "**-automorphism*". Finally, any *-homomorphism of the *-algebra A into $\mathcal{B}(H)$ is called a **-representation* of A on the (arbitrary) Hilbert space H; obviously this is a special case of the general concept of a representation of an algebra (Definition I.1.9).

§2.1. Algebras of bounded, compact and nuclear operators on a Hilbert space; their ideals and *-representations

Given any Banach space E, we already know about, besides $\mathcal{B}(E)$, the closed subalgebra $\mathcal{K}(E)$ consisting of the compact operators (Example I.2.11). The next most important example is of course the algebra of nuclear operators $(\mathcal{N}(E); \|\cdot\|_{\mathcal{N}})$ (see Chapter 0 §1.3) also with composition as the multiplication. It is obvious that this is also a Banach operator algebra and moreover, when $E = H$, it is self-adjoint but it is not a realized C^*-algebra like $\mathcal{B}(H)$ and $\mathcal{K}(H)$.

Exercise (III.2.1) (cf. Example I.2.33). Show that there is an a.i. but no b.a.i. in $\mathcal{N}(H)$.

Hint. Consider orthogonal projections onto finite-dimensional subspaces of H.

Let us put to our three algebras the question in our "questionnaire" (cf. introduction to the chapter) about the structure of closed ideals. Let A be any of them. Taking as a model the family of ideals in Example I.3.11, we put $I(E_0) = \{a \in A : a|_{E_0} = 0\}$ for each subspace E_0 of E.

Proposition (III.2.2). *$I(E_0)$ is a closed left ideal in A. Moreover if E_0 is finite-dimensional, it is automatically a modular ideal, and if E_0 is one-dimensional, it is a maximal ideal.*

▸ Any operator from A which is the identity on E_0 is a modular identity for $I(E_0)$. The rest is clear. ∎

If we combine Proposition 2.2 and Theorem I.3.53(Ia), we immediately get the following.

Corollary (III.2.3). *The algebras $\mathcal{B}(E)$, $\mathcal{K}(E)$ and $\mathcal{N}(E)$ are semisimple.*

---

[10] We postpone the axiomatic approach to the "star" operation and to the concepts derived from it until the second half of Chapter IV (beginning with §4).

A more complicated assertion is that the algebras of compact and nuclear operators in sufficiently nice spaces (we restrict ourselves to Hilbert spaces) have no other closed left ideals except those mentioned in Proposition 2.2. We first consider a subsidiary geometric fact.

Lemma (III.2.4). *Let* Λ *be a linearly ordered set and let* H_ν; $\nu \in \Lambda$, *be a decreasing family of closed subspaces in* H. *Put* $H_\infty := \bigcap_\nu H_\nu$, *and let* P_ν *be the orthogonal projection onto* H_ν, P_∞ *the orthogonal projection onto* H_∞. *Put* $Q_\nu := 1 - P_\nu$, $Q_\infty := 1 - P_\infty$. *Then, when* $A = \mathcal{K}(H)$ *or* $\mathcal{N}(H)$, *we have* $aQ_\infty = \lim_\nu aQ_\nu$ *(in* A*) for any* $a \in A$.

▶ Because of the obvious inequality $\| aQ_\infty - aQ_\nu \|_A \leq 2\| a - b\|_A + \| bQ_\infty - bQ_\nu\|_A$; $a, b \in A$, and the fact that the set of finite-dimensional (= linear combinations of one-dimensional) operators is dense in A, it is sufficient to verify the required limit when a is one-dimensional, that is, $a : x \longmapsto \langle x, y\rangle z$, where y and z are fixed vectors in H. In this case $aQ_\infty : x \longmapsto \langle Q_\infty x, y\rangle z = \langle x, Q_\infty y\rangle z$, and similarly $aQ_\nu : x \longmapsto \langle x, Q_\nu y\rangle z$; hence $\| aQ_\infty - aQ_\nu\|_A \leq \| Q_\infty y - Q_\nu y\|_H \|z\|_H$. But, obviously, $H_\infty^\perp$ is the closure of $\bigcup_\nu H_\nu^\perp$, so it follows that $Q_\infty y = \lim_\nu Q_\nu y$. The rest is clear. ■

Theorem (III.2.5) (Kaplansky). *When* $A = \mathcal{K}(H)$ *or* $\mathcal{N}(H)$ *any closed left ideal* I *of* A *is of the form* $I(H_I)$, *where* $H_I := \{x \in H : a(x) = 0$ *for all* $a \in I\}$. *As a consequence, any maximal modular left ideal in* A *is of the form* $I = \{a : a(x_0) = 0\}$, *where* $x_0 \in H \backslash \{0\}$ *is a certain fixed vector.*

▶ We consider the set $\mathfrak{U}$ of all closed subspaces H' of H such that $I(H') \subseteq I$; since it is obvious that $I \subseteq I(H_I)$, our problem is to establish that $H_I \in \mathfrak{U}$. We give $\mathfrak{U}$ the decreasing order (H' precedes H'' if $H' \supseteq H''$) and consider any linearly ordered subset $\{H_\nu : \nu \in \Lambda\}$. If, in the notation of Lemma 2.4, $a \in I(H_\infty)$, then $aQ_\infty = a$ and $aQ_\nu \in I(H_\nu) \subseteq I$; therefore since I is closed, Lemma 2.4 implies that $I(H_\infty) \subseteq I$. So $\{H_\nu ; \nu \in \Lambda\}$ has an upper bound, namely H_∞; hence, by Zorn's lemma, $\mathfrak{U}$ contains a maximal element, H_0 say. All that remains is to show that $H_0 = H_I$.

Suppose, on the contrary, that there exists $x \in H_0 \backslash \{0\}$ orthogonal to H_I. By the choice of H_I, there is some $a \in I$ with $y := a(x) \neq 0$. We put $a_0 := aP_0$, where P_0 is the orthogonal projection onto H_0; then $a_0 - a \in I(H_0) \subseteq I$, and therefore $a_0 \in I$; in addition, $a_0(x) = y$ and $a_0(z) = 0$ when $z \perp H_0$. Now we take $p_y : z \longmapsto \langle z, y\rangle y$ and $a_1 := a_0^* p_y a_0$; then $a_1 \in I$, $a_1^* = a_1$, $a_1(z) = 0$ when $z \perp H_0$ and, in addition, a_1 is one-dimensional: the latter follows from the fact that p_y is one-dimensional and $\langle a_1 x, x\rangle = \langle y, y\rangle^2 \neq 0$.

Hence, if we put $p := a_1/\|a_1\|$, we get an orthogonal projection onto a one-dimensional subspace of H_0, which belongs to I.

Finally, we take $H_1 := H_0 \cap \mathrm{Ker}(p)$ and the orthogonal projections Q_k onto $H_k^{\perp}$; $k = 0, 1$; then it is obvious that $Q_1 = p + Q_0$. Hence any $b \in I(H_1)$ has the form $b = bQ_1 = bp + bQ_0$, which, since $p \in I$ and $bQ_0 \in I(H_0) \subseteq I$, implies that $b \in I$. But this means that $I(H_1) \subseteq I$ and, because $H_1 \subsetneqq H_0$, this contradicts the fact that H_0 is maximal in $\mathfrak{U}$. ■

Since none of the ideals described in Theorem 2.5 is a right ideal, we get the following corollary.

Corollary (III.2.6) (cf. I.3.12). *There are no proper closed bi-ideals in $\mathcal{K}(H)$ and $\mathcal{N}(H)$.*

Exercise (III.2.7) (cf. I.3.11). Describe all the closed right ideals and all the maximal modular right ideals in these algebras.

Hint. The "star" establishes a bijection between the left and right ideals.

Remark (III.2.8). Non-closed bi-ideals do, of course, exist in both algebras — an example is the ideal of finite-dimensional operators. In $\mathcal{K}(H)$ there are, in addition, many bi-ideals that are Banach operator algebras with respect to special norms. The most popular of them, apart from $\mathcal{N}(H)$, is the algebra $\mathfrak{S}_2(H)$ of *Hilbert-Schmidt operators* — this is the name given to those operators $a \in \mathcal{K}(H)$, for which $a^*a \in \mathcal{N}(H)$ (when H is separable and infinite-dimensional these are precisely the operators that are unitarily equivalent to the classic integral operators in $L^2[0, 1]$; (see [79]). $\mathfrak{S}_2(H)$ is a Banach algebra with respect to the norm $\|a\|_2 := \sqrt{\|a^*a\|_{\mathcal{N}}}$. In fact $\mathfrak{S}_2(H)$ is just one of a sequence of Banach algebras $\mathfrak{S}_p(H)$; $1 \le p < \infty$, which are dense bi-ideals in $\mathcal{K}(H)$; here $\mathfrak{S}_1(H) = \mathcal{N}(H)$. Some information about the algebra $\mathfrak{S}_p(H)$ which, not without justification, is considered to be the "non-commutative analogue of l_p" (see §1.3) can be found in [24], [92].

In the case of $\mathcal{B}(H)$, its closed ideals can no longer be described so clearly: it is sufficient to note that, if we take any maximal left ideal containing $\mathcal{K}(H)$, then the intersection of the kernels of its operators is trivial. (This observation — and there will be others — already tells us that when we investigate $\mathcal{B}(H)$ it is going to turn out to be, unlike $\mathcal{K}(H)$, a very complicated algebra.) In general, the problem of a reasonably satisfactory description of the closed one-sided ideals in $\mathcal{B}(H)$ is at present fairly hopeless.

The picture is quite different with bi-ideals.

Lemma (III.2.9). *Let I be a bi-ideal in $\mathcal{B}(H)$, where H is separable, which contains a non-compact orthogonal projection p. Then $I = \mathcal{B}(H)$.*

▶ Since $\operatorname{Im} p$ is infinite-dimensional and H is separable, there exists a unitary map $v : H \longrightarrow H$ sending H onto $\operatorname{Im} p$. But then $p \in I$ and, obviously, $v = pv$ and $v^* v = 1$ which implies that I contains the identity and so also contains any operator in H. ■

Theorem (III.2.10) (Calkin). *Let H be separable and infinite-dimensional. Then $\mathcal{B}(H)$ has a unique closed proper non-trivial bi-ideal, namely $\mathcal{K}(H)$.*

▶ Our problem is to show that a bi-ideal I with the properties mentioned coincides with $\mathcal{K}(H)$. We take any $a \in I \setminus \{0\}$ and $x_0, y_0 \in H$ with $a(x_0) = y_0$ and $\|y_0\| = 1$. Then for any $x, y \in H$ the operator $q : z \longmapsto \langle z, x\rangle y$ can be presented in the form $q = ras$, where $s : z \longmapsto \langle z, x\rangle x_0$, $r : z \longmapsto \langle z, y_0\rangle y$. In this way any one-dimensional, and hence any finite-dimensional operator also, belongs to I; which, I being closed, means that $\mathcal{K}(H) \subseteq I$. Let us establish the reverse inclusion.

Suppose, on the contrary, that $a \in I \setminus \mathcal{K}(H)$; then it is easy to verify that $b := a^* a$ is also non-compact and in addition is self-adjoint. We apply Hilbert's theorem (Chap.0 §1.3) to it; let (X, μ) be the measure space and f be the function featuring in this theorem such that the diagram

$$\begin{array}{ccc} H & \xrightarrow{\ b\ } & H \\ {\scriptstyle u}\downarrow & & \downarrow{\scriptstyle u} \\ L^2(X, \mu) & \xrightarrow{\ \tilde{b}\ } & L^2(X, \mu) \end{array}$$

is commutative, where u is unitary and $\tilde{b} : g \longmapsto fg$. We put $\tilde{I} = \{ucu^{-1} : c \in I\}$; obviously $\tilde{I}$ is a proper (because I is) bi-ideal in $\mathcal{B}(L^2(X, \mu))$.

We now put $X_n := \{t \in X : f(t) \geq {}^1\!/\!_n\}$ for each $n \in \mathbb{N}$ and consider the functions χ_n, f_n and h_n that are equal to 1, $f(t)$ and $f(t)^{-1}$, respectively, on X_n and are zero on $X \setminus X_n$; the operators of multiplication by these functions are in $\mathcal{B}(L^2(X, \mu))$ and are denoted by $\tilde{p}_n$, $\tilde{b}_n$ and $\tilde{d}_n$, respectively. Since $\tilde{b} \in \tilde{I}$, then also $\tilde{p}_n = \tilde{b}\tilde{d}_n \in \tilde{I}$ for all n. Further the $\tilde{p}_n$ are obviously orthogonal projections, therefore applying Lemma 2.9 to $\tilde{I}$ and $\mathcal{B}(L^2(X, \mu))$, it follows that they are all compact, and hence the $\tilde{b}_n = \tilde{p}_n\tilde{b}_n$ are also compact. But the fact that f_n converges uniformly to f obviously implies that $\lim_{n\to\infty} \|\tilde{b} - \tilde{b}_n\| = 0$; hence $\tilde{b}$ is compact and so also is b which is unitarily equivalent to it. This contradiction establishes that $I \subseteq \mathcal{K}(H)$. ■

We now pass from the structure of (bi-)ideals to other questions about the algebras under discussion. Since the kernel of a non-trivial character on a Banach algebra is a closed bi-ideal of codimension 1, Theorem 2.10 and Corollary 2.6 imply that the algebras featuring there have no non-trivial characters. However, the reader may establish this fact more simply by verifying that the subalgebra of A generated by $\{v \in A : v^2 = 0\}$ is A itself for $\mathcal{B}(H)$ and is dense in A for $A = \mathcal{K}(H)$ or $\mathcal{N}(H)$.

The question about the structure of the representations of our algebras has more content. Here, thanks to the possibilities provided by the operation $a \longmapsto a^*$, it is far easier to work with $*$-representations; we restrict ourselves to these. It turns out that, for $\mathcal{K}(H)$ and $\mathcal{N}(H)$, all such representations can be described in fairly simple terms; informally speaking, they consist of a certain family of copies of the identity map, plus the zero representation. For simplicity we shall assume that all the Hilbert spaces featuring here are separable.

We first introduce one technically convenient concept. For a given H we let $\mathcal{H}$ denote either the space consisting of the row vectors $\underline{x} = \{x_k \in H : 1 \leq k \leq n\}$ for fixed n, or the sequences $\underline{x} = \{x_k \in H : k \in \mathbb{N}\}$, such that $\sum_{k=1}^{\infty} \|x_k\|^2 < \infty$. If we furnish $\mathcal{H}$ with the scalar product $\langle \underline{x}, \underline{y} \rangle := \sum_k \langle x_k, y_k \rangle$, we obviously get a Hilbert space which we shall call a (finite or, according to the sense, countable) *multiple of* H. Let A be a certain subalgebra of $\mathcal{B}(H)$.

Definition (III.2.11). The homomorphism $\mathfrak{m} : A \longrightarrow \mathcal{B}(\mathcal{H}) : a \longmapsto \underline{a}$ where $\underline{x} = \{x_k\} \longmapsto \underline{a}(\underline{x}) := \{a(x_k)\}$ is called the *replicating homomorphism of the algebra* A.

Theorem (III.2.12). *Let* $A = \mathcal{K}(H)$ *or* $\mathcal{N}(H)$ *and let* T *be a continuous*[11] *$*$-representation on a certain* H' *(*H *and* H' *are separable). Then* H' *decomposes into an orthogonal direct sum of subspaces* H_1 *and* H_0 *having the following properties:*

(I) *there exists a unitary operator* $u : H_1 \longrightarrow \mathcal{H}$, *where* $\mathcal{H}$ *is a multiple (finite or countable) of* H, *such that, for any* $a \in A$ *the diagram below is commutative;*

$$\begin{array}{ccc} H_1 & \xrightarrow{\;T_a|_{H_1}\;} & H_1 \\ {\scriptstyle u}\downarrow & & \downarrow{\scriptstyle u} \\ \mathcal{H} & \xrightarrow{\;\mathfrak{m}(a)\;} & \mathcal{H} \end{array} \qquad (1)$$

(II) $T_a|_{H_0} = 0$ *for any* $a \in A$.

---

[11] Later on, Theorem IV.5.25 shows us that the continuity assumption is automatically satisfied.

Thus, at an intuitive level, what has been said means that H' can be decomposed into an orthogonal direct sum of a finite or countable number of subspaces H_k; $k = 1, 2, \cdots$, isomorphic to H, and of one other space H_0 so that if we identify H with these H_k then, for any $a \in A$, the operator T_a may be written as a block matrix

$$\begin{pmatrix} a & & & 0 \\ & a & & \\ & & \ddots & \\ 0 & & & \boxed{0} \end{pmatrix}$$

▶ We introduce the brief notation $\tilde{a}$ instead of T_a. Let $\{e_n\}$ be an orthonormal basis in H; then A contains the one-dimensional operators $p_{mn} : x \longmapsto \langle x, e_n \rangle e_m$. Obviously, for all $k, l, m, n \in \mathbb{N}$, $p_{kl}p_{mn} = \delta^l_m p_{kn}$ (δ is the Kronecker symbol), and $p^*_{mn} = p_{nm}$. In view of the properties of T, these same relations are satisfied for $\tilde{p}_{mn}$. Hence the $\tilde{p}_{nn}$ are all orthogonal projections with pairwise orthogonal images and each map $\tilde{p}_{mn}$ is a partial isometry from $\operatorname{Im} \tilde{p}_{nn}$ onto $\operatorname{Im} \tilde{p}_{mm}$ which takes $\operatorname{Ker} \tilde{p}_{nn}$ to zero.

We fix on an orthonormal basis $\tilde{e}^k_1$; $k = 1, 2, \cdots$ (finite or countable) in $\operatorname{Im} \tilde{p}_{11}$, and for all k we put $\tilde{e}^k_n := \tilde{p}_{n1}(\tilde{e}^k_1)$; $n = 1, 2, \cdots$. The above properties of $\tilde{p}_{mn}$ imply that $\{\tilde{e}^k_n; k, n = 1, 2, \cdots\}$ is an orthonormal system in H'. We put $H_1 = \overline{\ell.h.}\{\tilde{e}^k_n;\ k, n = 1, 2, \cdots\}$ and $H_0 = H_1^{\perp}$.

Now let $\mathcal{H}$ be the multiple of H consisting of row vectors of length the dimension of $\operatorname{Im} \tilde{p}_{11}$, whatever this may be. Then there is an orthonormal basis $\{\underline{e}^k_n = (0, \cdots, e_n, 0, \cdots)$; $n, k = 1, 2, \cdots\}$ in $\mathcal{H}$, where in this row vector (or sequence) e_n occurs in the k-th place. In this way we define a unique unitary operator $u : H_1 \longrightarrow \mathcal{H} : \tilde{e}^k_n \longmapsto \underline{e}^k_n$.

With the help of the relations for $\tilde{p}_{mn}$ we can immediately verify that $[\mathfrak{m}(p_{lm}) \circ u](\tilde{e}^k_n) = [u \circ \tilde{p}_{lm}](\tilde{e}^k_n)$ for all possible indices, which implies that diagram (1) is commutative for all a of the form p_{lm}. But $\overline{\ell.h.}\{p_{lm}\} = A$; therefore, as T is linear and continuous, the diagram is commutative for all $a \in A$. All that remains is to note that, since $\tilde{p}_{lm}|_{H_0} = 0$ for all l, m, this implies precisely that $\tilde{a}|_{H_0} = 0$ for any $a \in A$. ∎

The situation is more complicated for the algebra $\mathcal{B}(H)$: if we try to prove an analogous theorem for this, we are soon convinced that a *-representation does not, generally speaking, reduce to "copies of the identity". Some information about their structure — at present still incomplete — is provided by the following proposition.

Proposition (III.2.13). *Theorem* 2.12 *remains valid for* $A = \mathcal{B}(H)$ *except that, instead of* (II) *it is now only asserted that* H_0 *is an invariant subspace for all* T_a; $a \in \mathcal{B}(H)$ *such that* $T_a | H_0 = 0$ *for* $a \in K(H)$ *(and that is all!).*

▶ We consider the restriction of T to $\mathcal{K}(H)$; then Theorem 2.12 provides the corresponding H_1 and H_0. We first show that diagram (1) is commutative for any $a \in \mathcal{B}(H)$. For this, it is obviously sufficient to check that $[T_a \circ u^{-1}](\underline{x}) = [u^{-1} \circ \mathfrak{m}(a)](\underline{x})$ for all $\underline{x} \in \mathcal{H}$ of the form $(0, \cdots, 0, x, 0, \cdots)$; $x \in H$. But for such an $\underline{x}$ one can easily see that there exists a one-dimensional orthogonal projection q in H such that $\underline{x} = \mathfrak{m}(q)\underline{x}$. Therefore, since $q, aq \in \mathcal{K}(H)$, according to Theorem 2.12 we get $T_a u^{-1}(\underline{x}) = T_a u^{-1}\mathfrak{m}(q)(\underline{x}) = T_a T_q u^{-1}(\underline{x}) = T_{aq} u^{-1}(\underline{x}) = u^{-1}\mathfrak{m}(aq)(\underline{x}) = u^{-1}\mathfrak{m}(a)\mathfrak{m}(q)(\underline{x}) = u^{-1}\mathfrak{m}(a)(\underline{x})$. All that remains is to note that, since H_1 is an invariant space for T_{a*}, it follows that $H_0 = H_1^{\perp}$ is an invariant subspace for $(T_{a*})^* = T_{a**} = T_a$, for any $a \in \mathcal{B}(H)$. ■

The result obtained, as can be seen, enables the study of arbitrary *-representations of the algebra $\mathcal{B}(H)$ to be reduced to the case $H = H_0$, that is, to representations taking all the compact operators to zero — but this case is the stumbling block. For the moment we only note that representations of this form induce (cf. Proposition I.3.44) representations of the quotient algebra $\mathcal{B}(H)/\mathcal{K}(H)$ which deserves special consideration.

§2.2. Calkin algebras, Fermion algebras, local and quasi-local observable spinor systems. Remarks about nest algebras

The quotient algebra $\mathcal{B}(H)/\mathcal{K}(H)$ has turned up in our account. In view of its prominent role in a whole range of questions[12] it is given a special name — the *Calkin algebra*. Theorem 2.10 and Proposition I.3.5 imply that this algebra, often denoted by $\mathcal{C}(H)$, is simple; recall (Remark I.3.69) that an algebra is said to be simple if it has no proper bi-ideals, closed or otherwise. As a consequence, any homomorphism from it ("into anywhere") is injective. Obviously $\mathcal{C}(H)$ is a Banach algebra having, furthermore, a star operation $a + \mathcal{K}(H) \longmapsto (a + \mathcal{K}(H))^* := a^* + \mathcal{K}(H)$. Since any *-representation T of the algebra $\mathcal{B}(H)$ for which $T|_{\mathcal{K}(H)} = 0$ induces a representation of the Calkin algebra, which is obviously a

[12] Especially in K-theory and the theory of C^*-algebra extensions [32], [227], [228].

*-representation, then, if we knew all the *-representations of $\mathscr{C}(H)$, we would know, by Proposition 2.13, all the representations of $\mathscr{B}(H)$. However, up to now only one sequence of specific *-representations of the algebra $\mathscr{C}(H)$ is known; this was constructed by Calkin. It shows, in particular, that $\mathscr{C}(H)$, although not formally consisting of operators, is nevertheless a realized C^*-algebra "in disguise" up to isometric *-isomorphism. Later we get this same fact by other methods, in the context of the general theory (see Corollary IV.7.82).

Returning to $\mathscr{B}(H)$ — our first example of an operator algebra, we can now observe the following phenomenon: there is in this algebra a bi-ideal ($\mathscr{K}(H)$), for which one can give a complete description of its *-representations, while at the same time the quotient algebra with respect to this bi-ideal ($\mathscr{C}(H)$) somehow does not wish to give any information about its *representations. Indeed, what has been said reflects a general fact which holds for all realized (and subsequent abstract) C^*-algebras. Namely, they may be separated into two classes, in many respects having contrasting properties. The first is formed by algebras "with a nice representation theory". Oversimplifying somewhat, these are the algebras that are obtained from $\mathscr{K}(H)$ with the help of a series of strictly stipulated constructions; they are called *postliminal* and may be characterized from various points of view (cf. later Chapter VI §2.2). Algebras of the other class — the antiliminal[13] algebras, are on the contrary those with representations about which extremely little is known. Moreover, it turns out that in every C^*-algebra there is a largest postliminal bi-ideal and the quotient algebra by this bi-ideal is antiliminal.

We refer the reader, who is interested in a rigorous account of the results mentioned, to books devoted entirely to C^*-algebras; see [31, 4.3.6] or [58, 6.2.7]. We note only that, if one has to work with antiliminal algebras, one has to put up with their bad behaviour: some of them are very important in theoretical physics. Our second example (after the Calkin algebra) of an antiliminal algebra is just such an algebra; unlike $\mathscr{C}(H)$, this algebra can be constructed directly as an operator algebra, and furthermore it is separable.

Let H be a separable Hilbert space. We fix on an orthonormal basis and let Φ_n; $n = 0, 1, 2, \cdots$ denote the subalgebra of $\mathscr{B}(H)$ consisting of the operators given by block matrices

[13] In this part of operator theory everything is beautiful, except the terminology: instead of "postiminal algebras" some authors speak of "GCR-algebras" or "algebras of type I" and instead of "antiliminal algebras" they talk about "$NGCR$-algebras".

$$\begin{pmatrix} M & & & 0 \\ & \ddots & & \\ & & M & \\ 0 & & & \ddots \end{pmatrix},$$

where the diagonal is made up of the same $2^n \times 2^n$ matrix M. Obviously Φ_n is isometrically *-isomorphic to the algebra of operators on 2^n-dimensional Hilbert space. There is a chain of subalgebras $\ell.h\{1\} = \Phi_0 \subset \Phi_1 \subset \Phi_2 \subset \cdots$; the closure in $\mathcal{B}(H)$ of its union is called the *Fermion algebra* and is denoted by Φ. Obviously Φ is a unital realized C^*-algebra; in addition, when we consider the operators made up of block matrices with rational elements down the diagonal, we see that it is separable.

Proposition (III.2.14). *The Fermion algebra is simple.*

▶ Suppose, on the contrary, that Φ has a proper non-zero bi-ideal I which, by Proposition I.3.5, we may assume to be closed. We take $x \in I$; $\|x\| = 1$; then there exist n and $a \in \Phi_n$; $\|a\| = 1$, such that $\theta := \|a - x\| < 1$. We put $y := a^*x$ and $b := a^*a$. Since b and all its powers are self-adjoint, then, by the C^*-property (Theorem 0.1.13), $\|b^{2^m}\| = \|b^{2^{m-1}}\|^2 = \cdots = \|b\|^{2^m} = \|a\|^{2^{m+1}} = 1$ for any m. Therefore, as the unit sphere in the (finite-dimensional) space Φ_n is compact, there exists a sequence $(b^{k_l};\ l \in \mathbb{N})$, converging to a certain $z \in \Phi_n$; $\|z\| = 1$.

We now put $z_m := b^m - (b - y)^m$; if we remove the brackets, we see that, since y belongs to I, so do all the z_m. Further $\|b^m - z_m\| \le \|b - y\|^m \le \|a^*\|^m \|a - x\|^m \le \theta^m$, hence z_{k_l} converges to the same z as b^{k_l}; consequently, $z \in I$.

Finally, we consider the bi-ideal $J := \Phi_n \cap I$ of Φ_n. Since $z \in J$ and $z \neq 0$, and $\Phi_n \simeq \mathcal{K}(H')$ when $\dim H' = 2^n$, then Corollary 2.6 implies that $J = \Phi_n$. Hence $1 \in J$, and hence $1 \in I$, consequently the bi-ideal I is not proper. Contradiction! ■

There is another method of constructing, to within isometric *-isomorphism, the Fermion algebra. It has the advantage that it enables an important additional structure to be defined naturally in the algebra that arises — the so-called action of (a certain) group of automorphisms. In the given case we are talking about the group $\mathbb{Z}^N$ (the "integer lattice"), where N is a fixed natural number.

Let Λ be the set of finite subsets of $\mathbb{Z}^N$ directed by inclusion. With every $\mathbf{m} \in \mathbb{Z}^N$ (a lattice point) we associate a two-dimensional Hilbert space $H_{\mathbf{m}}$ and put $\mathfrak{A}_{\mathbf{m}} := \mathcal{B}(H_{\mathbf{m}})$. For each $\lambda = (\mathbf{m}_1, \cdots, \mathbf{m}_k) \in \Lambda$ we consider the 4^k-dimensional algebra $\mathfrak{A}_\lambda := \bigotimes_{l=1}^{k} \mathfrak{A}_{\mathbf{m}_l}$ (see Chap.I §2.3; where here and later on we write $\bigotimes_{l=1}^{k}(\cdot)$ instead of $(\cdot)\otimes\cdots\otimes(\cdot)$ as a short notation for tensor products, elementary tensors and so on). We note that, when $\nu=\lambda\amalg\mu$ the associativity property of the tensor product (see *ibid*) enables $\mathfrak{A}_\lambda$ to be identified with a subalgebra of $\mathfrak{A}_\nu$ — the image of the injective homomorphism $i_\nu^\lambda : \mathfrak{A}_\lambda \longrightarrow \mathfrak{A}_\nu : a \longmapsto a\otimes 1_{\mathfrak{A}_\mu}$.

We now consider, in the set $\bigcup\{\mathfrak{A}_\lambda : \lambda\in\Lambda\}$ an equivalence relation, by declaring $a_\lambda\in\mathfrak{A}_\lambda$ and $a_\mu\in\mathfrak{A}_\mu$ to be equivalent, if there exists $\nu\supseteq\lambda\cup\mu$, such that $i_\nu^\lambda a_\lambda = i_\nu^\mu a_\mu$; this definition is justified by the obvious identity $i_\omega^\nu\circ i_\nu^\lambda = i_\omega^\lambda$, when $\lambda\subseteq\nu\subseteq\omega$. The corresponding set of equivalence classes is denoted by $\mathfrak{A}$. Further, if we take $\lambda\in\Lambda$ and to each $a_\lambda\in\mathfrak{A}_\lambda$ associate its equivalence class in $\bigcup\{\mathfrak{A}_\lambda : \lambda\in\Lambda\}$, we obviously get an injective map $i^\lambda : \mathfrak{A}_\lambda\longrightarrow\mathfrak{A}$; where it is easy to verify that $i^\nu\circ i_\nu^\lambda = i^\lambda$ for every $\nu\supseteq\lambda$.

Proposition (III.2.15). *There exists a unique structure of a unital algebra in* $\mathfrak{A}$ *for which* i^λ *is a unital homomorphism for every* $\lambda\in\Lambda$.[14]

▸ The last equation implies that, for any $a,b\in\mathfrak{A}$, there exist $\lambda\in\Lambda$ and $a_\lambda, b_\lambda\in\mathfrak{A}_\lambda$ with $a=i^\lambda a_\lambda$ and $b=i^\lambda b_\lambda$. We put $ab := i^\lambda(a_\lambda b_\lambda)$ and we define the linear operations in $\mathfrak{A}$ similarly. We leave the reader to verify that i^λ has the required properties. ■

Our next aim is to introduce a norm in $\mathfrak{A}$. We fix $\lambda=(\mathbf{m}_1,\cdots,\mathbf{m}_k)\in\Lambda$ and consider the 2^k-dimensional space $H_\lambda := \bigotimes_{l=1}^{k} H_{\mathbf{m}_l}$. This is a Hilbert space with scalar product that is uniquely defined by $\langle\bigotimes_{l=1}^{k}x_l, \bigotimes_{l=1}^{k}y_l\rangle := \prod_{l=1}^{k}\langle x_l, y_l\rangle$ (check that this is well defined and, for a better understanding, describe an orthonormal basis in H_λ starting from bases in the $H_{\mathbf{m}_l}$). Proposition I.2.29 obviously implies that the operator $\mathrm{ott} : \mathfrak{A}_\lambda = \bigotimes_{l=1}^{k}\mathcal{B}(H_{\mathbf{m}_l})\longrightarrow\mathcal{B}(H_\lambda)$ taking $\bigotimes_{l=1}^{k}a_{\mathbf{m}_l}$ to the operator "$\bigotimes_{l=1}^{k}x_l\longmapsto\bigotimes_{l=1}^{k}a_{\mathbf{m}_l}(x_l)$" is an algebra isomorphism. Therefore we are justified in defining a norm in $\mathfrak{A}_\lambda$ by putting

---

[14] The reader who is acquainted with the general categorical concept of a direct (= injective) limit, will see that the algebra $\mathfrak{A}$ is none other than the direct limit of the system $(\Lambda, \{\mathfrak{A}_\lambda\}, \{i_\mu^\lambda\})$ in the category of algebras.

$\|a\| := \|\mathbf{ott}(a)\|$, and also (at the same time) defining the operation "$*$" by putting $a^* := \mathbf{ott}^{-1}[\mathbf{ott}(a)]^*$.

Proposition (III.2.16). *There exist a unique (multiplicative) norm* $\|\cdot\|$ *in* $\mathfrak{A}$ *and a unique operation* $a \longmapsto a^*$ *(continuous with respect to the norm) such that, for any* $\lambda \in \Lambda$, $i^\lambda : \mathfrak{A}_\lambda \longrightarrow \mathfrak{A}$ *is an isometric* **-homomorphism.*

▶ $i^\lambda_\mu : \mathfrak{A}_\lambda \longrightarrow \mathfrak{A}_\mu$ is obviously an isometric *-homomorphism for $\lambda, \mu \in \Lambda$; $\lambda \subseteq \mu$. Therefore, given any $a \in \mathfrak{A}$, if $\lambda \in \Lambda$ and $a_\lambda \in \mathfrak{A}_\lambda$ are such that $a = i^\lambda a_\lambda$, then $\|a\| := \|i^\lambda a_\lambda\|$ and $a^* := i^\lambda(a^*_\lambda)$ do not depend on the choice of λ. The rest is clear. ■

If we take the completion of $\mathfrak{A}$ with respect to the norm we have introduced, we get a Banach algebra $\overline{\mathfrak{A}}$ (see Proposition I.2.7) furnished, in addition, with an operation $a \longmapsto a^*$ by taking the continuous extension of the corresponding operation in $\mathfrak{A}$. We now show that this method of giving the algebras $\mathfrak{A}$ and $\overline{\mathfrak{A}}$ naturally allows us to distinguish a certain group of automorphisms.

Identifying all the $H_{\mathbf{m}}$; $\mathbf{m} \in \mathbb{Z}^N$ (if we want, we can think of them as copies of the same space) we let $\mathrm{id}^{\mathbf{m}}_{\mathbf{n}} : \mathfrak{A}_{\mathbf{m}} \longrightarrow \mathfrak{A}_{\mathbf{n}}$ denote the corresponding *-isomorphism between the "algebras sitting over the lattice points $\mathbf{m}$ and $\mathbf{n}$". We fix $\mathbf{k} \in \mathbb{Z}^N$ and for any $\lambda = (\mathbf{m}_1, \cdots, \mathbf{m}_n)$ we put $\lambda + \mathbf{k} := (\mathbf{m}_1 + \mathbf{k}, \cdots, \mathbf{m}_n + \mathbf{k})$ and $\alpha_\lambda(\mathbf{k}) := \bigotimes_{l=1}^{n} \mathrm{id}^{\mathbf{m}_l}_{\mathbf{m}_l + \mathbf{k}} : \mathfrak{A}_\lambda \longrightarrow \mathfrak{A}_{\lambda+\mathbf{k}}$; obviously $\alpha_\lambda(\mathbf{k})$ is an isometric *-algebra *-isomorphism. It is easy to verify the following proposition.

Proposition (III.2.17). *There exist unique isometric* **-automorphisms* $\alpha(\mathbf{k}) : \mathfrak{A} \longrightarrow \mathfrak{A}$ *and* $\overline{\alpha}(\mathbf{k}) : \overline{\mathfrak{A}} \longrightarrow \overline{\mathfrak{A}}$ *such that, for any* $\lambda \in \Lambda$ *the diagram*

$$\begin{array}{ccc}
\mathfrak{A}_\lambda & \xrightarrow{\alpha_\lambda(\mathbf{k})} & \mathfrak{A}_{\lambda+\mathbf{k}} \\
{\scriptstyle i^\lambda}\downarrow & & \downarrow{\scriptstyle i^{\lambda+\mathbf{k}}} \\
\mathfrak{A} & \xrightarrow{\alpha(\mathbf{k})} & \mathfrak{A} \\
{\scriptstyle \mathbf{in}}\downarrow & & \downarrow{\scriptstyle \mathbf{in}} \\
\overline{\mathfrak{A}} & \xrightarrow{\overline{\alpha}(\mathbf{k})} & \overline{\mathfrak{A}}
\end{array}$$

is commutative. ▶■

Obviously the automorphisms of the form $\alpha(\mathbf{k})$ (or $\overline{\alpha}(\mathbf{k})$ respectively); $\mathbf{k} \in \mathbb{Z}^N$, form a subgroup of the group $\mathrm{Aut}(\mathfrak{A})$ (or $\mathrm{Aut}(\overline{\mathfrak{A}})$) of all *-algebra *-automorphisms of $\mathfrak{A}$ (or $\overline{\mathfrak{A}}$) and the map $\alpha : \mathbb{Z}^N \longrightarrow \mathrm{Aut}(\mathfrak{A})$ (or $\overline{\alpha} : \mathbb{Z}^N \longrightarrow \mathrm{Aut}(\overline{\mathfrak{A}})$) is obviously a group homomorphism mapping $\mathbb{Z}^N$ isomorphically onto this subgroup; it is called the *natural action of the group* $\mathbb{Z}^N$ *on* $\mathfrak{A}$ *(or* $\overline{\mathfrak{A}}$*)*.

Definition (III.2.18). The normed *-algebra $\mathfrak{A}$ (Banach *-algebra $\overline{\mathfrak{A}}$, respectively) furnished with the natural action of the group $\mathbb{Z}^N$ is called the *algebra of local (quasi-local, respectively) observable spinor systems.*

We note that the pair $(\overline{\mathfrak{A}}, \overline{\alpha})$ is one of the simplest examples of an important class of structures that are very interesting to study — the so-called C^*-dynamical systems, which the reader can learn about in, for example, [58]. See, for example [68], [97] to find out about the role and applications of $\mathfrak{A}$ and $\overline{\mathfrak{A}}$ in quantum statistical mechanics.

But now, to explain the comment made before we began constructing the algebras of observables, we show that by "forgetting about the action of $\overline{\alpha}$" we can identify $\overline{\mathfrak{A}}$ with the Fermion algebra we already know.

Proposition (III.2.19). *There exists an isometric *-isomorphism between Φ and $\overline{\mathfrak{A}}$.*

▶ We take an arbitrary bijection $\sigma : \mathbb{N} \xrightarrow{\simeq} \mathbb{Z}^N$ and introduce the following notation: $\lambda(n) := (\sigma(1), \cdots, \sigma(n)) \in \Lambda$, $\{e_1^n, e_2^n\}$ is an orthonormal basis in $H_{\sigma(n)}$, $q_{lm}^n \in \mathcal{B}(H_{\sigma(n)})$ is the one-dimensional operator $x \longmapsto \langle x, e_m^n \rangle e_l^n$, $l, m = 1, 2$, and finally a_M is the operator in Φ_n with M down the diagonal of a block matrix, where M is a $2^n \times 2^n$-matrix.

We use induction to construct isometric *-isomorphisms $j_n^0 : \Phi_n \longrightarrow \mathfrak{A}_{\lambda(n)}$; $n \in \mathbb{N}$. For $n = 1$ we associate to the operator a_M, where M is the 2×2-matrix $[z_{lm}]$, the operator $\sum_{l,m=1}^{2} z_{lm} q_{lm}^1 \in \mathcal{B}(H_{\sigma(1)}) \simeq \mathfrak{A}_{\lambda(1)}$. Suppose that j_k^0 has been constructed for $k = 1, \cdots, n$. We represent each $2^{n+1} \times 2^{n+1}$ matrix M as a block matrix $[M_{lm}]$, where M_{lm}; $l, m = 1,2$, is a $2^n \times 2^n$ matrix and using the representation $\mathfrak{A}_{\lambda(n+1)} = \mathfrak{A}_{\lambda(n)} \otimes \mathcal{B}(H_{\sigma(n+1)})$, we get $j_{n+1}^0 : \Phi_{n+1} \longrightarrow \mathfrak{A}_{\lambda(n+1)}$ $: a_M \longmapsto \sum_{l,m=1}^{2} j_n^0(a_{M_{lm}}) \otimes q_{lm}^{n+1}$.

We now introduce $j_n = i^{\lambda(n)} \circ j_n^0 : \Phi_n \longrightarrow \mathfrak{A}$; then, obviously, all the j_n are isometric *-homomorphisms such that $j_{n+1}|_{\Phi_n} = j_n$. This means that by using the j_n we can define in an obvious way an isometric *-isomorphism $j : \bigcup_n \Phi_n \longrightarrow \mathfrak{A}$ between *-algebras which are dense in Φ and $\overline{\mathfrak{A}}$ respectively. The rest is clear. ■

Exercise (III.2.20). Show that the group of *-automorphisms of the Fermion algebra contains an isomorphic image of any countable group G.

Hint. Construct the analogue $\overline{\mathfrak{A}}(G)$ of the algebra $\overline{\mathfrak{A}}$ using G instead of $\mathbb{Z}^N$ as the original group; show that $\overline{\mathfrak{A}}(G)$ is furnished with a unique action of the group G and at the same time (cf. Proposition 2.19) it is isometrically *-isomorphic to Φ.

* * *

As we have already said, among the operator algebras on a Hilbert space it is the self-adjoint algebras that are studied most; all the examples that we have met up to now were concerned with this class. Recently, however, there has been a growing interest in algebras which are not self-adjoint and, furthermore, have, in a well defined sense, the opposite structure. Among such algebras the simplest are the so-called nest algebras.

A system of closed subspaces of H is said to be *nested* if it is linearly ordered by inclusion. A (necessarily closed) subalgebra of $\mathcal{B}(H)$ is called a *nest algebra* if it consists of all those operators which leave invariant every subspace in a given nested system. Typical examples of nested systems are the family $H_n = \{\xi : \xi_m = 0$ when $m \geq n\}$; $n \in \mathbb{Z}$, in $l_2(\mathbb{Z})$ and its continuous analogue $H_t = \{f : f(s) = 0$ when $s \geq t\}$; $t \in \mathbb{R}$, in $L^2(\mathbb{R})$. The corresponding nest algebras are denoted by $N(\mathbb{Z}) \subset \mathcal{B}(l_2(\mathbb{Z}))$ and $N(\mathbb{R}) \subset \mathcal{B}(L^2(\mathbb{R}))$.

Unlike the self-adjoint subalgebras of $\mathcal{B}(H)$ which are always semisimple (see later Chap. IV §5.3), the nest algebras have a non-trivial radical, whose general characterization was given in a pioneering paper of Ringrose [220]. In particular, $\operatorname{Rad} N(\mathbb{Z}) = \{a \in N(\mathbb{Z}) : \lim_{n\to-\infty} p_n a p_n = 0, \quad \lim_{n\to\infty} (1-p_n)a(1-p_n) = 0$ and $(p_{n+1}-p_n)a(p_{n+1}-p_n) = 0$ for all $n \in \mathbb{Z}\}$, where p_n is the orthogonal projection onto H_n.

Many results and unsolved questions concerning various kinds of non-self-adjoint algebras — nest, triangular, reflexive and so on — are given in the survey by Erdos [300], where there is an extensive list of references and also in a concise "propaganda" survey by Lance [179]. Let us note the papers [177], [202], [116], [120] which are devoted to the homological properties of non-self-adjoint algebras.

§2.3. Some operator topologies and the corresponding polynormed algebras. Von Neumann algebras and the von Neumann bicommutant theorem. Remark on Lassner algebras

Operators provide a whole series of polynormed algebras, besides Banach algebras. We indicate four of the most important types, limiting ourselves for simplicity to operators acting on an infinite-dimensional Hilbert space H.

We fix a (for the moment arbitrary) subalgebra A of $\mathcal{B}(H)$ and recall the operator $\mathfrak{m} : \mathcal{B}(H) \longrightarrow \mathcal{B}(\mathcal{H})$ (see Definition 2.11), where $\mathcal{H}$ is a countable multiple of H. Every system of seminorms $\beta = \{\|\cdot\|_\nu;\ \nu \in \Lambda\}$ in $\mathfrak{m}(A)$ determines a system of seminorms $\mathfrak{m}\beta := \{\|\cdot\|'_\nu;\ \nu \in \Lambda : \|a\|'_\nu := \|\mathfrak{m}(a)\|_\nu\}$ in A, which we shall call a *multiple of the system* β.

Definition (III.2.21). The system of seminorms $\underline{s} := \{\|\cdot\|_x;\ x \in H\}$ in A, where $\|a\|_x := \|a(x)\|$, is called the *strong system of seminorms* (in A). Its multiple $\mathfrak{m}\underline{s}$ is called the *ultrastrong system of seminorms*. The system $\underline{w} := \{\|\cdot\|_{x,y};\ x,\ y \in H\}$ in A, where $\|a\|_{x,y} := |\langle a(x), y\rangle|$ is called the *weak system of seminorms* (in A). Its multiple $\mathfrak{m}\underline{w}$ is called the *ultraweak*[15] *system of seminorms*. The topologies that these give in A (denoted by the same symbols) are called the *strong, ultrastrong,*[16] *weak and ultraweak operator topologies* respectively.

(Obviously, the topology $\underline{s}$ in $\mathcal{B}(H)$ is a particular case of the strong operator topology in $\mathcal{B}(E)$ that was defined for general E in Chapter 0 §3.1.)

If we write out the operators as matrices in terms of various orthonormal bases, we see that the norms of the columns, treated as elements of l_2, are seminorms in $\underline{s}$, while the moduli of the elements of the matrix belong to $\underline{w}$.

Proposition (III.2.22). *The seminorms in* $\mathfrak{m}\underline{s}$ *may be indexed by the sequences* $\xi = \{\xi_n \in H : n \in \mathbb{N}\} \in \mathcal{H}$ *so that* $\|a\|_\xi := \left(\sum_{n=1}^{\infty} \|a\xi_n\|^2\right)^{1/2}$. *The seminorms in* $\mathfrak{m}\underline{w}$ *may be indexed by pairs of sequences* $\xi = \{\xi_n\}$, $\eta = \{\eta_n\} \in \mathcal{H}$ *so that* $\|a\|_{\xi,\eta} := |\sum_{n=1}^{\infty} \langle a\xi_n, \eta_n\rangle|$. ▶■

Remark (III.2.23). The reader can hardly have doubted that the strong and weak operator topologies are natural but he may be completely puzzled as to why it is necessary to take "multiples" of them. In fact, the ultraweak topology is the most important of the operator topologies (if we discount the uniform topology) and the basic source of its usefulness is that it coincides with

[15] The term "ultraweak" is unsuitable in that this topology is not weaker, on the contrary, it is stronger, than the "merely" weak. However it is so widely used that it is apparently too late to change it now.

[16] The ultrastrong topology is also sometimes known as the "strongest" operator topology.

the weak* topology of the dual space (Chapter 0 §1.2) if we identify $\mathcal{B}(H)$ with $\mathcal{N}(H)^*$ (cf. Exercise 0.3.45); for details see, for example, [75, II.1]. With regard to the ultrastrong topology, we need it later in the proof of the bicommutant theorem.

└

In the next proposition the symbol $\underline{u}$ denotes the uniform topology and the notation $\alpha < \beta$ means that the first topology is strictly stronger than the second.

Proposition (III.2.24). *Let A contain all the one-dimensional operators. Then we have the relations*

$$\underline{u} \prec \mathfrak{m}\underline{s} \prec \underline{s} \prec \underline{w}$$

$$\underline{u} \prec \mathfrak{m}\underline{s} \prec \mathfrak{m}\underline{w} \prec \underline{w}$$

where the topologies $\underline{s}$ and $\mathfrak{m}\underline{w}$ are not comparable.

▶ The relations are obviously true if the sign $\prec$ is understood to mean "not weaker". Further, let us fix on an orthonormal system $e_n \in H$; $n \in \mathbb{N}$, and put $q_{kl} : H \longrightarrow H : x \longmapsto \langle x, e_l \rangle e_k$. Then for $n \longrightarrow \infty$ the sequence q_{nn} converges to zero in $\mathfrak{m}\underline{s}$, but not in $\underline{u}$, and q_{n1} converges to zero in $\mathfrak{m}\underline{w}$ and in $\underline{w}$ but not in $\underline{s}$. So it is easy to see that all we have to do is to present a seminorm in $\mathfrak{m}\underline{w}$ which does not majorize any seminorm from the saturation of $\underline{s}$.[17]

We put, in the notation of Proposition 2.22, $\xi = \{\frac{1}{n} e_n;\ n \in \mathbb{N}\}$ and consider $\|\cdot\|_{\xi,\xi}$. Let $\|\cdot\|_{x_k}$; $k = 1, \cdots, m$, be a collection of seminorms in $\underline{s}$. Obviously there exists $y \in (\ell.h\{x_1, \cdots, x_m\})^{\perp}$ with $\langle y, e_n \rangle \neq 0$ for a certain $n \in \mathbb{N}$. Then for $a : H \longrightarrow H : z \longmapsto \langle z, y \rangle e_n$, $\max\{\|a\|_{x_1}, \cdots, \|a\|_{x_m}\} = 0$, while $\|a\|_{\xi,\xi} = |\sum_{l=1}^{\infty} \frac{1}{l^2} \langle ae_l, e_l \rangle| = \frac{1}{n^2} |\langle e_n, y \rangle| \neq 0$. The rest is clear. ■

Proposition (III.2.25). *Any subalgebra A of $\mathcal{B}(H)$ is a Hausdorff polynormed algebra with respect to each of the seminorm systems $\underline{s}$, $\underline{w}$, $\mathfrak{m}\underline{s}$ and $\mathfrak{m}\underline{w}$.*

▶ Our problem is to check that for fixed $a \in A$ the operators m_a and m_a^{op} (see I.2.2) are continuous in each of the four polynormed spaces. Using Proposition 0.1.9, we see that the assertion for $\underline{s}$ follows from $\|m_a(b)\|_x \le \|a\| \|b\|_x$ (here $\|\cdot\|$ is the operator norm) and $\|m_a^{op}(b)\|_x = \|b\|_{a(x)}$, and the assertion for $\underline{w}$ follows from $\|m_a(b)\|_{x,y} = \|b\|_{x,a^*(y)}$ and

---

[17] We cannot argue here, as we did before, in the language of sequences: this becomes clear from Exercise 2.27.

$\| m_a^{op}(b) \|_{x,y} = \| b \|_{a(x),y}$. For $\mathfrak{m}\underline{s}$ and $\mathfrak{m}\underline{w}$ the assertion is implied by similar inequalities and equations relating the "multiple" operators. ■

However for algebras that are fairly rich in operators, none of the four polynormed algebras that occur is jointly polynormed.

Exercise (III.2.26). Prove this fact for $A = \mathcal{B}(H)$.

Hint. For any neighbourhood of zero U in $(A; \underline{s})$ construct $a \in U$ with $a^2 = 1$.

We give a few further facts, restricting ourselves to the case $A = \mathcal{B}(H)$. First of all, none of the four polynormed algebras we have constructed is complete: for each of them one can easily construct a Cauchy net whose hypothetical limit would automatically be a discontinuous operator. Further, these algebras are not metrizable: none of the corresponding seminorm systems has a countable subsystem giving the same topology (cf. Chapter 0 §1.1).

All the same our algebras are not that bad: in practice we have to work with their bounded subsets (here and later — bounded with respect to the operator norm) and they have a much better structure. For example, the $\underline{u}$-closed bounded subsets of our four algebras are always complete. Further, such sets are always compact in the $\underline{w}$ and $\mathfrak{m}\underline{w}$-topologies, from which it is not difficult to deduce that on any bounded set these two topologies coincide (however, this fact, and also the fact that the $\underline{s}$ and $\mathfrak{m}\underline{s}$-topologies coincide on bounded sets, can be verified immediately).

Exercise (III.2.27). Prove that any sequence of operators which converges to zero in the $\underline{s}$-topology converges to zero in the $\mathfrak{m}\underline{w}$-topology also.

Hint. The Banach-Steinhaus theorem shows that such a sequence is bounded.

Finally, for separable H, in all the four cases, every bounded set is, as a topological space, metrizable.

Exercise (III.2.28). Let $\underline{S}$ be the unit ball in $(\mathcal{B}(H); \| \cdot \|_0)$, and let $(x_n, n \in \mathbb{N})$, be a dense subset in $\underline{S}$. Prove that the strong topology in $\underline{S}$ is given by the metric $\rho(a, b) := \sum_{n=1}^{\infty} 2^{-n} \| a(x_n) - b(x_n) \|$, and the weak topology is given by the metric $\rho(a, b) := \sum_{m,n=1}^{\infty} 2^{-m-n} | \langle a(x_m), x_n \rangle - \langle b(x_m), x_n \rangle |$.

* * *

For any subset M in $\mathcal{B}(H)$ we let $\overline{M}_{\underline{w}}$, $\overline{M}_{\underline{s}}$ and so on denote its closure with respect to the corresponding topology; then Proposition 2.24 implies the inclusions

$$\overline{M}_{\mathfrak{m}\underline{s}} \subseteq \overline{M}_{\underline{s}} \subseteq \overline{M}_{\underline{w}}$$
$$\overline{M}_{\mathfrak{m}\underline{s}} \subseteq \overline{M}_{\mathfrak{m}\underline{w}} \subseteq \overline{M}_{\underline{w}} \tag{1}$$

each of which, when considering arbitrary M, may turn out to be proper. A fundamental fact in the theory of operator algebras, discovered by von Neumann, is the following: if M is a self-adjoint algebra, then all four closures coincide and may be characterized in purely algebraic terms.

Definition (III.2.29). Let E be a Banach space. The set $M^! := \{a \in \mathcal{B}(E) : ab = ba$ for all $b \in M\}$ is called the *commutant* of the subset M of $\mathcal{B}(E)$ and the set $M^{!!} := (M^!)^!$ is called the *bicommutant*.

Proposition (III.2.30). (I) $M^!$ *is a unital subalgebra in* $\mathcal{B}(H)$ *which is closed in the* $\underline{w}$- *(and as a consequence),* $\mathfrak{m}\underline{w}$-, $\underline{s}$-, $\mathfrak{m}\underline{s}$- *and* $\underline{u}$-*topologies, for any* $M \subseteq \mathcal{B}(H)$.

(II) *If, whenever* M *contains* a *it also contains* a^*, *then* $M^!$ *is a self-adjoint algebra.*

(III) $M \subseteq M^{!!}$, $(M^!)^{!!} = M^!$ *for any* M, *and (as a consequence)* $(M^{!!})^{!!} = M^{!!}$.

▶ If the net $(a_\nu \in \mathcal{B}(H);\ \nu \in \Lambda)$, converges to a in the topology $\underline{w}$, then by Proposition 2.25, for any $b \in \mathcal{B}(H)$ in the same topology, $\lim_\nu (a_\nu b - ba_\nu) = ab - ba$. Therefore, if all $a_\nu \in M^!$ and $b \in M$, then $ab - ba = 0$, in other words, $a \in M^!$. This proves (I); the rest is clear. ■

Corollary (III.2.31). *If* A *is a self-adjoint algebra, then* $A^!$ *is a realized* C^*-*algebra, which is* $\underline{w}$-*closed in* $\mathcal{B}(H)$.

Recalling the space $\mathcal{H}$ (a "multiple of H"), for each $n \in \mathbb{N}$ we consider the operators $i_n : H \longrightarrow \mathcal{H} : x \longmapsto (0, \cdots, 0, x, 0, \cdots)$, where x is at the n-th place and $j_n : \mathcal{H} \longrightarrow H : \xi = \{\xi_k\} \longmapsto \xi_n$. For $\underline{b} \in \mathcal{B}(\mathcal{H})$ we put $b_{mn} := j_m \circ \underline{b} \circ i_n$ (intuitively speaking, this is an element of the operator block matrix depicting $\underline{b}$). It is easy to verify the following.

Proposition (III.2.32). *If* $a \in \mathcal{B}(H)$ *and* $\underline{b} \in \mathcal{B}(\mathcal{H})$, *then* $\mathfrak{m}(a)$ *commutes with* $\underline{b}$ $\Leftrightarrow$ a *commutes with all* b_{mn}; $m, n \in \mathbb{N}$. *As a consequence,* $\mathfrak{m}(M^{!!}) \subseteq (\mathfrak{m}(M))^{!!}$ *for any* $M \subseteq \mathcal{B}(H)$. ▶■

Exercise (III.2.33). Prove that, in fact, $\mathfrak{m}(M^{!!}) = (\mathfrak{m}(M))^{!!}$.

Theorem (III.2.34) ("von Neumann's bicommutant theorem"). *Let A be a self-adjoint algebra containing* **1**. *Then each of the sets $\overline{A}_{\underline{w}}$, $\overline{A}_{\mathfrak{m}\underline{w}}$, $\overline{A}_{\underline{s}}$ and $\overline{A}_{\mathfrak{m}\underline{s}}$ coincides with $A^{!!}$. As a consequence, the property that an algebra of the class indicated be closed with respect to any of the topologies under consideration is equivalent to its coinciding with its bicommutant.*

▶ In view of $\overline{A}_w \subseteq A^{!!}$ (Proposition 2.30) and the relations (1), our problem is to show that $A^{!!} \subseteq \overline{A}_{\mathfrak{m}\underline{s}}$. According to Proposition 2.22, for any two seminorms $\|\cdot\|_\xi$; $\xi = \{\xi_n\}$ and $\|\cdot\|_\eta$; $\eta = \{\eta_n\}$ in the system $\mathfrak{m}\underline{s}$,, there is a third, for example, $\|\cdot\|_\zeta$ with $\zeta = (\xi_1, \eta_1, \xi_2, \eta_2, \cdots)$, which majorizes each of them. Therefore it is sufficient for us to fix $a_0 \in A^{!!}$, $\underline{x} = \{x_n\} \in \mathcal{H}$ and, letting $\mathcal{H}_0$ denote the closure in $\mathcal{H}$ of elements of the form $\mathfrak{m}(b)\underline{x}$; $b \in A$, to establish that $\mathfrak{m}(a_0)\underline{x} \in \mathcal{H}_0$.

We let $\underline{p} \in \mathcal{B}(\mathcal{H})$ denote the orthogonal projection onto $\mathcal{H}_0$. Obviously, $\underline{p}\mathfrak{m}(c)\underline{p} = \mathfrak{m}(c)\underline{p}$ for any $c \in A$, and, as A is self-adjoint (this is where it is used!), $\underline{p}\mathfrak{m}(c^*)\underline{p} = \mathfrak{m}(c^*)\underline{p}$; hence $\underline{p}\mathfrak{m}(c) - \mathfrak{m}(c)\underline{p} = \underline{p}\mathfrak{m}(c) - \underline{p}\mathfrak{m}(c)\underline{p} = (\mathfrak{m}(c^*)\underline{p} - \underline{p}\mathfrak{m}(c^*)\underline{p})^* = 0$. Thus $\underline{p} \in (\mathfrak{m}(A))^!$ and, according to Proposition 2.32, the operators $\mathfrak{m}(a_0)$ and $\underline{p}$ commute. All that remains is to note that, by the condition $1_H \in A$, $\underline{x} = \mathfrak{m}(1)\underline{x} \in \mathcal{H}_0$ and therefore $\mathfrak{m}(a_0)\underline{x} = \mathfrak{m}(a_0)\underline{p}(\underline{x}) = \underline{p}(\mathfrak{m}(a_0)\underline{x}) \in \mathcal{H}_0$. ■

Let us look into the question of whether the two conditions of this theorem, that A be self-adjoint and $1_H \in A$, are of equal weight. The first is absolutely essential; with regard to the second, the exercises suggested below show that it can be weakened, and, by making the formulation of the theorem somewhat more complicated, it may be dropped entirely.

Exercise (III.2.35). Show that the assertion of the theorem is true for any self-adjoint subalgebra A of $\mathcal{B}(H)$ such that $A \cdot H := \overline{\ell.h}\{a(x) : a \in A, x \in H\} = H$.

Hint. If, in the notation of the preceding proof, $\underline{q} := 1_{\mathcal{H}} - \underline{p}$, then $\mathfrak{m}(a)\underline{q}(\underline{x}) = 0$ for all $a \in A$. Hence, using $(A \cdot H)^\perp = 0$, it is easy to show that $\underline{q}(\underline{x}) = 0$, that is $\underline{x} \in \mathcal{H}_0$.

From this it is not difficult to establish a general form of the bicommutant theorem.

Exercise (III.2.36). Let A be an arbitrary self-adjoint subalgebra in $\mathcal{B}(H)$ and let $p \in \mathcal{B}(H)$ be the orthogonal projection onto $\overline{A \cdot H}$. Prove that $\overline{A}_{\underline{w}} = \overline{A}_{\mathfrak{m}\underline{w}} = \overline{A}_{\underline{s}} = \overline{A}_{\mathfrak{m}\underline{s}} = \{a \in A^{!!} : ap = pa = a\}$ and further $A^{!!} = \ell.h\{\overline{A}_{\underline{w}}; 1\}$.

Thus we see from 2.34 - 2.36 that the various approaches yield the same class of operator algebras. On these grounds we adopt the following definition.

Definition (III.2.37). A self-adjoint subalgebra in $\mathcal{B}(H)$ which is closed with respect to the weak operator topology, is called a *von Neumann algebra.*

Of course, von Neumann algebras furnished with the operator norm are realized C^*-algebras; in addition, it is appropriate to consider them simultaneously also as polynormed algebras (especially with the ultraweak seminorm system). They can be characterized as Banach *-algebras in purely intrinsic terms (without mentioning $\mathcal{B}(H)$); see Assertion IV.7.60 about this.

In conclusion we formulate another very useful assertion about closures which coincide in different topologies (watch out, this does not follow from von Neumann's theorem).

Assertion (III.2.38) ("Kaplansky density theorem"). *Let A be a self-adjoint subalgebra of $\mathcal{B}(H)$; let $\underline{S}$ be the unit ball in A and let $\underline{S}_w$ be the unit ball in $\overline{A}_{\underline{w}}$. Then $\underline{S}$ is dense in $\underline{S}_w$ with respect to any of the topologies $\underline{s}$, $m\underline{s}$, $\underline{w}$ and $m\underline{w}$.*

* * *

We would like to mention another important class of polynormed algebras, this time provided by the theory of unbounded operators on a Hilbert space. When speaking of them below, we shall assume that the reader is acquainted with the foundations of this theory (see, for example [63, Chap.VIII]).

Let H be a Hilbert space and let $\tilde{H}$ be a fixed dense subspace of H. We let $\tilde{\mathcal{B}}(\tilde{H})$ denote the set of those "operators" a in H with domain of definition $\tilde{H}$,[18] for which $a(\tilde{H}) \subseteq \tilde{H}$ and, in addition, having a dual a^*, also determined on $\tilde{H}$ and such that $a^*(\tilde{H}) \subseteq \tilde{H}$; obviously this is an algebra if we take composition as the multiplication. Let $A(\tilde{H})$ be a certain unital subalgebra of $\tilde{\mathcal{B}}(\tilde{H})$, which if it contains an operator also contains its dual. We introduce in it the system of seminorms $\|\cdot\|_M$ defined by $\|a\|_M = \sup\{|\langle x, a(y)\rangle| : x, y \in M\}$, where M runs over the collection of those subsets of $\tilde{H}$ for which, for any $a \in A(\tilde{H})$, the set $\{a(x) \in \tilde{H} : x \in M\}$ is bounded with respect to the norm in $\tilde{H}$. It is not difficult to verify that $A(\tilde{H})$ is a polynormed algebra having, furthermore, an additional structure — the operation $a \longmapsto a^*$.

---

[18] Recall that in the theory of unbounded operators, an operator means a map, defined, generally speaking, not on the whole of H but only on a certain, as a rule, dense, subspace.

Polynormed algebras of the form $A(\tilde{H})$ were intensively studied in the works of the Leipzig school, and we shall call them *Lassner algebras.*[19] The topology in these algebras is called the *uniform topology*; it is, as was shown in [184], a natural generalization of the uniform topology in realized C^*-algebras (in particular $\tilde{\mathcal{B}}(\tilde{H})$ generalizes $(\mathcal{B}(H); \|\cdot\|_0)$). However, Lassner algebras have a lot of other topologies which are of interest in themselves [243].

Concerning the maximal Lassner algebra $\tilde{\mathcal{B}}(\tilde{H})$, it has, under certain natural conditions, a unique closed bi-ideal $\tilde{\mathcal{K}}(\tilde{H})$ which plays the same role for $\tilde{\mathcal{B}}(\tilde{H})$ as $\mathcal{K}(H)$ does for $\mathcal{B}(H)$. Moreover, as Löffler and Timmermann [190] showed, the quotient algebra $\tilde{\mathcal{B}}(\tilde{H})/\tilde{\mathcal{K}}(\tilde{H})$ is topologically isomorphic to a certain Lassner algebra; it is therefore "realized by unbounded operators" in the same way as the Calkin algebra is "realized by bounded operators" (cf. what was said at the beginning of §2.2).

Timmermann [241] studied a lot of other bi-ideals of $\tilde{\mathcal{B}}(\tilde{H})$ including "unbounded" analogues of the bi-ideals $\mathfrak{S}_p(H)$ of $\mathcal{B}(H)$ (see 2.8). He undertook a systematic investigation of the states (this kind of functional will be defined in IV.5.1) on these algebras (see [242] and the references contained in it).

The question of being able to give an abstract description of the Lassner algebras (see the discussion in [184]) was solved positively by Schmüdgen [293] in terms of an order structure; this was preceded by a partial result due to Lassner which will be formulated later (see Chap.V Assertion 3.47).

For applications of the algebras under discussion to quantum mechanics see [186], [85].

§3. GROUP ALGEBRAS

In abstract algebra we know, in practice, only one reasonable method of constructing a complex associative algebra starting from a given group G: we are referring to the algebra $\mathbb{C}(G)$, which is already known to the reader from Chap.I §1.1.[20] When we are dealing with analysis, there is a whole range of distinct Banach algebras connected with one and the same group, which is usually a

---

[19] They were first considered by Lassner [184] in 1969 who christened these algebras, and all algebras that are topologically *-isomorphic to them, $\tilde{AO}^*$-algebras; later, Powers arrived at this same class of algebras independently [206].

[20] Using the same principle, one can construct "abstract" group algebras over other base fields and, more generally, over commutative rings; see, for example, [78, Vol.I, p.291].

topological group and locally compact as well, and if it is a Lie group, there are also polynormed algebras. It is difficult to say which of these algebras is "best": each has its own advantages. All the same, one class of group algebras, the so-called L^1-algebras, are met perhaps more often than others. We shall begin with examples of these algebras.

§3.1. L^1-algebras of discrete groups, the line, and the circle. Their approximate identities, ideals and characters

Let G be a group, considered first of all without a topology, or if appropriate, with the discrete topology. We let $l_1(G)$ denote the set of those functions $f: G \longrightarrow \mathbb{C}$ that have a finite sum $\|f\|_1 := \sum_{s \in G} |f(s)|$ (for which, of course, it is necessary that the set $\{s \in G : f(s) \neq 0\}$ be at most countable). Obviously $(l_1(G); \|\cdot\|_1)$, is a Banach space with respect to pointwise operations. Furthermore, we introduce the function $f * g$, defined by $f * g(s) := \sum_{t \in G} f(t)g(t^{-1}s)$ which is called the *convolution* of the functions $f, g \in l_1(G)$. A routine check, using the rearrangement theorem for absolutely convergent series, shows that $l_1(G)$, with convolution as multiplication, is a Banach algebra with multiplicative norm. It is called the *l_1-algebra of the group* G (recall that we are already aquainted with one of these algebras — $l_1(\mathbb{Z})$; see Example I.2.12).

For $s \in G$ we let δ_s denote the function on G taking s to one, and the other points to zero. If we identify s with δ_s, we are justified in assuming that G is a subset of $l_1(G)$. We note that $\delta_{st} = \delta_s * \delta_t$ for all $s, t \in G$; this observation implies a result which should convince the reader that the convolution multiplication is natural.

Proposition (III.3.1). (I) *Convolution is the unique continuous multiplication in* $l_1(G)$ *that coincides with the original group multiplication on* G.

(II) *The Banach algebra* $l_1(G)$ *is (up to isometric isomorphism) the completion of* $\mathbb{C}(G)$ *with respect to the norm* $\|\sum_{k=1}^{n} \lambda_k s_k\| := \sum_{k=1}^{n} |\lambda_k|$.

▶ If we associate to each $\sum_{k=1}^{n} \lambda_k s_k \in \mathbb{C}(G)$ the function on G that is equal to λ_k at s_k; $1 \leq k \leq n$, and to zero at the remaining points, we get an isometric isomorphism from $\mathbb{C}(G)$ onto a dense subalgebra in $l_1(G)$. The rest is clear. ■

We now pass to examples of group algebras for continuous (= non-discrete) groups; for simplicity we limit ourselves to the two most easily visualized groups: $\mathbb{R}$ with the operation of addition ("the line") and $\mathbb{T}$ with the operation of multiplication ("the circle"). We let m denote the usual Lebesgue measure on the line and the Lebesgue measure normalized by the factor $(2\pi)^{-1}$ on the circle (so that $m(\mathbb{T}) = 1$), and we let $L^1(\mathbb{R})$ and $L^1(\mathbb{T})$, as usual, denote the corresponding spaces (of equivalence classes) of integrable functions. We consider, in $L^1(\mathbb{R})$, the classic convolution $f * g(s) = \int_{\mathbb{R}} f(t)g(s-t)dt$, and in $L^1(\mathbb{T})$ we consider the analogous operation $f * g(s) = \int_{\mathbb{T}} f(t)g(t^{-1}s)dt$, (we write dt instead of $dm(t)$). Just as for $l_1(G)$ (although the role of the rearrangement theorem for series is now played by Fubini's theorem) it is easy to ascertain that both spaces are Banach algebras and moreover both have a multiplicative norm with respect to the convolution multiplication.

We first note that elementary calculations using Fubini's theorem show that $L^1(\mathbb{R})$ and $L^1(\mathbb{T})$ are commutative. With regard to the algebra $l_1(G)$; obviously, it is commutative $\Leftrightarrow$ G is commutative.

Just like $l_1(\mathbb{Z})$ (see Example I.2.14), $L^1(\mathbb{R})$ and $L^1(\mathbb{T})$ can be considered as "disguised function algebras". In the case of the line we use the classical Fourier transformation $\Phi : L^1(\mathbb{R}) \longrightarrow C_0(\mathbb{R})$ and put $\Phi(\mathbb{R}) := \operatorname{Im}\Phi$. The uniqueness theorem in the theory of Fourier transformations enables the norm $\|\Phi(f)\| := \|f\|_1$ to be well defined in $\Phi(\mathbb{R})$: then the familiar formula $\Phi(f * g) = \Phi(f)\Phi(g)$ implies that $(\Phi(\mathbb{R}); \|\cdot\|)$, is a Banach function algebra on $\mathbb{R}$ which is isometrically isomorphic to $L^1(\mathbb{R})$. Similarly, if we associate to each $f \in L^1(\mathbb{T})$ the (two-sided) sequence of its Fourier coefficients and use the uniqueness theorem of the theory of Fourier series, we get an isometric isomorphism from $L^1(\mathbb{T})$ onto a certain Banach function algebra on $\mathbb{Z}$, denoted by $\Phi(\mathbb{Z})$. The algebras $\Phi(\mathbb{R})$, $\Phi(\mathbb{Z})$ and with them the Wiener algebra, denoted in this context by $\Phi(\mathbb{T})$, are called the *Fourier algebras* of the groups $\mathbb{R}$, $\mathbb{Z}$ and $\mathbb{T}$ respectively.

In future χ_M will everywhere denote the characteristic function of the subset M of the group under consideration.

Proposition (III.3.2). (I) *For any G, $l_1(G)$ is a unital algebra with identity $\delta_{\underline{e}}$, where $\underline{e}$ is the identity of the group G.*

(II) *There is no identity in $L^1(\mathbb{R})$ and $L^1(\mathbb{T})$, however there is a metric approximate identity. One can take this to be $e_U := \|\chi_U\|_1^{-1}\chi_U$, where U runs over a basis of neighbourhoods of zero in $\mathbb{R}$, or, according to the sense, neighbourhoods of the identity in $\mathbb{T}$, directed by inverse inclusion ($U \ll V$ means $U \supseteq V$).*

▶ (I) is obvious. We deal with (II), limiting ourselves to the case $\mathbb{R}$ and leaving the reader to construct analogous arguments for $\mathbb{T}$.

We first prove the assertion concerning the a.i. Since $\|e_U\|_1 = 1$ for all U, then for this it is obviously sufficient to check that $\lim_U f * e_U = f$ for all f in a certain set N such that $\overline{\ell.h.\{N\}} = L^1(\mathbb{R})$. As a suitable set N it is convenient to consider the characteristic functions of intervals.

We take arbitrary $\alpha, \beta \in \mathbb{R}$; $\alpha < \beta$, and put $X := X_{[\alpha,\beta]}$, then

$$X * e_U(s) = \int_\alpha^\beta e_U(s-t)dt = (m(U))^{-1}m(\{s-t : t \in U\} \cap [\alpha, \beta])$$

for any U and $s \in \mathbb{R}$. Hence, if we put $d_U := \sup\{|t| : t \in U\}$ and note that, as the system U is a neighbourhood basis, we are justified in restricting attention to U with $d_U \le \frac{1}{2}(\beta - \alpha)$, we see that our convolution takes values in $[0, 1]$, and moreover it is automatically equal to zero for $s \le \alpha - d_U$ and $s \ge \beta + d_U$, and to one for $\alpha + d_U < s < \beta - d_U$. Hence

$$\|X * e_U - X\|_1 = \int_{\mathbb{R}} |X * e_U(s) - X(s)|\, ds \le$$

$$\le m[\alpha - d_U, \alpha + d_U] + m[\beta - d_U, \beta + d_U] = 4d_U,$$

and, since it is obvious that $\lim_U d_U = 0$, then $\lim_U X * e_U = X$.

All that remains is to show that there is no "real" identity in $L^1(\mathbb{R})$. It is easy to see that for any U in our system there automatically exists $V \gg U$ such that $\|e_U - e_V\|_1 \ge 1$. This means that the net e_U does not converge in $L^1(\mathbb{R})$ and the required fact is implied by Proposition I.2.36. ■

Exercise (III.3.3). Exhibit a sequence in $L^1(\mathbb{R})$ and $L^1(\mathbb{T})$ consisting of continuous functions and forming a metric a.i.

For any group G, any function f on G and any $r \in G$, the function ${}^r f : s \longmapsto f(rs)$ (or $f^r : s \longmapsto f(sr)$) will be called the *left* (*right*, respectively) *translate* of f by r. Obvious calculations show that, for all the convolutions defined above, we have ${}^r f * g = {}^r(f * g)$ and $f * g^r = (f * g)^r$. This, taking Proposition 3.2 into account, easily implies the following.

Proposition (III.3.4). (I) *For any* G, $f \in l_1(G)$ *and* $r \in G$, *we have* ${}^r f = \delta_{r^{-1}} * f$ *and* $f^r = f * \delta_{r^{-1}}$.

(II) *For* $G = \mathbb{R}$ *or* $\mathbb{T}$ *and any* $f \in L^1(G)$, $r \in G$, *we have* ${}^r f = \lim_U h_U * f$, *where* U *runs over an arbitrary neighbourhood basis for the inverse element of* r (i.e. $-r \in \mathbb{R}$ *or* $r^{-1} \in \mathbb{T}$), *directed by inverse inclusion and where* $h_U := \|X_U\|_1^{-1} X_U$. ▶■

We are now in a position to establish an important characteristic property of the closed ideals of our algebras.

Theorem (III.3.5). *A closed subspace in* $l_1(G)$ *(for any* G*),* $L^1(\mathbb{R})$ *or* $L^1(\mathbb{T})$ *is a left (right, respectively) ideal* $\Leftrightarrow$ *whenever it contains a function it also contains all its left (right) translates.*

▶ "$\Rightarrow$" is obviously implied by Proposition 3.4. In proving "$\Leftarrow$" we again limit ourselves to the case $L^1(\mathbb{R})$, leaving the reader to carry out analogous arguments for $L^1(\mathbb{T})$ and the considerably simpler argument for $l_1(G)$.

So, our problem is to show that, for any "translation-invariant" I, $g * f \in I$ for any $f \in I$ and $g \in L^1(\mathbb{R})$. According to Corollary B of the Hahn–Banach theorem (Chap.0 §1.2), it is sufficient to check that $\langle \varphi, g * f \rangle = 0$ for any $\varphi \in L^1(\mathbb{R})^*$ with $\varphi|_I = 0$. In view of the familiar identification of $L^1(\mathbb{R})^*$ with $L^\infty(\mathbb{R})$, there exists $\tilde{\varphi} \in L^\infty(\mathbb{R})$ such that $\langle \varphi, h \rangle = \int_{\mathbb{R}} \tilde{\varphi}(s)h(s)ds$ for all $h \in L^1(\mathbb{R})$. But according to Fubini's theorem

$$\int_{\mathbb{R}} \tilde{\varphi}(s)\, g*f(s)ds = \int_{\mathbb{R}} \tilde{\varphi}(s)\left(\int_{\mathbb{R}} g(t)f(s-t)dt\right) =$$

$$= \int_{\mathbb{R}}\left(\int_{\mathbb{R}} \tilde{\varphi}(s)(^{-t}f(s))ds\right)g(t)dt = \int_{\mathbb{R}} \langle \varphi, {}^{-t}f \rangle g(t)dt.$$

Since $^{-t}f \in I$ for all $t \in \mathbb{R}$, the integrand is identically zero. The rest is clear. ■

Example (III.3.6). Theorem 3.5 implies that in each of the algebras being considered the set $I := \{f : \sum_{s \in G} f(s) = 0$ (or, according to the sense, $\int_G f(s)ds = 0)\}$ is a closed bi-ideal of codimension 1. This ideal (very important, for example, for homology theory [38], [30], [282]) is called the *augmentation ideal*.

Can all closed ideals of the algebras under consideration be described in an explicit form? We have already seen in §1.3 in the example of the Wiener algebra — an isometrically isomorphic copy of $l_1(\mathbb{Z})$ — that for some of them such a problem may turn out to be extremely difficult. Nevertheless it can be done for $L^1(\mathbb{T})$ (but not for $L^1(\mathbb{R})$; see below) and moreover in fairly simple terms.

We consider in $L^1(\mathbb{T})$ functions of the form $w(z) = z^n$; $n \in \mathbb{Z}$. The definition of convolution obviously implies the following lemma.

Lemma (III.3.7). (I) *For each* n *the one-dimensional subspace* $\ell.h.\{z^n\}$ *is an ideal in* $L^1(\mathbb{T})$.[21]

(II) *The function* $z^m * z^n$ *is equal to* z^n *when* $m = n$ *and to zero when* $m \neq n$. ▶■

We now let I_M denote, for an arbitrary subset M of $\mathbb{Z}$, the set of those $f \in L^1(\mathbb{T})$ that have m-th Fourier coefficient $\int_{\mathbb{T}} f(s)s^{-m}ds$ (this is just $\frac{1}{2\pi}\int_{-\pi}^{\pi} f_1(t)e^{-imt}dt$, where $f_1(t) := f(e^{it})$) equal to zero when $m \in M$.

Lemma (III.3.8). $I_M = \{f : f * z^m = 0$ for all $m \in M\} = \overline{\ell.h.}\{z^n : n \in \mathbb{Z}\backslash M\}$.

▶ Lemma 3.7(II) implies that the third of these sets is contained in the second. Since the Féjer sums of any $f \in L^1(\mathbb{T})$ converge to f it follows that the first lies in the third. Finally, it is obvious that the first coincides with the second. ■

Theorem (III.3.9). *A subspace* I *of* $L^1(\mathbb{T})$ *is a closed ideal of this algebra* $\Leftrightarrow$ $I = I_M$ *for a certain* $M \subseteq \mathbb{Z}$.

▶ "$\Leftarrow$" is obvious. Let I be a closed ideal, we put $M := \{m \in \mathbb{Z}: z^m \notin I\}$. The form of the third set in Lemma 3.8 implies that $I_M \subseteq I$ and the form of the second implies that, to show $I \subseteq I_M$, it is sufficient to show that $f * z^m = 0$ for $f \in I$ and $m \in M$. But $f * z^m \in I$, while, according to Lemma 3.7(I), $f * z^m$ is proportional to $z^m \notin I$. The rest is clear. ■

Corollary (III.3.10). (I) *Every maximal ideal in* $L^1(\mathbb{T})$ *is determined by an integer* m *and consists of those* f *for which* $\int_{\mathbb{T}} f(s)s^{-m}ds = 0$.

(II) *Every minimal ideal in* $L^1(\mathbb{T})$ *is determined by an integer* n *and consists of all functions that are multiples of* z^n.

Obviously Corollary 3.10(I) may also be formulated thus:

Corollary (III.3.11). *Spectral synthesis holds in the function algebra* $\Phi(\mathbb{Z})$.

For a finite group G the question of constructing the ideals of $l_1(G) \simeq \mathbb{C}(G)$ (like all the questions about these algebras) is solved in principle by the following classic assertion.

Assertion (III.3.12) ("Mashke theorem", see for example, [40, §10.1]). *If* G *is finite, then* $l_1(G)$ *is classically*

---

[21] So there are minimal ideals in $L^1(\mathbb{T})$ — not such a common phenomenon for the "algebras of analysis"!

semisimple; in other words (cf. Assertion I.3.68) $l_1(G)$ *is isomorphic to the Cartesian product of a finite number of complete matrix algebras.*

(Burnside made this theorem more precise by showing that the number of Cartesian factors is equal to the number of conjugacy classes in the group G.)

With regard to $l_1(G)$ when G is infinite and $L^1(\mathbb{R})$, the question of describing their closed ideals is at present a long way from being completely solved. A deep — and yet disheartening — result is that there exists in a commutative infinite-dimensional $l_1(G)$ and in $L^1(\mathbb{R})$ a closed ideal that is not an intersection of maximal ideals. (This fact is a consequence of a general theorem of Malliavin that we shall talk about in §3.2.)

The situation is considerably better when it comes to the question of describing the characters (and therefore, the maximal modular ideals in the commutative case — see later Theorem IV.1.1). The meaning of the relevant theorem is, loosely speaking, that the two meanings of the word "character" coincide. Recall that the latter comes originally from group theory: a *character of the group* G is any homomorphism into $\mathbb{T}$ — in other words, a function $\tilde{\chi}: G \longrightarrow \mathbb{C}$ such that $|\tilde{\chi}(s)| = 1$ and $\tilde{\chi}(st) = \tilde{\chi}(s)\tilde{\chi}(t)$ for all $s, t \in G$. It is known (if this is new to you — you must carry out the appropriate argument!) that each continuous character on $\mathbb{R}$ is determined by some $x \in \mathbb{R}$ and it takes t into e^{-itx}, and each continuous character of $\mathbb{T}$ ("itself") is determined by an integer n and takes t into t^{-n} (there is no reason for the minus sign, of course; it is only written "by tradition").

Theorem (III.3.13). *Let* G *denote an arbitrary discrete group, or* $\mathbb{R}$ *or* $\mathbb{T}$ *and let* $L^1(G)$ *be* $l_1(G)$, $L^1(\mathbb{R})$ or $L^1(\mathbb{T})$ *respectively. Then there exists a bijection between the set of continuous characters of the group* G *and the set of non-trivial characters of the algebra* $L^1(G)$, *associating the ("group") character* $\tilde{\chi}$ *with the ("algebra") character* χ *by the equation* $\chi(f) := \sum_{s\in G} f(s)\tilde{\chi}(s)$ *for discrete groups and* $\chi(f) := \int_G f(s)\tilde{\chi}(s)ds$, *for continuous groups. As a consequence, each character on* $L^1(\mathbb{R})$ *is determined by a real number* x *and takes* f *into* $\int_{\mathbb{R}} f(s)e^{-ixs}ds$, *and on* $L^1(\mathbb{T})$ *by an integer* n *and takes* f *into* $\int_{\mathbb{T}} f(s)s^{-n}ds$.

▸ Firstly, it can easily be verified that the given rule does give a map between the "group" and the "algebra" characters in all cases and moreover that it is injective. Therefore it only remains to show that every "algebra" character can be constructed by the above method starting from a certain "group" character.

If G is discrete, then it is sufficient to put $\tilde{\chi}(s) := \chi(\delta_s)$ for all $s \in G$; all the required properties are easily verified. Concerning the continuous groups, we limit ourselves to $\mathbb{R}$, leaving the reader to carry out an analogous argument for $\mathbb{T}$ (it may also be noted that the necessary fact for $\mathbb{T}$ follows easily from Corollary 3.10(I)).

So let $\chi \in X\alpha\rho\,(L^1(\mathbb{R}))$. According to the general form of functionals on $L^1(\mathbb{R})$ there exists some $\tilde{\chi} \in L^\infty(\mathbb{R})$ such that for all $f \in L^1(\mathbb{R})$

$$\chi(f) = \int_{\mathbb{R}} \tilde{\chi}(s) f(s) ds \tag{1}$$

Since $\chi \neq 0$, and the step functions are dense in $L^1(\mathbb{R})$, there exist $\alpha, \beta \in \mathbb{R}$; $\alpha < \beta$, such that for $X := X_{[\alpha,\beta]}$ the number $\theta := \chi(X) \neq 0$. Hence $\chi(f) = \chi(f)\chi(\theta^{-1}X) = \chi(f * \theta^{-1}X)$, and therefore Fubini's theorem implies that

$$\begin{aligned}\chi(f) &= \int_{\mathbb{R}} \tilde{\chi}(t)\left(\int_{\mathbb{R}} \theta^{-1} f(s) X(t-s) ds\right) dt = \\ &= \int_{\mathbb{R}}\left(\int_{\mathbb{R}} \theta^{-1}\tilde{\chi}(t) X(t-s) dt\right) f(s) ds = \\ &= \int_{\mathbb{R}}\left(\int_{\alpha}^{\beta} \theta^{-1}\tilde{\chi}(s+t) dt\right) f(s) ds.\end{aligned}$$

Since $\tilde{\chi}$ is essentially bounded, it obviously follows that the function in the last brackets is uniformly continuous with respect to s, so we are justified in assuming right from the start that the $\tilde{\chi}$ that gives our character by formula (1), is continuous. It remains to show that this is a character of the group $\mathbb{R}$.

Recall the net e_U in Proposition 3.2 and note that, for any $s \in \mathbb{R}$, there exists $\lim_U \chi({}^s e_U) = \lim_U \int_{\mathbb{R}} \tilde{\chi}(t) e_U(s+t) dt = = \lim_U (m(U))^{-1} \int_U \tilde{\chi}(t-s) dt = \tilde{\chi}(-s)$. Then, for any $f \in L^1(\mathbb{R})$ and $s \in \mathbb{R}$, Proposition 3.4(II) implies that

$$\begin{aligned}\int_{\mathbb{R}} \tilde{\chi}(s+t) f(t) dt &= \int_{\mathbb{R}} \tilde{\chi}(t)({}^{-s}f(t)) dt = \chi(\lim_U {}^{-s}e_U * f) = \\ &= \lim_U \chi({}^{-s}e_U)\chi(f) = \tilde{\chi}(s)\chi(f) = \int_{\mathbb{R}} \tilde{\chi}(s)\tilde{\chi}(t) f(t) dt.\end{aligned}$$

As f is arbitrary and $\tilde{\chi}$ is continuous, this means that $\tilde{\chi}(s+t) = \tilde{\chi}(s)\tilde{\chi}(t)$ for all $s, t \in \mathbb{R}$. As $\tilde{\chi}$ is bounded, this, in its turn, is possible only when $|\tilde{\chi}(s)| = 1$ for all $s \in \mathbb{R}$. ■

Proposition (III.3.14). *The algebras* $L^1(\mathbb{R})$ *and* $L^1(\mathbb{T})$ *are semisimple.*

▶ Since the kernel of any character of an algebra is a maximal modular ideal, Theorem I.3.53(Ia) and Theorem 3.13 imply that, for any $f \in \mathrm{Rad}(L^1(\mathbb{R}))$, we have $\int_{\mathbb{R}} f(s)e^{-ixs}ds = 0$ for all $x \in \mathbb{R}$. By the uniqueness theorem in the theory of (classical) Fourier transforms, this means that $f = 0$. Therefore $L^1(\mathbb{R})$ is semisimple; the same uniqueness theorem in the theory of Fourier series shows that $L^1(\mathbb{T})$ is semisimple. ■

Remark (III.3.15). The algebras $l_1(G)$ are also semisimple for all G but we can get this fact later (and simultaneously another proof of Proposition 3.14) as a corollary of a stronger assertion on the so-called star semisimplicity of the corresponding algebras (see IV.5.32 - IV.5.36).

§3.2. Some remarks about general L^1-algebras of locally compact groups and associated questions in harmonic analysis

The theory of L^1-algebras, that generalizes the results just obtained about $L^1(\mathbb{R})$ and the other specific algebras, forms, together with its applications, one of the most important chapters in harmonic analysis.[22]

Unfortunately a detailed account of this would require, even at the very beginning, giving a lot of space to the construction of the new machinery of invariant integration — and this would lead to an unwarranted distortion of the balance of the book. All the same we are sure that our reader ought to know at least the statements of the basic theorems: besides being beautiful and valuable in themselves, it is only with these as a background that the true place of the "experimental results" of §3.1 can be explained. Therefore we try to say, briefly and informally, something about certain concepts and results, referring the reader to the monographs [53], [87], [40], [48] for a complete and careful account.

When talking about measures on a set G (for the moment arbitrary), we always assume that they are countably additive and take values in $\mathbb{R}^{(+)} \cup \{\infty\}$. If G is furnished with a topology, then we usually require that the measure μ be *regular*; this means that: (I) the σ-ring on which it is defined contains all

[22] It should be noted that many authors use the term "group algebra" to mean exclusively the L^1-algebras.

the closed sets; (II) for any set M of finite measure and any $\varepsilon > 0$ one can find a compact set $K \subseteq M$ and an open set $U \supseteq M$ such that $\mu(U \setminus K) < \varepsilon$, and (III) the measure of any compact set is finite ("because it is small"). If, however, G is a group, then it is important to know how the measure is affected by translation; in particular, a measure μ is said to be *left-invariant* if, for any $r \in G$, $rM := \{rs : s \in M\}$ is measurable whenever $M \subseteq G$ is measurable and moreover $\mu(rM) = \mu(M)$.

Now let G be a topological group. Are there many measures on it compatible with the two structures? The following fundamental theorem, which has a long history, is connected with the names of Haar, von Neumann and Weil.

Assertion (III.3.16). *Let G be a locally compact group. Then up to a constant factor it has a unique left-invariant, regular measure.*[23]

The left-invariant measure featuring in Assertion 3.16 is called the *Haar measure* and is denoted by m. It is, in this way, implicitly contained in the algebraic-topological structure of G: the group, as it were, "knows its own measure".

Where there is a measure there is also, of course, integration with respect to this measure. According to the standard strategy, we get a Banach space of functions (more precisely, of equivalence classes) that are integrable with respect to the Haar measure on G, denoted by $L^1(G)$; moreover if G is compact, it is convenient to assume that $m(G) = 1$.

The following step transforms $L^1(G)$ into a Banach algebra. The appropriate multiplication is called *convolution* and it is defined by $f * g(s) = \int_G f(t)g(t^{-1}s)dt$ (we write dt instead of $dm(t)$). The Banach algebra $L^1(G)$ is called the *L^1-algebra* of the (locally compact) *group* G. Of course, our familiar $l_1(G)$, $L^1(\mathbb{R})$ and $L^1(\mathbb{T})$ are particular cases of L^1-algebras; in the first the Haar measure is equal to 1 on the one-point subsets (and therefore the integral becomes the sum) and in the second and — to within a normalization factor — third cases it coincides with the classical Lebesgue measure.

We note that $L^1(G_1) \hat{\otimes} L^1(G_2)$ is, to within an isometric isomorphism between Banach algebras, $L^1(G_1 \times G_2)$.

Exercise (III.3.17). Prove this fact for the case when G_1 and (independently) G_2 are either $\mathbb{R}$ or $\mathbb{T}$.

We now consider what general results lie behind the

---

[23] In addition, if the group is complete, then the local compactness condition is also necessary for the existence of such a measure.

observations made for specific algebras in §3.1 and say a bit more about them.

Proposition 3.2 becomes: *$L^1(G)$ is unital $\Leftrightarrow$ G is discrete* (in other words, $L^1(G) = l_1(G)$); however, *any* $L^1(G)$ *has a metric* a.i. (which is constructed, according to type, in the same way as those in Proposition 3.2, but using the characteristic functions of neighbourhoods of the identity in G).

The formula ${}^{r}f = \lim_{U} h_U * f$ from Proposition 3.4(II) carries over to any $L^1(G)$; however, the formula $f^{r} = \lim_{U} f * h_U$ is true only for those G that have Haar measures coinciding with their "right-invariant" analogue (for example, if G is commutative, discrete or compact). Nevertheless Theorem 3.5, which characterizes ideals in terms of translates, remains valid for left, as well as right, ideals in any $L^1(G)$.

Recall, further, that in the attempt to describe the closed ideals of the algebras in §3.1 we succeeded with $L^1(\mathbb{T})$ but suffered a fiasco with $l_1(\mathbb{Z})$ and $L^1(\mathbb{R})$. It turns out that it all depends on whether the group is compact or not: for compact G, and only for these, the closed ideals in $L^1(G)$ can be described fairly explicitly in terms of the so-called irreducible, finite-dimensional, representations of the group G (for details see [54, IV §3]). The simplest situation of all is when G is a compact commutative group: its continuous characters belong to $L^1(G)$ and any closed ideal in $L^1(G)$ is the closure of the linear hull of a certain family of them (cf. the role of z^n; $n \in \mathbb{N}$ in Theorem 3.9).

Finally, Theorem 3.13 is a particular case of a general theorem describing the characters of $L^1(G)$ in terms of the continuous characters of G: *there is a bijection* $\tilde{\chi} \longmapsto \chi$ *between the continuous characters of* G *and the characters of* $L^1(G)$ *defined by*

$$\chi(f) = \int_G f(s)\tilde{\chi}(s)ds; \quad f \in L^1(G). \tag{1}$$

Among arbitrary groups one can distinguish two classes where one can make considerably more progress, when studying their L^1-algebras (and other related function spaces), than in the general case. These are the compact groups (as we have already observed) and the commutative (= Abelian) groups; we now concentrate on the second. The two basic concepts for harmonic analysis on commutative groups are the character group and the ("abstract") Fourier transformation.

Thus, until we specifically say otherwise, G is a commutative, locally compact, group. We let $\hat{G}$ denote its set of continuous characters. We consider $\hat{G}$ to be a group with respect to pointwise multiplication and a simultaneously to be a topological space with the topology of uniform convergence on compact subsets of G; we note that both structures are induced by the natural embedding of $\hat{G}$ in $C(G)$ (cf. the beginning of

§1.1). It is easily verified that $\hat{G}$ is a topological group and that it is locally compact (so it is "certainly not worse" than the original group; so one can talk about the Haar measure on $\hat{G}$ and so on). Moreover the operation $G \longrightarrow \hat{G}$ turns discrete groups into compact groups and vice versa.

Definition (III.3.18). The *Fourier transform* of a function $f \in L^1(G)$ is the function $\hat{f} : \hat{G} \longrightarrow \mathbb{C}$ that sends the character $\tilde{\chi}$ to $\hat{f}(\tilde{\chi}) := \int_G f(s)\tilde{\chi}(s)ds$ (which we have already met as formula (1)).

Example (III.3.19). The general form, mentioned in §3.1, for the continuous characters on $\mathbb{R}$ and $\mathbb{T}$ easily implies that, up to a isomorphism of topological groups, $\hat{\mathbb{R}} = \mathbb{R}$ and $\hat{\mathbb{T}} = \mathbb{Z}$; in addition, we can immediately see that $\hat{\mathbb{Z}} = \mathbb{T}$. Hence it is clear that, making the appropriate identifications, the "abstract" Fourier transform in Definition 3.18 becomes, when $G = \mathbb{R}$ the classical Fourier transform; and when $G = \mathbb{T}$ it becomes the map taking a function to the sequence of its Fourier coefficients; when $G = \mathbb{Z}$ it takes a given set of coefficients of a trigonometric series to the sum of this series.

For any $f \in L^1(G)$ the function $\hat{f}$ is continuous and is zero at infinity; therefore we have a map $\underline{\Phi} : L^1(G) \longrightarrow C_0(\hat{G}) : f \longmapsto \hat{f}$, which is an algebra homomorphism of norm 1. It is still called the *Fourier transformation* (this time not just of an individual function) *of the whole algebra* $L^1(G)$. Later (Remark IV.3.14) we shall see that this is the special form taken for L^1-algebras of a very general construction in the theory of commutative Banach algebras — the Gel'fand transformation.

Up to now the fact that G is commutative has not been used in an essential way; this property comes to the fore in the following non-trivial assertion.

Assertion (III.3.20) ("uniqueness theorem for Fourier transforms"). *If the Fourier transforms of the two functions* $f_1, f_2 \in L^1(G)$ *are everywhere equal on* $\hat{G}$, *then* f_1 *and* f_2 *are themselves almost everywhere equal with respect to the Haar measure on* G (in other words, the map $\underline{\Phi}$ is injective).

Assertion 3.20 reduces, without special difficulty, to the following fact about the interrelation between G and $\hat{G}$.

Assertion (III.3.21) ("on the sufficiency of the set of characters"). *For any two distinct* $s, t \in G$ *there exists* $\tilde{\chi} \in \hat{G}$ *such that* $\tilde{\chi}(s) \neq \tilde{\chi}(t)$.

Recall the assertion, which is similar in form, about the sufficiency of continuous functionals on the Banach space E (a particular case of Corollary B of the Hahn–Banach theorem in Chap.0 §1.2) which implied that the canonical embedding $E \longrightarrow E^{**}$

is injective (Example 0.2.8). In exactly the same way Assertion 3.21 obviously implies that the corresponding map d from G into the group $\widehat{\widehat{G}}$ (of continuous characters on $\widehat{G}$), defined by $[d(s)](\widetilde{\chi}) := \widetilde{\chi}(s)$, is injective. However, and this is a very deep fact, there is a crucial difference between the properties of these two constructions. Unlike Banach spaces, the groups under consideration are always "reflexive".

Assertion (III.3.22) ("Pontryagin's duality principle"). *For any commutative, locally compact, group G the map $d : G \longrightarrow \widehat{\widehat{G}}$ is an isomorphism of topological groups.*

(For a proof that is fairly near to the original and uses the structure theory of the groups under consideration see [87, §24]; for a proof by the methods of the theory of Banach algebras see [53, §31].)

Returning to the Fourier transformation of the algebra $L^1(G)$ we put $\Phi(G) := \operatorname{Im} \Phi$. Obviously this is a Banach function algebra on $\widehat{G}$ with norm $\|\Phi(f)\| := \|f\|$ (the fact that it is well-defined comes from Assertion 3.20). This algebra $\Phi(G)$ is called the Fourier algebra of the group $\widehat{G}$; we have already met examples of it for $\widehat{G} = \mathbb{T}, \mathbb{R}$ and $\mathbb{Z}$. All maximal ideals of this algebra are fixed (this follows immediately from the form of the characters on $L^1(G)$ and Theorem IV.1.1 later on); in addition — what is not so trivial — all its point ideals are maximal. However, apart from when $\widehat{G}$ is discrete (= G is compact), the problem of describing all closed ideals in $\Phi(\widehat{G})$ — or, equivalently, in $L^1(G)$, is unsolved, and moreover it is known that $\Phi(\widehat{G})$ *automatically does not have spectral synthesis.* This is just the general form of Malliavin's theorem, already mentioned for $\widehat{G} = \mathbb{T}$ in §1.3 and (in an equivalent form) for compact $\widehat{G}$ and also for $\widehat{G} = \mathbb{R}$ in §3.1.

In conclusion, a few words about L^1-algebras that are in general non-commutative. In studying these we have in our repertoire no very powerful method like the Fourier transformation. This turns into a less effective construction (a special case of the homomorphism $\mathcal{G}$ that comes later in Chap. VI §2.2) which constructs from $f \in L^1(G)$ a certain operator-valued function taking values in algebras $\mathcal{B}(H)$, where the H are Hilbert spaces that are in general distinct at each point in the domain of definition. The role of the group characters is played by a generalization of them called the irreducible unitary representations of the group G (in these same H) which will be mentioned later (at the end of Chap. VI §2). Using these one can, however, solve a lot of important questions, especially in the case of compact groups when all these H are finite-dimensional (cf. Assertion VI.2.19 and Remark VI.2.86 later). In particular, a satisfactory duality theory can be constructed for compact groups (telling us how to reconstruct the group from its irreducible representations) and one can prove the "Tannaka-Krein duality principle" which plays a role for compact groups that is reminiscent of the role of the Pontryagin principle for commutative groups; for this see, for example [87, §30].

It is possible to prove by the methods of representation theory that $L^1(G)$ is *semisimple for any locally compact* G (to find out how this is done, see Remark IV.5.37 later). This, in its turn, has enabled us to obtain a meaningful generalization for all such G of Assertion 3.21 about the sufficiency of characters (the "Gel'fand-Raikov theorem", for a precise formulation of which see Assertion VI.2.85 later). There is also a meaningful analogue of the concept of a Fourier algebra for all such G, suggested by Eymard [297]. A lot of results on these algebras and certain generalizations of them are due to Lau (see [182] and the references mentioned there).

§3.3. Operator group algebras, measure algebras, certain group algebras of test functions and generalized functions

The L^1-algebras considered above are, possibly, the most natural and popular algebras but are a long way from being the only class of algebras that occur in harmonic analysis and closely related topics. In certain questions, for example, in the theory of group representations, an equally important role is played by those group algebras that consist of operators or that are isometrically isomorphic to operator algebras (thus §2 could be applied to these with the same success — or failure).

The most useful of these "operator-group" algebras are the so-called C^*-algebras of locally compact groups — the completions of the L^1-algebras with respect to a special, weaker, norm. It is, however, inadvisable to introduce them now: their definition will be accepted as being natural only after one is acquainted with the fundamental ideas of the general theory of C^*-algebras and we postpone it until Definition IV.7.67. However, the reader will have scarcely any doubt that it is natural to define another closely related class of algebras that is more intuitive and immediate than these operator algebras.

First let G be a discrete group. We consider the Hilbert space $l_2(G)$ consisting of those functions on G for which $\sum_{s\in G} |f(s)|^2 < \infty$; we put $\langle f, g\rangle := \sum_{s\in G} f(s)\overline{g(s)}$. We consider, for each $t \in G$, the operator $U_t : f \longrightarrow {}^t f$ in $l_2(G)$; it is obviously unitary. We consider further the subset $C^*_r(G) := \overline{\ell.h.}\{U_t : t \in G\}$ in the Banach algebra $\mathcal{B}(l_2(G))$; this is obviously closed and, since $U_t^* = U_{t^{-1}}$; $t \in G$, it is a self-adjoint subalgebra.

Definition (III.3.23). The realized C^*-algebra $C_r^*(G)$ (cf. the beginning of §2) is called the *reduced C^*-algebra of the (discrete) group* G.

The algebra we have introduced has many of the general features of $l_1(G)$; in particular its characters can be described in the same terms.

Exercise (III.3.24). Prove that there exists a bijection, between the characters of the algebra $C_r^*(G)$ and the characters of the group G, which takes χ into $\tilde{\chi} : t \longmapsto \chi(U_t)$.

For continuous groups such a method of defining the corresponding operator algebras "pays no heed" to their topology; therefore we start in another way. To be definite we take $G = \mathbb{R}$ and consider, besides the algebra $L^1(\mathbb{R})$, the Hilbert space $L^2(\mathbb{R})$. It is not difficult to check that the convolution formula gives a function $f * g \in L^2(\mathbb{R})$ for each pair $f \in L^1(\mathbb{R})$ and $g \in L^2(\mathbb{R})$. Therefore, any $f \in L^1(\mathbb{R})$ gives a map $T_f^{\mathbb{R}} : L^2(\mathbb{R}) \longrightarrow L^2(\mathbb{R}) : g \longmapsto f * g$, which is easily seen to be an operator of norm ≤ 1. The obvious "asssociativity" $(f_1 * f_2) * g = f_1 * (f_2 * g)$ immediately implies that the map $T^{\mathbb{R}} : L^1(\mathbb{R}) \longrightarrow \mathcal{B}(L^2(\mathbb{R})) : f \longmapsto T_f^{\mathbb{R}}$ is an algebra homomorphism and therefore a representation of $L^1(\mathbb{R})$ on $L^2(\mathbb{R})$.

The above construction is, of course, quite general and can be carried across, word for word, after changing the Lebesgue integral to an integral with respect to Haar measure, to any locally compact group G; thus we have a representation $T^G : L^1(G) \longrightarrow \mathcal{B}(L^2(G))$.

Definition (III.3.25). The subalgebra of $\mathcal{B}(L^2(G))$ that is the closure of the image T^G, is called the *reduced C^*-algebra of the group* G and is denoted (as in Definition 3.23) by $C_r^*(G)$.

We recommend the reader to consider Definition 3.25, having carefully carried out all the above arguments, in the case $G = \mathbb{T}$ and the case when G is discrete; in the latter case it should be shown that, as one would expect, both Definition 3.23 and Definition 3.25 provide the same algebra.

It is not difficult to show that $C_r^*(G)$ is a realized C^*-algebra for all, and not only for discrete, G (see Remark IV.5.38 later). We shall talk about the interrelations between the reduced C^*-algebra and the "simple" C^*-algebra mentioned above, sometimes called the complete C^*-algebra of a locally compact group, and the conditions for these to coincide (quite often satisfied) in Remark IV.7.68.

In the case of commutative groups their reduced C^*-algebras turn out to be, "after unmasking", objects that we know well.

Exercise (III.3.26). For $G = \mathbb{Z}$, $\mathbb{R}$ and $\mathbb{T}$ construct an isometric isomorphism between the algebras $C^*_r(G)$ and $C_0(\hat{G})$.

Hint. Consider the commutative diagram

$$\begin{array}{ccc} L^2(G) & \xrightarrow{T_f^G} & L^2(G) \\ \Big\downarrow{\scriptstyle \underline{\underline{\Phi}}} & & \Big\downarrow{\scriptstyle \underline{\underline{\Phi}}} \\ L^2(\hat{G}) & \xrightarrow{\varphi} & L^2(\hat{G}), \end{array}$$

in which φ is the operator of multiplication by $\underline{\Phi}(f) \in C_0(\hat{G})$, and $\underline{\underline{\Phi}}$ is the unitary operator coinciding with $\underline{\Phi}$ on $L^1(G) \cap L^2(G)$ (just as in Plancherel's classical theorem for $G = \mathbb{R}$).

In fact, the result of the exercise and the proof noted above carry across to any commutative locally compact G; in the appropriate argument a basic role is played by the "abstract Plancherel theorem" (which provides the necessary $\underline{\underline{\Phi}}$), for which see, for example, [53, §31].

If G is compact, then there is a whole family of algebras, besides $L^1(G)$ and $C^*_r(G)$, connected with it: these are $L^p(G)$; $1 < p < \infty$, with multiplication in all cases given by the same convolution. Of these the most useful is $L^2(G)$, called the *Hilbert algebra of the (compact) group* G. All the above algebras are isometrically isomorphic to the completions of a direct sum of exactly the same family of complete matrix algebras (generally speaking, distinct) but the completions are taken with respect to different norms. For $L^2(G)$ (and also for $C^*_r(G)$ but not for $L^1(G)$) the corresponding norm is fairly simply described. In particular, if G is, in addition, commutative, then all the matrices mentioned have dimension 1×1 (= scalars) and $L^2(G)$ is isometrically isomorphic to $l_2(\hat{G})$ with pointwise multiplication (prove this for $G = \mathbb{T}$). For more details about the algebras $L^2(G)$ see, for example, [53, §32].

* * *

Another algebra that deserves our attention is connected with an arbitrary locally compact group G. Although it is perhaps defined more simply than $L^1(G)$ and $C^*_r(G)$, without using the deep concept of Haar measure, its structure is nevertheless far more complicated and in many respects it remains a mysterious object even for the simplest continuous groups.

The underlying Banach space of this algebra is $C_0(G)^*$; so it consists of continuous functionals on $C_0(G)$. In order to introduce a multiplication, we note that, for any $f \in C_0(G)$, $\varphi, \psi \in C_0(G)^*$, the function $\psi \odot f : G \longrightarrow \mathbb{C} : s \longmapsto \psi({}^s f)$ belongs to $C_0(G)$ (check it!). Therefore the formula

$$\langle \varphi * \psi, f \rangle := \langle \varphi, \psi \odot f \rangle \tag{2}$$

is a good definition of the functional $\varphi * \psi \in C_0(G)^*$, called the *convolution* (this time of *functionals*) of φ and ψ. It is easy to show that $C_0(G)^*$ is a Banach algebra with convolution as multiplication; in this capacity it is called the *measure algebra on* G and is denoted by $M(G)$. The algebra we have introduced is unital: the "δ-function" $\delta : f \longmapsto f(\underline{e})$ serves as an identity, where $\underline{e}$ is the identity in G.

In what way are these measures? Recall Markov's theorem generalizing the classical theorem of Riesz about functionals on $C[0, 1]$; it asserts the following. Let Ω be a locally compact space; a complex measure on Ω is a function of subsets of Ω which has the same regularity property as the measures in §3.2 but with the difference that it takes values in $\mathbb{C}$ (not in $[0, \infty]$). Then there exists a bijection, between the set of complex measures on Ω and the set $C_0(\Omega)^*$, which takes the measure μ to the functional φ defined by

$$\langle \varphi, f \rangle = \int_\Omega f(s) d\mu(s); \quad f \in C_0(\Omega).$$

From what has been said we are justified, returning to G, in identifying the elements of $M(G)$ with complex measures on G, and from now on talking about the *convolution of two measures*; then a more suitable way of writing formula (2) is

$$\int_G f(s) d(\mu * \nu)(s) = \int_G \int_G f(st) d\nu(t) d\mu(s) \tag{3}$$

The above interpretation of $M(G)$ enables one to use the fairly well developed machinery of measure theory to study it. As a first application, Fubini's theorem immediately implies that; $M(G)$ is commutative $\Leftrightarrow$ G is commutative.

The convolution of measures not only reminds one of the convolution of functions in §3.2 but it can also be considered as a generalization of it. To each $g \in L^1(G)$ we associate the functional φ_g on $C_0(G)$ defined by $\langle \varphi_g, f \rangle := \int_G f(s) g(s) ds$ and we put $i_G : L^1(G) \longrightarrow M(G) : g \longmapsto \varphi_g$. It is easy to verify that if G is discrete, then i_G is an isometric isomorphism (of

Banach algebras) and thus $M(G)$ is just the same $L^1(G) = l_1(G)$ but "in a new wrapper". In general, we have the following assertion.

Assertion (III.3.27). *For any* G, i_G *is an isometric homomorphism whose image is the closed bi-ideal in* $M(G)$ *consisting of all measures that are absolutely continuous with respect to Haar measure* (= equal to zero on all sets of Haar measure zero).

Exercise (III.3.28). Prove this fact for $G = \mathbb{R}$ and $\mathbb{T}$.

So $L^1(G)$ can be identified with a bi-ideal in $M(G)$ and then the convolution of functions becomes the convolution of measures.

We shall exhibit another bi-ideal in $M(G)$ and at the same time an interesting subalgebra which is not a bi-ideal. We let $M_c(G)$ denote the set of continuous measures (= equal to zero on finite sets) and $M_d(G)$ denote the atomic measures (= concentrated on a countable subset). One immediately verifies the following proposition.

Proposition (III.3.29). (I) $M_c(G)$ *is a closed bi-ideal of* $M(G)$ *which, when* G *is continuous, contains* $L^1(G)$.

(II) $M_d(G)$ *is a closed subalgebra of* $M(G)$ *which is isometrically isomorphic to* $l_1(G_d)$, *where* G_d *is the group* G *with the discrete topology.* ▸∎

Further, the classical theorem on the representation of a function of bounded variation on $\mathbb{R}$ as the sum of a continuous function and a jump function can be naturally generalized to the following assertion.

Assertion (III.3.30). $M(G) = M_c(G) \oplus M_d(G)$ *and* (as a consequence) $M_d(G) \simeq M(G)/M_c(G)$.

The reader has already been warned that $M(G)$ has a more complicated structure than (its bi-ideal) $L^1(G)$. In fact, if we apply the standard questionnaire (cf. the beginning of the chapter) we come up against an explicit "refusal to answer" far more often. Here is how matters stand on the question, say, of the structure of the characters.

Recall the idyllic relation between the characters of $L^1(G)$ and the continuous characters of G (Theorem 3.13 and formula (1) in §3.2). Each character χ on $M(G)$ also generates a group character $\tilde{\chi} : G \longrightarrow \mathbb{T}$ and, moreover, in a simpler way: $\tilde{\chi}$ sends $s \in G$ into $\chi(\delta_s)$, where $\delta_s : C_0(G) \longrightarrow \mathbb{C} : f \longmapsto f(s)$. The trouble, however, is that $\tilde{\chi}$ is not necessarily continuous and that, worse still, it is in general impossible — even if it is continuous — to resurrect the original χ if we are given $\tilde{\chi}$.

Example (III.3.31). Consider $\chi_1 : M(G) \longrightarrow \mathbb{C} : \mu \longmapsto \mu(G)$ and $\chi_2 : M(G) \longrightarrow \mathbb{C} : \mu \longmapsto \sum_{s \in G} \mu(\{s\})$. Obviously these are characters that are distinct when G is continuous. Nevertheless they generate the same group character: $\tilde{\chi}(s) \equiv 1$.

In fact we have given here the simplest characters of $M(G)$ from two natural classes. The characters of the first class — they are called *absolutely continuous* — are constructed from the continuous characters on G on the same principles as one constructs the characters on $L^1(G)$. They are the most obvious (but, according to Example 3.31, not unique) extensions of the latter.

Exercise (III.3.32). Show that each continuous character $\tilde{\chi} : G \longrightarrow \mathbb{T}$ gives a character $\chi : M(G) \longrightarrow \mathbb{C} : \mu \longmapsto \int_G \tilde{\chi}(s) d\mu(s)$.

The characters of the second class, called *discrete*, are constructed quite differently; they are described by the following assertion.

Assertion (III.3.33). *For every (even discontinuous) character* $\tilde{\chi}$ *of the group* G *there exists a character* χ *of the algebra* $M(G)$ *defined uniquely by the following conditions:*

(I) $\chi(\mu) = 0$ *for every continuous measure* μ (and hence a *fortiori* $\chi|_{L^1(G)} = 0$);

(II) $\chi(\mu) = \sum_{n=1}^{\infty} \tilde{\chi}(s_n)\mu(\{s_n\})$ *for an (atomic) measure* μ *concentrated on the countable set* $\{s_n \in G\}$.

Exercise (III.3.34). Prove this assertion for $G = \mathbb{R}$ and $\mathbb{T}$.

From what has been said above it is clear that the set of absolutely continuous characters on $M(G)$ forms, generally speaking, only a small part of the whole of $\mathrm{X\alpha\rho}\, M(G)$: its cardinality is the same as the cardinality of the set of continuous characters on G, while, according to Assertion 3.33, the cardinality of $\mathrm{X\alpha\rho}\, M(G)$ is at least the cardinality of the set of all — even discontinuous — characters of our group. For example, for $G = \mathbb{R}$, the first cardinality is, of course, that of the continuum while the second is at least that of the exponent of the continuum [87, §25.6].

On $M(G)$ there are many characters that are not much like either an absolutely continuous character or a discrete character; for examples of these when $G = \mathbb{R}$ see, say, [20, §31]. As regards the problem of describing all the characters of $M(G)$ (even for $G = \mathbb{R}$ or $\mathbb{T}$), there is as yet no satisfactory solution, at least no transparent one. The strongest result is still that obtained by Shreider in 1950 (see [20, §33] or [294]) who succeeded in expressing the characters of $M(G)$ in terms of objects, far more closely connected with G itself — the so-called generalized characters of G.

Some further information about the algebra of measures will be given in Chap. IV §3.2.

* * *

Group algebras that do not belong to the class of Banach algebras arise, basically, when we are talking about Lie groups: they "react" in the proper way to the structure of a smooth manifold. Among these, we consider most often either the algebras of test, or basic, functions (infinitely smooth functions with some condition or other about them decreasing at infinity), or algebras of generalized functions (= continuous functionals on spaces of test functions). It is not difficult to see that, for compact G, such algebras contain either far fewer elements than $L^1(G)$, or, conversely, far more than $M(G)$. We deal briefly with only two specific examples connected with the simplest group $\mathbb{T}$, referring the reader to the books [40], [33], [34], [35] and their references.

Our first algebra, denoted by $\mathcal{E}(\mathbb{T})$, consists, like $C^\infty(\mathbb{T})$, of all infinitely smooth functions on the circle and is furnished with the same norms; however, we introduce another multiplication in it — convolution. The second, and perhaps the more important and interesting algebra, is denoted by $\mathcal{E}(\mathbb{T})^*$ and, correspondingly, consists of continuous functionals on $\mathcal{E}(\mathbb{T})$ ("generalized functions on $\mathbb{T}$"); it is furnished with a system of seminorms indexed by the bounded (with respect to each seminorm) sets $N \subset \mathcal{E}(\mathbb{T})$ and defined by $\|\varphi\|_N := \sup\{|\varphi(f)| : f \in N\}$ (Proposition 0.1.9 implies that this is well-defined). The multiplication in $\mathcal{E}(\mathbb{T})^*$ is again called *convolution* and is defined (in a similar way to that on $M(G)$) by $\langle\varphi * \psi, f\rangle := \langle\varphi, \psi \odot f\rangle$, where $\psi \odot f \in \mathcal{E}(\mathbb{T})$ takes s into $\langle\psi, {}^s f\rangle$; $f \in \mathcal{E}(\mathbb{T})$. We note that we have a chain of embeddings $\mathcal{E}(\mathbb{T}) \subset L^1(\mathbb{T}) \subset M(G) \subset \mathcal{E}(\mathbb{T})^*$.

Exercise (III.3.35). Prove that $\mathcal{E}(\mathbb{T})$ is an Arens-Michael algebra whereas $\mathcal{E}(\mathbb{T})^*$ is complete but is not an Arens-Michael algebra.

Remark (III.3.36). It is not difficult to see that the generalized functions concentrated at $1 \in \mathbb{T}$ (= linear combinations of the δ-function and its derivatives of all orders) form a subalgebra in $\mathcal{E}(\mathbb{T})^*$ which is isomorphic to the algebra of polynomials $\mathcal{P}$. This is a particular case of Schwartz's theorem, characterizing anew a well-known object connected with a Lie group G — the so-called enveloping algebra of the Lie algebra of this group: it is isomorphic to the algebra of generalized functions on G concentrated at the identity of the group (for details see [40]).

Both of the algebras we have introduced can arise in the guise of sequence algebras.

Exercise (III.3.37). Prove that $\mathcal{E}(\mathbb{T})^*$ is topologically isomorphic to the algebra c_p in §1.3 and $\mathcal{E}(\mathbb{T})$ is the algebra of "rapidly decreasing sequences" c_s consisting of those $\xi = \{\xi_n\}$ that have $\|\xi\|_k := \sup_n |n^k \xi_n| < \infty$ for all $k \in \mathbb{N}$, furnished with the system of norms: $\|\cdot\|_k$ for all k.

Hint. Go across from a function (or a generalized function) to the sequence of its Fourier coefficients.

Using the result of this exercise one can give a description of the closed ideals and characters of the algebras $\mathcal{E}(\mathbb{T})$ and $\mathcal{E}(\mathbb{T})^*$ analogous to that obtained for $L^1(\mathbb{T})$ in Theorems 3.9 and 3.13. We suggest the reader gives a precise formulation of, and proves, the appropriate assertions.

Chapter IV

Commutative and star Banach algebras

Having considered the basic examples, we return to questions of a more general nature. In this chapter we shall speak of results that historically form the core of our subject and reflect that interplay of analysis and algebra which is characteristic of it.

Here we shall describe, for the most part, the achievements of the "classical period" of Banach algebras (approximately 1939–1953).[1] The results, as a rule, refer to two classes of algebras: (I) commutative and (II) those having some intrinsic symmetry — the so-called star algebras. Accordingly, two closely-related theories are given (the second depends on the first). In each of them there is a climax — when there appears a construction for making "something concrete and tangible", within the given class, from a given (abstract) algebra: in the first case this is a function algebra, in the second it is an operator algebra. The constructed algebra contains extensive information about the original algebra and, under certain additional assumptions, it is an isometrically isomorphic copy of it.

The most important of the results in this chapter can be expressed thus: up to a certain (explicitly measured) imprecision, commutative Banach algebras are algebras of continuous functions and star Banach algebras are algebras of operators on a Hilbert space.

---

[1] Except for certain facts about Hermitian algebras; cf. the footnote near beginning of §6.

§1. THE SPECTRUM (= MAXIMAL IDEAL SPACE) OF A COMMUTATIVE BANACH ALGEBRA

Warning. Throughout this and the next two sections, when we speak of the algebra A we always mean a commutative Banach algebra.

§1.1. Identification of maximal ideals with characters. Applications to function algebras and trigonometric series (theorems of Wiener and Levy)

Recall the set of non-trivial characters of the algebra A, denoted by $X\alpha\rho$ (Chap.I §2.5). We consider, along with these, the set of maximal modular ideals which are denoted (for the time being) by Max. (We have already met something similar for the special case of function algebras in Chap.III §1.1.)

Theorem (IV.1.1). *Every maximal modular ideal in* A *is the kernel of a uniquely defined character* (in other words, ω in Chap.III §1.1 is bijective).

▶ We take $I \in \text{Max}$ and consider the quotient algebra A/I; it is, by Proposition I.3.40(II), a Banach field. Therefore, according to the Gel'fand-Mazur theorem (II.1.17), there exists a topological isomorphism $i : A/I \xrightarrow{\cong} \mathbb{C}$. We put $\chi := i \circ \mathbf{pr}$, where $\mathbf{pr} : A \longrightarrow A/I$ is the natural projection; then it is obvious that χ is a character of the algebra A with kernel I.

We have shown that ω is surjective. Further, for any $\chi \in X\alpha\rho$, it is obvious that $\chi(u) = 1$, where u is a modular identity for $\omega(\chi) = \text{Ker}\,\chi \in \text{Max}$. In view of $A = \text{Ker}\,\chi \oplus \ell.h.\{u\}$, this means that χ is uniquely defined by its kernel, that is, ω is injective. ■

Let us give a number of corollaries of the theorem we have just proved. First of all, if we combine it with the purely algebraic Theorem I.3.19, we get the following corollary.

Corollary (IV.1.2). *The element* a *of a unital commutative Banach algebra is invertible* $\Leftrightarrow$ $\chi(a) \neq 0$ *for each character of this algebra.*

In turn it is obvious that this implies the following:

Corollary (IV.1.3). *The spectrum of an element* a *of a commutative Banach algebra* A *is the set* $\{\chi(a) : \chi \in X\alpha\rho\}$ *if* A *is unital, and this same set, together with zero, if* A *is not unital.*

Further, Theorem 1.1 combined with Theorem I.3.53 provides the following corollary.

Corollary (IV.1.4). *The radical of a commutative Banach algebra is the intersection of the kernels of all its non-trivial characters.*

In its turn this implies the following corollary.

Corollary (IV.1.5). *For any element* $a \neq 0$ *of a semisimple commutative Banach algebra there exists a character* χ *of this algebra with* $\chi(a) \neq 0$.

This last fact is often expressed by saying that semisimple commutative Banach algebras have "sufficiently many characters" (cf. a similar statement about functionals in Corollary C after 0.1.12).

Having proved Theorem 1.1, we immediately get the description of maximal modular ideals, promised in Chap.III, for certain of the specific algebras considered there. Namely, Theorem 1.1 in combination with III.1.14, III.1.23 and III.3.13 imply that these ideals for the algebras W, $\mathcal{A}(\overline{\mathbb{D}})$, $L^1(\mathbb{R})$ and $L^1(\mathbb{T})$ are given by the points of $\mathbb{T}$, $\overline{\mathbb{D}}$, $\mathbb{R}$ and $\mathbb{Z}$ respectively; if we believe what was said in Chap.III §3.2, then, for the L^1-algebra of any commutative locally compact group G these ideals are given by the points of $\hat{G}$; we leave the reader to formulate carefully all the appropriate assertions.

With regard to W and $\mathcal{A}(\overline{\mathbb{D}})$, a description of their maximal ideals is a particular case of an assertion that also follows immediately from Theorem 1.1:

Corollary (IV.1.6). *If all the non-trivial characters of a certain Banach function algebra are fixed, then all the maximal modular ideals of this algebra are also fixed.*

We note also that Corollary 1.3 implies the following.

Corollary (IV.1.7). *If* A *is a unital Banach function algebra with fixed characters, then the spectrum of any function in* A *(as an element* A*) is its set of values.*

Theorem 1.1 and its corollaries enable a lot of questions, at first glance of a technical analytical nature, to be reduced to studying the characters of suitably chosen algebras. With the aim of showing how this can happen, we first satisfy the promise given in Remark II.2.5.

Theorem (IV.1.8). *Let* A *be a unital Banach function algebra, with all the characters fixed. Let* g *be a function belonging to* A *and let* U *be a domain in* $\mathbb{C}$ *containing its set of values. Then given any holomorphic function* w *on* U *the composition* $w \circ g$ *belongs to* A *and the map* $w \longmapsto w \circ g$ *is a holomorphic calculus for* g *in* U.

▶ By Theorem II.2.15 and Corollary 1.7, a holomorphic calculus for g exists in U since g is an element of A; let $w(g)$ be the function corresponding to $w \in \mathcal{O}(U)$. Then, if s is any point in the domain of definition of this function in A and χ_s is the character it determines, Theorem II.2.18 implies that

$$[w(g)](s) = \chi_s(w(g)) = w(\chi_s(g)) = w(g(s)).$$

Thus the function $w(g)$ coincides with $w \circ g$. ■

One of the applications of this result is a proof, suggested by Gel'fand, of a well-known theorem about trigonometric series, first obtained by Wiener and then generalized by a theorem of Levy.

Theorem (IV.1.9). *Suppose the function* $f : \mathbb{R} \longrightarrow \mathbb{C}$ *can be expanded as an absolutely convergent Fourier series. Then*

(I) (Wiener) *if* f *is nowhere zero, then* $1/f$ *can also be expanded as an absolutely convergent Fourier series*;

(II) (Levy) *if* w *is a function of a complex variable which is holomorphic in a certain domain containing the set of values of the function* f, *then* $w \circ f$ *can also be expanded as an absolutely convergent Fourier series.*

▶ Let $\sum_{n=-\infty}^{\infty} c_n e^{int}$; $t \in \mathbb{R}$ be the Fourier series of the function f; then $g(z) := \sum_{n=-\infty}^{\infty} c_n z^n$; $z \in \mathbb{T}$, belongs to the Wiener algebra W. Therefore, under the conditions of (I), Theorem III.1.14 implies that $0 \notin \mathrm{Sp}_W(g)$ and so $1/g$ can be represented as a certain absolutely convergent series $\sum_{n=-\infty}^{\infty} d'_n z^n$, and hence $1/f$ can be represented by the series $\sum_{n=-\infty}^{\infty} d'_n e^{int}$. If, however, the conditions of (II) are satisfied, then Theorem 1.8, applied with $A = W$, implies that $w \circ g$ can be represented by a certain absolutely convergent series $\sum_{n=-\infty}^{\infty} d''_n z^n$, and hence, $w \circ f$ can be represented as the series $\sum_{n=-\infty}^{\infty} d''_n e^{int}$. ■

Many years later Katznelson [156] proved the following assertion, which is evidence of the deep distinction between W and algebras such as $C^1(\mathbb{T})$ or $C(\mathbb{T})$, say.

Assertion (IV.1.10). *Let* w *be a function defined on the interval* $[-1, 1]$ *such that* $w \circ f \in W$ *for any* $f \in W$ *with* $\mathrm{Im}\, f \in [-1, 1]$. *Then* w *may be extended to a function that is holomorphic in a certain neighbourhood, in* $\mathbb{C}$, *of the interval* $[-1, 1]$.

This means, of course, that in formulating Levy's theorem the class of holomorphic functions cannot be changed to any wider class.

§1.2. The spectrum of an algebra and its topology

From now on we shall identify — by using the map $\omega : \chi \longmapsto \operatorname{Ker} \chi$ — the set $\mathrm{X}\alpha\rho$ and Max and call both of them the *spectrum of the algebra* A. It is psychologically convenient to think of the spectrum as a point set, with points indexing both the non-trivial characters and the maximal modular ideals. We shall denote the spectrum of the algebra A by $\underline{\Omega}(A)$ or, if there is no danger of confusion, by Ω, and the points of the spectrum by s, t, and so on; moreover for any $s \in \Omega$ the corresponding character will be denoted by χ_s and the ideal by I_s.

Recall the (weak*) topology defined in Chap.I §2.5 on the set of continuous characters of any polynormed algebra. In the case of a commutative Banach algebra we are justified, in view of what has just been said, in talking about the corresponding topology on the spectrum; here it is called the *Gel'fand topology.*[2] In future it will be this topology that we have in mind everywhere, except in certain specified cases.

If we reword Proposition I.2.45(II), we immediately get the following proposition.

Proposition (IV.1.11). *For every* $s \in \Omega$ *the sets of the form* $U_{s,a,\varepsilon} = \{t \in \Omega\ ;\ |\chi_t(a) - \chi_s(a)| < \varepsilon\}$, *taken over all* $a \in A$ *and* $\varepsilon > 0$, *form a subbasis for the neighbourhoods of this point.* ▸■

Theorem (IV.1.12). (I) *The spectrum of a commutative Banach algebra is locally compact; and in the case of a unital algebra it is compact.*

(II) *The one-point compactification* $\underline{\Omega}(A)_+$ *of the space* $\underline{\Omega}(A)$ (cf. Chap.0 §0.3°) *coincides, to within a homeomorphism, with* $\underline{\Omega}(A_+)$.

▸ The first assertion follows immediately from Proposition I.2.45(II) and Theorem I.2.50(II), and the second by Corollary I.2.52. ■

We note also the following useful proposition.

Proposition (IV.1.13). *The spectrum of the algebra* A *coincides, to within a homeomorphism, with the spectrum of* $A/\operatorname{Rad} A$.

▸ Let χ be a non-trivial character of A; then its kernel, being a maximal modular ideal, contains the radical. Hence χ induces (in the sense of Chap.0 §0.2°) a certain

---

[2] The spectrum with its Gel'fand topology is also often known as the "carrier space" of the algebra.

non-trivial character $\hat{\chi}$ of the algebra $\hat{A} := A/\mathrm{Rad}\, A$; in this way one defines a map $\beta : \underline{\Omega}(A) \longrightarrow \underline{\Omega}(\hat{A})$, which is obviously continuous.

Further, for any character $\hat{\chi}$ of the algebra $\hat{A}$, the map $\chi := \hat{\chi} \circ \mathbf{pr}$, where $\mathbf{pr} : A \longrightarrow \hat{A}$ is the natural projection, is obviously a character of A. The map $\underline{\Omega}(\hat{A}) \longrightarrow \underline{\Omega}(A) : \hat{\chi} \longmapsto \chi$ is also obviously continuous and is the inverse to β. ∎

Let us illustrate the concept we have introduced by two exercises. The first of these justifies using the term "spectrum" in two different senses.

Exercise (IV.1.14). Let A be a unital Banach algebra with a single topological generator a (this means that the set of polynomials in a is dense in A). Show that the spectrum of the algebra A coincides, to within a homeomorphism, with the spectrum of a as an element of A.

Hint. The map $\alpha : \Omega \longrightarrow \mathrm{Sp}\, a \ : s \longmapsto \chi_s(a)$ is well defined and satisfies the conditions of Alexandrov's theorem in Chap.0 §0.3°. All that is needed is to explain surjectivity — the existence, given $\lambda \in \mathrm{Sp}\, a$, of a character χ with $\chi(a) = \lambda$. Since $p(\lambda) \in \mathrm{Sp}\,(p(a))$; $p \in \mathcal{P}$, and the spectral radius is bounded by the norm, the character we need is well defined and is continuous on the subalgebra of polynomials in a, which is dense in A.[3]

Exercise (IV.1.15). Let A be a unital Banach algebra with n topological generators $a_1, \cdots, a_n$ (this means that the smallest subalgebra in A that contains these elements and the identity is dense in A). Show that the spectrum of A is homeomorphic to a compact subset of $\mathbb{C}^n$.

Hint. Consider the map $\Omega \longrightarrow \mathbb{C}^n : s \longmapsto \{\chi_s(a_1), \cdots, \chi_s(a_n)\}$.

* * *

Thus with every commutative Banach algebra there is associated a topological space — its spectrum. We find ourselves in one of those situations when we ought to follow the general rule formulated in their time by Eilenberg and MacLane. They taught us to look for a functor![4]

The appropriate definition, reminiscent of the definition of the functor (*) in Chap.0 §2.2, looks simplest in the case of

[3] Alternatively, use Corollary 1.3.

[4] "...whenever new abstract objects are constructed in a specified way out of given ones, it is advisable to regard the construction of the corresponding induced mappings on these new objects as an integral part of their definition." [295]

unital algebras; we restrict ourselves to this case as it is sufficient for our applications. We let $\mathcal{UCBA}$ denote the category of unital commutative Banach algebras and their continuous unital homomorphisms and let $\mathcal{CTop}$ denote the category of compact topological spaces and their continuous maps. Let $\varphi : A \longrightarrow B$ be a morphism in the first category and let χ be a non-trivial character of B; then $\chi \circ \varphi$ (that is, $\varphi^*(\chi)$, where $\varphi^* : B^* \longrightarrow A^*$ is the operator conjugate to φ) is obviously a non-trivial character of A. In this way, interpreting the points of the spectrum as characters, we construct a map $\underline{\Omega}(\varphi) : \underline{\Omega}(B) \longrightarrow \underline{\Omega}(A) : \chi \longmapsto \chi \circ \varphi$, where the fact that φ^* is continuous with respect to the weak* topologies in B^* and A^* implies immediately that $\underline{\Omega}(\varphi)$ is continuous. If we interpret the points of the spectrum as maximal ideals, it is obvious that the map $\underline{\Omega}(\varphi)$ takes a maximal ideal I in B to its preimage under φ (which, as it is not difficult to verify, is a maximal ideal in A).

The association $A \longmapsto \underline{\Omega}(A)$; $\varphi \longmapsto \underline{\Omega}(\varphi)$ is obviously a contravariant functor, denoted later by $\underline{\Omega} : \mathcal{UCBA} \longrightarrow \mathcal{CTop}$. We call it the *spectrum functor*.

Remark (IV.1.16). Since characters are automatically continuous (I.2.6), the spectrum functor may obviously be extended from $\mathcal{UCBA}$ to the category with the same objects but with arbitrary — even discontinuous — unital homomorphisms as morphisms.

Remark (IV.1.17). In the definition of the spectrum functor one can also relinquish the requirement that the algebras and homomorphisms under consideration be unital. Here, however, we need to be somewhat more precise in defining the morphisms of the category in which the images of the corresponding functor lie (the objects are, of course, locally compact spaces). See, for example, [254, p.122] for more about this category.

It is quite interesting to observe the way in which the spectrum functor establishes a connection between significant relations between "Banach algebras" and "topological" relations between objects of the corresponding category. Here are a few illustrations.

Exercise (IV.1.18). Let $\mathbf{pr} : A \longrightarrow A/I$ be the natural projection. Show that $\underline{\Omega}(\mathbf{pr})$ is a topologically injective map; and if $I = \operatorname{Rad} A$, then $\underline{\Omega}(\mathbf{pr})$ is a homeomorphism (cf. 1.13).

Exercise (IV.1.19). Let I be a closed ideal in A, let $I_1 = \{a + \lambda e;\ \ a \in I,\ \ \lambda \in \mathbb{C}\}$, and let $\mathbf{in} : I_1 \longrightarrow A$ be the natural embedding. Show that $\underline{\Omega}(\mathbf{in})$ maps a certain closed set $\Delta \subset \underline{\Omega}(A)$ (which one?) to a certain point $s_\infty \in \underline{\Omega}(I_1)$ (which one?) and determines a homeomorphism between $\underline{\Omega}(A)\setminus\Delta$ and $\underline{\Omega}(I_1)\setminus\{s_\infty\}$.

Hint. Interpreting the points of the spectrum as maximal ideals, every $J \in \underline{\Omega}(I_1)$, apart from I, has a unique preimage under $\underline{\Omega}(\mathbf{in})$ namely $M = \{ae' + x : a \in A,\ x \in I \cap J\}$, where e' is a modular identity for the ideal $I \cap J$ in the algebra J.

Finally, Proposition I.2.53, translated into the language of spectra, gives us the following.

Proposition (IV.1.20). *Let* $A, B \in \mathcal{UCBA}$. *Then* $\underline{\Omega}(A \hat{\otimes} B)$ *coincides, to within a homeomorphism, with* $\underline{\Omega}(A) \times \underline{\Omega}(B)$. ▶■

In fact one can extract something more from Proposition I.2.53: the identification established there between the two spaces is compatible with the morphisms of the appropriate categories. Here is one of the methods for expressing this important fact in a formal way.

Exercise (IV.1.21). Define carefully the functors, from $\mathcal{UCBA}$ into $\mathcal{CTop}$, denoted symbolically by $\underline{\Omega}(A \hat{\otimes} ?)$ and $\underline{\Omega}(A) \times \underline{\Omega}(?)$, where A is fixed, and show that they are naturally equivalent.

* * *

Besides the topology under consideration, there is another useful topology that can be introduced on the spectrum of a commutative Banach algebra, and which is a version of the structure space topology familiar in ring theory [29].

Definition (IV.1.22). Let M be a subset of A. Its *hull* is the subset $h(M)$ of $\underline{\Omega}(A)$ consisting of all characters that are zero on M (or, equivalently, consisting of all maximal ideals containing M).

Obviously $h(M) = h(I)$, where $I = \overline{\ell.h.}\{ax : a \in A, x \in M\}$ is an ideal of A.

Exercise (IV.1.23). Show that the complements of the hulls of all possible subsets of the algebra A form a topology on the spectrum that is, generally speaking, weaker than the Gel'fand topology.

This new topology is called the *hull-kernel topology* or the *Jacobson topology*. It is, in general, not Hausdorff (the reader can check this by taking the algebra $\mathcal{A}(\overline{\mathbb{D}})$ as an example); however, for unital algebras, if it is Hausdorff, then it coincides with the Gel'fand topology [64, III §7]. The class of commutative Banach algebras for which this last property is satisfied (it is easy to verify that these include $C_0(\Omega)$, $C^n[0,1]$ and W), is a very interesting one and it is important in many questions in the theory of functions and harmonic analysis. This is the class of so-called regular, or Shilov, algebras, that are studied in detail in [290].

Finally, another, third, topology on the spectrum deserves to be remembered: this is the topology of the spectrum as a subset of the normed space A^*. In contrast, it is in general stronger than the Gel'fand topology (for example, in this topology the spectrum of the algebra $C_0(\Omega)$ is discrete). This topology is often called the *Gleason topology* and it is used, primarily, in studying Banach algebras of analytic functions [218], [212].

§2. TRANSFORMATION OF A COMMUTATIVE BANACH ALGEBRA INTO AN ALGEBRA OF CONTINUOUS FUNCTIONS (THE GEL'FAND TRANSFORMATION)

Warning. Here, as before, A is a commutative Banach algebra with spectrum $\Omega = \underline{\Omega}(A)$, which is given the Gel'fand topology.

§2.1. The functions that represent the elements

Let a be an element of the algebra A (for the moment fixed).

Definition (IV.2.1). The function $\hat{a} : \Omega \longrightarrow \mathbb{C}$ that associates to the point $s \in \Omega$ the number $\chi_s(a)$ is called the *Gel'fand transform of the element* a.

The Gel'fand transform of a, considered as an element of A_+ is denoted by $\hat{a}_+$; in view of the the fact that the spectrum of A_+ can be identified with Ω_+ (Theorem 1.12(II)), we are justified in considering $\hat{a}_+$ as a function on Ω_+. The set of values of the function $\hat{a}$ is denoted by $\hat{a}(\Omega)$ and that of the function $\hat{a}_+$ by $\hat{a}_+(\Omega_+)$. It is obvious that the following holds.

Proposition (IV.2.2). *For any* $s \in \Omega$, *we have* $\hat{a}_+(s) = \hat{a}(s)$; *in addition* $\hat{a}_+(\infty) = 0$. *As a consequence* $\hat{a}_+(\Omega_+) = \hat{a}(\Omega) \cup \{0\}$. ▸■

Theorem (IV.2.3). *If* A *is unital, then the spectrum of the element* a *coincides with the set of values of the function* $\hat{a}$ *(that is,* $\mathrm{Sp}\, a = \hat{a}(\Omega)$*); if however* A *is not unital, then* $\mathrm{Sp}\, a = \hat{a}_+(\Omega_+)$. *As a consequence, for non-empty* Ω [5] *we have in both cases* $\underline{r}(a) = \sup\{|\hat{a}(s)| : s \in \Omega\} = \|\hat{a}\|_0$.

[5] When $\Omega = \varnothing$, that is, when A is radical, it is obvious that $\underline{r}(a) = 0$.

▶ This follows from Corollary 1.3, taking into account the preceding proposition. ■

We note that the description, given in 1.11, of a topology on the spectrum, takes the following form in the new terminology:

Proposition (IV.2.4). *For any* $s \in \Omega$ *the sets* $U_{s,a,\varepsilon} = \{t \in \Omega : |\hat{a}(t) - \hat{a}(s)| < \varepsilon\}$, *taken over all* $a \in A$ *and* $\varepsilon > 0$, *form a subbasis for the neighbourhoods in* Ω *of this point.* ▶■

Theorem (IV.2.5). *For any* $a \in A$ *the function* $\hat{a}(s); s \in \Omega$, *is continuous and vanishes at infinity* (in other words, $\hat{a} \in C_0(\Omega)$).

▶ The assertion about continuity follows immediately from Proposition 2.4; hence, in particular, the function $\hat{a}_+$, which is zero at the adjoined point (see Proposition 2.2), is continuous at this point. The rest is clear. ■

If we combine 2.4 and 2.5 we immediately get the following.

Corollary (IV.2.6). *The Gel'fand topology is the weakest topology on* Ω *among all those topologies for which all the functions* $\hat{a}$; $a \in A$, *are continuous.*

We turn our attention to the fact that the above properties of $\hat{a}$ often facilitate a description of the spectra of specific algebras. As a basis we have the following proposition.

Proposition (IV.2.7). *Any topology on* Ω *with respect to which* Ω *is locally compact and all the functions* $\hat{a}$; $a \in A$, *are continuous and vanish at infinity, coincides with the Gel'fand topology.*

▶ Let Ω'_+ be the one-point compactification of the spectrum with respect to the given topology. As the set Ω'_+ coincides with Ω_+, and moreover, by hypothesis, all $\hat{a}_+$; $a \in A$, are continuous as functions on Ω'_+; so then, obviously, all $\hat{b}$; $b \in A_+$ are continuous as functions on Ω'_+. But then Corollary 2.6 (replacing A by A_+ and Ω by Ω_+) implies that the identity map Ω'_+ on Ω_+ is a continuous bijection from one compact space to another, and therefore, by Alexandrov's theorem, it is a homeomorphism. ■

Here is a typical application.

Corollary (IV.2.8) (a "topological supplement" to Corollary 1.6). *Let a certain Banach function algebra on a locally compact space* Ω *be such that all its functions vanish at infinity. Then, if all its non-trivial characters are fixed, the map* $t \longmapsto \chi_t$ *is a homeomorphism from* Ω *onto the spectrum of this algebra.*

Remark (IV.2.9). This corollary explains why we usually use the notation Ω for the domain of definition of function algebras as well as for spectra. If there is danger of confusion, of course, we have to use a different notation.

§2.2. The transformation of an algebra in the large. Its functorial properties and its compatibility with the holomorphic calculus

Definition (IV.2.10). The map $\Gamma = \Gamma_A : A \longrightarrow C_0(\Omega) : a \longmapsto \hat{a}$ is called the *Gel'fand transformation of the algebra* A.[6]

Theorem (IV.2.11). (I) Γ *is a continuous homomorphism between Banach algebras and moreover if the norm in* A *is multiplicative, then* $\|\Gamma\| \leq 1$.

(II) $\operatorname{Ker}\Gamma = \operatorname{Rad} A$; *consequently, if* A *is semisimple,* Γ *is injective.*

(III) $\operatorname{Im}\Gamma$ *separates the points of* Ω (in the sense of Chap.II §1.1).

(IV) *if* A *is unital, then* $\Gamma(e) \equiv 1$.

▶ To prove (I), we are justified in supposing that the norm is multiplicative; then $\underline{r}(a) \leq \|a\|$ is satisfied for any $a \in A$ (II.1.24) and the assertion is implied by Theorem 2.3. Further (II) is implied by the characterization of the radical given in Corollary 1.4 and the other assertions are obvious. ■

We let Γ_+ denote the Gel'fand transformation of the algebra A_+. Then it is easy to see that the two homomorphisms are compatible:

Proposition (IV.2.12). *The diagram*

$$\begin{array}{ccc} A & \xrightarrow{\ \Gamma\ } & C_0(\Omega) \\ {\scriptstyle \mathbf{in}_1}\downarrow & & \downarrow{\scriptstyle \mathbf{in}_2} \\ A_+ & \xrightarrow{\ \Gamma_+\ } & C(\Omega_+) \end{array}$$

in which $\mathbf{in}_1$ *and* $\mathbf{in}_2$ *are the natural embeddings, is commutative.* ▶■

Let us consider the image of Γ, denoted in future by $\hat{A}$; obviously, this is a subalgebra of $C_0(\Omega)$. For any $f \in \hat{A}$ we put $\|f\|^{\wedge} := \inf\{\|a\| : a \in A;\ \Gamma(a) = f\}$.

Proposition (IV.2.13). $(\hat{A};\ \|\cdot\|^{\wedge})$ *is a function algebra on* Ω *which is isometrically isomorphic to* $A/\operatorname{Rad} A$. *Moreover its*

---

[6] As in the literature on Fourier transforms (which, as we shall see in §3.2, turn out to be a particular case of the concept we have introduced), the term "Gel'fand transform" is used both for an individual function, formed using the given construction, and for the construction as a whole. [However, in this English translation, the word "transform" tends to be used for the individual function and "transformation" for the process.]

spectrum coincides, up to a homeomorphism, with the spectrum of A.

▶ This follows immediately from Theorem 2.11(II) and Proposition 1.13. ■

The Gel'fand transformation is the construction promised at the beginning of the chapter that makes a function algebra out of a given (abstract) commutative Banach algebra. We give the appropriate summary.

Theorem (IV.2.14) (Gel'fand). *Let* A *be a commutative Banach algebra. Then there exists a continuous surjective homomorphism from* A *onto a certain Banach function algebra, defined on the (locally compact) space* $\underline{\Omega}(A)$, *whose kernel coincides with the radical of the algebra. As a consequence, a semisimple commutative Banach algebra is isometrically isomorphic to a certain Banach function algebra defined on its spectrum.* ▶■

In connection with this theorem, recall (I.3.62) which says that the elements of the radical of any algebra are mapped into zero under a homomorphism into a function algebra. Therefore the method described of getting a "functional representation" of the given algebra is, in a definite sense, the best possible: loosely speaking, the loss of information about the original algebra is reduced to a minimum.

We emphasize that the norm $\|\cdot\|^{\wedge}$ in the function algebra $\hat{A}$ is, in general, stronger than the uniform norm; in particular, if A is semisimple this means that its Gel'fand transformation, although injective, is not necessarily topologically injective. How can one characterize those algebras, for which the latter property is satisfied? The following gives the answer.

Theorem (IV.2.15). *The following properties of a commutative Banach algebra* A *are equivalent*:

(I) *there exists a norm* $\|\cdot\|^{0}$ *on* A *that is equivalent to the original norm and satisfies* $\|a^2\|^{0} = (\|a\|^{0})^2$ *for all* $a \in A$;

(II) $\Gamma : A \longrightarrow C_0(\Omega)$ *is a topologically injective map*;

(III) $\hat{A}$ *is a closed subalgebra of* $C_0(\Omega)$.

▶ If (I) is satisfied, then by Theorem II.1.28, taking into account Theorem 2.3, we have $\|\hat{a}\|_0 = \lim_{n\to\infty}\left(\|a^{2^n}\|^0\right)^{1/2^n} = \lim_{n\to\infty}\left(\|a^{2^{n-1}}\|^0\right)^{1/2^{n-1}} = \cdots = \|a\|^0$; in this way (I) implies (II). Conversely, if (II) is satisfied, then, putting $\|a\|^0 := \|\hat{a}\|_0$, we get (I). Finally, the equivalence of (II) and (III) obviously follows from Banach's theorem. ■

An algebra having the properties indicated in the theorem is called a *uniform algebra* or a *sup-norm algebra*. Research into these algebras has for years formed a special section in the theory of Banach algebras, having links with complex analysis and having a characteristic range of problems and results. The monograph [17] is devoted to uniform algebras.

We now show that the Gel'fand transformations of various algebras are mutually compatible. The appropriate precise assertion can be formulated most simply for unital algebras, to which we shall limit ourselves. We consider. along with the spectrum functor $\underline{\Omega} : \mathcal{UCBA} \longrightarrow \mathcal{CT}op$ (see §1.2), a functor $\underline{C}$, acting in the opposite direction, that sends the compact space Ω to the algebra $C(\Omega)$ and the continuous map between compact spaces $\omega : \Omega_1 \longrightarrow \Omega_2$ to the homomorphism $\underline{C}(\omega) : C(\Omega_2) \longrightarrow C(\Omega_1)$ $: f \longmapsto f \circ \omega$. Then one can immediately verify the following.

Proposition (IV.2.16). *The commutative diagram*

$$\begin{array}{ccc} A & \xrightarrow{\ \varphi\ } & B \\ {\scriptstyle \Gamma_A}\downarrow & & \downarrow{\scriptstyle \Gamma_B} \\ C(\underline{\Omega}(A)) & \xrightarrow{\ \underline{C}\circ\underline{\Omega}(\varphi)\ } & C(\underline{\Omega}(B)) \end{array}$$

holds for every morphism φ *in the category* $\mathcal{UCBA}$, in other words (cf. Chap.0 §2.2) *the collection* $\{\Gamma_A : A \in \mathcal{UCBA}\}$ *forms a natural transformation between the identity functor in* $\mathcal{UCBA}$ *and the composition of functors* $\underline{C} \circ \underline{\Omega} : \mathcal{UCBA} \longrightarrow \mathcal{UCBA}$. ▶■

A reader who is fairly familiar with category theory, probably feels that here "there is a whiff of conjugate functors". We suggest that such a reader strengthens the observation we have just made:

Exercise (IV.2.17). Show that the functors $\underline{\Omega}$ and $\underline{C}$ are conjugate.

The following theory shows that the Gel'fand transformation is compatible with the holomorphic calculus. For simplicity let A again be unital. We fix $a \in A$ and a function $w \in \mathcal{O}(U)$ such that the domain U contains $\mathrm{Sp}_A a$. Then, according to Theorem II.2.15, there is in A an element $w(a) := \Gamma o \lambda(w)$, where $\Gamma o \lambda$ is the holomorphic calculus for a in U, which by Theorem II.2.4 must be unique; in this way we can give a meaning to the Gel'fand transform $w(a)^{\wedge} = \Gamma_A(w(a))$. On the other hand, since $\hat{a}(s) \in \mathrm{Sp}_A\, a$ for any $s \in \underline{\Omega}(A)$, the composition $w \circ \hat{a} : \underline{\Omega}(A) \longrightarrow \mathbb{C} : s \longmapsto w(\hat{a}(s))$ is also well defined.

Theorem (IV.2.18). *The functions* $w(a)^{\wedge}$ *and* $w \circ \hat{a}$ *coincide.*

▶ For any $s \in \underline{\Omega}(A)$ and the corresponding character χ_s, it follows from Theorem II.2.18 that $w(a)^\wedge(s) = \chi_s(w(a)) = w(\chi_s(a)) = w(\hat{a}(s))$. ■

A similar theorem also holds for the holomorphic calculus in several elements of a commutative Banach algebra, as described in Chap.II §2.2: if w is a function that is holomorphic in the neighbourhood of the joint Shilov spectrum of the elements $a_1, \cdots, a_n \in A$, then we have $[w(a_1, \cdots, a_n)]^\wedge(t) = w(\hat{a}_1(t), \cdots, \hat{a}_n(t))$ for all $t \in \underline{\Omega}(A)$. One of the most important applications of this equation, virtually as important as the construction of the multi-dimensional holomorphic calculus itself, is the following assertion.

Assertion (IV.2.19) ("Shilov's idempotent theorem"; see, for example, [14, p.51]). *Let the spectrum of a commutative Banach algebra* A (not necessarily unital) *contain a subset* Ω_0 *that is simultaneously both compact and open. Then there exists an idempotent* p *in* A *such that* $\hat{p}(s) = 1$ *when* $s \in \Omega_0$ *and* $\hat{p}(s) = 0$ *for all remaining* s.

It is worth recalling a recent result of Putinar [214] showing that an analogous equation $[w(S_1, \cdots, S_n)]^\wedge(t) = w(\hat{S}_1(t), \cdots, \hat{S}_n(t))$ holds for the multi-operator holomorphic calculus of Taylor (in the sense of [250], cf. Chap.II §2.2) Here $\mathcal{S} = (S_1, \cdots, S_n)$ is a commutative system of operators on the Banach space E, w is any function which is holomorphic in a neighbourhood of its joint Taylor spectrum $\operatorname{Sp}\mathcal{S}$ and the hat "$\wedge$" denotes the Gel'fand transformation in any closed commutative subalgebra of $\mathcal{B}(E)$ containing all holomorphic functions of the system $\mathcal{S}$ (taken over all domains containing $\operatorname{Sp}\mathcal{S}$).

* * *

It is not surprising that after introducing the Gel'fand transformation a lot of effort has been spent on extending this construction in some form or other to important classes of algebras other than commutative or Banach algebras. Moreover there has been a measure of success in preserving some of the properties described above, although some properties have inevitably been lost. Several results of this kind about commutative polynormed algebras will be stated in the next chapter. On the other hand, for some special classes of non-commutative Banach algebras there also exists a satisfactory construction generalizing the Gel'fand transform: this is the so-called representation as cross-sections of continuous fields of Banach algebras. For this sort of question see, for example, [28] and [31 §10].

§3. REPRESENTATION BY MEANS OF FUNCTIONS OF SPECIFIC CLASSES OF ALGEBRAS

When we are considering specific classes of commutative Banach algebras with applications in some scientific domain or other, the Gel'fand transformation usually takes a specific form corresponding to a set of concepts in the given discipline. It turns out moreover that many of the constructions devised earlier in various domains of analysis and topology, and to all appearances in no way interrelated, are, in fact, some specialized form or other of this transformation.

§3.1. Function algebras: transferring functions to a "revised" domain of definition. Application: maximal compact extensions

We begin with algebras that from the outset are given as algebras of continuous functions. For them, the general idea behind the Gel'fand transformation, informally speaking, is that it shows how the functions in question should appear on a domain of definition that is a version of the original one but revised by removing "superfluous" points and adding "omitted" points.

Example (IV.3.1) ("Let well alone"). Let A be a Banach function algebra on the compact set Ω that separates the points of the latter and is such that all its non-trivial characters are fixed. Then, in view of Corollary 2.8, the spectrum of this algebra coincides (to within a homeomorphism) with Ω and $\Gamma : A \longrightarrow C(\Omega)$ is a natural embedding. Thus, for such algebras (and this includes, as we have seen, $C(\Omega)$ itself, and also $C^n[0, 1]$, $\mathcal{A}(\overline{\mathbb{D}}^n)$, W) the specific form of Γ tells us that the first choice of the domain of definition for the functions was the "correct" one.

Example (IV.3.2). Let A be the algebra of continuous 2π-periodic functions on $\mathbb{R}$ with pointwise operations and uniform norm. Then its spectrum, obviously, is homeomorphic to the circle $\mathbb{T} \simeq \mathbb{R}/2\pi\mathbb{Z}$ and $\Gamma : A \longrightarrow C(\mathbb{T})$ takes $f \in A$ into the well-defined function $\hat{f} : e^{ix} \longmapsto f(x)$. Thus the revision of the domain of definition — in this case of the real line — consists of identifying ("gluing") points that are distant from one another by some multiple of 2π.

We suggest the reader considers a natural generalization of the given example.

Exercise (IV.3.3). Let $\Omega = \coprod_{\nu \in \Lambda} \Omega_\nu$ be a partitioning of a certain compact space into closed subsets, let A be the subalgebra of $C(\Omega)$ consisting of those functions that are constant on each Ω_ν; $\nu \in \Lambda$. Find the spectrum and the Gel'fand transformation of this algebra.

Hint. "Glue" each of the Ω_ν to a point.

A more interesting situation is where the revision of the original domain of definition of the functions consists of adjoining new points. Here different kinds of behaviour are possible among which we can distinguish, roughly speaking, "continuous extension" and "analytic extension".

We first consider an arbitrary, completely regular, (= Tikhonov) space Δ and the algebra $C_b(\Delta)$ consisting of all bounded continuous functions on Δ furnished with the uniform norm (cf. Chap.III §1.1). Let Ω be the spectrum of this algebra. We note that the two spaces are connected by an injective map $\omega : \Delta \longrightarrow \Omega$ associating to every $t \in \Delta$ the point of the spectrum corresponding to the character $\chi : f \longmapsto f(t)$. For brevity we put $\Delta' = \operatorname{Im} \omega$ and $\omega' = \omega|^{\Delta'}$.

Proposition (IV.3.4). (I) *$\omega' : \Delta \longrightarrow \Delta'$ is a homeomorphism.*
(II) *Δ' is dense in Ω.*

▶ (I) The form of a neighbourhood base in Ω and the continuity of the functions $f \in C_b(\Delta)$ immediately implies that ω, and hence also ω', is continuous. Further, since Δ is completely regular, it follows that that any point t in Δ has a neighbourhood base consisting of sets of the form $V_{t,f,\varepsilon} = \{s \in \Delta : |f(s) - f(t)| < \varepsilon\}$; $f \in C_b(\Delta)$, $\varepsilon > 0$. Since, in the notation of 2.4, $\omega'(V_{t,f,\varepsilon}) = \Delta' \cap U_{\omega(t),f,\varepsilon}$, this means that ω' is open.

(II) By Proposition 2.4, it is sufficient to show that, for any $t' \in \Omega$, any neighbourhood of it of the form $U = \bigcap_{k=1}^{n} U_{t',f_k,\varepsilon}$; $f_1, \cdots, f_n \in C_b(\Delta)$, $\varepsilon > 0$, contains at least one point from Δ'. Replacing f_k, if necessary, by $f_k - \hat{f}_k(t')$, we may assume that, for all f_k, we have $\hat{f}_k(t') = 0$, that is, $f_k \in I_{t'}$. But then the function $g = \sum_{k=1}^{n} f_k\overline{f}_k$, being an element of this ideal $I_{t'}$, is not invertible in $C_b(\Delta)$; the latter, obviously, means that $\inf\{g(s); s \in \Delta\} = 0$. All that remains is to take $s \in \Delta$ such that $g(s) < \varepsilon^2$, and to check directly that $s' = \omega(s) \in \Delta'$ belongs to U. ■

From now on we shall, using the homeomorphism ω', identify Δ with Δ' and thus think of Δ as a dense subset of Ω.

Corollary (IV.3.5). *$\Gamma : C_b(\Delta) \longrightarrow C(\Omega)$ associates with each function a continuous extension and is an isometric isomorphism between the two algebras.*

In topology any compact space which contains a given topological space as a dense subset, is customarily called a *compactification* of this space. Proposition 3.4 thus provides a certain special compactification of the space Δ; we shall now see that it may be distinguished among the others by its exclusive properties.

A compactification of the space Δ denoted (traditionally) by $\beta\Delta$ is said to be *maximal* if, for any compactification $K \supset \Delta$, there exists a (necessarily unique) continuous map from $\beta\Delta$ into K leaving the points of Δ fixed; we note that, as the continuous image of a compact space is itself compact, such a map is necessarily surjective. Obviously, any two maximal compactifications coincide to within a homeomorphism that is the identity on Δ.

Proposition (IV.3.6). *Let $\tilde{K}$ be a compactification of the space Δ such that any continuous bounded function on Δ admits a continuous extension to the whole of $\tilde{K}$. Then $\tilde{K}$ is a maximal compactification.*

▶ Let K be another compactification of Δ. For any $f \in C(K)$ we let $\tilde{f}$ denote the continuous extension of $f|_{\Delta}$ to the whole of $\tilde{K}$; obviously $\varphi : C(K) \longrightarrow C(\tilde{K}) : f \longmapsto \tilde{f}$ is a continuous homomorphism. Then the spectrum functor (see §1.2) provides a continuous map $\alpha = \underline{\Omega}(\varphi) : \tilde{K} \longrightarrow K$, where, if $s \in \Delta$, then $f(s) = \tilde{f}(s)$ implies that $\alpha(s) = s$. ■

What has been proved, taking into account Proposition 3.4, implies the following.

Theorem (IV.3.7). *Any completely regular space Δ has a maximal compactification:*[7] *namely, the spectrum of the algebra $C_b(\Delta)$.* ■

The discovery of this fact, that every completely regular space has a maximal compactification, is associated with the names of Tikhonov, Čech and Stone, who used purely topological methods [244], [281]. The theorem that says that such an extension may be exhibited as $\underline{\Omega}(C_b(\Delta))$ was proved by Gel'fand and Shilov [115]. (We note that the requirement of complete regularity is not only sufficient but is also a necessary condition for the existence of compactifications: any compact space is completely regular [95] and hence so is any subspace.)

For certain specific spaces the maximal compactification can be constructed fairly simply. For example, if Δ is the space of countable ordinals with the order topology [95, 3.1.27], then $\beta\Delta$ coincides with the one-point compactification of Δ and the adjoined point may be identified with the first uncountable ordinal [95, 3.6.10]. In many cases, however, the maximal compactification may turn out to be considerably more complicated

[7] We note for comparison that "minimal" compactifications — that is, one-point compactifications — exist only for locally compact spaces.

than the original space. An example is $\beta(\mathbb{N})$, where $\mathbb{N}$ is taken with the discrete topology: this space contains 2^c points (here c is the cardinality of the continuum) and it contains not a single convergent sequence of distinct points [95, 3.6.12 and 15]. We recall that, according to Theorem 3.7, $\beta(\mathbb{N})$ is the spectrum of the algebra of the bounded sequences $c_b = C_b(\mathbb{N})$, introduced in Chap. III §1.3.

For various other aspects of the "phenomenon of continuous extension" that occur when we take the Gel'fand transformation of certain significant classes of uniform algebras other than $C_b(\Delta)$, see, for example [20 §43] (where, in particular, they mention algebras of almost periodic functions) and also [67, p.273] (where the algebra $L^\infty[0, 1]$ is discussed).

Let us say a word about the other kind of phenomenon, about what we called analytic extension: we limit ourselves to the simplest examples of this.

Example (IV.3.8). Consider the algebra $\mathcal{A}(\mathbb{T})$ consisting of those functions on the circle that are uniform limits of polynomials. Then, as we know from Chap.III §1.4, $\mathcal{A}(\mathbb{T})$ is isometrically isomorphic to the "disc algebra" $\mathcal{A}(\overline{\mathbb{D}})$; therefore Proposition III.1.23, describing the characters of the last algebra, implies that the spectrum of $\mathcal{A}(\mathbb{T})$ is the whole of the disc $\overline{\mathbb{D}}$. Loosely speaking, in this case the additional points, which were not in the original domain of definition, form an "analytic membrane stretched over the circle". As for $\Gamma : \mathcal{A}(\mathbb{T}) \longrightarrow C(\overline{\mathbb{D}})$, this associates to each function its analytic continuation over the whole disc.

A somewhat more general situation is described by the following.

Exercise (IV.3.9). Let K be a compact subset of $\mathbb{C}^n$, let $\mathcal{A}(K)$ be the algebra of functions on K which are the uniform limits of polynomials (with the sup-norm). Prove that the spectrum of this algebra coincides, to within a homeomorphism, with the set $\{z \in \mathbb{C}^n : |p(z)| \le \max_{\lambda \in K} |p(\lambda)|$ for any polynomial $p\}$ (that is, with the so-called polynomially convex hull of the set K).

This fact is important in approximation theory for functions of several complex variables; for details of this see [55].

§3.2. Group algebras: the Fourier transform

From function algebras we pass to group algebras. We begin with the algebra $L^1(\mathbb{R})$ that serves as a typical example.

Theorem (IV.3.10). *There exists a homeomorphism between the spectrum Ω of the algebra $L^1(\mathbb{R})$ and the line $\mathbb{R}$ (with the usual topology) such that, after making the corresponding* *identification of* Ω *with* $\mathbb{R}$, *the Gel'fand transformation* $\Gamma : L^1(\mathbb{R}) \longrightarrow C_0(\mathbb{R})$ *takes the function* f *into* $\hat{f}(x) = \int_{\mathbb{R}} f(s)e^{-isx}ds$. (In other words, *the Gel'fand transformation of* $L^1(\mathbb{R})$ *is none other than the classical Fourier transformation.*)

▶ The description of the characters of the algebra $L^1(\mathbb{R})$ given in Theorem III.3.13 immediately implies two things: (I) Ω may be identified, as a set, with $\mathbb{R}$, and (II) under this identification the Gel'fand transform of each function in $L^1(\mathbb{R})$ coincides with its Fourier transform. Since the Fourier transform of every integrable function belongs to $C_0(\mathbb{R})$, Proposition 2.7 ensures that Ω and $\mathbb{R}$ coincide as topological spaces. ■

With the help of such arguments, using the remaining assertions of Theorem III.3.13, we can rediscover some more "old acquaintances". We suggest the reader examines two cases.

Exercise (IV.3.11). Show that the spectrum of the algebra $l_1(\mathbb{Z})$ is homeomorphic to the circle $\mathbb{T}$ and, after making the corresponding identification, $\Gamma : l_1(\mathbb{Z}) \longrightarrow C(\mathbb{T})$ takes $a = \{a_n;\ n \in \mathbb{Z}\}$ into $\hat{a}(z) = \sum_{n=-\infty}^{\infty} a_n z^n$; $z \in \mathbb{T}$. (In other words, $\Gamma = i \circ j$, where j is the isometric isomorphism between $l_1(\mathbb{Z})$ and the Wiener algebra W described in Example I.2.12 while i is the natural embedding of W in $\mathbb{C}(\mathbb{T})$.)

Exercise (IV.3.12). Show that the spectrum of the algebra $L^1(\mathbb{T})$ is homeomorphic to $\mathbb{Z}$ (with the discrete topology) and, after making the corresponding identification, $\Gamma : L^1(\mathbb{T}) \longrightarrow C_0(\mathbb{Z})$ associates to each integrable function on the circle the sequence of its Fourier coefficients.

Starting from the examples we have considered, one can guess what form the Gel'fand transformation will take for the L^1-algebra of an arbitrary group (see III §3.2).

Assertion (IV.3.13) [20, p.135]. *Let* G *be a locally compact Abelian group. Then the spectrum of the algebra* $L^1(G)$ *is homeomorphic to the set* $\hat{G}$ *of characters of the group* G, *furnished with the Pontryagin topology. If we make the appropriate identification, then* $\Gamma : L^1(G) \longrightarrow C_0(\hat{G})$ *is none other other than the (abstract) Fourier transformation.*

Remark (IV.3.14). This assertion, together with Theorem 2.11 (II), means that the algebra $L^1(G)$ is semisimple $\Longleftrightarrow$ the "uniqueness theorem for Fourier transforms" holds (cf. III.3.20). In this way, for all the familiar groups — for example $\mathbb{R}$ and $\mathbb{T}$ — the semisimplicity of the L^1-algebras follows immediately from a classical theorem on Fourier integrals or series. It is interesting that in the general case, where G

is arbitrary, it is more convenient to do quite the reverse: first, using a completely different approach, we show that the algebras $L^1(G)$ are semisimple and then as corollaries we obtain the uniqueness theorem together with the theorem on the sufficiency of the set of characters (III.3.21). We shall talk later about a suitable method for proving semisimpicity; see 5.32 - 5.37.

As before, let G be a locally compact Abelian group. Although, as we have just shown, the Gel'fand transformation of the algebra $L^1(G)$ can be described explicitly in the terminology of harmonic analysis, if we consider another algebra connected with G — the measure algebra $M(G)$ — the picture is drastically more complicated. First of all, the domain of definition itself of the functions that arise — that is, the spectrum of our algebra — is a very complicated object both from the point of view of how its elements are defined (essentially we have already spoken about this in Chap.III §3.3) and from a topological point of view. In general, there are far more negative results — counterexamples to various conjectures — about the Gel'fand transformation of the measure algebra than there are positive ones. Here we only cite, and moreover, without producing any proof, several of the most important facts that are relevant here, referring the reader for details to the monograph by Taylor [321]. (The survey paper by Hewitt [278] gives a good presentation of the situation in this area at a comparatively early stage.)

We put $\Omega = \underline{\Omega}(M(G))$. In Chap.III §3.3 we described the set of points of Ω that correspond to absolutely continuous characters of the measure algebra; recall that this is the name given to those characters defined by

$$\chi(\mu) = \int_G \tilde{\chi}(s)d\mu(s); \quad \mu \in M(G),$$

in which $\tilde{\chi}$ is some (continuous) character of the group G. In this way, using the above formula, one constructs a map $\hat{G} \longrightarrow \Omega : \tilde{\chi} \longmapsto \chi$ which, as it is not difficult to verify, gives a homeomorphism of $\hat{G}$ onto an open subset of Ω (we recommend the reader to show that the inverse of this homeomorphism is a particular case, namely, for $I = L^1(G)$ and $A = M(G)$, of what we were talking about in Exercise 1.19). Thus we are justified in assuming that Ω contains $\hat{G}$ as an open subset.

Take an arbitrary measure $\mu \in M(G)$ and consider the restriction of its Gel'fand transform $\hat{\mu}(t)$; $t \in \Omega$ to $\hat{G}$; in other words, we consider the function on $\hat{G}$

$$\hat{\mu}(t) = \int_G \tilde{\chi}_t(s)d\mu(s),$$

where $\tilde{\chi}_t$ is the character of the group G corresponding to the point $t \in \hat{G}$ (if you like, you can say that it is this point). The function thus constructed $\hat{\mu}(t)$; $t \in \hat{G}$, is called the *Fourier-Stieltjes transform* of the measure μ; this is, so to speak, the "visible part" of the Gel'fand transform of our measure. Theorems 2.5 and 2.11(I) imply immediately that this function is continuous and $\sup\{|\hat{\mu}(t)| : t \in \hat{G}\} \le \|\mu\|$.

As we have already said in Chap.III §3.3 (in terms of characters), $\hat{G}$ forms a "very small" part of Ω. There, we drew attention to the difference in their cardinalities; here we add the remark that what was said can also be understood from the topological point of view: generally speaking, $\hat{G}$ is not even dense in Ω [20]. It is interesting, however, that, in spite of this last fact, any function $\hat{\mu} : \Omega \longrightarrow \mathbb{C}$ (that is, the Gel'fand transform of a measure μ) is uniquely defined by its restriction to $\hat{G}$ (that is, by the Fourier-Stieltjes transform of this measure) — in exactly the same way as analytic functions are uniquely defined by their restrictions to a set of uniqueness. This follows immediately from the following important theorem in harmonic analysis.

Assertion (IV.3.15) ("uniqueness theorem for Fourier-Stieltjes transforms"). *If the Fourier-Stieltjes transforms of two measures from* $M(G)$ *coincide* (are identical on $\hat{G}$), *then the measures themselves coincide.*

Of course, this result generalizes the uniqueness theorem for Fourier transforms (III.3.20), which corresponds to the case of measures that are absolutely continuous with respect to Haar measure. It is true the proof becomes more complicated and requires new ideas; see, for example, [87, 31.5]. An obvious corollary of Assertion 3.15 is that the algebra $M(G)$ is semisimple.

Remark (IV.3.16). In fact the functions $\hat{\mu}$; $\mu \in M(G)$ not only behave like analytic functions (cf. above) but actually are analytic functions with respect to a certain infinite-dimensional analytic structure that exists on part of Ω. For the appropriate exact definitions and results see, for example, [111], [247].

Of the "positive" results about the structure of the spectrum of the measure algebra, it is worth mentioning the description of its integral cohomology groups due to Taylor [248]; for zero-dimensional cohomology it gives Cohen's theorem [164], which was known before, that characterizes the idempotents in $M(G)$ (= the measures whose Gel'fand transform takes only the values 0 and 1).

└

* * *

The Gel'fand transformation plays no less a role for the third traditional class of algebras in analysis — the operator algebras — than it does for function algebras and group algebras. Here, as we shall see, this construction turns into the so-called continuous functional calculus for an operator and, in the particular case when the algebra is generated by some self-adjoint operator, it turns out to be an equivalent way of writing the spectral decomposition of this operator.

However, we are not yet completely ready, in our exposition, for these questions: they belong to a set of questions that involve an additional algebraic structure — involution. Therefore the relevant results appear towards the end of this chapter in §7.2. We now deal with involution in real earnest.

§4. INVOLUTIONS AND STAR ALGEBRAS

§4.1. Initial definitions and properties

As has already been noted at the beginning of the chapter, an involution is a structure which automatically gives a certain intrinsic symmetry in the algebra under consideration. Here is the precise definition.

Definition (IV.4.1). Let A be an algebra. A map $*: A \longrightarrow A$ is called an *involution* in A if, writing a^* instead of $*(a)$, for all $a, b \in A$, $\lambda \in \mathbb{C}$, the following are satisfied:

(I) $(a+b)^* = a^* + b^*$, (II) $(\lambda a)^* = \bar{\lambda} a^*$,
(III) $(ab)^* = b^* a^*$, (IV) $a^{**} = a$.

An algebra furnished with an involution is called an *involutive algebra* or briefly a **-algebra*.

The simplest *-algebra (excluding, of course, the zero algebra) is $\mathbb{C}$ with the operation of complex conjugation as the involution. A basic stock of examples will be given in the next section.

A homomorphism $\kappa: A \longrightarrow B$ between *-algebras is called an *involutive homomorphism* or a **-homomorphism* if $\kappa(a^*) = \kappa(a)^*$ for all $a \in A$. A similar meaning is also given to more general concepts like **-map* and **-operator* and also very often to concepts like **-isomorphism* and **-character* (in the last case $\mathbb{C}$ is considered to have the involution introduced above).

If a is an element of the *-algebra A, then a^* is called the *adjoint* of a. An element that is equal to its adjoint is called *self-adjoint* or *Hermitian*. The set of self-adjoint elements in A is called the *self-adjoint part* of A and is denoted by A_{sa}; obviously this is a (real) subspace of the underlying real linear space of the algebra A. We also note that the product of two elements from A_{sa} remains in A_{sa} $\iff$ these elements commute.

Proposition (IV.4.2). *Any element of a* **-algebra* A *can be represented uniquely as* $h + ik$, *where* $h, k \in A_{sa}$. *Here* $h = \frac{1}{2}(a + a^*)$ and $k = \frac{1}{2i}(a - a^*)$. ▶■

Corollary (IV.4.3). *A character* χ *of the* **-algebra* A *is a* **-character* $\iff$ $\chi(h) \in \mathbb{R}$ *for all* $h \in A_{sa}$ $\iff$ *(when* A *is unital)* $\chi(h) \neq i$ *for any* $h \in A_{sa}$.

We note that one can construct another pair of self-adjoint elements from a given $a \in A$: these are obviously a^*a and aa^*.

An element $p \in A$ is said to be *positive* if (I) it is self-adjoint and (II) $\mathrm{Sp}_A(p) \subseteq [0, \infty)$. The set of positive elements in A is denoted by A_{pos}; obviously, if it contains a, it contains the whole ray $\{ta : t \geq 0\}$. (We shall see later that the question of the relation between the sets A_{pos} and $\{a^*a : a \in A\}$ turns out to be extremely important.)

In a unital *-algebra, it is obvious that $(a^{-1})^* = (a^*)^{-1}$ for any invertible a; in particular $e^* = e$. An element u of such an algebra is said to be *unitary* if it is invertible and $u^{-1} = u^*$. The involution in an arbitrary *-algebra A may be uniquely extended to the involution $* : a + \lambda e_+ \longmapsto a^* + \overline{\lambda} e_+$ in A_+.

An element a of a *-algebra is said to be *normal* if $a^*a = aa^*$. Thus self-adjoint elements and — if there is an identity — unitary elements are particular cases of normal elements.

If M is a subset of a *-algebra A, we put $M^* = \{a^* : a \in M\}$. Here M is called a *self-adjoint subset* or briefly a **-subset* (in A) if $M^* \subseteq M$ — which obviously implies $M^* = M$. A similar meaning is given to such terms as **-subalgebra*, **-ideal* and so on. Here is an example of their use: it is obvious that the kernel of any *-homomorphism is a *-bi-ideal of the appropriate *-algebra.

Proposition (IV.4.4). *If* I *is a left ideal of the* **-algebra* A, *then* I^* *is a right ideal; moreover, if* I *is maximal or modular, then so also is* I^*. *Further, the map* $I \longmapsto I^*$ *is a bijection between the set of left and right ideals of* A *and also between the corresponding sets of maximal (and) modular ideals.* ▶■

Remark (IV.4.5). It is very obvious from this proposition that algebras that are furnished with an involution have a great advantage compared with others: concepts and facts expressed in terms of right and left multiplication can be interchanged. As it is not difficult to see, the basis of all such results is that involution realizes an isomorphism between A and A^{op} as algebras over $\mathbb{R}$.

Proposition (IV.4.6). *If there exists a one-sided (left or right) identity in a *-algebra, then this algebra is unital.* ▶■

Proposition (IV.4.7). *The radical of a *-algebra is a *-bi-ideal.*

▶ If a belongs to all maximal modular left ideals of our algebra, then a^*, on the basis of Proposition 4.4, belongs to all maximal modular right ideals. All that remains is to use Theorem I.3.53. ■

Any *-subalgebra (in particular, any *-ideal) of a given *-algebra is obviously a *-algebra itself. Further, the quotient algebra of a *-algebra A with respect to a *-bi-ideal I is a *-algebra with respect to the involution that is well-defined by $(a + I)^* = a^* + I$; $a \in A$; we shall call it the *quotient-*-algebra*. Finally, the Cartesian product of any family of *-algebras is a *-algebra with respect to the "coordinatewise" definition of involution.

§4.2. Star algebras and basic examples; *-representations

Definition (IV.4.8). A Banach algebra furnished (as a "pure" algebra) with an involution is called a *star Banach algebra* if this involution is continuous.[8]

The more general concept of a *star polynormed algebra* is defined in a similar way and also a star multiplicatively polynormed algebra and so on; the meaning of such terms is obvious. We note that property (IV) in Definition 4.1 immediately implies the following.

Proposition (IV.4.9). *Let a *-algebra A be allotted an (arbitrary) topology. Then if the involution is continuous, it gives a homeomorphism from A onto itself.* ▶■

Let us discuss the way in which the property that the involution is continuous may be expressed in terms of the norm (or, according to the sense, the family of seminorms).

Definition (IV.4.10). Let A be a *-algebra. The (semi)norm $\|\cdot\|$ in A is called a *star (semi)norm* if $\|a^*\| = \|a\|$ for all $a \in A$.

Proposition (IV.4.11). *A Banach algebra furnished with an involution is a star Banach algebra $\Longleftrightarrow$ there exists in it a multiplicative star norm equivalent to the original (if A has an identity, this norm may also be assumed to be unital).*

---

[8] [Thus, in the author's terminology, a star Banach algebra must be distinguished from a Banach *-algebra.]

▶ Let the involution be continuous; then by Proposition 4.9 this is a topological automorphism of A as a real linear space. Consequently, for some $c, C > 0$, $c\|a\| \le \|a^*\| \le C\|a\|$ for all $a \in A$. Hence a new norm that is obviously a star norm is given, in A, by $\|a\|' = \max\{\|a\|, \|a^*\|\}$ and this is equivalent to the original norm. We also note that, if $\|\cdot\|$ is multiplicative and (for unital A) unital, then so also is $\|\cdot\|'$. The rest is clear. ■

The analogous assertion about polynormed algebras is suggested as an exercise.

Exercise (IV.4.12). Let A be a polynormed, respectively, multiplicatively polynormed algebra, furnished with an involution. Prove that A is a star algebra (of the given type) $\iff$ its topology may be given with the help of a system of star, or star and multiplicative, respectively, seminorms.

We leave the reader to explain what are the cases for which some class or other of star polynormed algebras allow the operation of taking *-subalgebras, quotient-*-algebras and Cartesian products (cf. the analogous questions about algebras without involution as considered in Chap.I §2.3 and §3.3). We pick out only the obvious fact that for a star Banach algebra any closed sub-*-algebra and any Hausdorff quotient-*-algebra of it are themselves star Banach algebras.

There are also a number of constructions involving the tensor product that can be made without going out of the class of star polynormed algebras. In a purely algebraic situation, as can easily be shown, the tensor product $A \otimes B$ of two *-algebras A and B is itself a *-algebra with respect to the well-defined involution $(\sum_{k=1}^{n} a_k \otimes b_k)^* := \sum_{k=1}^{n} a_k^* \otimes b_k^*$; $a_k \in A$, $b_k \in B$. The corresponding "Banach" result requires somewhat more delicate arguments:

Proposition (IV.4.13). *Let A and B be star Banach algebras. Then the Banach algebra $A \hat{\otimes} B$ (see I.2.26) is a star Banach algebra with respect to the involution determined uniquely by $(a \otimes b)^* := a^* \otimes b^*$ for all $a \in A$, $b \in B$. Moreover, if the norms in A and B are star norms, than so is the norm in $A \hat{\otimes} B$.*

▶ For any linear space E we let E^- denote the linear space with the same underlying additive group as E but with a new multiplication by scalars $\lambda \circ x$, equal to the "previous" $\bar{\lambda}x$. Obviously, if E is a Banach space, then so also is E^- (with respect to the same norm). In these terms the required involution in $A \hat{\otimes} B$ may be given by the commutative diagram

$$\begin{array}{ccc} A \hat{\otimes} B & \xrightarrow{\;*\;} & A \hat{\otimes} B \\ \Big\downarrow i & & \Big\uparrow *\otimes* \\ A \hat{\otimes} B & \xrightarrow{\;j\;} & A^{-} \hat{\otimes} B^{-} \end{array}$$

in which i is the operator associated with the bi-operator $(a, b) \longmapsto a \otimes b$ (and of course this is an isometric isomorphism), j is the identity map on the corresponding sets (and therefore an isometric anti-isomorphism of the corresponding Banach spaces), and finally $*\otimes*$ is the operator associated with a continuous bi-operator which must obviously be $(a, b) \longmapsto a^* \otimes b^*$.

It is immediately obvious from this construction that the map $*$ is continuous, that it has properties (I) and (II) from Definition 4.1 and that it acts on elementary tensors in the way indicated. This, taking into account the fact that $A \hat{\otimes} B = \overline{\ell.h.}\{a \otimes b : a \in A,\ b \in B\}$, obviously implies the remaining properties in the definition of an involution and also the assertion about uniqueness. The assertion about star norms follows easily from Theorem 0.3.29. ∎

Of course, this proposition carries over easily to more general classes of polynormed algebras with different versions of the topological tensor product (cf. Chap.I §2.1). Details of the corresponding formulations and proofs are left to the reader.

* * *

We give a few examples. The first of these is the most important in that it served in its time as the motivation for introducing the concept of an abstract involution.

Example (IV.4.14). Let H be a Hilbert space. Then $\mathcal{B}(H)$ is a star Banach algebra with respect to the operator norm and the involution given by taking the adjoint (= dual) operator in the usual sense. More generally, any self-adjoint Banach operator algebra in H (see the beginning of Chap.III §2) is a star algebra with respect to this same involution; in particular, $\mathcal{K}(H)$ and $\mathcal{N}(H)$ are star algebras with respect to this same involution.[9] They are both *-bi-ideals in $\mathcal{B}(H)$ and moreover $\mathcal{K}(H)$ is closed. Further the quotient-*-algebra $\mathcal{B}(H)/\mathcal{K}(H)$ is the Calkin algebra $\mathcal{C}(H)$ with involution $a + \mathcal{K}(H) \longmapsto a^* + \mathcal{K}(H)$ (cf. Chap.III §2.2).

---

[9] So also are the algebras $\mathfrak{S}_p(H)$; $1 < p < \infty$, mentioned in III.2.8.

Exercise (IV.4.15). Let σ denote one of the systems of seminorms, $\underline{s}$, $\mathfrak{m}\underline{s}$, $\underline{w}$ or $\mathfrak{m}\underline{w}$, defined in $\mathcal{B}(H)$ (see Chap.III §2.3). Show that the polynormed algebra $(\mathcal{B}(H); \sigma)$ is a star algebra when $\sigma = \underline{w}$ or $\sigma = \mathfrak{m}\underline{w}$ and is not a star algebra in the two other cases.

Remark (IV.4.16). For this reason, we sometimes consider, instead of $\underline{s}$ (that is the strong system of seminorms), the so-called strong* system $\|S\|'_x := \max\{\|Sx\|, \|S^*x\|\}$; $x \in H$ and the system $\mathfrak{m}\underline{s}$ is dealt with in a similar way. For details see, for example, [75, II.2].

Recall the concept of a representation of an algebra on a linear space (Chap.I §1.1). Example 4.14 enables an analogous "involutive" concept to be introduced which turns out to be very important in what follows.

Definition (IV.4.17). Any *-homomorphism from A into $\mathcal{B}(H)$ is called a **-representation of the *-algebra A on the Hilbert space H.*

Recall that we have already met special classes of *-representations earlier in Chap.III §2.2.

As a rule we shall associate with *-representations the same notation T and T_a as for ordinary representations (cf. Chap.I §1.1). If A is a star Banach algebra, then we say a *-representation T of it is *isometric* if $\|T_a\| = \|a\|$ for all $a \in A$. An isometric representation is of course injective.

A one-dimensional *-representation (corresponding to the case $\dim H = 1$, that is, $\mathcal{B}(H) \simeq \mathbb{C}$) of a given *-algebra is obviously the same as a *-character.

If we continue down the list of examples, we can ascertain that the majority of the specific algebras considered earlier have natural involutions.

Example (IV.4.18). Let Ω be a locally compact space. Then $C_0(\Omega)$ is a star Banach algebra with respect to the uniform norm and the involution given by taking the complex-conjugate of a function. More generally, any of the function algebras considered in Chap.III §1, which if it contains a function also contains its complex-conjugate, is a star polynormed algebra of some class or other with respect to this same involution; in particular, these include $C(\Omega)$, $C^n[0, 1]$; $n = 1, 2, \cdots, \infty$ and W.

Example (IV.4.19). The Banach group algebras $l_1(G)$, where G is a discrete group, $L^1(\mathbb{R})$ and $L^1(\mathbb{T})$ are star algebras with respect to the involution given, using the multiplicative notation for the inverse of an element, by $f^*(s) = \overline{f(s^{-1})}$. Involution is defined analogously for the L^1-algebras of all those locally compact groups in which the left- and right-invariant measures coincide. In the general case the formula is somewhat more complicated due to the appearance of the so-called modular function (see, for example, [31, §13.2]).

The algebra $M(G)$, where G is an arbitrary locally compact group, is a star algebra with respect to the involution $\mu \longmapsto \mu^*$, uniquely defined by $\int_G f(t)d\mu^*(t) = \int_G f(t^{-1})d\mu(t)$ for all $f \in C_0(G)$. Moreover $L^1(G)$ with the above involution may be identified with a *-bi-ideal in $M(G)$ (cf. Chap.III §3.3).

Involutions can also be introduced in some analytic function algebras.

Example (IV.4.20). The Banach algebra $\mathcal{A}(\overline{\mathbb{D}})$ is a star algebra with respect to the involution $w \longmapsto w^* : w^*(z) = \overline{w(\overline{z})}$; $z \in \overline{\mathbb{D}}$. The same formula gives an involution in $\mathcal{O}(U)$, where U is a domain in $\mathbb{C}$, which is symmetric with respect to the real axis; moreover this is, obviously, a Fréchet star algebra.

Note that among the star algebras listed, some turn out to be better than others from the point of view of the subsequent general theory. Our last example provides a "quite bad" star algebra.

Example (IV.4.21). We furnish $\mathbb{C}^2$ (with coordinatewise multiplication and arbitrary norm) with a new involution — unlike the one introduced for $\mathbb{C}^M$ in Example 4.18 — by putting $(\xi_1, \xi_2)^* = (\overline{\xi}_2, \overline{\xi}_1)$; $\xi_1, \xi_2 \in \mathbb{C}$. The star algebra thus constructed is denoted by $\widetilde{\mathbb{C}}^2$. We note that one can take $C(\Omega)$, where $\Omega = \{i, -i\}$, (i being here the imaginary unit), furnished with the involution given by the same formula as in Example 4.20, as another realization (to within a *-isomorphism) of this *-algebra.

The following useful observations show the dissimilarity in the properties of our algebras. They are of a preliminary character; we shall see later what are the general facts that lie behind them.

Exercise (IV.4.22). Which of the maximal ideals in $C(\Omega)$ (Ω is a compact space), $C^n[0, 1]$, $L^1(\mathbb{R})$, $\mathcal{A}(\overline{\mathbb{D}})$ and $\widetilde{\mathbb{C}}^2$ are *-ideals?

Answer. In $\widetilde{\mathbb{C}}^2$ — none of them, in $\mathcal{A}(\overline{\mathbb{D}})$ only those that "lie on the real axis", in the other algebras — all of them.

Exercise (IV.4.23). Show that (I) $C[-1, 1]$ has an isometric *-representation; (II) $C^n[-1, 1]$ and $\mathcal{A}(\overline{\mathbb{D}})$ have an injective but do not have an isometric *-representation; (III) $\widetilde{\mathbb{C}}^2$ has only the zero *-representation.

Hint. To construct the representations, consider the operation of multiplication by a function on $H = L^2[-1, 1]$. Then use the fact that, when $S \in \mathcal{B}(H)$, $\|S^*S\| = \|S\|^2$; in particular $S^* = S$ implies $\|S^2\| = \|S\|^2$ and $S \neq 0$ implies $S^*S \neq 0$.

§5. POSITIVE FUNCTIONALS, THE GNS-CONSTRUCTION AND THE UNIVERSAL REPRESENTATION

§5.1. Positive functionals and associated GNS-representations

Everywhere in this section A is a certain *-algebra.

The following class of functionals plays an outstanding role in the theory of star algebras; comparable to the role played by characters in the theory of commutative algebras.

Definition (IV.5.1). A functional f on A is said to be *positive* if $f(a^*a) \geq 0$ for each $a \in A$. If A is unital, then a positive functional f on A, such that $f(e) = 1$, is called a *state.*[10]

We shall denote the set of positive functionals on A by PF(A) and, in the unital case, the set of states by <u>St</u>(A).

Example (IV.5.2). Obviously, every *-character of an algebra is a positive functional and every non-trivial *-character of a unital algebra is a state.

Let us indicate an important generalization of this example. Let $T : A \longrightarrow \mathcal{B}(H)$ be a certain *-representation. For any $x \in H$ we let $f_{T,x}$ denote the functional on A defined by $f_{T,x}(a) := \langle T_a x, x\rangle$. A functional of the form $f_{T,x}$ is called a *vector functional* (for each possible T and x).

Proposition (IV.5.3). *For fixed* T

(I) $f_{T,x}$ *is positive for all* $x \in H$, *and*

(II) $\|T_a\|^2 = \sup\{f_{T,x}(a^*a) : x \in H, \|x\| = 1\}$ *for any* $a \in A$.
If A *is unital and* $H_1 := \operatorname{Im} T_e$, *then, in addition,*

(III) $f_{T,x}$ *is a state for all* $x \in H_1$ *with* $\|x\| = 1$, *and*

(IV) $\|T_a\|^2 = \sup\{f_{T,x}(a^*a) : x \in H_1, \|x\| = 1\}$ *for any* $a \in A$.

▶ $\langle T_{a*a}x, x\rangle = \langle T_a^*T_a x, x\rangle = \langle T_a x, T_a x\rangle$ for all $a \in A$ and $x \in H$; hence $f_{T,x}(a^*a) = \|T_a x\|^2$. So, if A is unital, $\|T_a x\|^2 = \|T_a T_e x\|^2 = f_{T,x_1}(a^*a)$, where $x_1 := T_e x$. The rest is clear. ■

---

[10] This term is inherited from physics and is closely connected with the (physical) concept of the state of a system in quantum-mechanics; see, for example, [10, pp.6–7].

Thus, starting from a given *-representation, we get a certain set of positive functionals. Conversely, the basic content of this and the next section is a description of a method that, for a wide class of *-algebras, enables one to construct *-representations starting from given positive functionals. Moreover it turns out, in particular, that Proposition 5.3 actually describes all positive functions on these algebras (that is, for such A, PF(A) consists entirely of vector functionals). It is usual to call the construction set out below the GNS-*construction* in honour of its creators Gel'fand, Naimark, and Segal [114], [237].

From now on, until further notice, f is a fixed positive functional on A. Naturally, we should first of all get a Hilbert space from somewhere or other — this will be the field of action of the operators that arise. With this in mind, we note the following.

Proposition (IV.5.4). $f(b^*a) = \overline{f(a^*b)}$ *for any* $a, b \in A$.

▶ $f((\lambda a + b)^*(\lambda a + b)) \geq 0$ for all $\lambda \in \mathbb{C}$; therefore, $\lambda f(b^*a) + \overline{\lambda} f(a^*b) \in \mathbb{R}$ for all λ. This is possible only when $f(b^*a)$ and $f(a^*b)$ are complex conjugates. ■

We consider the complex function on $A \times A$ that sends the pair $a, b \in A$ to the number $f(b^*a)$. Definition 5.1 and Proposition 5.4 imply that this function has all the properties of a scalar product, except that the "scalar square" $f(a^*a)$ being zero does not, in general, imply that $a = 0$. The following follows immediately from this.

Proposition (IV.5.5) (the "Cauchy–Bunyakovskii inequality"). $|f(b^*a)|^2 \leq f(a^*a)f(b^*b)$ *for all* $a, b \in A$. ▶■

Corollary (IV.5.6). *For any* $a \in A$ *the conditions* (I) $f(a^*a) = 0$; (II) $f(ba) = 0$ *for all* $b \in A$ *and* (III) $f(a^*b) = 0$ *for all* $b \in A$, *are equivalent.*

If A has no identity, then we shall in future only use those positive functionals on A that may be extended to a positive functional on its unitization. Such functionals have some more good properties.

Proposition (IV.5.7). *Let* f *extend to a positive functional* f_1 *on* $A_\times$.[11] *Then for any* $a \in A$

(I) $f(a^*) = \overline{f(a)}$; *in particular,* $f(a) \in \mathbb{R}$ *when* $a \in A_{sa}$;
(II) $|f(a)|^2 \leq f_1(e)f(a^*a)$.

▶ This is obtained from Propositions 5.4 and 5.5 by the substitution $b = e$. ■

---

[11] Recall (cf. Chap.I §1.2) that $A_\times$ is the notation used for A if it is unital and for its unitization A_+ otherwise.

Example (IV.5.8). Let $A \neq \mathbb{O}$ be a *-algebra with the zero mutliplication (any of the specific *-algebras we know, say, after replacing the original multiplication by the zero multiplication). Then Proposition 5.7(II) implies that no non-trivial functional on A can be extended to a positive functional on $A_\times$; at the same time it is obvious that every functional on A is positive.

Exercise (IV.5.9). Prove that the relations (I) and (II) from Proposition 5.7 — if we replace $f_1(e)$ by a constant that does not depend on a — are (not only necessary but also) sufficient conditions for one to be able to extend f to a positive functional on $A_\times$.

From now on, the notation f is used for a given fixed positive functional on the *-algebra $A_\times$.

We put $J^f = \{a \in A_\times : f(a^*a) = 0\}$. By Corollary 5.6, J^f is a left ideal of $A_\times$. We let $\widetilde{H}^f$ denote the quotient space $A_\times/J^f$. Again by applying Corollary 5.6 we get that the number $f(b^*a)$; $a, b \in A$, does not alter when a and b are replaced by elements belonging to the same coset with respect to J^f. This means that there is a well-defined complex function on $\widetilde{H}^f \times \widetilde{H}^f$ that sends the pair $x = a + J^f$, $y = b + J^f$ to the number $f(b^*a)$, which we shall in future call $\langle x, y\rangle$. Obviously, $\langle\cdot, \cdot\rangle$ already has all the properties, without exception, of a scalar product; in other words, $(\widetilde{H}^f, \langle\cdot, \cdot\rangle)$ is a pre-Hilbert space. We let H^f denote the Hilbert space which is its completion. In future H^f ($\widetilde{H}^f$ respectively) will be called the *(pre-)Hilbert space associated with* f.

Since it is a left ideal in $A_\times$, J^f is an invariant subspace for all the operators $m_a : A_\times \longrightarrow A_\times : b \longmapsto ab$; $a \in A$. In this way each m_a; $a \in A$, induces (in the sense of Chap.0 §0.2°) an operator acting in $\widetilde{H}^f$, denoted in future by $\widetilde{T}^f_a$, that is well defined by $\widetilde{T}^f_a(b + J^f) = ab + J^f$. Consequently, we get a map $\widetilde{T}^f : A \longrightarrow \mathcal{L}(\widetilde{H}^f) : a \longmapsto \widetilde{T}^f_a$.

Proposition (IV.5.10). *$\widetilde{T}^f$ is a representation of the algebra A on the linear space $\widetilde{H}^f$ which is unital when $A = A_\times$. In addition, $\langle\widetilde{T}^f_a x, y\rangle = \langle x, \widetilde{T}^f_{a^*} y\rangle$ for any $a \in A$, $x, y \in \widetilde{H}^f$.* ▶■

The reader probably expects that our next step will be to go from operators in $\widetilde{H}^f$ to operators in H^f; it is soon clear, however, that this is possible only if we make additional assumptions about A. Nevertheless we are already in a position to indicate the final result we want.

Definition (IV.5.11). The GNS-*representation of a *-algebra A associated with* $f \in \mathrm{PF}(A_\times)$ is the *-representation $T^f : A \longrightarrow \mathcal{B}(H^f) : a \longmapsto T^f_a$ for which $T^f_a x = \widetilde{T}^f_a x$ for any $x \in \widetilde{H}^f$.

It is obvious from this definition what is *a priori* the only possible obstacle to the construction of GNS-representations:

Proposition (IV.5.12). *A GNS-representation T^f, associated with $f \in \mathrm{PF}(A_X)$ exists $\Longleftrightarrow$ all the operators $\widetilde{T}^f : \widetilde{H}^f \longrightarrow \widetilde{H}^f$, $a \in A$, are bounded. If this holds, then T^f is uniquely defined and, for any $a \in A$, T^f_a is the (unique) continuous extension of the operator $\widetilde{T}^f_a$ to H^f.* ▶■

The following is evidence that GNS-representations do not always exist.

Example (IV.5.13). Consider the algebra c_p of slowly increasing sequences (see the end of Chap.III §1.3) with the involution $\xi = \{\xi_n\} \longmapsto \xi^* := \{\bar{\xi}_n\}$. Put $f : c_p \longrightarrow \mathbb{C}$ $: \xi \longmapsto \sum_{n=1}^{\infty} \xi_n e^{-n}$. Then, obviously, f is positive, $(\widetilde{H}^f; \langle\cdot, \cdot\rangle)$ is c_p with $\langle\xi, \eta\rangle = \sum_{n=1}^{\infty} \xi_n \bar{\eta}_n e^{-n}$ and $\widetilde{T}^f_\xi : c_p \longrightarrow c_p$ is the operator of coordinatewise multiplication by ξ. The action of this operator on a sequence of the form $(0, \cdots, 0, 1, 0, \cdots)$ implies immediately that for unbounded ξ it is not bounded in $\widetilde{H}^f$.

We now show, in the notation of Proposition 5.3, that the connection between f and T^f is of a reciprocal nature. Let $\hat{e}$ be the coset of the identity of the algebra A_X in $\widetilde{H}^f = A_X/J^f$.

Proposition (IV.5.14). *Let a GNS-representation exist for $f \in \mathrm{PF}(A_X)$. Then*

(I) $f = f_{T^f, \hat{e}}$ *and thus the functional f is a vector functional;*

(II) $\operatorname{Ker} T^f \subseteq J^f$;

(III) *if f is, in addition, a state, then* $\|T^f_a\| \geq \sqrt{f(a^*a)}$.

▶ (I) is implied by the obvious equation $\langle T^f_a\hat{e}, \hat{e}\rangle = f(a)$; $a \in A$. Further, if $T^f_a = 0$, then $T^f_a\hat{e} = a + J^f$ is the zero coset in A_X/J^f and therefore $a \in J^f$; this proves (II). Finally, if f is a state, then $\|\hat{e}\| = \sqrt{f(e^*e)} = 1$; and therefore (I) and the expression for $\|T_a\|^2$ in Proposition 5.3 imply (III). ■

§5.2. Automatic continuity of positive functionals and GNS-representations of star Banach algebras

We turn to considering conditions that are sufficient for the existence of GNS-representations. We first indicate such a condition in the context of abstract algebra.

Definition (IV.5.15). Let A be a *-algebra. The element $a \in A$ is said to be *star-bounded* if there exists $C \geq 0$ such that $g(a^*a) \leq Cg(e)$ for all $g \in \mathrm{PF}(A_X)$. The number $B^*(a) \geq 0$, whose square is the smallest C satisfying this inequality, is called the *star bound* of this element. The

*-algebra is said to be *star-bounded* if all its elements are star-bounded.

Obviously the star bound $B^*(a)$ may also be defined as $\sqrt{\sup\{g(a^*a) : g \in \underline{\mathrm{St}}(A_X)\}}$.

Proposition (IV.5.16). *Let A be a star-bounded algebra. Then for any $f \in \mathrm{PF}(A_X)$ there exists an associated GNS-representation T^f. Moreover $\|T^f_a\| \le B^*(a)$ for any $a \in A$.*

▶ We fix f and for any $b \in A_X$ we put $f_b : A \longrightarrow \mathbb{C} : c \longmapsto f(b^*cb)$; then it is easy to see that $f_b \in \mathrm{PF}(A_X)$. Hence, from the construction of $\tilde{T}^f_a$, $\langle \tilde{T}^f_a x, \tilde{T}^f_a x\rangle = f((ab)^*ab) = f_b(a^*a) \le B^*(a)^2 f_b(e) = B^*(a)^2 f(b^*b) = B^*(a)^2\langle x, x\rangle$ for all $a \in A$ and $x \in \tilde{H}^f$; $x = b + J^f$; $b \in A_X$. Consequently $\|\tilde{T}^f_a\| \le B^*(a)$, and all that remains is to apply Proposition 5.12. ■

Let us now concentrate on Banach algebras: it turns out that, for these, the above condition is automatically satisfied. At the root of this fact is a theorem, which is of interest in its own right, about automatic continuity (cf. the comments to Theorem I.2.6). It, in its turn, is based on the following lemma.

Lemma (IV.5.17). *Let A be a unital star Banach algebra and let a be a self-adjoint element with $\underline{r}(a) < 1$. Then there exists $b \in A_{sa}$ such that $e - a = b^2$.*

▶ We put $U := \{z + 1 : z \in \mathbb{D}\}$; then $\underline{r}(a) < 1$ implies that $\mathrm{Sp}_A(e - a) \subset U$, and therefore we are justified in considering $b := \Gamma o\lambda(w)$, where $\Gamma o\lambda$ is the holomorphic calculus for $e - a$ in U, with $w(z) = \sqrt{z}$ (cf. II.2.20). If we expand w as a Taylor series in powers of $z - 1$, we see that it is approximated in $\mathcal{O}(U)$ by polynomials in z with real coefficients. Hence, as $\Gamma o\lambda$ is continuous, b is approximated in A by the "same" polynomials in a. But the latter, obviously are self-adjoint and the involution is continuous; therefore b is also self-adjoint. ■

In the rest of the section when we talk about star Banach algebras we shall always assume that the given norm is multiplicative, star, and, if there is an identity, unital. According to Proposition 4.11, there is no loss of generality here.

Theorem (IV.5.18). *Let A be a star Banach algebra, and let g be an arbitrary positive functional on A_X. Then g is continuous and moreover $\|g\| = g(e)$ and, as a consequence, $\underline{\mathrm{St}}(A_X) = \{g \in \mathrm{PF}(A_X) : \|g\| = 1\}$.*

▶ It is sufficient to show that $\|a\| < 1$; $a \in A_X$, implies the inequality $|g(a)| \le g(e)$. According to Proposition 5.7(II), $|g(a)|^2 \le g(e)g(a^*a)$. But $\underline{r}(a^*a) \le \|a^*a\| \le \|a\|^2 < 1$ and $a^*a \in (A_X)_{sa}$; therefore the lemma implies that $e - a^*a = h^2$ for a certain $h \in (A_\times)_{sa}$. But then, as $g \in \mathrm{PF}(A_\times)$, $g(e - a^*a) = g(h^*h) \geq 0$, that is, $g(a^*a) \leq g(e)$. The rest is clear. ■

Remark (IV.5.19). The positive functionals on A itself, if it is not unital, are no longer necessarily continuous: it is sufficient to take a discontinuous linear functional on an infinite-dimensional Banach *-algebra with zero multiplication (cf. Example 5.8). Nevertheless one can be sure that the functionals in $\mathrm{PF}(A)$ are continuous if, say, A has a bounded approximate identity (see, for example [9, p.201]).

Theorem (IV.5.20). *Let A be a star Banach algebra. Then it is star-bounded and (as a consequence) for any $f \in \mathrm{PF}(A_\times)$ there exists an associated* GNS-*representation T^f. Moreover $\|T^f_a\| \leq B^*(a) \leq \|a\|$ for any $a \in A$.*

▶ The preceding theorem implies that $f(a^*a) \leq \|f\|\,\|a^*a\| \leq f(e)\|a\|^2$ for any $a \in A$; hence a is star-bounded and $B^*(a) \leq \|a\|$. The rest follows from Proposition 5.16. ■

We give a few illustrations.

Example (IV.5.21). Let $A = C[0, 1]$, $f : A \longrightarrow \mathbb{C} : a \longmapsto \int_0^1 a(t)dt.$

Obviously, $f \in \underline{\mathrm{St}}(A)$, $H^f = L^2[0, 1]$ and T^f_a; $a \in C[0, 1]$, is the operator of pointwise multiplication by the function $a(t)$ on $L^2[0, 1]$. Thus, in this case the GNS-representation is a realization of the isometric isomorphism between $C[0, 1]$ and $B \subset \mathcal{B}(L^2[0, 1])$ described in Example I.2.13.

Exercise (IV.5.22). Let $A = \mathcal{B}(H)$. Prove that for any $x \in H$ and $f : \mathcal{B}(H) \longrightarrow \mathbb{C} : a \longmapsto \langle a(x), x\rangle$ the space H^f is isometrically isomorphic to H and T^f is an isometric *-isomorphism between $\mathcal{B}(H)$ and $\mathcal{B}(H^f)$.

Hint. Consider the map $\widetilde{H}^f \longrightarrow H : a + J^f \longmapsto a(x)$.

Exercise (IV.5.23). Let Ω be a discrete uncountable topological space. Prove that any GNS-representation of the algebra $A = C_0(\Omega)$ has a non-zero kernel.

Hint. Since $C_0(\Omega)^* \simeq l_1(\Omega)$, for each $f \in \mathrm{PF}(A_\times)$ the kernel contains all functions that are zero on some countable set depending only on f.

We conclude this section with a useful inequality concerning arbitrary *-representations. Given some *-representation $T : A \longrightarrow \mathcal{B}(H)$ we let $T_\times : A_\times \longrightarrow \mathcal{B}(H)$ denote T itself if A is unital and the extension of T to $A_\times$ that takes e_+ to 1 otherwise.

Proposition (IV.5.24). *Let* A *be a star-bounded algebra and let* T *be an arbitrary* **-representation. Then* $\|T_a\| \le B^*(a)$ *for any* $a \in A$.

▶ Since $\|T_a\| = \|T_{\chi_a}\|$, the required bound is implied by Proposition 5.3 (III) and (IV) applied to T_χ. ■

This, taking into account the bound $B^*(a) \le \|a\|$ contained in Theorem 5.20, immediately implies another theorem about automatic continuity — this time about continuity of *-representations.

Theorem (IV.5.25). *Let* A *be a star Banach algebra and let* T *be an arbitrary* **-representation of* A. *Then* T *is continuous and* $\|T\| \le 1$; *if* T *is also unital, then* $\|T\| = 1$. ▶■

§5.3. Universal representation, star radical and star semisimplicity

Exercise 5.23 warns us that, in contrast with the favourable situations described in Example 5.21 and Exercise 5.22, even good algebras[12] may not have injective GNS-representations; informally speaking, part of the information about such an algebra is lost with each separate GNS-representation. Our next problem is to "reduce the loss of information to a minimum" and, in particular, to learn how to construct representations with the smallest possible kernels. With this in mind we proceed as follows.

Let $\{H_\nu;\ \nu \in \Lambda\}$ be a certain family of Hilbert spaces. We take the Cartesian product $\bigtimes_\nu H_\nu$ for whose elements ("rows") we use a notation like $\bar{x} = (\cdots, x_\nu, \cdots)$; $x_\nu \in H_\nu$, and in this product we consider the subspace consisting of those rows $\bar{x}$ for which $\sum_\nu \|x_\nu\|^2 < \infty$. The latter, since it has the scalar product $\langle \bar{x}, \bar{y}\rangle = \sum_\nu \langle x_\nu, y_\nu\rangle$, is obviously a Hilbert space. We call it the *Hilbert sum of the family of spaces* $\{H_\nu;\ \nu \in \Lambda\}$ and denote it by $\boxplus_\nu H_\nu$.

Further, let A be a *-algebra and $T^\nu : A \longrightarrow \mathcal{B}(H_\nu)$; $\nu \in \Lambda$, be a family of *-representations of this algebra, such that, for any $a \in A$, $\sup_\nu \|T^\nu_a\| < \infty$ (by Proposition 5.24, an example would be any family of *-representations of a star bounded algebra). Then for every $a \in A$, there is a well-defined operator, acting on $\boxplus_\nu H_\nu$, that sends the row $(\cdots, x_\nu, \cdots)$ into $(\cdots, T^\nu_a x_\nu, \cdots)$; let us call it $\boxplus_\nu T^\nu_a$.

---

[12] The algebra in this exercise belongs to the class of C^*-algebras that was considered to be the "best" type of *-algebra in §4.2.

In addition, it is easy to see that the map $a \longmapsto \boxplus_{\nu} T^{\nu}_{a}$ is a *-representation of the algebra A on the space $\boxplus_{\nu} H_{\nu}$. This representation is called the *Hilbert sum of the family of* *-representations $\{T^{\nu}, \nu \in \Lambda\}$ and is denoted by $\boxplus_{\nu} T^{\nu}$.

From now on up to the end of the section A is a certain star-bounded algebra.

We consider the set of states of the algebra $A_{\times}$ and for every state f we take the Hilbert space H^{f} and the GNS-representation T^{f} associated with it. We put $\mathcal{H}^{A} = \boxplus_{f} H^{f}$ and $\mathcal{T}^{A} = \boxplus_{f} T^{f}$, where the Hilbert sums are taken over all $f \in \underline{\mathrm{St}}(A_{\times})$. We shall often in future assume that A is fixed and write $\mathcal{H}$ instead of $\mathcal{H}^{A}$ and $\mathcal{T}$ instead of $\mathcal{T}^{A}$.

Definition (IV.5.26). The representation $\mathcal{T}^{A} : A \longrightarrow \mathcal{B}(\mathcal{H}^{A})$ is called the *universal representation* of the (star-bounded) algebra A.

We draw attention to the obvious relation $\mathcal{T}^{A} = \mathcal{T}^{A_{\times}}|_{A}$.

The universal representation is a construction enabling one to manufacture an operator algebra out of a given "nice" *-algebra in the same way as the Gel'fand transformation makes a function algebra out of a commutative Banach algebra (cf. what was promised at the beginning of the chapter). The next few results show that this representation really does give the maximum possible information, compared with others, about the original algebra.

The class of all *-representations of the algebra A will in future be denoted by $\mathcal{R}ep\, A$.

Theorem (IV.5.27). *For any* $a \in A$

$$\|\mathcal{T}_{a}\| = \sup\{\|T^{f}_{a}\| : f \in \underline{\mathrm{St}}(A_{\times})\} = \sup\{\|T_{a}\| : T \in \mathcal{R}ep\, A\} = B^{*}(a).$$

▶ In the chain of inequalities $\|\mathcal{T}_{a}\| \le \sup\{\|T_{a}\| : T \in \mathcal{R}ep\, A\} \le B^{*}(a) \le \sup\{\|T^{f}_{a}\| : f \in \underline{\mathrm{St}}(A_{\times})\} \le \|\mathcal{T}_{a}\|$ the first is obvious, the second is implied by Proposition 5.24, the third by Proposition 5.14(III) and finally the fourth follows from Definition 5.26 itself.■

Proposition (IV.5.28). *Let A be a star Banach algebra. Then* $\|\mathcal{T}_{a}\|$ (generally speaking, this is less than $\|a\|$; see Theorem 5.25) *is, among all* $\|T_{a}\|$; $T \in \mathcal{R}ep\, A$, *that which is closest to* $\|a\|$.

▶ This follows from Theorem 5.25 (with $T = \mathcal{T}$) and Theorem 5.27.■

We deal with the kernel of the universal representation.

Theorem (IV.5.29). *The following properties of the element* $a \in A$ *are equivalent:*

(I) $\mathcal{T}_a = 0$;

(II) $T_a = 0$ *for all* **-representations of the algebra* A;

(III) $T_a = 0$ *for all* GNS-*representations of the algebra* A;

(IV) $f(a^*a) = 0$ *for all* $f \in \mathrm{PF}(A_\times)$ ($\Leftrightarrow$ *for all* $f \in \underline{\mathrm{St}}(A_\times)$);

(V) $f(a) = 0$ *for all* $f \in \mathrm{PF}(A_\times)$ ($\Leftrightarrow$ *for all* $f \in \underline{\mathrm{St}}(A_\times)$).

▶ The equivalence of properties (I) - (IV) follows immediately from Theorem 5.27. Further, by Proposition 5.7(II), (IV) implies (V). Finally, if (V) is satisfied, then Proposition 5.3(I), applied to the representation $\mathcal{T}^{A_\times}$, implies that $\langle \mathcal{T}_a\overline{x}, \overline{x}\rangle = 0$ for any $\overline{x} \in \mathcal{H}$. But then for any $\overline{x}, \overline{y} \in \mathcal{H}$ and $\overline{z}_1 = \overline{x} + \overline{y}$, $\overline{z}_2 = \overline{x} - \overline{y}$, $\overline{z}_3 = \overline{x} + i\overline{y}$, $\overline{z}_4 = \overline{x} - i\overline{y}$, we have

$$\langle \mathcal{T}_a x, y\rangle = \frac{1}{4}\langle \mathcal{T}_a\overline{z}_1, \overline{z}_1\rangle - \frac{1}{4}\langle \mathcal{T}_a\overline{z}_2, \overline{z}_2\rangle + \frac{i}{4}\langle \mathcal{T}_a\overline{z}_3, \overline{z}_3\rangle - \frac{i}{4}\langle \mathcal{T}_a\overline{z}_4, \overline{z}_4\rangle = 0.$$

This means that $\mathcal{T}_a = 0$ and so (V) implies (I). ■

Remark (IV.5.30). A result that is deeper than Theorems 5.27 and 5.29 is that in the case of a star Banach algebra we could, in both theorems, replace the class of all GNS-representations by the considerably narrower class of so-called topologically irreducible *-representations. We shall talk about these representations and the ("pure") states corresponding to them in Chap. VI §2.4.

One could say that, in Theorem 5.29 we are talking about those elements of a *-algebra that "prevent it being an operator algebra" — in the same way as the presence of a radical "prevents a commutative Banach algebra being a function algebra". This motivates the following definition.

Definition (IV.5.31). The set of elements of a (star-bounded) algebra A having any of the properties indicated in Theorem 5.29 is called the *star radical* of this algebra and it is denoted by *-Rad A. The algebra A is said to be *star semisimple* if *-Rad $A = \{0\}$ and *star radical* if *-Rad $A = A$.

Since it is the kernel of a *-homomorphism (namely $\mathcal{T}$), the star radical of an algebra is a *-bi-ideal.

We thus see that "star semisimple algebra" is just another name for an algebra that is *-isomorphic to some self-adjoint algebra of operators on a Hilbert space. Moreover a star Banach algebra A of this class is topologically *-isomorphic to some self-adjoint Banach operator algebra on Hilbert space (compare §2.2) which we may suppose to be $\operatorname{Im} \mathcal{T} \subset \mathcal{B}(\mathcal{H})$ with norm $\|\cdot\|'$ well defined by $\|\mathcal{T}_a\|' = \|a\|$.

What has been said of course implies that $\mathcal{B}(H)$ and the other operator algebras in Example 4.14 are star semisimple. It is useful to look at the other specific algebras that we know.

Proposition (IV.5.32). (I) *$C_0(\Omega)$, $C^n[0, 1]$ and $\mathcal{A}(\overline{\mathbb{D}})$ (see 4.18 and 4.20) are star semisimple.*

(II) *$l_1(G)$, $L^1(\mathbb{R})$ and $L^1(\mathbb{T})$ (see 4.19) are star semisimple.*

(III) *$\widetilde{\mathbb{C}}^2$ (see 4.21) is star radical.*

▸ (I) Let A be one of the three algebras indicated and let us write Δ for Ω_+, $[0, 1]$ or $\mathbb{D} \cap \mathbb{R}$ respectively. We take $t \in \Delta$ and $\chi_t : A_X \longrightarrow \mathbb{C} : a \longmapsto a(t)$; obviously $\chi_t \in \mathrm{PF}(A_X)$. Therefore, if $a \in {}^*\text{-Rad}\, A$, then, by Theorem 5.29(IV), $0 = \chi_t(a^*a) = |a(t)|^2$ for all $t \in \Delta$. In all cases (for $\mathcal{A}(\overline{\mathbb{D}})$ we use the uniqueness theorem for analytic functions) this means that $a = 0$.

(II) For each of the L^1-algebras under consideration we proceed in the same way, namely, we produce a special injective *-representation and then use Theorem 5.29(II). This turns out to be the homomorphism $T^G : L^1(G) \longrightarrow \mathcal{B}(L^2(G)) : a \longmapsto T^G_a$, where $T^G_a : b \longmapsto a * b$, introduced in Chap.III §3.3. In fact for discrete G

$$\begin{aligned}\langle T^G_{a*}b, c\rangle &= \sum_s a^* * b(s)\overline{c(s)} = \sum_s\sum_t \overline{a(t^{-1})}b(t^{-1}s)\overline{c(s)} = \\ &= \sum_t \left(\sum_s b(ts)\overline{c(s)}\right)\overline{a(t)} = \sum_t\sum_s b(s)\overline{c(t^{-1}s)}\overline{a(t)} = \\ &= \sum_s b(s)\overline{a * c(s)} = \langle b, T^G_a c\rangle; \quad a \in L^1(G), \quad b, c \in L^2(G).\end{aligned}$$

hence $T^G_{a*} = (T^G_a)^*$. This conclusion is implied by a similar calculation when $G = \mathbb{R}$ and $G = \mathbb{T}$ but with an integral instead of the sum.

Thus, T^G is a *-representation; all that remains is to show that $\operatorname{Ker} T^G = 0$. Let $T^G_a = 0$ for $a \in L^1(G)$. This means, in particular, that $a * (\lambda b_U) = 0$ for any $\lambda \in \mathbb{C}$ and the characteristic function b_U of any measurable set $U \subset G$. But the set $\{\lambda b_U\}$ lies in $L^1(G)$ and contains an approximate identity (in the discrete case the ordinary identity) of the algebra (Proposition III.3.2). Hence clearly $a = 0$.

(III) It is obvious that $(\widetilde{\mathbb{C}}^2)_{sa} = \{(z, \overline{z}) : z \in \mathbb{C}\} = \{(z, 1)^* \cdot (z, 1) : z \in \mathbb{C}\}$. Hence any $f \in \underline{\mathrm{St}}(\widetilde{\mathbb{C}}^2)$ is zero on $(\widetilde{\mathbb{C}}^2)_{sa}$ and hence on the whole of $\widetilde{\mathbb{C}}^2$. The rest is clear. ■

Essentially, part of this proposition was considered earlier (Exercise 4.23), where another argument was suggested.

The following proposition helps clarify the interrelation between the star radical and the usual radical.

Proposition (IV.5.33). *Let h be a self-adjoint element in A with zero spectrum. Then $h \in {}^*\text{-Rad}\, A$. As a consequence, an element of a star semisimple algebra with these properties must be zero.*

▶ Let $T : A \longrightarrow \mathcal{B}(H)$ be an arbitrary *-representation of our algebra. As this is a homomorphism, Proposition II.1.6 implies that, for every $a \in A$, $\text{Sp}\, T_a \subseteq \text{Sp}_A\, a \cup \{0\}$. Hence T_h is a self-adjoint operator on H with zero spectrum; from $\|T_h^2\| = \|T_h\|^2$ (0.1.13) and the formula for the spectral radius (II.1.28), it follows that T_h is zero. All that remains is to apply Theorem 5.29(II). ■

Corollary (IV.5.34). *Let $p \neq 0$ be a positive element of a star semisimple algebra. Then $-p$ is not positive.*

Theorem (IV.5.35). $\text{Rad}\, A \subseteq {}^*\text{-Rad}\, A$ *for any (star-bounded) algebra A. As a consequence a star semisimple algebra is semisimple.*

▶ Let $r \in \text{Rad}\, A$; then, by Proposition 4.7, $r^* \in \text{Rad}\, A$ also. According to Proposition 4.2, this means that the elements $r_1, r_2 \in A_{sa}$ such that $r = r_1 + ir_2$ also belong to the radical and so, like all elements of the radical, they have zero spectrum. But then, by Proposition 5.33, r_1 and r_2 lie in ${}^*\text{-Rad}\, A$, and hence so also does r. ■

Corollary (IV.5.36). *The algebra $L^1(G)$ where G is a discrete group* (so $L^1(G)$ is $l_1(G)$), *is semisimple.*

(Recall that this is the last of the algebras appearing in Proposition 5.32, concerning which we did not yet know whether it was semisimple.)

Remark (IV.5.37). The arguments in Proposition 5.32(II) can in practice be carried across unchanged to the case of an arbitrary locally compact group G. This was precisely the way, using star semisimplicity, in which it was proved that all the group L^1-algebras are (plain) semisimple; for Abelian groups this reduces to the corollary mentioned in Remark 3.14. For a complete proof, due to Raikov, see, for example, [53, §28].

That the algebra of measures $M(G)$ is star semisimple (and so plain semisimple) is proved by similar arguments for any locally compact group G [87, §22.11].

Remark (IV.5.38). The fact that, for any locally compact G, the homomorphism T^G is involutive, has another important corollary: the algebra $C_r^*(G)$ (see III.3.25), being the closure of $\text{Im}\, T^G$ with respect to the operator norm, is self-adjoint and therefore it is a realized $\mathbb{C}^*$-algebra.

* * *

The reader will no doubt have noticed that the space of the universal representation $\mathcal{H}$ is, as a rule, not separable; the universal representation, unlike the Gel'fand transformation, almost never has a dense image (in $\mathcal{B}(\mathcal{H})$), even before we ask about surjectivity. Nevertheless, it has gradually (but not immediately) become clear that it is precisely this, and no other possibly more economical, construction that plays an outstanding role in the theory of operator algebras, especially in modern topics. Basically, this role (and the term "universal" itself) is connected with special properties of the co-restriction $\mathcal{T}|^{A!!} : A \longrightarrow A^{!!}$, where $A^{!!}$ denotes the weak operator closure (or, what is equivalent, the bicommutant; see Chap.III §2.3) of the image of $\mathcal{T}$ in $\mathcal{B}(\mathcal{H})$. In particular, if A is a C^*-algebra (see §7 later for something about this), then $A^{!!}$ may be identified, as a Banach space, with the second dual A^{**} and $\mathcal{T}|^{A!!}$ may be identified with the canonical embedding of A in A^{**}. This range of these questions, however, is beyond the scope of the initial background that is given in this book; for this see, for example, [31, §12] and [58, Chap.3].

§6. HERMITIAN ALGEBRAS

It is natural to expect that when we are studying some class or other of *-algebras, primarily Banach *-algebras, the results will turn out to be stronger the more the involution is compatible with the other structures. In this sense the C^*-algebras we are going to talk about in the next section are best. But now we take an important step on the road to C^*-algebras and consider a wider class of algebras with involution which is quite interesting in itself.[13] This is based on the following purely algebraic definition.

Definition (IV.6.1). An involutive algebra A is said to be *Hermitian* if $\mathrm{Sp}(h) \subseteq \mathbb{R}$ for any $h \in A_{sa}$.

From the definition it obviously follows that A is Hermitian $\Longleftrightarrow$ $A_{\times}$ is Hermitian.

---

[13] Many results in this section were obtained earlier for C^*-algebras (in the period 1939-1953; cf. the beginning of the chapter). However later on it became clear that it is precisely the Hermitian algebras that form the natural class in which they are valid.

We suggest that the reader checks that both the star algebras $L^1(\mathbb{R})$ and $L^1(\mathbb{T})$ in Example 4.19 are Hermitian and that the algebras in Examples 4.20 and 4.21 are not Hermitian.

We give, without proof, some information about the other examples. The algebras $L^1(G)$, where G is a commutative or compact group, are automatically Hermitian [225], however for arbitrary G this is, generally speaking, not so: a counterexample is the group of 2×2-matrices with complex elements and with determinant equal to one [20, §52, p.268]. The algebra of measures $M(G)$, as a rule, is not Hermitian; in the commutative case this is connected with the existence of an analytic structure in its spectrum (see [294], [247]).

Proposition (IV.6.2). *Let the* **-algebra* A *be such that all elements of the form* a^*a; $a \in A_\times$, *are positive (see* §4.1). *Then* A *is Hermitian.*

▶ Our problem is to show that, for $h \in A_{sa}$ and $\lambda = s + it$, $s, t \in \mathbb{R}$, $t \neq 0$, $h - \lambda e$ is invertible in $A_\times$. We put $k = h - se$; then k is self-adjoint and, by hypothesis, k^2 is positive. Hence $k^2 + t^2e = (k + ite)(k - ite)$ is invertible and hence so are both its commuting factors. But the second of them is just $h - \lambda e$. ■

Exercise (IV.6.3). Verify that the above sufficient condition for A to be Hermitian is equivalent to the following: for any $a \in A$, $e + a^*a$ is invertible.

Let us consider what advantages Hermitian algebras have over other algebras.

In the next proposition A is a Hermitian star Banach algebra and B a closed Hermitian *-subalgebra of A. The case when B is unital but the identity in B is not an identity in A is, for brevity, said to be special.

Proposition (IV.6.4). $\mathrm{Sp}_B(a) = \mathrm{Sp}_A(a)$ *is satisfied for each* $a \in B$, *except for the special case when* $\mathrm{Sp}_B(a) \cup \{0\} = \mathrm{Sp}_A(a)$.

▶ Obviously, in the "non-special" case we can identify $B_\times$, and in the special case we can identify B_+, with a subalgebra of $A_\times$ so that both algebras contain a common identity. Therefore, taking into account $\mathrm{Sp}_{B_+}(a) = \mathrm{Sp}_B(a) \cup \{0\}$ (II.1.5) we are justified in assuming that A and B themselves have a common identity e.

Let $b := a - \lambda e$; $\lambda \in \mathbb{C}$, be invertible in A; then so are b^*, b^*b and bb^*. As B is Hermitian, $(b^*b - ite)^{-1} \in B$ exists for any $t > 0$. But, since the map taking an element to its inverse is continuous (I.2.43), $\lim_{t \to 0} (b^*b - ite)^{-1} = (b^*b)^{-1}$.

Hence, as B is closed, $(b^*b)^{-1} \in B$; in exactly the same way $(bb^*)^{-1} \in B$. Therefore b has in B a left and right inverse, and therefore also a ("complete") inverse.

We have shown that $\lambda \notin \mathrm{Sp}_A(a)$ implies $\lambda \notin \mathrm{Sp}_B(a)$. The converse is obvious. ■

Warning. From now on up to the end of this section A is some Hermitian star Banach algebra with star multiplicative norm.

Proposition (IV.6.5). *$\underline{r}(a)^2 \leq \underline{r}(a^*a)$ is satisfied for any $a \in A$.*

▶ Assume otherwise; then, by replacing a by an element proportional to it, we are justified in assuming that $\underline{r}(a^*a) < 1$ and $1 \in \mathrm{Sp}(a)$. We note that, because $a - e$ is not invertible in A_X, at least one of the elements $b := (e + a^*)(e - a)$ or $c := (e - a)(e + a^*)$ is not invertible in A_X.

If we consider Lemma 5.17 with a^*a instead of a, we get that $e - a^*a = h^2$ for a certain self-adjoint and, obviously, invertible $h \in A_X$. But then $b = h^2 + a^* - a = (-ih)(k + ie)h$, where $k := ih^{-1}(a^* - a)h^{-1}$. As k is self-adjoint and A_X is Hermitian, all three factors in the expression for b are invertible; consequently b itself is also invertible. A similar argument, taking into account the equality $\underline{r}(aa^*) = \underline{r}(a^*a)$ (arising from Proposition II.1.8), shows that c is also invertible. Contradiction! ■

Proposition (IV.6.6). *$\underline{r}(hk) \leq \underline{r}(h)\underline{r}(k)$ holds for any $h, k \in A_{sa}$.*

▶ Proposition 6.5 and Proposition II.1.8 imply that, for any n,

$$\underline{r}(h^n k^n) \leq \sqrt{\underline{r}((h^n k^n)^*(h^n k^n))} = \sqrt{\underline{r}(k^n h^{2n} k^n)} = \sqrt{\underline{r}(h^{2n} k^{2n})}.$$

Hence, putting $n := 2^m$; $m = 0, 1, 2, \cdots$, we see that for these m

$$\underline{r}(hk) \leq \sqrt[2^m]{\underline{r}(h^{2^m} k^{2^m})} \leq \sqrt[2^m]{\|h^{2^m}\| \|k^{2^m}\|}.$$

All that remains is to let m tend to infinity and apply Theorem II.1.28. ■

The following property of Hermitian algebras turns out to be very useful later on.

Theorem (IV.6.7). *If the set A_{pos} contains two elements, then it contains their sum; consequently, it is convex.*

▶ Let $p, q \in A_{pos}$. Then $e + p$ and $e + q$ are invertible in A_X and it is easy to verify that the element $e + p + q$ can be written as $(e + p)(e - hk)(e + q)$, where $h := (e + p)^{-1}p$, $k := q(e + q)^{-1}$. But by Theorem II.1.10,

$\mathrm{Sp}(h) = \{\lambda(1+\lambda)^{-1} : \lambda \in \mathrm{Sp}(p)\}$; which means that $\mathrm{Sp}(h) \in [0, 1)$. Consequently, $\underline{r}(h) < 1$, and in exactly the same way $\underline{r}(k) < 1$. Therefore, by Proposition 6.6, $\underline{r}(hk) < 1$ and hence $e - hk$ is invertible.

So, for any $p, q \in A_{\mathrm{pos}}$ the element $e + p + q$, being the product of three invertible elements, is itself invertible. But then, for any $t < 0$, by replacing p and q by sp and sq, where $s := -t^{-1}$, we get that $p + q - te = -t(e + sp + sq)$ is invertible. The rest is clear. ■

Recall Proposition 6.2. It turns out that the converse is true for Hermitian algebras, although this is not so simple.

Lemma (IV.6.8). *Let* $a \in A$; $m := \sup \mathrm{Sp}(a^*a)$. *Then* $m \geq 0$ *and* $\mathrm{Sp}(a^*a) \subseteq [-m, m]$.

▶ We put $a = h + ik$; $h, k \in A_{\mathrm{sa}}$; then, obviously $me + aa^* = 2h^2 + 2k^2 + (me - a^*a)$. Since $h^2 + te = (h + i\sqrt{t}e)(h - i\sqrt{t}e)$ is invertible when $t > 0$, then $h^2 \in A_{\mathrm{pos}}$; in exactly the same way $k^2 \in A_{\mathrm{pos}}$; finally the definition of m implies that $me - a^*a \in A_{\mathrm{pos}}$ as well. Hence, by Theorem 6.7, $me + aa^*$ is positive; this means precisely that $\inf \mathrm{Sp}(aa^*) \geq -m$. All that remains is to use the fact that $\mathrm{Sp}(aa^*) \cup \{0\} = \mathrm{Sp}(a^*a) \cup \{0\}$ (II.1.8). ■

Lemma (IV.6.9). *Let* $a \in A$ *be such that* $\underline{r}(a^*a) < 1$. *then* $r(b^*b) < 1$ *for* $b := 2a(e + a^*a)^{-1}$. *Moreover, if* $t_0 \in \mathrm{Sp}(a^*a)$, *then* $s_0 := 4t_0(1 + t_0)^{-2} \in \mathrm{Sp}(b^*b)$.

▶ Since a^*a and $(e + a^*a)^{-1}$ commute, then $b^*b = 4a^*a(e + a^*a)^{-2}$. Hence, by Theorem II.1.10 (or II.2.23), $\mathrm{Sp}(b^*b) = \{4\lambda(1+\lambda)^{-2} : \lambda \in \mathrm{Sp}(a^*a)\}$. This already proves the second assertion; in addition, as $t < 1$ implies $4t(1+t)^{-2} < 1$, it follows from $\underline{r}(a^*a) < 1$ and the expression for $\mathrm{Sp}(b^*b)$ that $\sup \mathrm{Sp}(b^*b) < 1$. By Lemma 6.8, this means that $\underline{r}(b^*b) < 1$. ■

Theorem (IV.6.10) (Shirali–Ford).[14] *A star Banach algebra* A *is Hermitian* $\Leftrightarrow$ *for any* $a \in A_{\times}$ *the element* a^*a *is positive.*

▶ In view of Proposition 6.2 it is sufficient to establish "$\Rightarrow$". Assume otherwise; then there exists $a_0 \in A_{\times}$ such that $\mathrm{Sp}(a_0^*a_0)$ contains some $t_0 < 0$; where we may assume that $\underline{r}(a_0^*a_0) < 1$. Using Lemma 6.9 we construct by induction the sequence $a_n \in A_{\times}$; $a_{n+1} := 2a_n(e + a_n^*a_n)^{-1}$ and $t_n \in \mathbb{R}$; $t_{n+1} := 4t_n(1+t_n)^{-2}$; $n = 0, 1, 2, \cdots$. We note that $4t(1+t)^{-2} < 4t$ is satisfied for any $t \in (-1, 0)$; therefore

---

[14] For the original proof see [322]; our argument, with certain modifications, follows [9, §41]. An alternative proof was given by Ptak [351].

$t_0 \in (-1, 0)$ implies the existence of N such that $t_N \leq -1$. At the same time, by Lemma 6.9, $\underline{r}(a_n^* a_n) < 1$ and $t_n \in \mathrm{Sp}(a_n^* a_n)$ for all n, that is $t_n > -1$. A contradiction. ∎

Exercise (IV.6.11). Construct a proof for the case when a is normal.

Recall the relation between the star radical and the usual radical established in Theorem 5.35. We now return to this question for the class of Hermitian algebras in the light of Theorem 6.7.

Lemma (IV.6.12). *Let A be a unital Hermitian star Banach algebra and let I be a proper left ideal of A. Then there exists $f \in \underline{\mathrm{St}}(A)$ such that I is contained in the ideal* $J^f := \{a \in A : f(a^*a) = 0\}$ (see just after IV.5.9).

▶ We put $I_{(sa)} := I \cap A_{sa}$ and consider the subspace $I^+_{(sa)} := \{te + h : h \in I_{(sa)}, t \in \mathbb{R}\}$ of the real linear space A_{sa} and also the functional $g_0 : I^+_{(sa)} \longrightarrow \mathbb{R} : te + h \longmapsto t$.

Further, we put $I^+_{(pos)} := I^+_{(sa)} \cap A_{pos}$. Let $p \in I^+_{(pos)}$; then $p - g_0(p)e$, by the definition of g_0, belongs to $I_{(sa)} \subset I$ and hence it is not invertible. Hence $g_0(p) \in \mathrm{Sp}(p)$ and therefore $g_0(p) \geq 0$. But, according to Theorem 6.7, A_{pos} is a convex set and an obvious consequence of A being Hermitian is that e is an interior point. In this way we find that we have the conditions for the "geometric version" of the Hahn-Banach theorem — Theorem 0.1.12, where one must put $E := A_{sa}$, $E_0 := I^+_{(sa)}$ and $M := A_{pos}$. As a consequence there exists a functional $g : A_{sa} \longrightarrow \mathbb{R}$ extending g_0 and such that $g(p) \geq 0$ for all $p \in A_{pos}$.

Now for any $a \in A$; $a = h + ik$, $h, k \in A_{sa}$, we put $f(a) = g(h) + ig(k)$. Obviously, $f : A \longrightarrow \mathbb{C}$ is a well-defined functional such that $f(p) = g(p) \geq 0$ for any $p \in A_{pos}$. Further, Theorem 6.10 implies that f is positive and, in view of $f(e) = g_0(e) = 1$, it is a state. All that remains is to note that $a^*a \in I_{(sa)}$ when $a \in I$ and therefore $f(a^*a) = g_0(a^*a) = 0$ ∎

Theorem (IV.6.13). *Let A be a Hermitian star Banach algebra. Then $^*\text{-Rad}\, A = \text{Rad}\, A$. As a consequence, a semisimple algebra of this kind is star semisimple.*

By Theorem 5.35 it is sufficient to show that the star radical is contained in the ordinary radical. Let $a \in {}^*\text{-Rad}\, A$ and let I be a certain maximal left ideal of $A_\times$. According to the previous lemma, there exists a state f on $A_\times$ such that $I \subseteq J^f$. Since I is maximal and J^f is a proper ideal, this means that $I = J^f$ and, by Theorem 5.29(IV) and Definition 5.31, $a \in I$. Since I is arbitrary, all that remains is to use the characterization of the (ordinary) radical given in Theorem I.3.53(I). ∎

In conclusion we dwell a moment on the case of commutative algebras; here it turns out that there are new equivalent definitions of Hermitian.

Theorem (IV.6.14). *The following properties of a commutative star Banach algebra A are equivalent:*

(I) *A is Hermitian;*
(II) *any character of the algebra A is a *-character;*
(III) *any maximal modular ideal of A is a *-ideal;*
(IV) *the Gel'fand transformation $\Gamma : A \longrightarrow C_0(\Omega)$ is a *-homomorphism.*

▸ Let A be Hermitian and χ its character corresponding to the point $s \in \Omega$ (see §1.2). Then, for any $h \in A_{sa}$, $\chi(h) = \hat{h}(s)$; hence, according to Theorem 2.3, $\chi(h) \in \operatorname{Sp} h \subset \mathbb{R}$. This means, in view of Corollary 4.3, that χ is a *-character.

Thus (I) implies (II) and a similar argument proves the converse. The equivalence of (II) and (III) follows immediately from Theorem 1.1 about identifying characters with maximal ideals and the equivalence of (II) and (IV) follows immediately from the form of the homomorphism Γ. ■

Later we require the well-known *Stone-Weierstrass theorem*, according to which, *if K is a compact space, every subalgebra of $C(K)$ which, if it contains a function, also contains its complex-conjugate, which contains the constants and which separates the points of K, is dense in $C(K)$ with respect to the uniform norm* (see, for example, [67, Chap.5.7]).

Theorem (IV.6.15). *Let A be a star commutative Banach algebra with spectrum Ω. If A is Hermitian, then the image of its Gel'fand transformation is dense in $C_0(\Omega)$.*

▸ Recall the commutative diagram in Proposition 2.12 connecting the Gel'fand transformations of the algebras A and A_+. By Theorem 6.14 ((I) $\Leftrightarrow$ (IV)), if the algebra $\hat{A} := \operatorname{Im} \Gamma \subseteq C_0(\Omega)$ contains a function it also contains its complex conjugate. Therefore, taking into account Theorem 2.11(III), the subalgebra $\hat{A}_+ := \operatorname{Im} \Gamma_+$ in $C(\Omega_+)$ satisfies all the conditions of the Stone-Weierstrass theorem with $K := \Omega_+$ and consequently it is dense in $C(\Omega_+)$. This obviously implies that $\hat{A}$ is dense in $C_0(\Omega)$. ■

§7. C^*-ALGEBRAS

§7.1. Definition and simplest properties. Description of commutative C^*-algebras as the algebras $C_0(\Omega)$

Let A be a *-algebra. A multiplicative seminorm $\|\cdot\|$ in A is called a *C^*-seminorm* (and when it is a norm, then a C^*-norm) if it satisfies

$$\|a^*a\| = \|a\|^2; \quad a \in A,$$

We shall call this identity the *C^*-property.*

Proposition (IV.7.1). (I) *Let* $\|\cdot\|$ *be a multiplicative seminorm in* A *such that* $\|a\|^2 \le \|a^*a\|$ *for all* $a \in A$. *then this is a* C^*-*seminorm.*

(II) *Every* C^*-*seminorm is a star seminorm.*

(III) *Every* C^*-*seminorm in a unital algebra, apart from the zero seminorm, is unital.*

▶ Under the conditions (I) it is obvious that $\|a\|^2 \le \|a^*\|\|a\|$ and therefore $\|a\| \le \|a^*\|$ for all $a \in A$. Since replacing a by a^* gives the opposite inequality, then $\|a^*\| = \|a\|$. This proves (II) already and $\|a^*a\| \le \|a^*\|\|a\| = \|a\|^2$ immediately implies (I). Finally (III) obviously follows from $\|e\| = \|e^*e\| = \|e\|^2$ and the bound $\|a\| \le \|a\|\|e\|$ for any $a \in A$. ■

The following class of algebras is the most important in the whole theory of involutive Banach algebras.

Definition (IV.7.2). A Banach *-algebra is called a *C^*-algebra* if its norm is a C^*-norm.[15]

This emphasizes that C^*-algebras form a special class of star Banach algebras.

Example (IV.7.3). By Theorem 0.1.13, the operator norm in $\mathcal{B}(H)$, where H is a Hilbert space, has the C^*-property. Therefore $\mathcal{B}(H)$ is a C^*-algebra and so also is any self-adjoint closed subalgebra of $\mathcal{B}(H)$ (with the same norm).

Example (IV.7.4). The algebra $C_0(\Omega)$, where Ω is a locally compact space, is also, obviously, a C^*-algebra.

Let us announce immediately that the meaning of the main results of this section is that there are no other C^*-algebras besides those described in these examples. Meantime we suggest the following exercise to the reader.

Exercise (IV.7.5). Show that $\mathcal{N}(H)$, $C^n[0, 1]$; $n > 0$, and $\mathcal{A}(\overline{\mathbb{D}})$ are not C^*-algebras with any norm equivalent to the original norm.

Hint. In each of these algebras there is a sequence of elements h_n such that $\|h_n\| = 1$ but $\lim_{n\to\infty} h_n^*h_n = 0$.

---

[15] In fact, it was shown in [98] that we could require less: if a norm in a Banach *-algebra has the C^*-property, then it is multiplicative.

Of course, any closed *-subalgebra of a C^*-algebra is itself a C^*-algebra. (The analogous fact about quotient algebras is more complicated and we shall establish it later after the necessary preparation; see Theorem 7.81.) Regarding the adjoining of an identity, it is easy to discover that, for any normed *-algebra $A \neq \mathbb{O}$, the standard norm in A_+ given in Chap.I §2.3 does not have the C^*-property. This inconvenience, however, does not turn out to be important.

Namely, let A be a non-unital C^*-algebra. We introduce a new seminorm in $A_\times$ (which is A_+) by putting $\|a\|' := \sup\{\|ab\| : b \in A, \ \|b\| \leq 1\}$ for $a \in A_\times$. In other words, $\|a\|'$ is the norm in $\mathcal{B}(A)$ of the operator $m_a : b \longmapsto ab$.

Proposition (IV.7.6). *$\|\cdot\|'$ is a C^*-norm in $A_\times$ coinciding on A with the original norm $\|\cdot\|$.*

▶ The equation $m_{ab} = m_a m_b$; $a, b \in A_\times$ obviously implies that $\|\cdot\|'$ is a multiplicative seminorm. Further, if $a \in A$, then the properties of the original norm imply that $\|a\|' \leq \|a\|$, and taking into account Proposition 7.1(II) they also imply that $\|a\|^2 = \|a^*\|^2 = \|aa^*\| = \|m_a(a^*)\| \leq \|m_a\|\,\|a^*\| = \|a\|'\,\|a\|$, that is $\|a\| \leq \|a\|'$. This means that the seminorm we introduced coincides on A with the original norm.

Let us show that $\|\cdot\|'$ is a norm. Suppose otherwise that there exists $a \in A_\times \backslash \{0\}$; $a := b + \lambda e$; $b \in A$, $\lambda \in \mathbb{C}$ such that $\|a\|' = 0$. Then $\|m_a\| = 0$, that is $ac = bc + \lambda c = 0$ for all $c \in A$. From what we did before, $a \notin A$ and therefore $\lambda \neq 0$. Hence $(-\lambda^{-1}b)c = c$ for any $c \in A$ and thus A has a left identity. By Proposition 4.6, this would mean that A, notwithstanding the original hypothesis, is unital.

All that remains is to verify the inequality in Proposition 7.1(I) that ensures the C^*-property. We take $a := b + \lambda e$; $b \in A$, $\lambda \in \mathbb{C}$. From the properties of the original norm and Proposition 7.1(II), we have $\|m_a(c)\|^2 = \|ac\|^2 = \|(ac)^*(ac)\| = \|c^*(a^*a)c\| \leq \|c^*\|\,\|m_{a*a}c\| \leq \|m_{a*a}\|\,\|c\|^2$ for any $c \in A$, hence $\|m_a\|^2 \leq \|m_{a*a}\|$. The rest is clear. ■

In future when speaking of $A_\times$, where A is a C^*-algebra without an identity, we shall always have in mind the corresponding C^*-algebra with the norm constructed as above.

Exercise (IV.7.7). Show that the algebra $A \times B$, where A and B are C^*-algebras, is a C^*-algebra with the norm $\|(a, b)\| := \max\{\|a\|, \|b\|\}$; $a \in A$, $b \in B$. Deduce from this that in the algebra A_+, where A is a C^*-algebra, a C^*-norm can be introduced that extends the original norm even in the case when A is unital.

We turn to studying the special properties of various elements in C^*-algebras and their consequences.

Proposition (IV.7.8). *If a is a normal element of a C^*-algebra, then* $\|a\| = \sqrt{\|a^2\|} = \underline{r}(a)$.

▶ For every element of a C^*-algebra we have $\|(a^*a)^2\| = \|a^*a\|^2 = \|a\|^4$ and for a normal element we have in addition $\|(a^*a)^2\| = \|(a^2)^*a^2\| = \|a^2\|^2$. This implies the first equality. Further, if a is normal so too are all powers of a. Therefore, from what we have proved it follows that $\|a^{2^n}\| = \|a^{2^{n-1}}\|^2 = \cdots = \|a\|^{2^n}$ for all n. All that remains is to apply the spectral radius formula. ■

Proposition (IV.7.9). *Every C^*-algebra is semisimple.*

▶ If a belongs to the radical of our algebra so too does a^*a; hence $\underline{r}(a^*a) = 0$ and, by Proposition 7.8, $\|a^*a\| = \|a\|^2 = 0$. ■

Proposition (IV.7.10). *If u is a unitary element of a C^*-algebra with identity, then* $\mathrm{Sp}(u) \subseteq \mathbb{T}$.

▶ By Proposition 7.1(III), $\|u\|^2 = \|u^*u\| = \|e\| = 1$; therefore Proposition 7.8 implies that $\underline{r}(u) = 1$ and in exactly the same way $\underline{r}(u^{-1}) = \underline{r}(u^*) = 1$. Thus $|\lambda| \le 1$ for $\lambda \in \mathrm{Sp}(u)$ and since $\lambda^{-1} \in \mathrm{Sp}\, u^{-1}$ (II.1.10), $|\lambda^{-1}| \le 1$. Hence $\lambda \in \mathbb{T}$. ■

Lemma (IV.7.11). *Let A be a unital star Banach algebra. Then for any $a \in A$ the element $e^{(a^*)}$* (see II.2.10) *coincides with $(e^a)^*$.*

▶ By Theorem II.2.7, the first of the two elements is $\sum_{n=0}^{\infty} \frac{(a^*)^n}{n!} = \sum_{n=0}^{\infty} \left(\frac{a^n}{n!}\right)^*$ and the second is $\left(\sum_{n=0}^{\infty} \frac{a^n}{n!}\right)^*$. The rest is clear. ■

Theorem (IV.7.12). *If h is a self-adjoint element of a C^*-algebra, then* $\mathrm{Sp}(h) \subset \mathbb{R}$. *(In other words, every C^*-algebra is Hermitian.)*

▶ We consider the element $u := e^{ih}$ in $A_{\times}$, where A is our algebra; then by Proposition II.2.11 and the preceding lemma, $u^* = u^{-1} = e^{-ih}$. This means that u is unitary. But by Theorem II.2.23, $e^{i\lambda} \in \mathrm{Sp}(u)$ holds for any $\lambda \in \mathrm{Sp}(h)$ and hence, by Proposition 7.10, $e^{i\lambda} \in \mathbb{T}$. But this is, of course, only possible when $\lambda \in \mathbb{R}$. ■

All is now ready to prove one of the most important results in the whole of this chapter.

Theorem (IV.7.13) ("First Gel'fand–Naimark theorem"). *Let A be a commutative C^*-algebra with spectrum Ω. Then its Gel'fand transformation $\Gamma : A \longrightarrow C_0(\Omega)$ is an isometric *-isomorphism.*

▶ Since $\|a\| = \underline{r}(a) = \|\hat{a}\|_0$ for any $a \in A$ (Propositions 7.8 and 2.3), Γ is an isometric embedding; hence, in particular, the subalgebra $\hat{A} = \mathrm{Im}\,\Gamma$ is closed in $C_0(\Omega)$. but Theorems 7.12, 6.14(IV), and 6.15 imply that Γ is a

*-homomorphism with a dense image. Therefore $\hat{A}$, being both closed and dense in $C_0(\Omega)$, coincides with the latter algebra. ∎

The case of operator algebras is worthy of particular mention in connection with what was promised at the end of §3.

Corollary (IV.7.14). *Let A be a closed self-adjoint subalgebra of $\mathcal{B}(H)$, where H is a Hilbert space. If A is commutative, then its Gel'fand transformation $\Gamma : \underline{A \longrightarrow} C_0(\Omega)$ is an isometric isomorphism and moreover $(S^*)^\wedge(t) = \overline{\hat{S}(t)}$ for any $S \in A$ and $t \in \Omega$.*

Recall Example I.2.13 that we considered earlier. The above corollary obviously shows us what assertion of a general character was behind the observation made there.

Let us note the interpretation of Theorem 7.13 in category theory.

Exercise (IV.7.15). Show that the category of unital commutative C^*-algebras and their continuous[16] *-homomorphisms coincides, to within an isomorphism, with the category that is dual to the category of compact topological spaces and their continuous maps.

Hint. The required isomorphism is a composition of the spectrum functor in §1.2, more precisely, its restriction to the first of the two categories and the functor of taking duals.

A more general assertion holds: the category of all commutative C^*-algebras is isomorphic to the dual of the category of locally compact spaces. However, it is true that one needs to be more precise in defining the morphisms of the second category; see, for example, [254, p.122].

Remark (IV.7.16). Theorem 7.13 shows, roughly speaking, the following: everything that can be done with continuous functions, can also be done with elements of commutative C^*-algebras. In particular, this relates to families of operators that happen to be included in some commutative self-adjoint subalgebra of $\mathcal{B}(H)$ — for example, to a (single) normal operator. In the remainder of the section we shall repeatedly meet specific applications of this principle.

Remark (IV.7.17). On the other hand, Theorem 7.13 enables us to consider arbitrary C^*-algebras as "non-commutative generalizations of $C_0(\Omega)$" and often to guess their properties,

---

[16] In fact *-homomorphisms between C^*-algebras are automatically continuous (see later 7.62 and 7.83); in the commutative case the reader may easily deduce this fact from Theorem 7.13.

starting from known properties of algebras of continuous functions.[17] Such an ambitious programme turned out to be extremely fruitful and led to such very useful (and applicable) concepts as "non-commutative integration", "non-commutative order" and even new independent disciplines such as "non-commutative algebraic topology".[18] An account of this range of questions goes beyond the scope of the aims we have set; among the papers that up to now have been devoted to this (their number is rapidly increasing) we recommend [304], [305], [227], [323].

§7.2. Continuous functions and Borel functions of certain elements of a C^*-algebra. Application: spectral theorem for a normal operator

The reader has long known that one can "apply holomorphic functions" to elements of Banach algebras provided their domain of definition contains the spectrum of the given element. Now, thanks to Theorem 7.13, we can in several important cases, considerably widen the class of functions (of a complex variable) f for which the expression $f(a)$ has a meaning: instead of holomorphic functions we shall be talking first about "continuous" functions and then about a "Borel" functional calculus. Here is the first formal definition.

Let A be a star polynormed algebra (for the time being arbitrary), let a be a normal element and let Ω be a compact set in $\mathbb{C}$.

Definition (IV.7.18). (Any) continuous unital *-homomorphism $\mathcal{C}ont\,;\ (C(\Omega);\ \|\cdot\|_0) \longrightarrow A_\times$ taking the function $f(z) = z$ into a, is called a *continuous functional calculus* or simply *a continuous calculus*.

We let $P_{z,\bar{z}}$ ($P^0_{z,\bar{z}}$ respectively) denote the smallest unital subalgebra (or just, subalgebra, respectively) of $C(\Omega)$ containing the functions $f(z) = z$ and $f(z) = \bar{z}$. Obviously the first consists of all the polynomials in z and $\bar{z}$ and the second of polynomials in z and $\bar{z}$ without a constant term.

---

[17] It is possible that the letter C appears in the very name "C^*-algebra", introduced by Segal [237], just because of this.

[18] In view of the category theoretic interpretation of Theorem 7.13 mentioned above, the subject of algebraic topology, that is the study of homotopy functors on the category of topological spaces, can be considered, almost entirely, as the "commutative" part of the discipline of studying the analogously defined functors on the category of all C^*-algebras. Hence the name "non-commutative algebraic topology" (Effros [305]).

Theorem (IV.7.19) (uniqueness). *If A is Hausdorff, then for given Ω and $a \in A$ there exists at most one continuous calculus.*

▶ By virtue of its algebraic properties, a continuous calculus, must be unique on $P_{z,\bar{z}}$ and this subalgebra, as a consequence of the Stone–Weierstrass theorem, is dense in $C(\Omega)$. The rest is clear. ∎

Just as for a holomorphic calculus, the spectrum of an element inevitably crops up when we try to clarify the conditions for the existence of a continuous calculus.

Proposition (IV.7.20). *A necessary condition for a continuous calculus for $a \in A$ to exist in Ω is that* $\mathrm{Sp}_A(a)$ *be contained in Ω.*

▶ The proof of Proposition II.2.6 can obviously be carried over from the "holomorphic" case to the "continuous" case. ∎

In particular, if A is not unital, then Proposition 7.20 implies that, if there exists a continuous calculus in the compact set Ω, it must automatically contain zero. A natural question about which "continuous functions in a" still belong to A ($A \neq A_\times$ now) is solved by the following proposition.

Proposition (IV.7.21). *In this situation* $\mathcal{C}ont(f) \in A \iff f(0) = 0$.

▶ "$\Leftarrow$". According to the Stone–Weierstrass theorem f is $\lim_{n\to\infty} g_n$, where $g_n \in P_{z,\bar{z}}$; moreover, as $f(0) = 0$, we may assume also that $g_n(0) = 0$, in other words, that $g_n \in P^0_{z,\bar{z}}$; $n \in \mathbb{N}$. But then, obviously $\mathcal{C}ont(g_n) \in A$; $n \in N$, and therefore, by virtue of the continuity of $\mathcal{C}ont$, $\mathcal{C}ont(f) \in A$.

"$\Rightarrow$". We put $g = f - f(0)1$; then $g(0) = 0$ and by what has already been proved, $\mathcal{C}ont(g) \in A$. Therefore, if $\mathcal{C}ont(f) \in A$, then $\mathcal{C}ont(f(0)1) = f(0)e \in A$, and this means that $f(0) = 0$. ∎

Let us turn to sufficient conditions for a continuous calculus to exist. First of all we turn our attention to the fact that for the class of star Banach algebras the necessary condition given in 7.20 is not sufficient (cf. Theorems II.2.6 and II.2.15 about the holomorphic calculus).

Example (IV.7.22). Obviously there does not exist in $\Omega = [0, 1]$ a continuous calculus for the function $f(t) = t$ as an element of the algebra $A = C^1[0, 1]$; on the other hand $[0, 1]$ is the spectrum of f.

However, if we restrict ourselves to considering C^*-algebras, then the picture changes.

Let A be a C^*-algebra and let a be a normal element. We take the C^*-algebra $A_\times$ (see §7.1) and put

$C^*(a) := \overline{\ell.h.}\{a^m(a^*)^n;\quad m, n = 0, 1, 2, \cdots\}$; obviously $C^*(a)$ is a commutative algebra which is the smallest C^*-subalgebra of A_X containing a and e.

We are again in the situation where the two meanings of the word "spectrum" coincide (cf. 1.14). For brevity we let Ω_a denote $\underline{\Omega}(C^*(a))$ and we associate to each $s \in \Omega_a$ the number $\chi_s(a)$ or, what is the same thing, $\hat{a}(s)$; here χ_s is the character on $C^*(a)$ corresponding to s and $\hat{a} := \Gamma_{C*(a)}(a)$.

Proposition (IV.7.23). *$\chi_s(a) \in \mathrm{Sp}_A(a)$ for any $s \in \Omega_a$ and the map $\omega : \Omega_a \longrightarrow \mathrm{Sp}_A(a) : s \longmapsto \chi_s(a)$ is a homeomorphism.*

▶ If we combine Proposition 6.4 and Theorem 7.12, we see that $\mathrm{Sp}_A(a) = \mathrm{Sp}_{C*(a)}(a)$. Therefore by Corollary 1.3 (or Theorem 2.3) ω is well defined, surjective and coincides with $\hat{a}|^{\mathrm{Sp}_A(a)}$, and hence, by Theorem 2.5, it is continuous. Moreover, if $\chi_s(a) = \chi_t(a)$ for some $s, t \in \Omega_a$ then, since $C^*(a)$ is Hermitian, $\chi_s(a^*) = \chi_t(a^*)$ and hence χ_s and χ_t coincide on the whole of $C^*(a)$. All that remains is to use Alexandrov's theorem from Chap.0 §0.3°. ■

Theorem (IV.7.24). *Let A be a C^*-algebra, let a be a normal element of A and let Ω be a compact set in $\mathbb{C}$ containing $\mathrm{Sp}_A(a)$. Then there exists on Ω a (unique by 7.19) continuous functional calculus for $a \in A$ of norm ≤ 1, and moreover its image coincides with $C^*(a)$. In the case when $\Omega = \mathrm{Sp}_A(a)$ this calculus is an isometric operator.*

▶ By Theorem 7.13, the Gel'fand transformation $\Gamma = \Gamma_{C*(a)} : C^*(a) \longrightarrow C(\Omega_a)$ is an isometric *-isomorphism. Therefore we are justified in defining a certain map $\mathcal{C}ont$ by the commutative diagram

$$\begin{array}{ccccc}
C(\Omega) & & \xrightarrow{\quad\mathcal{C}ont\quad} & & A_X \\
\downarrow \tau & & & & \uparrow \mathrm{in} \\
C(\mathrm{Sp}_A(a)) & \xrightarrow{C(\omega)} & C(\Omega_a) & \xrightarrow{\Gamma^{-1}} & C^*(a)
\end{array}$$

in which $\tau : f \longmapsto f|_{\mathrm{Sp}_A(a)}$, and $\underline{C}(\omega) : g \longmapsto g \circ \omega$ (cf. the functor $\underline{C}$ in Proposition 2.16). As $\underline{C}(\omega)$ and Γ^{-1} are obviously isometric *-isomorphisms, and τ is a *-homomorphism of norm ≤ 1, $\mathcal{C}ont$ is also a *-homomorphism of norm ≤ 1 which is isometric when $\Omega = \mathrm{Sp}_A(a)$, in other words, when $\tau = 1$. Moreover the path of the element $f \in C(\Omega) : f(z) = z$ in the above diagram, may be depicted as $f \longmapsto f|_{\mathrm{Sp}_A(a)} \longmapsto \hat{a} \longmapsto a \longmapsto a$. The rest is clear. ■

The next proposition says something about the compatibility of the two functional calculi we know about.

Proposition (IV.7.25). *Let* A *be a unital* C^**-algebra, let* a *be a normal element of* A. *Let* U *be an open set and* Ω *a compact subset of* $\mathbb{C}$ *such that* $\mathrm{Sp}_A(a) \subseteq \Omega \subset U$. *Let* $\Gamma o\lambda$ *and* $\mathcal{C}ont$ *be the holomorphic calculus in* U *and the continuous calculus in* Ω*, respectively, for* $a \in A$. *Then* $\Gamma o\lambda\, w = \mathcal{C}ont(w|_\Omega)$ *for any* $w \in \mathcal{O}(U)$.

▶ We consider the map from $\mathcal{O}(U)$ into A that takes w into $\mathcal{C}ont(w|_\Omega)$; obviously it satisfies Definition II.2.1. So the required equality follows from the uniqueness theorem for a holomorphic calculus (II.2.4). ■

From now on we are justified, with no danger of getting confused with what was said in §2.2, in calling the element $\mathcal{C}ont\, f$ *the value of the function* f *at* a and denoting it by $f(a)$.

Example (IV.7.26). When the conditions of Theorem 7.24 are satisfied, one can define elements in A such as $\mathrm{Re}\, a$, $\mathrm{Im}\, a$ and $|a|$; for example, $|a|$ ("the modulus of a") is $\mathcal{C}ont\, f$ when $f(z) = |z|$ and so on.

Example (IV.7.27). Suppose that, in addition to the conditions of Theorem 7.24, a is positive and $\Omega = \mathrm{Sp}_A(a)$. Then one can define the element $a^{1/n} := \mathcal{C}ont(t^{1/n})$ in A, where $t^{1/n}$ is the "algebraic n-th root" function; correspondingly $a^{1/n}$ is called the *algebraic* n*-th root of* a.

We note that the elements we get, generally speaking, are no longer holomorphic functions of a; in particular $a^{1/n}$, unlike $\sqrt[n]{a}$ in II.2.20, is also defined when $0 \in \mathrm{Sp}_A(a)$. As an application of this, the algebraic roots enable the positive elements of C^*-algebras to be completely characterized in terms of the Shirali–Ford theorem (6.10).

Theorem (IV.7.28). *Let* p *be an element of the* C^**-algebra* A. *Then* $p \in A_{\mathrm{pos}} \iff p = h^2$ *for a certain* $h \in A_{\mathrm{pos}} \iff p = a^*a$ *for a certain* $a \in A$.

▶ The second of these properties implies, of course, the third and the third, by Theorem 6.10, implies the first; we only have to deduce the second from the first. We put $h := p^{1/2}$. Then Proposition 7.21 implies that $h \in A$, whether there exists an identity in A or not; at the same time the algebraic properties of $\mathcal{C}ont$ imply that $h \in (A_X)_{sa}$ and, taking into account Proposition II.1.6, $\mathrm{Sp}_A h \in \mathbb{R}^{[+)}$. The rest is clear. ■

Exercise (IV.7.29). Formulate and prove, for a continuous calculus, a spectral mapping theorem analogous to II.2.23.

* * *

For certain questions in analysis, primarily in the theory of operators, it turns out to be very important that in certain situations one can go further and talk about more general functions of the elements of algebras than continuous functions.

Let Ω be a compact metrizable space, for the moment arbitrary. Recall that a subset $\Delta \subset \Omega$ is called a *Borel subset* if it belongs to the smallest σ-algebra in Ω containing all the open (or what is equivalent, all the closed) sets. A function $f : \Omega \longrightarrow \mathbb{C}$ is called a *Borel function* (or, some say Baire function, and also B-function) if the preimage of every open set in $\mathbb{C}$ is a Borel set in Ω; this same class of functions may be described also as the smallest class of functions on Ω that contains all the continuous functions and which is closed under the operation of taking pointwise limits of sequences (see, for example, [80, p.294]).

The set of bounded Borel functions on Ω is denoted by $B(\Omega)$; it is easy to see that these functions may also be described as the uniform limits of linear combinations of characteristic functions of Borel subsets in Ω. Obviously $\mathcal{B}(\Omega)$ is a C^*-algebra with respect to pointwise operations, the involution $f \longmapsto \overline{f}$ and the uniform norm; it contains $C(\Omega)$ as a C^*-subalgebra.

Our next aim is to introduce another structure in $B(\Omega)$: this time that of a star polynormed algebra. With this aim we note that $B(\Omega)$ is naturally identified with a certain closed subspace of $C(\Omega)^{**}$.

In fact, we have mentioned earlier (Chap.III §3.3) that $C(\Omega)^*$ can be identified with the Banach space $M(\Omega)$ of regular complex Borel measures on Ω. This enables us to define a map $B(\Omega) \longrightarrow C(\Omega)^{**} = M(\Omega)^*$ that takes f into the functional $\mu \longmapsto \int_\Omega f(t)d\mu(t)$; $\mu \in M(\Omega)$. It is easy to see that this is an isometric operator, whose restriction to $C(\Omega)$ is the canonical embedding into the second dual. Thus, we are justified in assuming that $C(\Omega) \subset B(\Omega) \subset C(\Omega)^{**}$.

Recalling what the seminorms are in $C(\Omega)^{**}$ that give the weak* topology, we consider the system of their restrictions to $B(\Omega)$ — in other words, the system $\|f\|_\mu := \left|\int_\Omega f(t)d\mu(t)\right|$; $f \in B(\Omega)$, where μ runs over $M(\Omega) = C(\Omega)^*$. This system of seminorms will be denoted by $\underline{\mathrm{m}}$ and we shall speak about the polynormed space $(B(\Omega); \underline{\mathrm{m}})$; the topology of this space will be called the $\underline{\mathrm{m}}$-*topology*.

Proposition (IV.7.30). $(B(\Omega); \underline{\mathrm{m}})$ *with pointwise multiplication and the involution* $f \longmapsto \overline{f}$ *is a star polynormed algebra and* $C(\Omega)$ *is a dense* **-subalgebra.*

▶ For any $f \in B(\Omega)$ and $\mu \in M(\Omega)$ we let $f \cdot \mu$ denote the measure identified with the functional $h \longmapsto \int_\Omega h(x)f(x)d\mu(x)$ on $C(\Omega)$. Then $\|fg\|_\mu = \|g\|_{f\cdot\mu}$ is obviously satisfied for any $g \in B(\Omega)$. This means, by Proposition 0.1.9, that the multiplication in $(B(\Omega); \underline{m})$ is separately continuous; as every seminorm in $\underline{m}$ is even a star seminorm, this proves the first assertion. All that remains is to use the fact that $C(\Omega)$, like any Banach space, is dense in its second dual with respect to the weak* topology. ■

Now let A, $a \in A$ and $\Omega \subset \mathbb{C}$ be as in Definition 7.18.

Definition (IV.7.31). A *Borel functional calculus*, or simply a *Borel calculus* on Ω for $a \in A$ is any continuous unital *-homomorphism $\mathcal{B}or : (B(\Omega); \underline{m}) \longrightarrow A_X$, taking the function $f(z) = z$ into a.

The specific requirement in Definition 7.31 — continuity with respect to the $\underline{m}$-topology — enables us to establish a uniqueness theorem analogous to Theorem 7.19.

Theorem (IV.7.32). *If A is Hausdorff, then for given Ω and $a \in A$ there exists at most one Borel calculus.*

▶ Proposition 7.30 and the Stone–Weierstrass theorem obviously imply that $P_{z,\bar{z}}$ is a dense subalgebra of $(B(\Omega); \underline{m})$ and (cf. the proof of Theorem 7.19) the Borel calculus is uniquely defined on this. The rest is clear. ■

Concerning the question of the existence of a Borel calculus, here again, besides the obvious necessary condition $\mathrm{Sp}_A\, a \subseteq \Omega$ (the proof, as in Proposition 7.20, is analogous to that given in II.2.6) special conditions on the algebra A are needed which arise in connection with the abundance of idempotent elements in $B(\Omega)$.

Exercise (IV.7.33). Suppose that a cannot be approximated in A by a linear combination of idempotents. Prove that there does not exist on any non-empty Ω a Borel calculus for $a \in A$.

In fact, the natural class of algebras which are suitable for constructing a Borel calculus for their normal elements are those which are topologically *-isomorphic to von Neumann algebras, taken with some topology that is no stronger than the ultraweak topology (see Chap.III §2.3). However, not wishing to make the exposition too cumbersome, we shall limit ourselves to only one of these algebras, namely $(\mathcal{B}(H); \underline{w})$; the reader can easily verify that the results we shall get still hold when this algebra is replaced by any other in the above class.

Everywhere from now on when we say "a *Borel calculus for the operator* a" we mean a to be taken as an element of $(\mathcal{B}(H); \underline{w})$.

We make some preliminary observations. Let a be a bounded normal operator on the Hilbert space H, let Ω be a compact set in $\mathbb{C}$ that contains its spectrum, and let $\mathcal{C}ont : C(\Omega) \longrightarrow (\mathcal{B}(H);\ \|\cdot\|_0)\ : f \longmapsto f(a)$ be the corresponding continuous calculus for a.

Proposition (IV.7.34). *For any* $x, y \in H$ *there exists a unique measure* $\mu_{x,y} \in M(\Omega)$ *such that* $\int_\Omega f(t) d\mu_{x,y}(t) = \langle f(a)x, y\rangle$ *for any* $f \in C(\Omega)$. *Moreover* $\|\mu_{x,y}\| \le \|x\|\,\|y\|$.

▶ The map $f \longmapsto \langle f(a)x, y\rangle$ is a functional on $C(\Omega)$ with norm ≤ 1, because of the bounds $\|f(a)\| \le \|f\|$ (see Theorem 7.24) and $|\langle f(a)x, y\rangle| \le \|f(a)\|\,\|x\|\,\|y\|$. We only have to apply Riesz's theorem to this functional. ■

Proposition (IV.7.35). *For any* $\alpha \in C(\Omega)^{**} = M(\Omega)^*$ *there exists a unique operator* $\alpha(a)$ *on* H *such that* $\langle \alpha(a)x, y\rangle = \alpha(\mu_{x,y})$ *for any* $x, y \in H$. *Moreover* $\|\alpha(a)\| \le \|\alpha\|$.

▶ We put $\mathcal{R}_\alpha : H \times H \longrightarrow \mathbb{C} : (x, y) \longmapsto \alpha(\mu_{x,y})$; obviously this is a sesquilinear functional on H; in addition the bounds in Proposition 7.34 imply that $|\mathcal{R}_\alpha(x, y)| \le \|\alpha\|\,\|\mu_{x,y}\| \le \|\alpha\|\,\|x\|\,\|y\|$ and therefore $\|\mathcal{R}_\alpha\| \le \|\alpha\|$. All that remains is to use the theorem about identifying sesquilinear functionals with operators (see the end of Chap.0 §1.3). ■

Thus, starting with $\mathcal{C}ont : C(\Omega) \longrightarrow \mathcal{B}(H)$ we have arrived at a certain map $\mathcal{A}r : C(\Omega)^{**} \longrightarrow \mathcal{B}(H);\ a \longmapsto \alpha(a)$;[19] obviously this is an operator.

Proposition (IV.7.36). (I) *As an operator between Banach spaces* $\mathcal{A}r$ *is continuous and has norm* ≤ 1.

(II) *As an operator between* $C(\Omega)^{**}$ *with the weak* topology and* $\mathcal{B}(H)$ *with the weak operator* ($\underline{w}$-)*topology* $\mathcal{A}r$ *is still continuous.*

(III) $\mathcal{A}r|_{C(\Omega)} = \mathcal{C}ont$.

(IV) $\operatorname{Im} \mathcal{A}r$ *lies in the closure of the subalgebra* $C^*(a)$ *of* $(\mathcal{B}(H);\ \underline{w})$.

▶ (I) is implied by the bound in Proposition 7.35.

(II) Given a seminorm $\|\cdot\|_{x,y}$ from the system $\underline{w}$ in $\mathcal{B}(H)$ (see III.2.21) we consider the seminorm $\|\cdot\|_{\mu_{x,y}}$ in $C(\Omega)^{**}$. Since $\|\alpha(a)\|_{x,y} = |\langle \alpha(a)x, y\rangle| = |\alpha(\mu_{x,y})| = \|\alpha\|_{\mu_{x,y}}$ is

---

[19] This notation is in honour of Arens who defined the structure of a Banach algebra in the second dual space of any Banach algebra [101]. It can be shown that $\mathcal{A}r$ is a homomorphism when we take precisely this "Arens product".

satisfied for any $\alpha \in C(\Omega)^{**}$, the required continuity is implied by 0.1.9.

(III) This is implied by the uniqueness in Proposition 7.35.

(IV) By (II) and the fact that $C(\Omega)$ is weak* dense in $C(\Omega)^{**}$, $\operatorname{Im} \mathcal{A}r$ lies in the closure of $\mathcal{A}r(C(\Omega))$ in $(\mathcal{B}(H);\ \underline{w})$. But by (III) and Theorem 7.24, $\mathcal{A}r(C(\Omega)) = \operatorname{Im} \mathcal{C}ont = C^*(a)$. ∎

Theorem (IV.7.37). *Let a be a normal operator on the Hilbert space H and let Ω be a compact set in $\mathbb{C}$ that contains its spectrum. Then there exists on Ω a Borel calculus for a (unique by Theorem 7.32) namely $\mathcal{B}or := \mathcal{A}r|_{B(\Omega)}$. In addition*

(I) *$\mathcal{B}or$ is also a continuous homomorphism of norm ≤ 1 between $B(\Omega)$ and $\mathcal{B}(H)$ as Banach algebras;*

(II) *the restriction of $\mathcal{B}or$ to $C(\Omega)$ is $\mathcal{C}ont$; finally,*

(III) *if a certain operator commutes with a and a^*, then it commutes with $\mathcal{B}or\, f$ for all $f \in B(\Omega)$* (in other words, $\operatorname{Im} \mathcal{B}or \subseteq \{a, a^*\}^{!!}$; see III.2.29).

▶ First of all, Proposition 7.36 (I, II) implies that the above map $\mathcal{B}or$ is an operator that is continuous in both the required senses and Proposition 7.36(III) implies that $\mathcal{B}or$ coincides with $\mathcal{C}ont$ on $C(\Omega)$; hence, in particular, $\mathcal{B}or\, f = a$ when $f(z) \equiv z$. Further, if we combine Proposition 7.36(IV) and von Neumann's bicommutant theorem (III.2.34), we see that the image of $\mathcal{B}or$ lies in $C^*(a)^{!!} = \{a, a^*\}^{!!}$. Thus we only have to show that $\mathcal{B}or$ is a *-homomorphism.

First let $f \in B(\Omega)$, $h \in C(\Omega)$. Then, using Proposition 7.30 and the properties of $\mathcal{B}or$ we already know about, we see that for any net $(f_\nu \in C(\Omega);\ \nu \in \Lambda)$, converging to f in $(B(\Omega);\ \underline{m})$ we have

$$\mathcal{B}or(fh) = \lim_\nu \mathcal{B}or(f_\nu h) = \lim_\nu \mathcal{C}ont(f_\nu h) = \lim_\nu (\mathcal{C}ont(f_\nu)\mathcal{C}ont\, h) =$$

$$= \lim_\nu (\mathcal{B}or(f_\nu)\mathcal{B}or\, h) = (\lim_\nu \mathcal{B}or\, f_\nu)\mathcal{B}or\, h = \mathcal{B}or\, f\ \mathcal{B}or\, h$$

(the limit is taken in the sense of the $\underline{w}$-topology in $\mathcal{B}(H)$). Hence for any $f, g \in B(\Omega)$, with the help of a similar calculation using a net $(g_\mu \in C(\Omega);\ \mu \in \Lambda')$ that converges (this time) to $g \in (B(\Omega);\ \underline{m})$ we get that $\mathcal{B}or(fg) = \mathcal{B}or\, f\ \mathcal{B}or\, g$. Finally, the fact that $\mathcal{C}ont$ is a *-map and $C(\Omega)$ is dense in $(B(\Omega);\ \underline{m})$ obviously implies that $\mathcal{B}or$ is also a *-map. ∎

We note that by Theorem 7.37(II) we can, with no danger of ambiguity, talk about the operator $\mathcal{B}or\, f$; $f \in B(\Omega)$, as a "Borel function f of a" and denote it by $f(a)$.

The continuity requirement entering into the definition of a Borel calculus could be formulated differently in the language of sequences:

Exercise (IV.7.38). Prove that, under the conditions of Theorem 7.37, a Borel calculus is the unique *-homomorphism from $B(\Omega)$ into $\mathcal{B}(H)$ taking $f(z) = z$ into a and having the following property: if g_n is a uniformly bounded sequence in $B(\Omega)$, pointwise convergent to g, then $g_n(a)$ converges to $g(a)$ in $(\mathcal{B}(H); \underline{w})$. Prove the analogous assertion obtained by replacing the weak operator topology ($\underline{w}$) by the strong operator topology ($\underline{s}$; see III.2.21).

Hint. The result about the $\underline{w}$-topology follows from Lebesgue's theorem on taking limits under the integral sign with respect to (any) measure in $M(\Omega)$. This, using

$$\|(g(a) - g_n(a))x\|^2 = \int_\Omega |g(t) - g_n(t)|^2 d\mu_{x,x}(t),$$

implies the result about the $\underline{s}$-topology.

Remark (IV.7.39). A non-trivial theorem due to Fuglede (see, for example, [67, p.300]) asserts that a bounded operator in H that commutes with the normal operator a also commutes with a^*. This implies that in Theorem 7.37(III) $\{a, a^*\}^{!!}$ in fact coincides with $\{a\}^{!!}$.

* * *

Is it impossible to describe some "analytic" means of getting $f(a)$ for a continuous or Borel function f such as the integral formula that featured in the construction of the holomorphic calculus in II.2.15? There actually does exist such a method for representing $f(a)$ by a certain formula involving a vector-valued integral — but quite different from the contour integral in Chap.II §2. The core of the construction is based on the observation that, since the Borel calculus for an operator is a *-homomorphism, it carries the self-adjoint idempotents of $B(\Omega)$, that is, the characteristic functions of Borel subsets, into self-adjoint idempotents in $\mathcal{B}(H)$, that is, into orthogonal projections. By axiomatizing certain properties of the family of operators that we get (although for the time being only the most obvious ones), we reach the following concept.

Let Ω be a certain compact space, let Σ be its collection of Borel subsets and let H be a Hilbert space.

Definition (IV.7.40). A *spectral measure* on Ω is a map $E : \Sigma \longrightarrow \mathcal{B}(H)$ taking values in the set of orthogonal projections such that

(I) $E(\emptyset) = 0$ and $E(\Omega) = 1$;

(II) if $\Delta = \Delta_1 \coprod \Delta_2$; $\Delta_1, \Delta_2 \in \Sigma$, then $E(\Delta) = E(\Delta_1) + E(\Delta_2)$ "finite additivity";

(III) if $\Delta = \Delta_1 \cap \Delta_2$; $\Delta_1, \Delta_2 \in \Sigma$, then $E(\Delta) = E(\Delta_1)E(\Delta_2)$.

Exercise (IV.7.41). Show that in fact (III) is implied by (II).

Hint. If $\|P_1 + P_2\| = 1$ for orthogonal projections P_1 and P_2, then $P_1P_2 = P_2P_1 = 0$.

We say that a Borel function on Ω is *simple* if it takes a finite number of values, in other words, if it is a linear combination of characteristic functions. The set of simple functions on Ω is denoted by $S(\Omega)$; obviously this is a *-subalgebra of $B(\Omega)$ which is dense with respect to the uniform norm as well as with respect to the m-topology.

Proposition (IV.7.42). *Let $E : \Sigma \longrightarrow \mathcal{B}(H)$ be a spectral measure. Then there exists a unique operator $I : B(\Omega) \longrightarrow \mathcal{B}(H)$ that is continuous as an operator between Banach spaces and is such that $I(f) = \sum_{k=1}^{n} \lambda_k E(\Delta_k)$ for $f = \sum_{k=1}^{n} \lambda_k \chi_{\Delta_k} \in S(\Omega)$. Moreover I is a *-homomorphism.*

▶ It is obvious that the map $S(\Omega) \longrightarrow \mathcal{B}(H)$ $: \sum_{k=1}^{n} \lambda_k \chi_{\Delta_k} \longmapsto \sum_{k=1}^{n} \lambda_k E(\Delta_k)$ is well defined and is a *-homomorphism of norm 1. All that remains is to take I to be the unique continuous extension to $B(\Omega)$. ■

Definition (IV.7.43). For $f \in B(\Omega)$ the operator $I(f) : H \longrightarrow H$ is called the *integral of the function* f *with respect to the spectral measure* E and it is also written as $\int_\Omega f(\lambda)dE_\lambda$.

Thus the definition implies that $\int_\Omega f(\lambda)dE_\lambda$ is the limit of the "approximating sums" $\sum_{k=1}^{n(m)} \lambda_k^{(m)} E(\Delta_k^{(m)})$ for any sequence of simple functions $f_m = \sum_{k=1}^{n(m)} \lambda_k^{(m)} \chi_{\Delta_k}^{(m)}$ converging uniformly to f.

If we rephrase Proposition 7.42, we get the following.

Proposition (IV.7.44). *Let $E : \Sigma \longrightarrow \mathcal{B}(H)$ be a spectral measure on Ω. Then the map $\varphi : B(\Omega) \longrightarrow \mathcal{B}(H) : f \longmapsto \int_\Omega f(\lambda)dE_\lambda$ is a continuous Banach algebra *-homomorphism. As a consequence, for $\Omega \subset \mathbb{C}$, the restriction of φ to $C(\Omega)$ is the continuous calculus in Ω for the operator $\int_\Omega \lambda dE_\lambda$* (that is, for $I(f)$ where $f(z) \equiv z$, $z \in \Omega$). ▶■

It is not difficult to see that all homomorphisms of the above sort have such integral representations:

Proposition (IV.7.45). *Let $\varphi : B(\Omega) \longrightarrow \mathcal{B}(H)$ be a continuous *-homomorphism of Banach algebras. Then there exists a unique* *spectral measure on* Ω, *namely* $E : \Delta \longmapsto \varphi(\chi_\Delta)$, *such that* $\varphi(f) = \int_\Omega f(\lambda)dE_\lambda$ *for any* $f \in B(\Omega)$.

▶ That E is a spectral measure follows obviously from the algebraic properties of φ. Further $I : f \longmapsto \int_\Omega f(\lambda)dE_\lambda$ coincides with φ on every χ_Δ; $\Delta \in \Sigma$, and hence also on $S(\Omega)$; therefore since they are both continuous (Proposition 7.42), $I = \varphi$. Finally, uniqueness is ensured by $E(\Delta) = \int_\Omega \chi_\Delta(\lambda)dE_\lambda$; $\Delta \in \Sigma$. ■

As a particular case of 7.45 we see that the Borel calculus and — since it is a restriction of this — the continuous calculus for any operator both have integral representations using a certain spectral measure. However, the results obtained cannot yet impress one as being conclusive. Firstly, the integral in Proposition 7.44, although it provides a continuous calculus, does not, in general, provide a Borel calculus: φ is not necessarily continuous in the sense of the appropriate polynormed algebras. Secondly, and this is more important, the spectral measure in Proposition 7.45, that provides the integral formula for the operators $\varphi(f)$, is uniquely defined by the values of φ on the whole of $B(\Omega)$, but, generally speaking, not by its values on the subalgebra $C(\Omega)$ and what is more — and this is what we would hope for! — not by its value on the "individual" function $f(z) \equiv z$ for $\Omega \subset \mathbb{C}$. In order to achieve the desired paradise, it is necessary to restrict ourselves to considering a special class of spectral measures which is extremely important in operator theory.

Definition (IV.7.46). A spectral measure $E : \Sigma \longrightarrow \mathcal{B}(H)$ on the compact space Ω is called a *resolution of the identity* if, for all $x, y \in H$, the set function $E_{x,y} : \Delta \longmapsto \langle E(\Delta)x, y\rangle$ is a regular complex measure on Ω (= belongs to $M(\Omega)$).

Exercise (IV.7.47). Show that the property of finite additivity in Definition 7.40 follows even if only one of the measures $E_{x,y}$ in Definition 7.46 is finitely additive.

Proposition (IV.7.48). *If* E *is a resolution of the identity, then* $\left\langle \left[\int_\Omega f(\lambda)dE_\lambda\right](x), y\right\rangle = \int_\Omega f(\lambda)dE_{x,y}(\lambda)$ *for any* $f \in B(\Omega)$ *and* $x, y \in H$ (in the usual sense of integration with respect to a regular measure). ▶■

We emphasize that not every spectral measure is a resolution of the identity.

Example (IV.7.49). Let μ be a finitely additive,[20] but not countably additive (= not regular), measure on $[0, 1]$. Then the map E taking Δ into the operator of multiplication by χ_Δ on $L^2([0, 1],\mu)$ is a spectral measure but is not a resolution of the identity on $[0, 1]$. (It is sufficient to verify that $E_{f,f}$, where $f(t) := 1$, coincides with μ.)

Let us consider what further possibilities there are in the new concept.

Proposition (IV.7.50) (cf. 7.44). *Let* $E : \Sigma \longrightarrow \mathcal{B}(H)$ *be a resolution of the identity on* Ω. *Then the map* $\varphi : (B(\Omega); \underline{m}) \longrightarrow (\mathcal{B}(H); \underline{w}) : f \longmapsto \int_\Omega f(\lambda)dE_\lambda$ *is a continuous *-homomorphism of polynormed algebras. As a consequence, for* $\Omega \subset \mathbb{C}$ *it is the Borel calculus on* Ω *for the operator* $\int_\Omega \lambda dE_\lambda$.

▸ By Proposition 0.1.9 it is sufficient, given $x, y \in H$, to produce a measure $\mu \in M(\Omega)$ such that $\left\| \int_\Omega f(\lambda)dE_\lambda \right\|_{x,y} = \| f \|_\mu$. But by 7.46 we are justified in putting $\mu = E_{x,y}$. The rest is clear. ■

All that remains is for us to combine the accumulated facts and establish the following result about the Borel calculus which in a particular case — applying it "personally" to the function $f(z) = z$ — turns into the classical spectral theorem.

Theorem (IV.7.51). *Let* a *be a normal operator on* H *and let* Ω *be a compact set in* $\mathbb{C}$ *containing its spectrum. Then*

(I) ("the spectral theorem for a normal operator") *there exists a unique resolution of the identity* E *on* Ω, *such that* $a = \int_\Omega \lambda dE_\lambda$;

(II) *the Borel calculus on* Ω *for* a *can be given by* $f(a) = \int_\Omega f(\lambda)dE_\lambda$; $f \in B(\Omega)$.

▸ We put $E(\Delta)$ equal to the value of the Borel function χ_Δ acting on a. Then Theorem 7.37(I) and Proposition 7.45 imply that E is a spectral measure giving the operators $f(a)$ from the equation indicated in the statement of the theorem; hence, in

---

[20] The reader who is acquainted with the concept of an invariant mean on a group [25] or at least a Banach limit [67, p.82] can easily construct such a measure.

particular, $a = \int_\Omega \lambda dE_\lambda$. Further from Proposition 7.35 and the form of the Borel calculus in 7.37 it obviously follows that $E_{x,y}(\Delta) = \int_\Omega \chi_\Delta(t) d\mu_{x,y}(t) = \mu_{x,y}(\Delta)$ for any $x, y \in H$ and Borel set Δ, where $\mu_{x,y}$ is the (automatically regular) measure given in Proposition 7.34; this means that E is a resolution of the identity.

All that remains is to prove uniqueness in (I). Let $F : \Delta \longmapsto F(\Delta)$ be a resolution of the identity in Ω such that $a = \int_\Omega \lambda dF_\lambda$. Then the map $f \longmapsto \int_\Omega f(\lambda) dF_\lambda$ is, by Proposition 7.50, a Borel calculus on Ω for a. As the latter is unique (7.32), this means that $\int_\Omega f(\lambda) dF_\lambda = \int_\Omega f(\lambda) dE_\lambda$ for all $f \in B(\Omega)$. If we take any Borel set Δ and put $f := \chi_\Delta$, we get $F(\Delta) = E(\Delta)$. ∎

Exercise (IV.7.52) (cf. 7.45). Let Ω be an arbitrary compact set and let φ be a continuous unital *-homomorphism between the polynormed algebras $(B(\Omega); \underline{m})$, and $(\mathcal{B}(H); \underline{w})$. Show that $E : \Delta \longmapsto \varphi(\chi_\Delta)$ is a resolution of the identity in Ω.

The resolution of the identity featuring in 7.51 is called a *resolution of the identity for the operator* a (on Ω). We show that it is in fact independent of the choice of $\Omega \supseteq \operatorname{Sp} a$.

Proposition (IV.7.53). *Let* E^0, E *be resolutions of the identity for the operator* a *on* $\operatorname{Sp} a$ *and on* Ω *respectively. Then* $E(\Delta) = E^0(\Delta \cap \operatorname{Sp} a)$ *for any Borel set* $\Delta \subseteq \Omega$. *Consequently,* $E(\Delta) = 0$ *when* $\Delta \subseteq \Omega \setminus \operatorname{Sp} a$.

▶ Obviously $E' : \Delta \longmapsto E^0(\Delta \cap \operatorname{Sp} a)$ is a resolution of the identity on Ω such that $\int_\Omega \lambda dE'_\lambda = \int_\Omega \lambda dE^0_\lambda = a$. Therefore the necessary equality is implied by the uniqueness property in Theorem 7.51(I). ∎

In conclusion we persist in recommending the reader, if he has not had to do this before, to find in an explicit form the resolution of the identity for at least the following operators: (I) an orthogonal projection; (II) the operator of pointwise multiplication by the bounded sequence $\{\lambda_n \in \mathbb{C};\ n \in \mathbb{N}\}$ in l_2; (III) the operator $g(t) \longmapsto tg(t)$ in $L^2[0, 1]$.

§7.3. Description of C^*-algebras as operator algebras

Now, armed with a complete understanding of commutative C^*-algebras, we turn to the basic results on general C^*-algebras.

Lemma (IV.7.54). *Let Ω be a compact space, let B be a unital commutative Banach algebra with multiplicative norm, and let $\kappa : C(\Omega) \longrightarrow B$ be an injective continuous*[21] *unital homomorphism. Then $\|\kappa(f)\| \geq \|f\|_0$ for any $f \in C(\Omega)$.*

▶ Recall the spectrum functor $\underline{\Omega}$ (see §1.2) and put $\alpha := \underline{\Omega}(\kappa) : \underline{\Omega}(B) \longrightarrow \Omega$ and $\Delta := \operatorname{Im} \alpha \subseteq \Omega$. We take $t \in \underline{\Omega}(B)$ and $s = \alpha(t)$; then $f(s) = f(\alpha(t)) = \kappa(f)^{\wedge}(t)$ for any $f \in C(\Omega)$. By Theorem 2.11(I), this means that $|f(s)| \leq \|\kappa(f)\|$.

By definition of the uniform norm $\|\cdot\|_0$, all that remains is for us to show that $\Delta = \Omega$. Since α is a continuous map between compact spaces, Δ is closed. Therefore, if $\Delta \neq \Omega$, Urysohn's theorem (see Chap.0 §0.3°) implies that there exist in $C(\Omega)$ functions g and h such that $g|_{\Delta} \equiv 1$, $h \neq 0$ and $gh = 0$. Since $\kappa(g)^{\wedge}(t) = g(\alpha(t)) = 1$ for any $t \in \underline{\Omega}(B)$, then by Corollary 1.2, $\kappa(g)$ is invertible in B. But then $\kappa(g)\kappa(h) = \kappa(gh) = 0$ implies that $\kappa(h) = 0$, but this contradicts the fact that κ is injective. ∎

Theorem (IV.7.55). *Let A be a C^*-algebra, let B be a star Banach algebra with multiplicative star norm, and let $\kappa : A \longrightarrow B$ be an injective continuous *-homomorphism. Then $\|\kappa(a)\| \geq \|a\|$ for any $a \in A$.*

▶ By Proposition 7.6, one may assume without loss of generality that κ is a unital homomorphism of unital algebras. We can choose arbitrarily $h \in A_{sa}$ and put $A^0 = \overline{\ell.h.\{e, h^n; n = 1, 2, \cdots\}}$. We also put $B^0 = \overline{\kappa(A^0)}$ and $\kappa^0 : A^0 \longrightarrow B^0 : a \longmapsto \kappa(a)$. Then A^0 is a commutative C^*-algebra, B^0 a commutative Banach algebra and, by Theorem 7.13, we find that we have satisfied the conditions of the preceding lemma. Hence $\|\kappa(h)\| = \|\kappa^0(h)\| \geq \|h\|$.

Finally we take any $a \in A$. Then, since $a^*a \in A_{sa}$, what has already been proved implies that $\|\kappa(a)\|^2 = \|\kappa(a)^*\|\,\|\kappa(a)\| \geq \|\kappa(a^*)\kappa(a)\| = \|\kappa(a^*a)\| \geq \|a^*a\| = \|a\|^2$. ∎

By combining this theorem and Theorem 5.25 we immediately get the following corollary.

Corollary (IV.7.56). *Any injective *representation of a C^*-algebra is an isometry.*

We now combine several of the results we have obtained.

---

[21] The assumption that κ be continuous here and in the next theorem could be waived; see Remark 1.16.

Flow diagram for the proof of Theorem IV.7.57

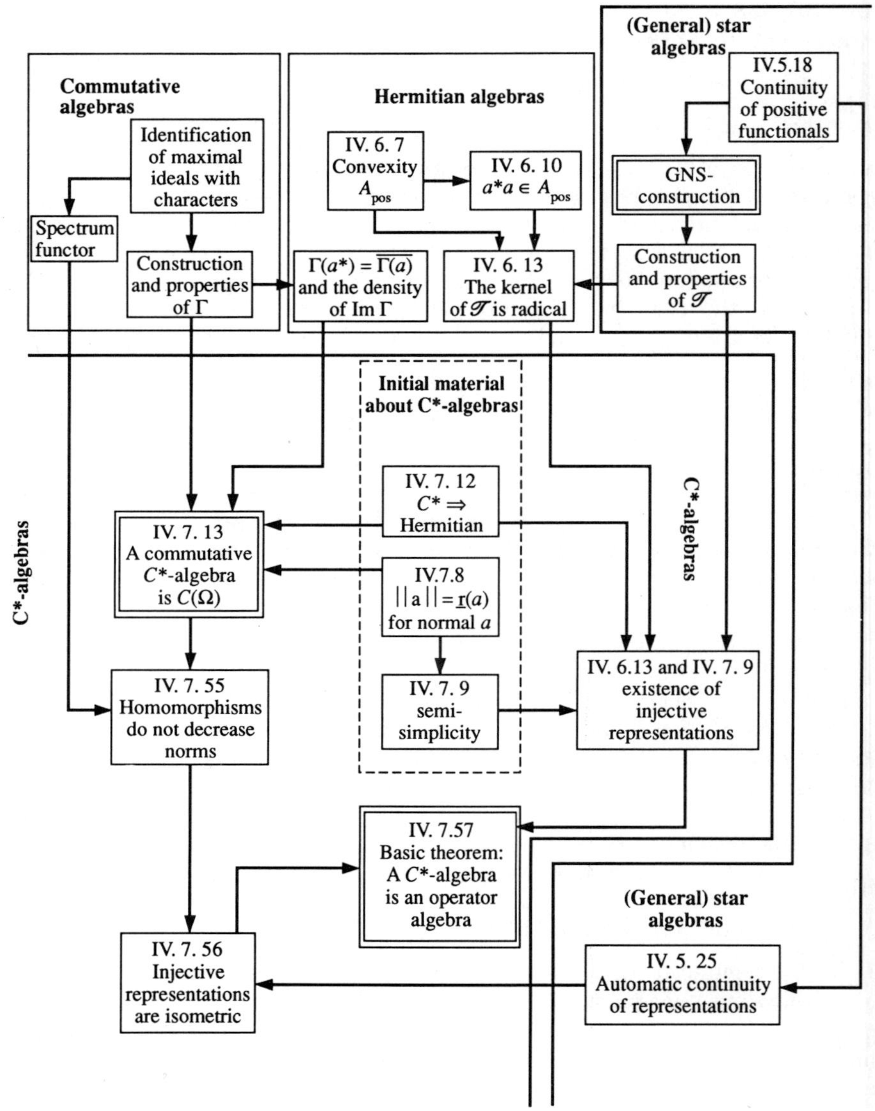

Theorem (IV.7.57) ("the second Gel'fand–Naimark theorem").[22] *Let A be a C^*-algebra. Then it has an isometric *-representation and moreover its universal representation is automatically such a representation. As a consequence, a star Banach algebra is a C^*-algebra $\Leftrightarrow$ it is isometrically *-isomorphic to a certain closed *-subalgebra of $\mathcal{B}(H)$ for a certain Hilbert space H.*

▶ As A is Hermitian (7.12) and semisimple (7.9), it is automatically star-semisimple (6.13) or, what is the same thing, its universal representation is injective. All we have to do now is to use the preceding corollary. ■

Of course, we now know why we used the term "realized C^*-algebra" in Chap. III §2.

The theorem we have proved is probably the main theorem in the theory of C^*-algebras. The reader may note that practically all the facts accumulated in this chapter (excluding the analysis of certain specific examples) was part of the preparation for this theorem. We illustrate, in a somewhat simplified form, the basic steps of the whole argument leading up to this theorem, with the help of the preceding flow diagram.

We note that all the proofs of Theorem 7.57 known to us at the present time are very similar and the differences between them are matters of detail only and not about principles [64], [9], [67]. For example, we included Theorem 6.13 and Corollary 7.56 in the argument as they are obviously of value in their own right. However, we could have reached the basic theorem without these "blocks" in our flow diagram by using instead an additional observation.

Exercise (IV.7.58). Let A be a unital C^*-algebra. Show that for any $a \in A$ there exists a state f on A such that $f(a^*a) = \|a\|^2$. Deduce Theorem 7.57 from this (omitting Theorem 6.13 and Corollary 7.56).

Hint. After applying Theorem 7.13 (taking into account Theorems 7.12 and 6.10), construct a state f_0 on the commutative *-subalgebra of A that contains a^*a and e, such that $f_0(a^*a) = \|a\|^2$. Then, using the Hahn–Banach theorem in its geometric form, extend f_0 to the required state on A. As this proves that $\|a\| \le B^*(a)$, Theorems 5.18 and 5.27 immediately imply that the universal representation of the algebra A is an isometric embedding.

---

[22] Gel'fand and Naimark proved this theorem [114] under the additional assumption that $e + a^*a$ be invertible for any $a \in A$ (cf. Exercise 6.3) and stated the conjecture that this last condition could be deduced from the others (that is, those in the definition of C^*-algebras). Later on Kelley and Vaught [158] and Fukamiya [262] showed that it could.

We also note that the proofs of some assertions that we made about general Hermitian algebras, may be simplified for the special case of C^*-algebras. We shall return to this question in the next section (see Remark 7.76).

The separable case opens up further possibilities.

Exercise (IV.7.59). Prove that a separable C^*-algebra has an isometric *-representation which moreover is a GNS-representation on a separable Hilbert space.

Hint. First show that the unit ball in $A^*_\times$, where A is our algebra, is separable with respect to the weak* topology and so $\underline{\mathrm{St}}(A_\times)$ must also be separable with respect to the weak* topology. Then, taking a dense subset $\{f_n; n = 1, 2, \cdots\}$ in $\underline{\mathrm{St}}(A_\times)$ consider the representation T^f, where $f = \sum_{n=1}^{\infty} \frac{1}{2^n} f_n$.

Theorem 7.57, when considered from the point of view of operator theory, gives an axiomatic description of those *-subalgebras of $\mathcal{B}(H)$ that are closed in the uniform topology. Let us formulate an important complement to this which provides a description, close in spirit, of the more restricted class of von Neumann algebras (= the *-subalgebras of $\mathcal{B}(H)$ that are closed in the weak operator topology; see III.2.37).

Assertion (IV.7.60) (Sakai's theorem). *The following properties of a C^*-algebra A are equivalent*:

(I) *A coincides, to within an isometric Banach space isomorphism, with E^* for a certain Banach space E (we say that A "has a predual")*;

(II) *A is isometrically *-isomorphic to a certain von Neumann algebra.*

For the proof see, for example, [58, 3.9]; this contains also other axiomatic descriptions of von Neumann algebras.

Remark (IV.7.61). In this identification of A with a suitable von Neumann algebra, the weak* topology in A, considered as a dual space, coincides with the topology induced from the ultraweak topology in $\mathcal{B}(H)$ (see III.2.21).

Here is a first application of the basic theorem.

Proposition (IV.7.62). (I) *Any *-homomorphism κ of a star Banach algebra into a C^*-algebra is continuous and moreover, if the norm in the first algebra is multiplicative, then $\|\kappa\| \le 1$.*

(II) *Any injective *-homomorphism between two C^*-algebras is isometric.*

▶ Let A and B be our algebras and let $\mathcal{T}$ be the universal representation of B. Then $T := \mathcal{T} \circ \kappa$ is a *-representation of the algebra A. Since $\mathcal{T}$ is an isometric operator, $\|\kappa(a)\| = \|\mathcal{T}(\kappa(a))\| = \|T(a)\|$ for any $a \in A$. Therefore (I) follows immediately from Theorem 5.25 and (II) follows from Corollary 7.56. ■

§7.4. Enveloping and group C^*-algebras. Remark about tensor products of C^*-algebras and nuclear C^*-algebras

We now show, guided by the basic theorem, that studying *-representations of any star bounded algebra A may be reduced to studying *-representations of some C^*-algebra, canonically associated with A.

Definition (IV.7.63). Let A be a *-algebra. The pair $(\Theta; \theta : A \longrightarrow \Theta)$, where Θ is a C^*-algebra and θ a *-homomorphism, is called an *enveloping C^*-algebra of the algebra A* if for any *-representation $T : A \longrightarrow \mathcal{B}(H)$ of the algebra A there exists a unique *-representation $T' : \Theta \longrightarrow \mathcal{B}(H)$ such that the diagram below commutes.

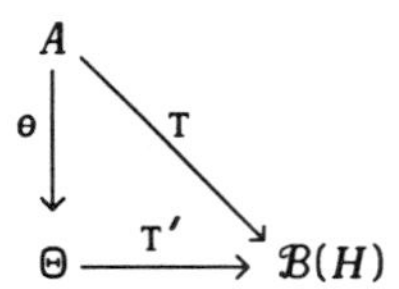

Let A be a star bounded algebra and let $\mathcal{T} : A \longrightarrow \mathcal{B}(\mathcal{H})$ be its universal representation. We let $C^*(A)$ denote the closure (in the operator norm) of the image of $\mathcal{T}$ and put $\vartheta = \mathcal{T}\,|^{\,C^*(A)}$. Obviously $C^*(A)$ is a C^*-algebra and ϑ a *-homomorphism.

Theorem (IV.7.64). *Any star bounded algebra A has an enveloping C^*-algebra, namely $(C^*(A), \vartheta)$.*

▶ Let $T : A \longrightarrow \mathcal{B}(H)$ be a *-representation. The bound $\|T_a\| \le \|\mathcal{T}_a\|$; $a \in A$ (Theorem 5.27), which ensures, in particular, that $\operatorname{Ker} \mathcal{T} \subseteq \operatorname{Ker} T$, implies that there exists a unique *-homomorphism $\varphi : \operatorname{Im} \mathcal{T} \longrightarrow \mathcal{B}(H)$ such that $\varphi(\mathcal{T}_a) = T_a$; $a \in A$, and that $\|\varphi\| \le 1$. Hence, taking the continuous extension of φ to the whole of $C^*(A)$ we get a *-representation T' that gives us (replacing (Θ, θ) by $(C^*(A), \vartheta)$) the commutative diagram in Definition 7.63. It is obvious from the density of $\operatorname{Im} \mathcal{T}$ in $C^*(A)$ and Theorem 5.25 that this representation is unique. ■

Let us describe another, basically very similar, method for constructing an enveloping algebra.

Exercise (IV.7.65). Let A be a star bounded algebra, let $B = A/{*}\text{-Rad}\,A$, and let $\tau : A \longrightarrow B$ be the natural projection. Prove that

(I) $\|a\| = B^*(a)$; $a \in A$, is a C^*-seminorm and moreover the largest possible C^*-seminorm on A;

(II) its quotient seminorm in B is a C^*-norm and moreover the largest possible C^*-norm on B;

(III) The completion of B with respect to this last norm is a C^*-algebra which is isometrically *-isomorphic to $C^*(A)$ and moreover τ, (with domain of values in the completion of B) becomes ϑ, after making the appropriate identification.

We suggest that the reader deduces from Definition 7.63 the following uniqueness property for the enveloping C^*-algebra.

Exercise (IV.7.66). Let (Θ_k, θ_k); $k = 1, 2$, be two enveloping C^*-algebras of the *-algebra A. Show that there exists an isometric *-isomorphism i such that the diagram

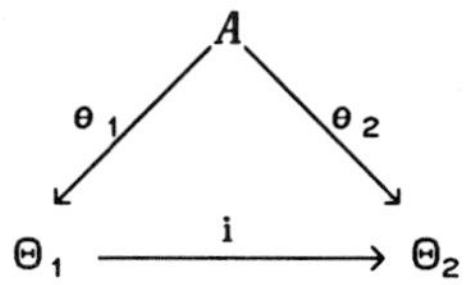

is commutative.

We are now finally in a position to fulfil the promise made at the beginning of Chap.III §3.3 to introduce the most important class of group algebras consisting of operators.

Let G be a locally compact group. We let $C^*(G)$ denote the algebra $C^*(A)$, where A is the star Banach algebra $L^1(G)$ (see 4.19).

Definition (IV.7.67). The algebra $C^*(G)$ is called the *C^*-algebra of the group G.*

Sometimes the algebra we have introduced is called the *complete* C^*-algebra of the group G in order to emphasize that it is different from the reduced C^*-algebra $C^*_r(G)$. Recall (Chap.III §3.3) that the latter was defined as the closure with respect to the operator norm of the image of a special homomorphism $T^G : L^1(G) \longrightarrow \mathcal{B}(L^2(G))$; $[T^G f](g) = f * g$. Since T^G is a *-representation (cf. 5.32(II) and 5.37), Definitions 7.63 and 7.67 obviously imply the existence of a *-homomorphism $\sigma : C^*(G) \longrightarrow C^*_r(G)$ uniquely defined by $\sigma(\mathcal{T}_f) := T^G_f$; $f \in L^1(G)$, and having a dense image.

Remark (IV.7.68). In fact, since the image of any *-homomorphism between C^*-algebras is closed (this will be proved below, see Theorem 7.83), σ is always surjective. Moreover, for the important class of so-called amenable groups, including all commutative and all compact groups, σ is an isometric *-isomorphism: however, for all other groups, for the free group F_n with $n > 1$ generators, say, this is not so. For details about these questions see [25 §3.5].

Remark (IV.7.69). In the theory of group representations complete C^*-algebras work better than reduced C^*-algebras. The fact is that all unitary representations of the group G (that is, all continuous homomorphisms of G into groups of unitary operators on Hilbert spaces; cf. later Chap. VI §1.3) may be found starting from *-representations of the algebra $C^*(G)$; see, for example, [31 §13.9]. Nevertheless there are generally fewer *-representations of the algebra $C^*_r(G)$; for example, when $G = F_2$, there are not enough of them even to describe the (huge number) of finite-dimensional representations of this group.

Another application of the construction of enveloping C^*-algebras concerns Banach tensor products. Using this construction one can propose a version of a Banach tensor product that stays within the class of C^*-algebras (in much the same way as we succeeded in doing this for the operation of adjoining an identity in Proposition 7.6).

Definition (IV.7.70). Let A and B be C^*-algebras. Their *maximal C^*-tensor product* is the enveloping C^*-algebra of the star Banach algebra $A \hat{\otimes} B$ (see 4.13).

This C^*-algebra is often denoted by $A \otimes_{\max} B$ and its norm by $\|\cdot\|_{\max}$.

There are, however, other versions of the concept of a C^*-tensor product besides the maximal C^*-tensor product; this is connected with the fact that there exist on the algebraic tensor product $A \otimes B$, generally speaking, a lot of different C^*-norms whose completions give distinct C^*-algebras.

Exercise (IV.7.71). Show that $\|\cdot\|_{\max}$ is the largest possible C^*-seminorm on the *-algebra $A \otimes B$.

It is curious that, in addition to the largest, there is also a smallest C^*-norm on $A \otimes B$, denoted by $\|u\|_{\min}$; it may be defined as the least upper bound of the set of $\|T_u\|_0$, where T runs over the *-representations of the algebra $A \otimes B$ — but taking, not all of them (which would, of course, give the norm $\|u\|_{\max}$), but only those which decompose into the so-called tensor products of *-representations of the algebras A and B. For the definition of these and other details see, for example, [75, IV.1].

A natural question arises: under what conditions on A does there exist only one C^*-norm — in other words, $\|\cdot\|_{\max} = \|\cdot\|_{\min}$. Studying this has led to the discovery of a very important class of algebras that Lance [324] calls the *nuclear C^*-algebras*:[23] these are, by definition, those C^*-algebras A for which $\|\cdot\|_{max}$ is the unique C^*-norm in $A \otimes B$ for any C^*-algebra B. It has been gradually discovered that nuclear C^*-algebras may be characterized from very different points of view, including one from the standpoint of their enveloping von Neumann algebras ("Connes injectivity"), another in the language of so-called completely positive operators and another in terminology rather like that of the approximation property. These will be mentioned later, in Chap.VII §1.5, with a description of them in terms of homology.

All postliminal C^*-algebras (see Chap.III §2.2 and also the end of Chap.VI §2.2) including all commutative algebras are nuclear C^*-algebras; on the other hand, $\mathcal{B}(H)$, for infinite-dimensional H and $C^*_r(F_n)$; $n>1$ (see Remark 7.69) are not nuclear C^*-algebras.

One can learn about the basic results on tensor products of C^*-algebras and nuclear C^*-algebras in the surveys [178], [304], [325] and the literature indicated there; there is some material in the monographs [56], [44, Vol.II].

§7.5. Positive elements, approximate identities, quotient algebras

All that remains is for us to set out a few more results from the beginning of the theory of C^*-algebras which are immediately related to the questions considered earlier. First of all, we deal more thoroughly with the set of positive elements A_{pos}.

Recall that the set K in a certain linear space is called a *cone* if (I) $x, y \in K \Rightarrow x+y \in K$; (II) $x \in K$, $t \geq 0 \Rightarrow tx \in K$ and (III) $x \in K$; $x \neq 0 \Rightarrow -x \notin K$. We have essentially already proved the following.

Proposition (IV.7.72). *The set of positive elements of a semisimple Hermitian star Banach algebra is a cone.*

▶ Property (I) forms the content of Theorem 6.7, (II) is obvious and (III) is implied by Theorems 5.34 and 6.13. ■

In the case of a C^*-algebra the set A_{pos} has important additional properties; besides, the definition of a positive element itself admits many equivalent formulations. We have already seen in Theorem 7.28 that the element p of a C^*-algebra A is positive $\Longleftrightarrow$ $p=h^2$ for a certain $h \in A_{sa}$ $\Longleftrightarrow$ $p=a^*a$

---

[23] By analogy with Grothendieck's nuclear polynormed spaces, whose definition involves comparing two systems of seminorms in $E \otimes F$ (see the end of Chap.0 §3).

for a certain $a \in A$. Here are another two characterizing properties:

Proposition (IV.7.73). *An element p of a C^*-algebra A is positive $\Leftrightarrow$ it is self-adjoint and $C \geq \|Ce - p\|$ (in A_X) is satisfied for all $C \geq \|p\|$ $\Leftrightarrow$ it is self-adjoint and the preceding inequality is satisfied for at least one $C \geq \|p\|$.*

▶ Let $p \in A_{sa}$; then $(Ce - p) \in (A_X)_{sa}$ for any $C \in \mathbb{R}$. According to Theorem 7.12, $\mathrm{Sp}(p), \mathrm{Sp}(Ce - p) \subseteq \mathbb{R}$, and according to Proposition 7.8, $\underline{r}(p) = \|p\|$ and $\underline{r}(Ce - p) = \|Ce - p\|$. Therefore, if $C \geq \|p\|$, then $p \in A_{pos} \Leftrightarrow \mathrm{Sp}(p) \subseteq [0, \|p\|] \Leftrightarrow \mathrm{Sp}(Ce - p) \subseteq [C - \|p\|, C] \Leftrightarrow \underline{r}(Ce - p) \leq C \Leftrightarrow \|Ce - p\| \leq C$. ■

Proposition (IV.7.74). *The set of positive elements of a C^*-algebra is a closed cone.*

▶ The fact that A_{pos}, where A is our algebra, is a cone, is implied by 7.72, taking into account Proposition 7.9 and Theorem 7.12. Further, let $p = \lim_{n\to\infty} p_n$, where $p_n \in A_{pos}$. Then there exists $C > 0$ such that $\|p\|, \|p_n\| \leq C$; $n \in \mathbb{N}$. Hence, as the first and second properties in 7.73 are equivalent, $C \geq \|Ce - p_n\|$ for all n. Passing to the limit with respect to n, we get that $C \geq \|Ce - p\|$ and, as the third and first properties in 7.73 are equivalent, $p \in A_{pos}$. The rest is clear. ■

Proposition (IV.7.75). *Let A be a C^*-algebra. Then $a^*pa \in A_{pos}$ for any $p \in A_{pos}$ and $a \in A$.*

▶ According to Theorem 7.28, $p = b^*b$ for a certain $b \in A$. Therefore the same Theorem 7.28 implies that $a^*pa = (ba)^*(ba) \in A_{pos}$. ■

Remark (IV.7.76). In proving Theorem 7.28, Propositions 7.74 and 7.75 as well as the basic Theorem 7.57, we have relied upon general results in the theory of Hermitian algebras; however, for the special case of C^*-algebras the proofs of these results may themselves be considerably simplified.

We show this first of all by taking the example of Theorem 6.7 (about the closure of A_{pos} under addition). Let p and q be positive elements of the C^*-algebra A and let $C > \|p\|, \|q\|$. Then, by the equivalences established in Proposition 7.73, $\|Ce - \frac{1}{2}(p + q)\| \leq \frac{1}{2}\|Ce - p\| + \frac{1}{2}\|Ce - q\| \leq C$ which immediately implies that $\frac{1}{2}(p + q)$, and hence also $p + q$, is positive.

Theorem 6.10 (that is, deducing the first property from the third in Theorem 7.28) may also be obtained more simply for C^*-algebras. In fact, let a be an element of the C^*-algebra A, let $p = a^*a$, and let B be any closed commutative *-subalgebra of A containing p. If, by Theorem 7.13, we identify B with $C_0(\Omega(B))$, we get that there exist $p_+, p_- \in B_{pos} \subseteq A_{pos}$ such that $p = p_+ - p_-$, $p_+p_- = 0$ and, in addition, $p_- = q^2$ for a certain $q \in B_{pos}$. Hence

$$(aq)^*(aq) = qa^*aq = q(p_+ - p_-)q = p_+p_- - p_-^2 = -p_-^2 .$$

Further, we represent aq in the form $h + ik$; $h, k \in A_{sa}$; then

$$(aq)(aq)^* = 2h^2 + 2k^2 - (aq)^*(aq) = 2h^2 + 2k^2 + p_-^2 .$$

Hence, as the sum still lies in A_{pos}, $(aq)(aq)^* \in A_{pos}$. As the spectrum of the product changes, under reversal of the factors, "at most by $\{0\}$" (II.1.8), this means that $(aq)^*(aq) \in A_{pos}$, that is, $-p_-^2 \in B_{pos}$. But then $p_- = 0$ and consequently $p = p_+ \in A_{pos}$.

Finally, we note that the proof of the third property of the cone ($p, -p \in A_{pos} \Rightarrow p = 0$) does not require Theorems 5.33 and 6.13 in the case of a C^*-algebra but follows immediately from Proposition 7.8.

The properties we have established for the set A_{pos} enable an additional structure to be introduced into the algebras we are considering — an *order relation*, which plays an outstanding role in the theory of C^*-algebras. Namely, we agree to write "$a \le b$"; $a, b \in A$, for any *-algebra A in which A_{pos} is a cone, if $b - a \in A_{pos}$.

Theorem (IV.7.77). *Let A be a *-algebra such that A_{pos} is a cone. Then the relation "$\le$" has the following properties:*

(I) $a \le b$ *and* $b \le c$ *implies* $a \le c$;

(II) $a \le b$ *and* $t \ge 0$ *implies* $ta \le tb$;[24]

(III) $a \le b$ *and* $b \le a$ *implies* $a = b$.

If, in addition, A is a C^*-algebra, then we can add to these properties

(IV) $a_n \le b_n$, $a = \lim_{n\to\infty} a_n$ *and* $b = \lim_{n\to\infty} b_n$ *implies* $a \le b$;

(V) *for any* $c \in A$, $a \le b$ *implies* $c^*ac \le c^*bc$;

(VI) $h \le \|h\| e$ *for any* $h \in A_{sa}$ (*in* $A_\times$);

(VII) $0 \le p \le q$ *implies* $\|p\| \le \|q\|$.

▶ (I) – (V). These properties follow immediately from the properties of A_{pos} established in Theorem 7.28 and Propositions 7.74 and 7.75.

---

[24] Properties (I) and (III) together with the obvious property $a \le a$ mean that this relation makes A into a partially ordered set.

(VI) If $h \in A_{sa}$, then $\operatorname{Sp} h \subseteq [-\underline{r}(h), \underline{r}(h)] = [-\|h\|, \|h\|]$; and $\operatorname{Sp}(-h)$ lies in the same interval. Hence it is obvious that $\|h\| e - h \in A_{pos}$.

(VII) If $0 \leq p \leq q$, then, by properties (VI) and (I), $\|q\| e - p \geq 0$, that is, $\|q\| - t \geq 0$ for all $t \in \operatorname{Sp}(p)$. As $\underline{r}(p) = \|p\|$, this is possible only when $\|p\| \leq \|q\|$. ■

As an application we establish a useful property of C^*-algebras — the presence of approximate identities (see Chap.I §2.3).

Proposition (IV.7.78). *Let* A *be a* C^**-algebra and let* I *be a* (not necessarily proper) *left ideal in* A. *Then* I *has a right* b.a.i. $(e_\nu;\ \nu \in \Lambda)$, *such that for all* ν, $0 \leq e_\nu \leq e$ (*in* $A_\times$) *and, as a consequence,* $\|e_\nu\| \leq 1$.

▶ Let Λ be the set of finite subsets of I directed with respect to inclusion. For $\nu = (a_1, \cdots, a_n) \in \Lambda$ we put $p_\nu = a_1^* a_1 + \cdots + a_n^* a_n$ and $e_\nu = (e + np_\nu)^{-1} np_\nu$ (e is the identity in $A_\times$); obviously $p_\nu, e_\nu \in I \cap A_{pos}$. We show that $(e_\nu;\ \nu \in \Lambda)$, is the required a.i.

We choose $a \in I$ and $\varepsilon > 0$ arbitrarily. We fix a certain subset $\nu_0 \in \Lambda$ containing more than ε^{-2} elements, among which is a. Let ν be a subset of n elements which contains ν_0. Then we put (in $A_\times$) $q = (e - e_\nu) a^* a (e - e_\nu)$ and $q_\nu = (e - e_\nu) p_\nu (e - e_\nu)$.

$a \in \nu$ and the form of p_ν imply that $0 \leq a^* a \leq p_\nu$; hence, by 7.77(V), $0 \leq q \leq q_\nu$. By 7.77(VII), this means that

$$\|a - ae_\nu\|^2 = \|(a(e - e_\nu))^* a(e - e_\nu)\| = \|q\| \leq \|q_\nu\| .$$

We now let B_ν denote any closed commutative *-subalgebra of $A_\times$ containing p_ν and e. Then Proposition 6.4 and Theorem 7.12 imply that B_ν contains $(e + np_\nu)^{-1}$ and with it e_ν and q_ν. We note that $q_\nu = p_\nu (e - e_\nu)^2 = p_\nu (e + np_\nu)^{-2}$; hence, by Theorem 7.13, $\|q_\nu\| = \left\| \dfrac{\hat{p}_\nu}{(1 + n\hat{p}_\nu)^2} \right\|_0$, where $\hat{p}_\nu(s)$; $s \in \underline{\Omega}(B_\nu)$, is the Gel'fand transform of p_ν as an element of B_ν. But according to the same theorem, $\hat{p}_\nu$ is a non-negative function; consequently

$$\|q_\nu\| \leq \left\| \frac{\hat{p}_\nu}{1 + 2n\hat{p}_\nu} \right\|_0 < \frac{1}{n} < \varepsilon^2 .$$

If we combine both the inequalities we have obtained, we see that for all ν containing ν_0, $\|a - ae_\nu\| < \varepsilon$. Thus, since a

and ε were arbitrary, we have verified that $(e_\nu;\ \nu \in \Lambda)$, is a right a.i. in I. All that remains is to note that the Gel'fand transform $\hat{e}_\nu$ of the element $e_\nu \in B_\nu;\ \nu \in \Lambda$, satisfies $0 \le \hat{e}_\nu(s) \le 1;\ s \in \underline{\Omega}(B_\nu)$; this immediately implies that $0 \le e_\nu \le e$ in A. ■

Theorem (IV.7.79). *Any closed bi-ideal in the C^*-algebra A is self-adjoint and has a* (two-sided) b.a.i. $(e_\nu;\ \nu \in \Lambda)$, *such that* $0 \le e_\nu \le e$ *(in* $A_\times$*). In particular, every* (generally speaking, non-unital) *C^*-algebra has a* b.a.i. *with this property*.

▶ Let I be our bi-ideal; then, by Proposition 7.78, it has a right b.a.i. $(e_\nu;\ \nu \in \Lambda)$, with $0 \le e_\nu \le e$. Therefore for any $a \in I$, $\|a^* - e_\nu a^*\| = \|(a - ae_\nu)^*\| = \|a - ae_\nu\|$ implies that $a^* = \lim\limits_\nu e_\nu a^*$. In view of $e_\nu \in I$ and the properties of I, this means that $a^* \in I$; thus I is self-adjoint. Hence any element in I is of the form a^* for some $a \in I$ and therefore the last equation means that e_ν is a left b.a.i. for I. The rest is clear. ■

This theorem immediately implies that the quotient algebra of a C^*-algebra A with respect to a closed bi-ideal I is a Banach *-algebra with a star multiplicative norm and involution well-defined by the formula $(a + I)^* = a^* + I;\ a \in A$. In fact, with the help of the next lemma we can prove rather more.

Lemma (IV.7.80). *Let A be a unital Banach algebra with a multiplicative norm, let I be a closed left ideal in A having a right* b.a.i. $(e_\nu;\ \nu \in \Lambda)$, *such that* $\|e - e_\nu\| \le 1$ *for all* ν. *Then* $\|a + I\| = \lim\limits_\nu \|a - ae_\nu\|$ *for any* $a \in A$.

▶ If $b \in I$, then $\lim\limits_\nu (b - be_\nu) = 0$ and therefore

$$\overline{\lim_\nu} \|a - ae_\nu\| = \overline{\lim_\nu} \|(a - ae_\nu) + (b - be_\nu)\| =$$

$$= \overline{\lim_\nu} \|(a + b)(e - e_\nu)\| \le \overline{\lim_\nu} \|a + b\| \|e - e_\nu\| \le \|a + b\|.$$

Taking the infimum over all $b \in I$, we get $\overline{\lim\limits_\nu} \|a - ae_\nu\| \le \|a + I\|$. On the other hand $\|a + I\| \le \inf\limits_\nu \|a - ae_\nu\| \le \underline{\lim\limits_\nu} \|a - ae_\nu\|$. The rest is clear. ■

Theorem (IV.7.81). *Let A be a C^*-algebra and let I be a closed bi-ideal in A. Then A/I* (with the quotient norm and the quotient involution) *is also a C^*-algebra.*

▶ By Theorem 7.79, there is a b.a.i. $(e_\nu;\ \nu \in \Lambda)$ in I, such that $0 \le e_\nu \le e$ for all ν (e, as usual, is the identity in $A_\times$), or what is equivalent, $0 \le e - e_\nu \le e$. Therefore by Theorem 7.77(VII) we have the conditions of the preceding lemma. Consequently, for $a \in A$

$$\|a + I\|^2 = \lim_\nu \|a - ae_\nu\|^2 = \lim_\nu \|(a(e - e_\nu))^* a(e - e_\nu)\| =$$

$$= \lim_\nu \|(e - e_\nu)a^* a(e - e_\nu)\| \le \lim_\nu \|a^* a - (a^* a)e_\nu\| = \|a^* a + I\|.$$

Thus the quotient norm in A/I satisfies the conditions of Proposition 7.1(I), on the strength of which it is a C^*-norm. ∎

We can now fulfil one of our long-standing promises, given in Chapter III §2.2.

Corollary (IV.7.82). *The Calkin algebra* $\mathcal{C}(H) = \mathcal{B}(H)/\mathcal{K}(H)$ *is a* C^**-algebra and is thus isometrically* **-isomorphic to a certain realized* C^**-algebra* (that is, one consisting of "genuine" operators; cf. Chap. III §2.2).

In conclusion let us show that one can construct all the *-homomorphisms between C^*-algebras in a fairly transparent form.

Theorem (IV.7.83). *Let* $\kappa : A \longrightarrow B$ *be a* **-homomorphism between* C^**-algebras. Then its kernel and image are closed and we have a commutative diagram of* C^**-algebras and* **-homomorphisms*

$$\begin{array}{ccc} A & \xrightarrow{\kappa} & B \\ {\scriptstyle\tau}\downarrow & & \uparrow{\scriptstyle \mathrm{in}} \\ A/\mathrm{Ker}\,\kappa & \xrightarrow{i} & \mathrm{Im}\,\kappa \end{array}$$

in which τ *is the natural projection,* in *is the natural embedding and* (the important thing!) i *is an isometric* **isomorphism.*

▶ By Proposition 7.62(I), κ is continuous and therefore the bi-ideal $\mathrm{Ker}\,\kappa$ is closed. Further, by Theorem 7.81, the operator $j : A/\mathrm{Ker}\,\kappa \longrightarrow B$, induced by the operator κ, is an injective *-homomorphism between C^*-algebras and, by Proposition 7.62(II), is it an isometric embedding; hence, in particular, $\mathrm{Im}\,\kappa$ is closed. The rest is clear. ∎

Corollary (IV.7.84). *Let* A *and* B *be* C^**-algebras and suppose further that there exists a* **-homomorphism from* A *into* B *with dense image. Then it is surjective, and* B *coincides, to within an isometric* **-isomorphism, with a certain quotient algebra of the algebra* A.

Hence we get, in particular, that for any locally compact group G, its reduced C^*-algebra is, after the appropriate identification, a quotient algebra of its complete C^*-algebra (see Remark 7.68 above).

Chapter V
Polynormed algebras that are near to Banach algebras

Historically, mathematicians became interested in general polynormed algebras (= locally convex algebras), as opposed to just Banach algebras, essentially for the same reasons which caused the theory of locally convex spaces to appear at a certain stage in the development of the theory of Banach spaces. The basic reason was, evidently, connected with applications: there had been an accumulation of significant "non-Banach" examples. Topologists needed to work with the algebras $C(\Omega)$ when Ω was non-compact, complex analysists needed to work with algebras of holomorphic functions defined on domains in $\mathbb{C}^n$ (specialists in operator theory needed to work with problems in functional calculus that also involved these same algebras), specialists in differential equations and harmonic analysis needed to work with various algebras of test functions and distributions. Another, this time intrinsic, stimulus was that studying structures that are more general than Banach algebras enables one to get a better understanding of the results of the "Banach" theory itself and to establish natural limitations to the scope of possible generalizations.

By now a considerable number of nice results have been obtained about various properties of specific (non-Banach) polynormed algebras that are typical of certain areas of analysis or topology (see, for example [7, Chap.I], [252], [184], [203], [40, §§10.4–5]). But as for results of a more general character, the overwhelming majority of them concern the class of multinormed algebras, and moreover they take a more definitive form for Arens–Michael algebras (sometimes with the additional assumption of metrizability). It can be said that Arens–Michael algebras (= complete Hausdorff multinormed algebras; cf. I.2.4) form that clearly defined class of polynormed algebras which is next, in degree of difficulty, after the class of Banach algebras. This is precisely the general sense of the results set out below and, primarily, Theorem V.2.17 later on.

There is some more detailed material about "algebras that are near to Banach algebras" in, for example, the monographs [50] and [7].[1]

It should be noted that there is, at the moment, another — quite different — class of algebras with an additional structure that is being studied and which may also be considered as rather nearer to the Banach algebras. These algebras are not, generally speaking, topological algebras: in them, one is given axiomatically the "bounded sets", and not the open sets. Whereas Arens-Michael algebras, as we shall see later (2.17 and 2.19) can be expressed in terms of Banach algebras by using the inverse limit construction of category theory, for a certain well-defined class of algebras "with a bounded structure" a similar role is played by the dual construction of the direct limit. However, this interesting range of questions goes beyond the scope of our book; for more about them see, for example, [7, Chap.VI], [15], [108], [326].

(The so-called metric generalizations of Banach algebras which are also related to the algebras under discussion, were investigated earlier by Żelazko [36]; they provide us with a significant class of "still topological" but "no longer polynormed" algebras.)

Warning. Throughout this chapter, when we talk about multinormed algebras with a fixed system of seminorms, we always assume that these seminorms are multiplicative (cf. I.2.3) and the system itself is saturated (cf. Chap.0 §1.1).

---

[1] We would like to recall for comparison that in the theory of polynormed spaces (without multiplication) we studied, alongside Banach spaces, spaces that, as measured by this theory, may be considered as "a long way away from being Banach spaces". There are such spaces in the theory of duality, especially the Grothendieck nuclear spaces mentioned at the end of Chap.0 §3: their properties are very unlike those of Banach spaces and, in certain respects, they are even better than the latter (cf. "the kernel theorem " [90, IV §9]). It must be confessed that in the theory of polynormed algebras there are, at present, no results that are comparable in significance on objects that are "far from being a Banach algebra"; possibly they will appear in the future.

§1. BANACH ALGEBRAS, CONCOMITANTS OF MULTINORMED ALGEBRAS. APPLICATION: A CHARACTERIZATION OF MULTINORMED SKEW FIELDS

We begin by describing a method that enables many questions about multinormed algebras to be reduced to analogous questions about certain Banach algebras canonically connected with them.

Let $(A;\ \|\cdot\|_\nu;\ \nu\in\Lambda)$, be a (for the moment arbitrary) multinormed algebra. For each ν we put $I_\nu = \{a\in A : \|a\|_\nu = 0\}$; the properties of $\|\cdot\|_\nu$ imply that I_ν is a bi-ideal in A. We consider the quotient seminorm $\|\cdot\|^\wedge_\nu$ of the seminorm $\|\cdot\|_\nu$ in the quotient algebra A/I_ν; then the definition of I_ν implies that this is a norm and the fact that $\|\cdot\|_\nu$ is multiplicative (in A) implies that $\|\cdot\|^\wedge_\nu$ is multiplicative (in A/I_ν). We let A_ν denote the completion (multiplicatively) of the normed algebra A/I_ν (see Chap.I §2.1) and we still use the notation $\|\cdot\|^\wedge_\nu$ for the extended norm. Together with A_ν we also consider the homomorphism $\tau_\nu : A \longrightarrow A_\nu : a \longmapsto a + I_\nu \in A/I_\nu \subseteq A_\nu$; which is obviously continuous.

Definition (V.1.1). The Banach algebras A_ν and the homomorphisms τ_ν are said to be *concomitant* with $(A;\ \|\cdot\|_\nu;\ \nu\in\Lambda)$.

We shall sometimes use more precise expressions like the "algebra A_ν concomitant with the seminorm $\|\cdot\|_\nu$", the meaning of which is obvious.

Some of the concomitant Banach algebras are also inter-related by homomorphisms. We introduce an order relation "$\mu < \nu$ if $\|\cdot\|_\mu \le \|\cdot\|_\nu$". In this way, since our system of seminorms is saturated, we get a directed set.

Proposition (V.1.2). *When* $\mu<\nu$; $\mu,\nu\in\Lambda$, *there exists a unique continuous homomorphism* $\tau^\nu_\mu : A_\nu \longrightarrow A_\mu$ *such that the diagram*

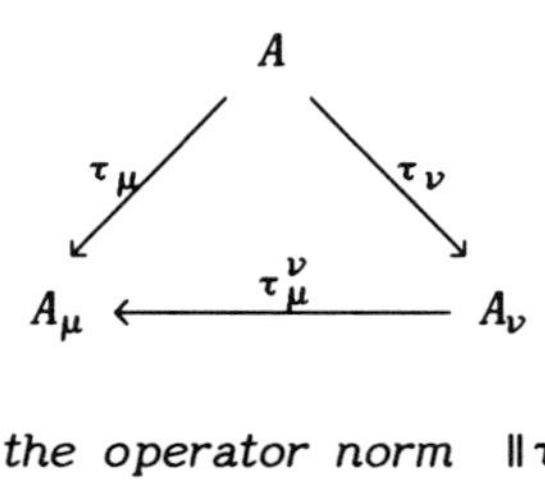

commutes. Moreover the operator norm $\|\tau^\nu_\mu\| \le 1$.

► $\mu<\nu$ implies that $I_\mu \supseteq I_\nu$. Therefore the map $a + I_\nu \longmapsto a + I_\mu$ is well defined and is a continuous homomorphism from $(A/I_\nu;\ \|\cdot\|^\wedge_\nu)$ to $(A/I_\mu;\ \|\cdot\|^\wedge_\mu)$. If we extend this by continuity, we get a homomorphism from A_ν into A_μ. The rest is clear. ■

Proposition (V.1.3). *Let* $\lambda < \mu < \nu$; $\lambda, \mu, \nu \in \Lambda$. *Then* $\tau_\lambda^\nu = \tau_\lambda^\mu \tau_\mu^\nu$.

▶ We consider three diagrams by applying 1.2 to the pairs (λ, μ), (μ, ν) and (λ, ν). Then the fact that the first two commute implies that the third is still commutative when τ_λ^ν is replaced by $\tau_\lambda^\mu \tau_\mu^\nu$. Hence, again by 1.2, these two homomorphisms coincide. ■

In connection with our constructions, let us recall a useful concept in general category theory. Let $\mathcal{K}$ be an arbitrary category and let Λ be some directed set with an order relation "$<$". Further, suppose that to each $\nu \in \Lambda$ there corresponds an object $X_\nu \in \mathcal{K}$ and to each pair $\mu, \nu \in \Lambda$ with $\mu < \nu$ there corresponds a morphism $\sigma_\mu^\nu : X_\nu \longrightarrow X_\mu$. "This set-up", that is the whole collection $\mathcal{X} = (\Lambda, \{X_\nu\}, \{\sigma_\mu^\nu\})$ is called a projective system or, as we shall more often say, an *inverse system* in $\mathcal{K}$ if $\sigma_\lambda^\nu = \sigma_\lambda^\mu \sigma_\mu^\nu$ whenever $\lambda < \mu < \nu$. Moreover the collection $(X, \{\varphi_\nu\})$ consisting of (yet another) object $X \in \mathcal{K}$ and morphisms $\varphi_\nu : X \longrightarrow X_\nu$, is called an *object over* $\mathcal{X}$ if $\varphi_\mu = \sigma_\mu^\nu \varphi_\nu$ whenever $\mu < \nu$.

Thus for our initial multinormed algebra $(A; \|\cdot\|_\nu; \quad \nu \in \Lambda)$, the totality $\mathcal{A} = (\Lambda, \{A_\nu\}, \{\tau_\mu^\nu\})$ is, according to Proposition 1.3, an inverse system in the category of Banach algebras and also in any category of polynormed algebras that contains it. In the spirit of Definition 1.1, we shall also say that this $\mathcal{A}$, and also the homomorphisms τ_μ^ν entering into it, is a *concomitant* of A. We note that the collection $(A, \{\tau_\nu\})$ is an object over $\mathcal{A}$ in the category of (this time) multinormed algebras or also in any ambient category. (These observations enable us later to use known facts in category theory "without re-inventing the wheel".)

Of course, if A is unital, then so are all its concomitant A_ν, τ_ν and τ_μ^ν.

Example (V.1.4). Let $A = (C(\Omega); \|\cdot\|_K; \quad K \in \Lambda)$, where Ω is a locally compact space and Λ is the collection of all its compact subsets (cf. the beginning of Chap.III §1.1). Here, obviously $C(K)$; $K \in \Lambda$, are the concomitant Banach algebras, to within an isometric isomorphism, and the concomitant homomorphisms (with one or two indices) are surjective and take a function to its restriction to the relevant compact subset. We note that when Ω is discrete (when $C(\Omega) = \mathbb{C}^\Omega$) all the concomitant Banach algebras are finite-dimensional.

Example (V.1.5). Let $A = (C^\infty[0, 1], \quad \|\cdot\|_n, \quad n = 0, 1, 2, \cdots)$ with the norms indicated in Chap.III §1.3. Here, as the set of indices is totally ordered, "all the concomitants of our algebra" may be depicted by a commutative diagram

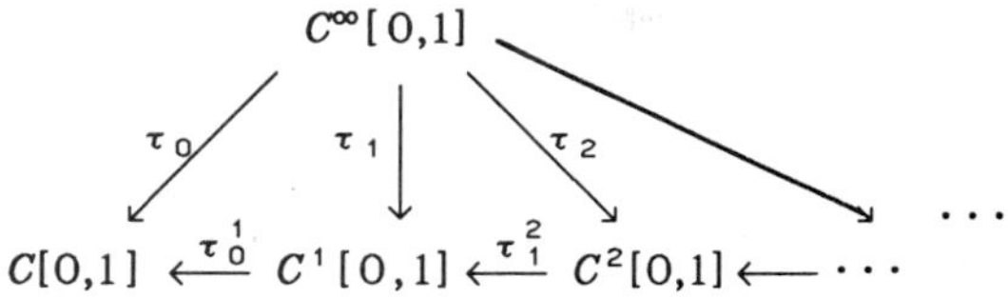

in which all the homomorphisms are natural embeddings (leaving the functions unaltered).

Here, finally, is the first application.

Proposition (V.1.6). *Every unital multinormed algebra* $(A;\ \|\cdot\|_\nu,\ \nu\in\Lambda)$, *is an algebra with continuous inverse.*

▶ Let $(a_\xi \in \mathrm{Inv}\,A;\ \xi\in\Xi)$ be a certain net that converges to the identity e in A; then for every $\nu\in\Lambda$ the net $\tau_\nu(a_\xi)$ converges to the identity e_ν in A_ν. But A_ν, being a Banach algebra, is an algebra with continuous inverse (I.2.43), hence $\tau_\nu(a_\xi^{-1}) = (\tau_\nu(a_\xi))^{-1}$ also converges to e_ν. In view of the obvious equation $\|a_\xi^{-1} - e\|_\nu = \|\tau_\nu(a_\xi^{-1}) - e_\nu\|_\nu^\wedge$, this means that a_ξ^{-1} converges to e with respect to each of the seminorms in A. ■

We are now in a position to obtain the following version of the Gel'fand–Mazur theorem.

Theorem (V.1.7). *Let* $(A;\ \|\cdot\|_\nu,\ \nu\in\Lambda)$, *be a multinormed skew field with any topology apart from the indiscrete topology* (in other words, there is at least one ν for which $\|\cdot\|_\nu \not\equiv 0$). *Then* A *is topologically isomorphic to* $\mathbb{C}$.

▶ From the hypotheses, the set $I^0 = \{a\in A : \|a\|_\nu = 0$ for all $\nu\}$ is obviously a bi-ideal in A and does not coincide with A. Since A is a skew field, Theorem I.3.20 implies that $I^0 = \{0\}$ and this means (see 0.1.4) that A is Hausdorff. All that remains is to combine Theorem II.1.17 and Proposition 1.6. ■

We pick out yet another fact that will prove useful to us later on.

Proposition (V.1.8). *Let* $(A;\ \|\cdot\|_\nu,\ \nu\in\Lambda)$, *be a multinormed algebra, and let* χ *be a continuous character of* A. *Then among the concomitant homomorphisms there exists* $\tau_\nu : A \longrightarrow A_\nu$ *such that* $\chi = \tilde{\chi}\tau_\nu$ *for some character* $\tilde{\chi}$ *of the algebra* A_ν.

▶ Since χ is continuous, there exist $\nu\in\Lambda$ and $C>0$ such that $|\chi(a)| \le C\|a\|_\nu$ for all $a\in A$ (see 0.1.9), hence $I_\nu \subseteq \mathrm{Ker}\,\chi$ and $\chi = \dot{\chi}\tau_\nu$, where $\dot{\chi}$ is the continuous character on A/I_ν induced by χ. Consequently $\chi = \tilde{\chi}\tau_\nu$, where $\tilde{\chi}$ is the (unique) continuous extension of $\dot{\chi}$ to A_ν. ■

§2. DESCRIPTION OF ARENS-MICHAEL ALGEBRAS IN TERMS OF BANACH ALGEBRAS

In this section it will be shown that for Arens-Michael algebras the concomitant inverse systems of Banach algebras carry a great deal of information about their structure. Here Theorem 2.10, which has a lot of applications, serves as the basic technique for extracting this information. Theorem 2.17 emerges as one important application of Theorem 2.10 which, together with its Corollary 2.19, completely characterizes Arens-Michael algebras in terms of Banach algebras.

Let $\mathcal{A} = (\Lambda, \{A_\nu\}, \{\sigma_\mu^\nu\})$ be a certain inverse system in the category of (for the moment all) polynormed algebras. The family $\{a_\nu \in A_\nu;\ \nu \in \Lambda\}$ (in other words, an element of the Cartesian product $\bigtimes_\nu A_\nu$) is called a *compatible family* (of this system) if $a_\mu = \sigma_\mu^\nu(a_\nu)$ whenever $\mu \prec \nu$. If $\mathcal{A}$ is the concomitant system of the multinormed algebra $(A;\ \|\cdot\|_\nu)$, then obviously every $a \in A$ provides a compatible family $\{a_\nu = \tau_\nu(a)\}$. Are there other compatible families for such an $\mathcal{A}$?

Example (V.2.9). We consider the same algebra of entire functions $\mathcal{O}(\mathbb{C})$ with two different systems of norms, $\|\cdot\|_n'$ and $\|\cdot\|_n''$, $n = 1, 2, \cdots$, obtained by putting $\|w\|_n' = \max\{|w(z)| : |z| \le n\}$ and $\|w\|_n'' = \max\{|w(t)| : t \in [-n, n]\}$. Then the n-th concomitant Banach algebra to $(\mathcal{O}(\mathbb{C});\ \|\cdot\|_n')$ is obviously the "disc algebra on $\{z : |z| \le n\}$" consisting of functions that are defined and continuous on this disc and analytic at its interior points (cf. the description of the "disc algebra" given in Proposition III.1.21). At the same time the n-th concomitant algebra to $(\mathcal{O}(\mathbb{C});\ \|\cdot\|_n'')$, is obviously $C[-n, n]$. Concerning compatible families of elements of the concomitant inverse systems, then for $(\mathcal{O}(\mathbb{C});\ \|\cdot\|_n')$, each of them consists, as is easily seen, of the restrictions (to the relevant disc) of the same $w \in \mathcal{O}(\mathbb{C})$ and for $(\mathcal{O}(\mathbb{C});\ \|\cdot\|_n'')$ they consist of the restrictions (to the corresponding intervals) of the same $f \in C(\mathbb{R})$; where the latter is not necessarily of the form $f = w|_{\mathbb{R}}$ for some $w \in \mathcal{O}(\mathbb{C})$. Thus in the first case all compatible families "originate from an element of the original algebra" and in the second they do not.

We turn our attention to the fact that the first of the multinormed algebras we have just considered is complete and the second is not. It turns out that the situation is as follows:

Theorem (V.2.10). *Let $(A;\ \|\cdot\|_\nu,\ \nu \in \Lambda)$, be an Arens-Michael algebra, let A_ν, τ_ν, τ_μ^ν be its concomitant Banach algebras and homomorphisms and let $\{\hat{a}_\nu \in A_\nu;\ \nu \in \Lambda\}$ be a certain compatible family of its concomitant inverse system. Then there exists a unique $a \in A$ such that $\hat{a}_\nu = \tau_\nu(a)$ for all $\nu \in \Lambda$.*

▶ If a and b are two elements in A with the required property, then $\tau_\nu(a-b)=0$ for all ν. From the definition of τ_ν, this means precisely that $\|a-b\|_\nu = 0$ for this particular ν. Since A is Hausdorff, this implies that $a-b=0$ (cf. 0.1.4). In this way we establish that the element we are interested in is unique; let us prove that it exists.

We consider $\Lambda \times \mathbb{N}$ to be a directed set with respect to the order "(μ, m) precedes (ν, n) if simultaneously $\mu \prec \nu$ and $m \le n$". Using the fact that A/I_ν is dense in A_ν, we get, for each pair (ν, n), some element $a_{\nu,n} \in A$ such that $\|\tau_\nu(a_{\nu,n}) - \hat{a}_\nu\|^\wedge_\nu < \frac{1}{n}$.

Let us show that the net $(a_{\nu,n};\ (\nu, n) \in \Lambda \times \mathbb{N})$ is a Cauchy net in A, that is, it is a Cauchy net in each seminormed space $(A;\ \|\cdot\|_{\nu_0})$, for fixed $\nu_0 \in \Lambda$. For this it is obviously sufficient to show that for each $\varepsilon > 0$, $\|a_{\lambda,l} - a_{\mu,m}\|_{\nu_0} < \varepsilon$ whenever $\nu_0 \prec \lambda, \mu$ and $\frac{2}{\varepsilon} < l, m$. But this in turn is implied by

$$\|a_{\lambda,l} - a_{\mu,m}\|_{\nu_0} = \|\tau_{\nu_0}(a_{\lambda,l}) - \tau_{\nu_0}(a_{\mu,m})\|^\wedge_{\nu_0} \le$$

$$\le \|\tau_{\nu_0}(a_{\lambda,l}) - \hat{a}_{\nu_0}\|^\wedge_{\nu_0} + \|\tau_{\nu_0}(a_{\mu,m}) - \hat{a}_{\nu_0}\|^\wedge_{\nu_0} =$$

$$= \|\tau^\lambda_{\nu_0}\tau_\lambda(a_{\lambda,l}) - \tau^\lambda_{\nu_0}(\hat{a}_\lambda)\|^\wedge_{\nu_0} + \|\tau^\mu_{\nu_0}\tau_\mu(a_{\mu,m}) - \tau^\mu_{\nu_0}(\hat{a}_\mu)\|^\wedge_{\nu_0} \le$$

$$\le \|\tau_\lambda(a_{\lambda,l}) - \hat{a}_\lambda\|^\wedge_\lambda + \|\tau_\mu(a_{\mu,m}) - \hat{a}_\mu\|^\wedge_\mu < \frac{1}{l} + \frac{1}{m} < \varepsilon,$$

using the definition of the norm in the relevant algebras, the property of a compatible family and both assertions in Proposition 1.2.

But we are assuming that A is complete so the net $a_{\nu,n}$, being a Cauchy net, converges to a certain element a in A. Now, having fixed any $\nu \in \Lambda$, we have to show that $\tau_\nu(a) = \hat{a}_\nu$.

We take $\varepsilon > 0$; then the construction of a implies that $\|a - a_{\mu,m}\|_\nu < \frac{\varepsilon}{2}$ for a certain $\mu \in \Lambda$, $\nu \prec \mu$ and $m > \frac{2}{\varepsilon}$. Hence using arguments similar to those used in the preceding calculations we see that

$$\|\tau_\nu(a) - \hat{a}_\nu\|^\wedge_\nu \le \|\tau_\nu(a) - \tau_\nu(a_{\mu,m})\|^\wedge_\nu + \|\tau_\nu(a_{\mu,m}) - \hat{a}_\nu\|^\wedge_\nu =$$

$$= \|a - a_{\mu,m}\|_\nu + \|\tau^\mu_\nu\tau_\mu(a_{\mu,m}) - \tau^\mu_\nu(\hat{a}_\mu)\|^\wedge_\nu < \frac{\varepsilon}{2} + \|\tau_\mu(a_{\mu,m}) - \hat{a}_\mu\|^\wedge_\mu < \varepsilon.$$

Since ε is arbitrary, this shows that the norm of the element $\tau_\nu(a) - \hat{a}_\nu \in A_\nu$ is zero. ■

As a first application of this theorem we prove the following, keeping the same notation.

Proposition (V.2.11) (Arens). *Let A be a unital Arens-Michael algebra. Then the element $a \in A$ is invertible $\Longleftrightarrow$ $\tau_\nu(a)$ is invertible in A_ν for every $\nu \in \Lambda$.*

▶ "$\Rightarrow$" is obvious; we deal with the inverse deduction. We put for brevity $\hat{a}_\nu = \tau_\nu(a)$ and $e_\nu = e_{A_\nu}$. We consider $\mu, \nu \in \Lambda$ with $\mu < \nu$; then $(\tau^\nu_\mu(\hat{a}_\nu^{-1}))\hat{a}_\mu = (\tau^\nu_\mu(\hat{a}_\nu^{-1}))(\tau^\nu_\mu(\hat{a}_\nu)) = \tau^\nu_\mu(\hat{a}_\nu^{-1}\hat{a}_\nu) = \tau^\nu_\mu e_\nu = e_\mu$. Hence $\tau^\nu_\mu(\hat{a}_\nu^{-1})$ is a left inverse and a similar argument shows that it is also a right inverse for $\hat{a}_\mu$. This means that $\tau^\nu_\mu(\hat{a}_\nu^{-1}) = \hat{a}_\mu^{-1}$, in other words, that the family $(\hat{a}_\nu^{-1} \in A_\nu;\ \nu \in \Lambda)$ is compatible. But then, by Theorem 2.10, there exists $b \in A$ such that $\tau_\nu(b) = \hat{a}_\nu^{-1}$ for all ν. Consequently $\tau_\nu(ba) = \tau_\nu(b)\tau_\nu(a) = \hat{a}_\nu^{-1}\hat{a}_\nu = e_\nu$ and in precisely the same way $\tau_\nu(ab) = e_\nu$; since at the same time $\tau_\nu(e_A) = e_\nu$, the uniqueness, again established in Theorem 2.10, of the element that "generates" the compatible family — in this case $\{e_\nu;\ \nu \in \Lambda\}$ — implies that $ba = ab = e_A$. ■

If, in this proposition, we consider $a - \lambda e;\ \ \lambda \in \mathbb{C}$, instead of a, we immediately get the following.

Corollary (V.2.12). *Let $(A;\ \|\cdot\|_\nu,\ \nu \in \Lambda)$, be an Arens-Michael algebra. Then $\mathrm{Sp}_A(a) = \bigcup_\nu \mathrm{Sp}_{A_\nu}(\tau_\nu(a))$ for any $a \in A$.*

We turn to the preparations for the second of the theorems promised at the beginning of the section. We first widen our knowledge of category theory by introducing the following important concept.

Let $\mathcal{K}$ be a certain category and let $\mathcal{X} = (\Lambda, \{X_\nu\}, \{\sigma^\nu_\mu\})$ be an inverse system in $\mathcal{K}$.

Definition (V.2.13). An object $(X, \{\sigma_\nu\})$ over $\mathcal{X}$ is called the *inverse limit* of this system if for any other object $(Y, \{\varphi_\nu\})$ over $\mathcal{X}$ there exists a unique morphism $\alpha : Y \longrightarrow X$ such that for all $\nu \in \Lambda$ the diagram below is commutative.

$$\begin{array}{ccc} Y & \xrightarrow{\ \alpha\ } & X \\ & {\scriptstyle\varphi_\nu}\searrow \quad \swarrow{\scriptstyle\sigma_\nu} & \\ & X_\nu & \end{array}$$

(Sometimes we shall call X itself the inverse limit of $\mathcal{X}$ intending that this is so when it is suitably "equipped" with morphisms into $X_\nu;\ \nu \in \Lambda$.)

In practice the concept of an inverse limit is used as one of the ways of reducing comparatively complicated objects (X) to a system of simpler ones (X_ν). We need to get to know just one specific situation.

Proposition (V.2.14) ("uniqueness theorem"). *Let* $(X^k, \{\sigma_\nu^k\})$, $k = 1, 2$, *be two inverse limits of the system* $\mathcal{X}$. *Then there exists an isomorphism* $i : X^1 \longrightarrow X^2$ *such that for any* $\nu \in \Lambda$ *the diagram below is commutative.*

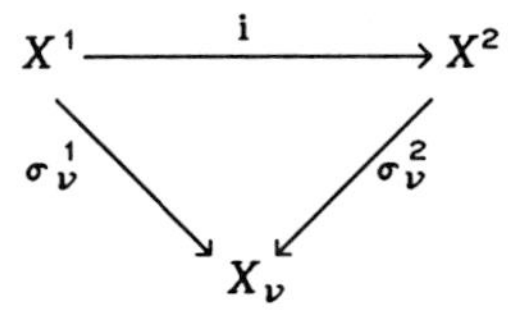

▶ If we put $(X^2, \{\sigma_\nu^2\}) = (X, \{\sigma_\nu\})$ and $(X^1, \{\sigma_\nu^1\}) = (Y, \varphi_\nu)$ in Definition 2.13, we get (as α) some morphism i and by exchanging the two given inverse limits we get a certain morphism j. Since for any $\nu \in \Lambda$ the diagram

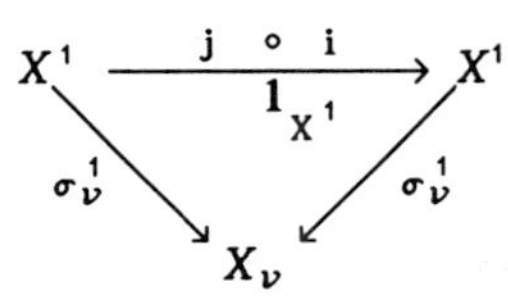

is obviously commutative, the uniqueness of the morphism α in 2.14 implies that $j \circ i = 1_{X^1}$. An analogous argument shows that $i \circ j = 1_{X^2}$. ■

Exercise (V.2.15). Defining the category of objects over $\mathcal{X}$ in a suitable way, show that Proposition 2.14 is a particular case of the uniqueness, up to an isomorphism, of the final object in any category (see Chap.0 §2.2).

We shall return to the specific categories of our account consisting of polynormed algebras. For them the "uniqueness theorem" means, in particular, that two algebras that are inverse limits of the same inverse system coincide to within a topological isomorphism. About the existence of inverse limits, it is not difficult to see that by no means every inverse system in the category of Banach algebras has them. However, in a lot of wider categories, as we shall now show, inverse limits always exist.

Let $\mathcal{A} = (\Lambda, \{A_\nu\}, \{\tau_\mu^\nu\})$ be an inverse system in the category of (for the moment) all polynormed algebras, denoted in future by $\mathcal{PNA}$. We consider the Cartesian product $\bigtimes_\nu A_\nu$ and let A^0 denote its subalgebra consisting of all those "rows" $\bar{a} = \{a_\nu;\ \nu \in \Lambda\}$ that are compatible families; since all the τ_μ^ν

are continuous, this subalgebra must be closed. Further we put $\tau_\nu^0 : A^0 \longrightarrow A_\nu : \bar{a} \longmapsto a_\nu;\quad \nu \in \Lambda$.

Proposition (V.2.16). *Let* $\mathcal{K}$ *be a complete subcategory in* $\mathcal{PNA}$ *which, if it contains any algebra, also contains any of its closed subalgebras, and, if it contains any family of algebras, also contains their Cartesian product* (for example, $\mathcal{PNA}$ itself or the category of Arens-Michael algebras). *Then any inverse system in* $\mathcal{K}$ *has an inverse limit in this category; namely, for* $\mathcal{A}$ *this is, in the notation we have just described,* $(A^0, \{\tau_\nu^0\})$.

▶ First of all, the condition implies that $A^0 \in \mathcal{K}$. Further, let $(B, \{\varphi_\nu\})$; $B \in \mathcal{K}$, be an arbitrary object over $\mathcal{A}$. For any $b \in B$ we consider the family $\{\varphi_\nu(b) \in A_\nu;\ \nu \in \Lambda\}$; as it is obviously compatible, it is an element of $A^0 \subseteq \bigtimes_\nu A_\nu$. Therefore the map $\alpha : B \longrightarrow A^0 : b \longmapsto \{\varphi_\nu(b)\}$ is well defined and it is easy to see that it is a continuous homomorphism. It remains to note that it follows from the way that τ_ν^0 is constructed that α is the unique map making the diagram in Definition 2.13 commutative, after the appropriate substitutions (A^0 instead of X and so on). ■

Our preparation is finally completed.

Theorem (V.2.17). *Let* $(A; \|\cdot\|_\nu;\ \nu \in \Lambda)$, *be an Arens-Michael algebra. Then*

(I) A *is the inverse limit of its concomitant inverse system of Banach algebras*;

(II) A *is, up to a topological isomorphism, a closed subalgebra of the Cartesian product of the family of its concomitant Banach algebras.*

▶ Let $\mathcal{A}$ be the concomitant system for A; let us show that the object $(A, \{\tau_\nu : A \longrightarrow A_\nu\})$ over $\mathcal{A}$ (τ_ν are the concomitant homomorphisms) is an inverse limit for $\mathcal{A}$. We take an arbitrary "alien" object $(B, \{\varphi_\nu\})$ over $\mathcal{A}$. We consider the family $\{\varphi_\nu(b) \in A_\nu\}$ for $b \in B$; obviously it is compatible. Therefore, by Theorem 2.10, there exists a unique $\alpha(b) \in A$ such that $\tau_\nu(\alpha(b)) = \varphi_\nu(b)$.

We put $\alpha : B \longrightarrow A : b \longmapsto \alpha(b)$; then, since the choice of $\alpha(b)$, given b, is unique, it is obvious that this implies that α is an algebra homomorphism and, since τ_ν is continuous, the fact that $\|\varphi_\nu(b)\|_\nu^\wedge = \|\alpha(b)\|_\nu$ for all ν implies that α is continuous. Since the construction of α implies that the diagram in the definition of an inverse limit is commutative, this proves (I). This implies (II), taking into account the specific form of this inverse image that was described in Proposition 2.16 and the uniqueness of this inverse image up to a category isomorphism (see 2.14). ■

Remark (V.2.18). If we consider algebras with zero multiplication, Theorem 2.17 becomes a theorem in the theory of topological vector spaces about representing complete Hausdorff polynormed spaces as the inverse limit of Banach spaces (see, for example, [90, II, §5.4]).

By combining the theorem we have proved with the explicit construction of an inverse limit preceding it (2.16) and the "uniqueness theorem" (2.14) we immediately get the following.

Corollary (V.2.19). *A polynormed algebra is an Arens–Michael algebra* $\Longleftrightarrow$ *it is the inverse limit of a certain inverse system of Banach algebras* $\Longleftrightarrow$ *it is topologically isomorphic to a closed subalgebra of the Cartesian product of a certain family of Banach algebras.*

Returning to the specific Arens–Michael algebras $C(\Omega)$, $C^{\infty}[0, 1]$ and $\mathcal{O}(\mathbb{C})$ in Examples 1.4, 1.5 and 2.9, we thus see that the first of them can be represented as the inverse limit of the system of Banach algebras $C(K)$, where K is a compact subset of Ω, the second is the inverse limit of the system of Banach algebras $C^{n}[0, 1]$; $n = 0, 1, \cdots$, and finally, the third is the inverse limit of the system of Banach "disc algebras on $\{z \in \mathbb{C} : |z| \le n\}$; $n = 1, 2, \cdots$".

In the case of an arbitrary (not necessarily either complete and/or Hausdorff) multinormed algebra A the inverse limit of its system of concomitant Banach algebras can be described as follows: we put $I_{\mathrm{A}}^{0} = \{a : \|a\|_{\nu} = 0$ for all $\nu \in \Lambda\}$; obviously this is the closed bi-ideal in A consisting of all the points "tied to zero" and the quotient algebra A/I_{A}^{0} is Hausdorff. We denote the completion of the latter (see I.2.23) by $\overline{\overline{A}}$ and put $\tau : A \longrightarrow \overline{\overline{A}} : a \longmapsto a + I_{\mathrm{A}}^{0}$.

Exercise (V.2.20). Show that the inverse limit of the system of concomitant Banach algebras of A is $(\overline{\overline{A}}, \{\overline{\overline{\tau}}_{\nu}\})$, where $\overline{\overline{\tau}}_{\nu}$; $\nu \in \Lambda$, is uniquely defined by the commutative diagram

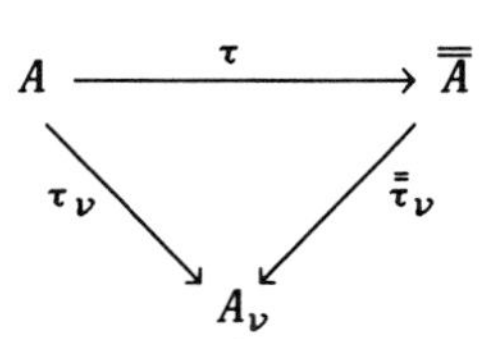

* * *

To complete this section we show that for an arbitrary polynormed algebra one can define, in a reasonable way, the Arens–Michael algebra that is "closest to it".

Definition (V.2.21). Let A be a polynormed algebra. The pair $(\widetilde{A}, \omega)$, where $\widetilde{A}$ is an Arens–Michael algebra and $\omega : A \longrightarrow \widetilde{A}$ is a continuous algebra homomorphism, is called *the Arens–Michael* *hull* of the algebra A if, for any Arens-Michael algebra B and continuous homomorphism $\kappa : A \longrightarrow B$, there exists a unique continuous homomorphism $\tilde{\kappa} : \tilde{A} \longrightarrow B$ such that the diagram

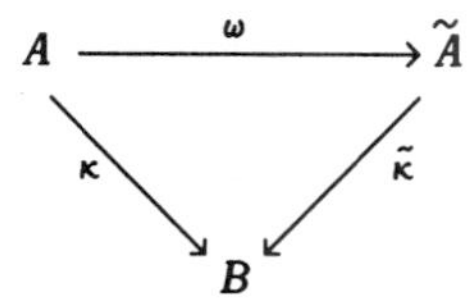

commutes.

We offer the reader, in the form of exercises, a few simple facts about Arens-Michael hulls.

Exercise (V.2.22). Define the concept we have introduced in terms of the final object of a certain category. As a consequence (cf.2.14 and 2.15) formulate and prove a "uniqueness theorem" for the Arens-Michael hull of a given algebra.

Let us turn to the question of the existence of Arens-Michael hulls. First we consider a preparatory result.

Exercise (V.2.23). Let A be a multinormed algebra, and let $\overline{\overline{A}}$ and τ be the algebra and homomorphism introduced above (and figuring in Exercise 2.20). Show that $(\overline{\overline{A}}, \tau)$ is the Arens-Michael hull of the algebra A.

Now let $(A; \|\cdot\|_\nu;\ \nu \in \Lambda)$, be an arbitrary polynormed algebra with a saturated system of seminorms. We let $(A_0, \|\cdot\|'_\mu,\ \mu \in \Lambda')$, denote the multinormed algebra with the same underlying pure algebra as A, but with a new system of seminorms, namely, that consisting of all multiplicative seminorms that are continuous, as functions $a \longmapsto \|a\|'_\mu$, in the topology of the original algebra A. (It is easy to verify that this last condition means precisely that $\|\cdot\|'_\mu \le C\|\cdot\|_\nu$ for certain $\nu \in \Lambda$ and $C > 0$.) Obviously the homomorphism $j : A \longrightarrow A_0$, leaving all elements fixed, is continuous, and moreover j is a topological isomorphism $\Longleftrightarrow$ A itself is a multinormed algebra.

If we substitute A_0 for the algebra in the previous exercise, the Arens-Michael hull that featured there is denoted by $(\tilde{A}, \tilde{\tau})$. We also put $\omega = \tilde{\tau} \circ j : A \longrightarrow \tilde{A}$.

Exercise (V.2.24) ("existence theorem"). Show that every polynormed algebra A has the pair $(\tilde{A}, \omega)$ as its Arens-Michael hull.

Hint. Every continuous homomorphism from A into any Arens-Michael algebra is also continuous with respect to the topology defined in A_0.

Here are a few illustrations.

Exercise (V.2.25). Show that when $A = \mathcal{P}$ (see I.2.17) $\widetilde{A}$ is $\mathcal{O}(\mathbb{C})$; when $A = c_p$ (see Chap.III §1.3) $\widetilde{A}$ is c_∞ (see I.2.16) and when $A = L^\omega[0, 1]$ (see I.2.22) $\widetilde{A} = \mathbb{O}$.

Another significant class of examples is contained in the paper [252, §7] by Taylor who, apparently, was the first to introduce (as an explicit construction and in other terminology [251, Definition 5.1]) the concept of an Arens-Michael hull.

We emphasize, however, that the construction we have described does not by any means give good results for all polynormed algebras: as we have already seen, there are those, including a few fairly important examples, for which the Arens-Michael hull is trivial. These algebras thus form the opposite class to the class of Arens-Michael algebras (in approximately the same vague sense in which radical algebras are the opposite to semisimple algebras; cf. Chap.I §3.4). Such examples crop up, in particular, in operator theory:

Exercise (V.2.26). Show that, when $A = \mathcal{B}(H)$; $\dim H = \infty$ with any of the systems of seminorms — strong, ultrastrong, weak and ultraweak — introduced in Chap.III §2.3, then $\widetilde{A} = \mathbb{O}$.

Hint. Using the hint to Exercise III.2.26 one can show that, for any neighbourhood of zero U in any of the four topologies, the set $\{ab : a, b \in U\}$ is the whole of $\mathcal{B}(H)$.

§3. COMMUTATIVE ARENS-MICHAEL ALGEBRAS AND THE LIKE

How much of the "classical" theory, which was the subject of the preceding chapter, have we succeeded in extending from the class of Banach algebras to a larger class — or at least to the class of Arens-Michael algebras? In this section we recall certain unavoidable and at first sight very notable losses, however attention is mainly concentrated on those relatively important facts that — sometimes after a suitable rethinking — may nevertheless be saved. A large part of these positive results concern commutative multinormed algebras in a varying degree of generality.

Recall that in the basic theory of commutative Banach algebras there is the Theorem IV.1.1 about identifying maximal ideals with characters. If we go outside the class of Banach algebras such an identification necessarily involves only a part of the two sets concerned.

Proposition (V.3.27). *Let A be a commutative multinormed algebra. Then there exists a bijection, between the set of its continuous non-trivial characters on the one hand and the set of* *its closed maximal modular ideals on the other, that associates each character with its kernel.*

▶ Let I be a closed maximal modular ideal in A, then A/I (with the system of quotient seminorms) is a Hausdorff multinormed algebra which, by Theorem I.3.34, is a field. Hence, by Theorem 1.7, $A/I \simeq \mathbb{C}$.

The rest of the proof is the same as in Theorem IV.1.1. ■

Example (V.3.28). For the polynomial algebra $\mathcal{P}$ its set of (non-trivial) characters as well as its maximal ideals are characterized precisely by the complex plane: the first are of the form $\chi_z : p \longmapsto p(z)$ and the second are of the form $I_z = \{p : p(z) = 0\}$; $z \in \mathbb{C}$. If we introduce the multiplicative norm $\| p\| = \max\{|p(t)| : t \in [0, 1]\}$ in $\mathcal{P}$, then the continuous characters, as well as the closed maximal ideals of our polynormed algebra, are obviously those that correspond to points of the interval $[0, 1] \subset \mathbb{C}$. The bijection in 3.27 is, for our algebra, the map $\chi_t \longmapsto I_t$; $t \in [0, 1]$, and thus is a "part" of the bijection between the set of all its characters and the set of all its maximal ideals.

Remark (V.3.29). In the given example every maximal ideal of the algebra under consideration is the kernel of some character or other. However, even in Arens–Michael algebras there are maximal ideals that are not kernels of characters, even discontinuous characters. In particular, one could take all the maximal ideals in c_∞ containing c_{00} (cf. Chap.I §3.2) or all the maximal ideals in $C(\mathbb{R})$ containing the ideal of functions of compact support. (The proof of these facts is based on the so-called "real-compactness", in the sense of Hewitt, of $\mathbb{N}$ and $\mathbb{R}$ as topological spaces; see, for example, [7, 1.5–1, 1.5–3 and 1.6–1].)

Thus, by Proposition 3.27 the problem of describing the closed maximal ideals of a given commutative multinormed algebra reduces to the problem of describing its continuous characters, which — and we have already noticed this — may turn out to be comparatively simple. For example, if we combine 3.27 and III.1.18, we get the following.

Corollary (V.3.30). *Every closed maximal ideal of the algebra $\mathcal{O}(U)$, where U is a domain of $\mathbb{C}$, is fixed.*

We suggest that the reader obtains a similar description of the closed maximal ideals in $C(\Omega)$, where Ω is a completely regular topological space, starting from 3.27 and the result of Exercise III.1.8.

* * *

Although the closed maximal ideals form, even in the case of Arens–Michael algebras, only part of the totality of maximal ideals, nevertheless in this case there are enough of them to solve two questions: to characterize the invertible elements and to characterize the radical.

Warning. From now on until the end of the chapter we shall, for simplicity, assume everywhere that the algebra under consideration is unital. The reader may, without difficulty, obtain analogues, for arbitrary algebras, of the following results by a routine replacement of A by $A_{\times}$.

Proposition (V.3.31) (cf. I.3.19(III) and IV.1.2). *Let A be a commutative Arens-Michael algebra. Then $a \in A$ is invertible $\Longleftrightarrow$ it does not belong to any closed maximal ideal in A $\Longleftrightarrow$ $\chi(a) \neq 0$ for each continuous non-trivial character χ on A.*

▶ Suppose that a is not invertible; then, by Proposition 2.11, there exist a Banach algebra A_ν (obviously commutative) and a homomorphism $\tau_\nu : A \longrightarrow A_\nu$, which are concomitant to A, such that $\hat{a}_\nu = \tau_\nu(a)$ is not invertible in A_ν. But then, by Corollary IV.1.2, $\hat{\chi}(\hat{a}_\nu) = 0$ for a certain character $\hat{\chi}$ on A_ν, hence $\chi(a) = 0$ for the character $\chi = \hat{\chi} \circ \tau_\nu$ on A. Moreover, since $\hat{\chi}$ is "automatically" continuous (see I.2.6), χ is also continuous.

We have shown that the third of the three properties implies the first, the purely algebraic Theorem I.3.19 ensures that the second property can be deduced from the first, and Proposition 3.27 shows that the third can be deduced from the second. ∎

Proposition (V.3.32) (cf. I.3.53 and IV.1.4). *The radical of a commutative Arens-Michael algebra is the intersection of all its closed maximal ideals and is the intersection of the kernels of all its continuous characters.*

▶ By 3.27, the two sets that we are claiming to be the radical, coincide; by Theorem I.3.53, the radical is contained in the first of them. All that remains is to prove that the second set is contained in the radical.

Let $a \notin \operatorname{Rad} A$, where A is our algebra. Then, by the definition of the radical, there is some $b \in A$ such that $e - ba$ is not invertible in A. Hence, by Proposition 3.31, there exists a continuous non-trivial character χ on A such that $\chi(e - ba) = 1 - \chi(b)\chi(a) = 0$, and therefore $\chi(a) \neq 0$. The rest is clear. ∎

It is useful to note that these two last propositions can also be reworded as follows:

Corollary (V.3.33). *In a commutative Arens-Michael algebra the union (intersection, respectively) of all its maximal ideals coincides with the union (intersection, respectively) of all its closed maximal ideals.*

Recall the concept of the spectrum of a commutative Banach algebra that was so important in the previous chapter. Proposition 3.27 suggests the following generalization of this concept: *the spectrum of a commutative multinormed algebra* A is the set of its continuous non-trivial characters (denoted, as usual, by $X\alpha\rho\, A$) or the set of its closed maximal ideals that are canonically identified with them. As before (cf. Chap.IV §1.2), the weak* topology on $X\alpha\rho\, A$ is considered to be the standard topology on the spectrum. As in the Banach case, we shall use $\underline{\Omega}(A)$ or Ω as the notation for the spectrum, and s, χ_s or I_s as the notation for a point in it — depending on how it is to be interpreted (cf. *ibid.*).

The situation is fundamentally worse than in the case of "Banach" algebras in that the spectra of (even unital) Arens-Michael algebras are no longer necessarily compact (cf. IV.1.12) or even locally compact.

Exercise (V.3.34). Show that the spectrum of the algebra $C(\Omega)$, where Ω is completely regular (see Chap.III §1.2), is homeomorphic to Ω.

Hint. Use the result of Exercise III.1.8.

Thus any completely regular topological space satisfying the not very burdensome conditions discussed at the end of Chap.III §1.2, could be the spectrum of a unital Arens-Michael algebra.

Let us note a certain (though fairly narrow) class of algebras for which the spectra are as good as before.

Exercise (V.3.35). Let A be an Arens-Michael algebra whose set of invertible elements is open (cf. I.2.6 and I.1.11). Prove that $\mathrm{Sp}(a)$ is a compact subset of $\mathbb{C}$ for any $a \in A$ and deduce from this that the spectrum of the algebra A is compact.

Hint. The map $\underline{\Omega}(A) \longrightarrow \mathop{\times}\limits_{a\in A} (\mathrm{Sp}(a)) : \chi \longmapsto \{\chi(a) : a \in A\}$ is topologically injective and its image is closed.

Let us turn to the question of realizing the elements of multinormed algebras as continuous functions on the spectrum. In view of the possible unpleasant topological properties of the spectrum, such a construction is not going to be so effective as for Banach algebras; all the same it retains certain attractive features.

Let a be an element of the multinormed algebra A. For $s \in \Omega$ ($= \underline{\Omega}(A)$) we put $\hat{a}(s) = \chi_s(a)$ (cf. IV.2.1). As in the Banach case, the *function that we get* $\hat{a} : \Omega \longrightarrow \mathbb{C}$ *is continuous* (this is implied by the definition of the topology on the spectrum) and, for an Arens-Michael algebra, $\mathrm{Sp}(a)$ *coincides with the set of its values* (the proof in IV.2.3 remains valid if we replace the reference to Corollary IV.1.3 by a reference to 3.31). The function $\hat{a}$, as before, is called the *Gel'fand transform of the element* a, and the map $\Gamma : A \longrightarrow C(\Omega)$ associating such an $\hat{a}$ to each $a \in A$ is the *Gel'fand transformation* of the (now multinormed) *algebra* A (cf. Definitions IV.2.1 and IV.2.10).

Theorem (V.3.36). (I) Γ *is a unital algebra homomorphism.*
(II) $\operatorname{Im}\Gamma$ *separates the points of* Ω.
(III) *If* A *is an Arens-Michael algebra, then* $\operatorname{Ker}\Gamma = \operatorname{Rad} A$.

▶ (I) and (II) are obvious, (III) is implied by 3.32. ■

Corollary (V.3.37). *Every semisimple commutative Arens-Michael algebra is isomorphic* (in a purely algebraic sense) *to a certain subalgebra of the algebra of continuous functions on its spectrum.*

On comparing Theorem 3.36 and its Banach counterpart IV.2.11, we see that one very essential detail is missing: the former says nothing about the continuity of the homomorphism Γ. In fact, when we consider $C(\Omega)$ to have the standard topology, — we mean by this the compact-open topology defined in Chap.0 §1.1, — this continuity is ensured only when certain additional conditions are satisfied.[2] Here is one such:

Proposition (V.3.38). *If* $(A; \|\cdot\|_\nu;\ \nu \in \Lambda)$, *is a metrizable Arens-Michael algebra* (= a multinormed Fréchet algebra), *then the homomorphism* $\Gamma : A \longrightarrow C(\Omega)$ *is continuous.*

▶ We fix a compact set $K \subset \Omega$ and consider the subset $M_K = \{a : \max_{s \in K} |\hat{a}(s)| \le 1\}$ of A. Since, for any $a \in A$, the continuous function $\hat{a}(s)$ is bounded on the compact set K, M_K is an absorbing set; in addition, it is obviously closed, balanced and convex. Since, in a Fréchet space, a set with such properties (a so-called "barrel") contains a neighbourhood of zero, there exist $\nu \in \Lambda$ and $\varepsilon > 0$ such that the "ball" $\{a : \|a\|_\nu \le \varepsilon\}$ lies inside M_K. In this way the seminorm $\|\cdot\|_K$ from the family defining the compact-open topology in Ω and the seminorm $\|\cdot\|_\nu$ in A found from it, are connected by the relation $\|\Gamma(a)\|_K \le \frac{1}{\varepsilon}\|a\|_\nu$ for any $a \in A$. As K is arbitrary, we have verified the conditions for the operator Γ to be continuous (see 0.1.9). ■

Exercise (V.3.39). Show that another condition ensuring that Γ is continuous is that $\operatorname{Inv} A$ is open in A.

---

[2] The reader who is acquainted with topological spaces of ordinals, can easily construct a suitable counterexample by taking A to be the algebra $C(\Omega)$, where Ω is the space of countable ordinals.

Now, since we have the "Banach" Theorem IV.2.15, naturally we pose the question: under what conditions can a given algebra be identified — algebraically and topologically — with a certain subalgebra of the algebra of all continuous functions on some topological space? A partial answer is given by the following.

Proposition (V.3.40). *The following properties of the commutative Arens-Michael algebra* A *are equivalent*:

(I) *there exists a defining system of seminorms* $(\|\cdot\|_\nu;\ \nu\in\Lambda)$ *in* A, *such that* $\|a^2\|_\nu = \|a\|_\nu^2$ *for all* $a\in A$ *and all* ν;

(II) A *is topologically isomorphic to a closed subalgebra of* $C(\Delta)$ *for some locally compact Hausdorff space* Δ.

If A *is, in addition, metrizable, then a further property is equivalent to both these properties*:

(III) *the Gel'fand transformation* $\Gamma : A \longrightarrow C(\Omega)$ *realizes a topological isomorphism between* A *and a closed subalgebra of* $C(\Omega)$.

(Thus, although the spectrum of our algebra does not necessarily have good topological properties — cf. 3.34 — all the same, in a given situation its elements may be realized as functions on another — and this time automatically locally compact — space Δ. True, in this new realization, the functions on Δ do not now necessarily separate the points of its domain of definition — in contrast to the Gel'fand transformation.)

▸ (I) ⇒ (II). Let $(A_\nu;\ \nu\in\Lambda)$, be the family of concomitant Banach algebras for A. The condition obviously implies that, for any $\nu\in\Lambda$ and $\hat{a}\in A_\nu$, $\|\hat{a}^2\|^\wedge_\nu = (\|\hat{a}\|^\wedge_\nu)^2$; therefore, by IV.2.15, the Gel'fand transformation $\Gamma_\nu ; A_\nu \longrightarrow C(\Delta_\nu)$, where Δ_ν is the (compact) spectrum of the algebra A_ν, is an isometric homomorphism. We put $\Delta = \coprod_\nu \Delta_\nu$ and introduce the topology of the disjoint union on this set; obviously the space thus constructed is locally compact and Hausdorff. We consider the homomorphism $\gamma : \bigtimes_\nu A_\nu \longrightarrow C(\Delta)$ sending the "row" $\{a_\nu \in A_\nu;\ \nu\in\Lambda\}$ to the function f such that $f(s) = [\Gamma_\nu(a_\nu)](s)$ for $s\in\Delta_\nu\subset\Delta$. Since all the Γ_ν are isometries it follows immediately from the form of the seminorms determining the topologies in $\bigtimes_\nu A_\nu$ and $C(\Delta)$ that γ is topologically injective.

Thus $C(\Delta)$ contains a subalgebra which is topologically isomorphic to $\bigtimes_\nu A_\nu$ and this, in its turn, by Theorem 2.17, contains a subalgebra topologically isomorphic to A. Hence, of course, A is topologically isomorphic to a subalgebra of $C(\Delta)$ which (since A is complete) is complete and therefore closed in $C(\Delta)$.

(II) ⇒ (III). We consider the defining system of seminorms $\{\|\cdot\|_K;\ K\in\Lambda\}$ in $A\subseteq C(\Delta)$, where Λ is the family of all compact subsets of Δ. We fix K and associate to each $t\in K$ the character $\chi: A\longrightarrow\mathbb{C}: f\longmapsto f(t)$. In this way we have obviously constructed a certain map α from K into the spectrum Ω of the algebra A; the fact that the function f is continuous and the definition of the topology in Ω immediately imply that α is continuous. Therefore its image L, being the continuous image of a compact set, is itself compact. This means that, given the seminorm $\|\cdot\|_K$ in A, there exists a seminorm in the system determining the topology in $C(\Omega)$ — namely $\|\cdot\|_L$ — such that $\|f\|_K=\|\Gamma(f)\|_L$ for any $f\in A$.

Recall that, provided A is metrizable, the homomorphism Γ, by Proposition 3.38, is continuous; at the same time the equation we have established between the seminorms shows that the corestriction Γ onto $\operatorname{Im}\Gamma$ is open; finally, A, being a function algebra, is semisimple (see I.3.62) and therefore, taking into account 3.36(III), Γ is injective. All this together means that Γ maps A topologically and isomorphically onto its image, which must thus be complete and is therefore a closed subalgebra of $C(\Omega)$.

(II) ⇒ (I) and (III) ⇒ (I). Both of these follow immediately from the form of the seminorms determining the (compact-open) topology in $C(\Delta)$ or, depending on which are talking about, the topology in $C(\Omega)$. ∎

* * *

The last group of questions which are left for us to discuss in this chapter concern the structure of multinormed algebras with involution. In future, when we talk about a system of seminorms in a star multinormed algebra, we shall invariably assume that all the seminorms are star and multiplicative, and that the system itself is saturated.[3] First we note a simple proposition.

Proposition (V.3.41). *Let* $(A;\ \|\cdot\|_\nu;\ \nu\in\Lambda)$, *be a star multinormed algebra. Then in each of its concomitant Banach algebras* A_ν *there exists a uniquely defined involution that turns* A_ν *into a star algebra and is such that all the concomitant* τ_ν *and* τ^ν_μ *turn out to be *-homomorphisms.*

▸ In this situation all the I_ν are *-bi-ideals in A and, as a consequence, all the A/I_ν are star normed algebras. Further, the involution in A/I_ν can be extended uniquely by continuity to an involution in A_ν. The rest is clear. ∎

A standard example of a commutative star multinormed algebra is provided, of course, by the algebra $C(\Delta)$, where Δ is a

---

[3] cf. the result of Exercise IV.4.12.

topological space and the algebra is taken with pointwise operations, with the system of seminorms giving the compact-open topology and with involution defined as taking the complex-conjugate of a function.

Proposition (V.3.42) (cf. IV.6.14). *The following properties of a commutative star Arens-Michael algebra* $(A; \|\cdot\|_\nu;\ \nu \in \Lambda)$, *are equivalent:*

(I) *A is a Hermitian algebra;*

(II) *all Banach algebras concomitant to A are Hermitian;*

(III) *the Gel'fand transformation $\Gamma : A \longrightarrow C(\Omega)$ is a *-homomorphism.*

▶ (I) ⇒ (II). We fix ν and take a self-adjoint $\dot{a} \in A_\nu$. Obviously $\dot{a}$ can be represented as $\lim_{n\to\infty} \dot{a}_n$, where $\dot{a}_n = \tau_\nu(a_n)$; $a_n \in A$; moreover, we can replace a_n, if necessary, by $\frac{1}{2}(a_n + a_n^*)$ and so we are justified in assuming that all the $\dot{a}_n$ are self-adjoint. But then, for all n, by 2.12, $\mathrm{Sp}_{A_\nu}(\dot{a}_n) \subseteq \mathrm{Sp}_A(a_n) \subseteq \mathbb{R}$ and therefore, by Theorem IV.2.3, the function $\Gamma_\nu(\dot{a}_n)$, where Γ_ν is the Gel'fand transformation of the algebra A_ν, takes real values. But Γ_ν being continuous (see IV.2.11) implies that the function $\Gamma_\nu(\dot{a}_n)$ converges uniformly to $\Gamma_\nu(\dot{a})$ as $n \longrightarrow \infty$, and therefore the latter also takes real values. Hence by Theorem IV.2.3, $\mathrm{Sp}_{A_\nu} \dot{a} \subset \mathbb{R}$ and we have checked the definition of a Hermitian algebra.

(II) ⇒ (III). We take $s \in \Omega$ and the corresponding character χ_s on A, then, by Proposition 1.8, $\chi_s = \tilde{\chi} \circ \tau_\nu$ for a certain concomitant homomorphism $\tau_\nu ; A \longrightarrow A_\nu$ and a certain character $\tilde{\chi}$ on A_ν. As A_ν is Hermitian, Theorem IV.6.14 implies that

$$\tilde{\chi}(\dot{b}^*) = \overline{\tilde{\chi}(\dot{b})}$$

for any $\dot{b} \in A_\nu$; hence, taking into account 3.41, for any $a \in A$

$$\chi_s(a^*) = \tilde{\chi}(\tau_\nu(a^*)) = \tilde{\chi}(\tau_\nu(a)^*) = \overline{\tilde{\chi}(\tau_\nu(a))} = \overline{\chi_s(a)}$$

or what is the same

$$\hat{a}^*(s) = \overline{\hat{a}(s)}\,.$$

We have verified that Γ is a *-homomorphism.

(III) ⇒ (I). If $a \in A$ is self-adjoint, then the condition on Γ implies that the set of values of the function $\hat{a}(s)$; $s \in \Omega$, lies in $\mathbb{R}$. All that remains is to recall that the set is precisely $\mathrm{Sp}_A(a)$. ■

Proposition (V.3.43) (cf. IV.6.15). *Let A be a commutative star Arens-Michael Hermitian algebra with spectrum Ω. Then the image of its Gel'fand transformation is dense in $C(\Omega)$.*

▸ In view of the definition of the compact-open topology in $C(\Omega)$, it is sufficient to show that, for any compact set $K \subset \Omega$, the set $\{\Gamma(a)|_K : a \in A\}$ is dense in $C(K)$ with respect to the uniform norm. But Γ, by 3.42, is a *-homomorphism; this immediately implies that the set is a self-adjoint subalgebra of $C(K)$; in addition it separates the points of K and contains the constants. This ensures, by the Stone-Weierstrass theorem (see the end of Chap.IV §6), that the above set is dense in $C(K)$. ■

We are now in a position to extend the first Gel'fand-Naimark theorem to a certain class of algebras beyond that of Banach algebras. Recall (see the beginning of Chap.IV §7) that a multiplicative seminorm $\|\cdot\|$ in the *-algebra A is called a C^*-seminorm if it satisfies $\|a^*a\| = \|a\|^2$; $a \in A$.

Definition (V.3.44). A complete Hausdorff polynormed *-algebra is called a *multi-C^*-algebra* [4] if it has a defining system of C^*-seminorms.

Proposition IV.7.1(II) obviously implies that every multi-C^*-algebra is a star Arens-Michael algebra.

Of course every complete[5] algebra $C(\Delta)$, where Δ is a topological space, is a multi-C^*-algebra with respect to the standard algebraic operations and topology.

Theorem (V.3.45) (cf. IV.7.13). *Let A be a metrizable commutative multi-C^*-algebra with spectrum Ω. Then its Gel'fand transformation $\Gamma : A \longrightarrow C(\Omega)$ is a topological *-isomorphism between these algebras.*

▸ Let $(\|\cdot\|_\nu;\ \nu \in \Lambda)$, be a defining system of C^*-seminorms for A; then, obviously, each of the concomitant Banach *-algebras A_ν is a C^*-algebra. Therefore Proposition IV.7.8 implies that, for any ν and $a \in A$, $\|a^2\|_\nu = \|\tau_\nu(a^2)\|^\wedge_\nu = \|(\tau_\nu(a))^2\|^\wedge_\nu = (\|\tau_\nu(a)\|^\wedge_\nu)^2 = \|a\|^2_\nu$. Hence, by 3.40 ((I) $\Rightarrow$ (III)), Γ is topologically injective. In addition, by combining Theorem IV.7.12 and Proposition 3.42 ((II $\Rightarrow$ (III)), we get that Γ is a *-homomorphism.

All that remains is to show that the image of Γ coincides with the whole of $C(\Omega)$ and for this it is sufficient to verify that it is dense and closed in $C(\Omega)$. Since Theorem IV.7.12 together with Proposition 3.42 ((II) $\Rightarrow$ (I)) ensures that it is Hermitian, 3.43 implies that it is dense. Finally, that the

---

[4] These algebras (under another name) were thoroughly studied in a quite recent paper by Fragoulopoulou [343].

[5] cf. the discussion at the end of Chap.III. §1.2.

image of Γ is closed follows from the fact that it is complete and this is a consequence of the fact that, as we have already proved, A is complete and Γ is topologically injective. ∎

Remark (V.3.46). Regarding a more general (not necessarily metrizable) commutative multi-C^*-algebra, such an algebra A is topologically isomorphic to the algebra $C(\Omega)$ on its spectrum, where this however is given a topology which, generally speaking, is weaker than the compact-open topology: this topology is given by the seminorms $\|\cdot\|_K$ only for those compact sets K that, as sets of characters on A, are equicontinuous [7, p.249 and p.263].

... Informally speaking, the above facts show that in the commutative case multi-C^*-algebras are the "correct" generalization of C^*-algebras: outside the context of the class of Banach algebras they play the same role as the commutative C^*-algebras within this class. For arbitrary, generally speaking, non-commutative, algebras such a view of things would be correct, of course, only if a meaningful analogue were obtained for the second ("non-commutative") Gel'fand-Naimark theorem. Up to now, however, it is not clear what specific polynormed algebras should be taken as the "realized multi-C^*-algebras, that is, should play the role of self-adjoint operator algebras on a Hilbert space in Theorem IV.7.57.

Recall the Lassner algebras (of unbounded operators) defined at the end of Chap.III §2.3; they are directly related to the question under discussion as is shown by the following assertion.

Assertion (V.3.47) (cf. [185, Theorem 1]). *Every barrelled*[6] *multi-C^*-algebra is topologically *-isomorphic to a certain Lassner algebra.*

The converse, however, is not true. Certainly not every Lassner algebra is a multi-C^*-algebra and the problem here is not only just that it may not be complete as in the commutative case (the algebras $C(\Delta)$): for the "majority" of pre-Hilbert spaces $\tilde{H}$ the multiplication in $\mathcal{B}(\tilde{H})$ is not even jointly continuous [184, Theorem 3.2 and Example 4.1].

---

[6] A polynormed space is said to be barrelled if every closed convex, balanced and absorbing subset contains a neighbourhood of zero; in particular, all Fréchet spaces are barrelled (cf. the proof of 3.38).

Chapter VI
Modules, they are also representations

Catherine the Great: "Tell me, General, what exactly is the difference between a howitzer and a mortar?"
Cossack-General Platov: "Oh, you see, Mother-Czarina, a howitzer is a specialized sort of thing and this is a mortar - an altogether specialized sort of thing!"

(From an old Russian anecdote.)

The reader has already been introduced to representations and he can judge from the results of Chap.IV §§5-7 how useful they can be. When we consider the representations of a given algebra it is as though we were taking photographs of it from different sides. The "operator" photos we get (which may also include, as in Theorem IV.7.57, three-dimensional photos) enable us to get a much better understanding of its structure and to form a foundation for its study using results and methods accumulated in operator theory.

This chapter pursues two aims. On the one hand, it will include a considerably wider range of material about representations and about facts in the general theory that are based on them; we are primarily talking about a set of questions connected with the fundamental concept of irreducibility. Here, the material in the chapter, while remaining an organic part of the theory of "the algebras of analysis", simultaneously joins another vast discipline — the theory of group representations ([40], [54], [49], [35], ...), and provides the latter with effective methods for research. In our account, however, we stop at the very threshold of the theory of group representations and restrict ourselves to only a few statements of the more impressive of the "neighbouring" results (cf. later 1.32, 1.34, 2.85).

On the other hand, representations will interest us not only as a means of studying algebras but also as a method of introducing new mathematical structures which are of great interest in their own right — the so-called modules.

A module, as the reader will see, is something like a linear space in which the role of the scalars in the base field is played by the elements of an arbitrary given algebra. It immediately becomes clear from the definition that the concept of a module and the concept of a representation are equivalent; in essence, this is a single concept considered from different points of view. (There is a psychological shift reflected in the notation: instead of $T_a(x)$, we now write $a \cdot x$.) Nevertheless by a certain established tradition, when we speak of modules, it is usual to have in mind one body of examples, exercises, methods and so on, and when we speak of representations, a somewhat different one. There has even appeared two parallel languages: "module", "morphism", "simplicity", ... and at the same time "representation", "intertwining operator", "irreducibility", ... Moreover in some questions (when using the methods of category or homology theory, say) the language of modules is more convenient and natural and in others the language of representations is more suitable; so that we shall use each in turn.[1]

Having introduced the reader to modules let us make the last important remark. In spite of the similarity of their definitions, modules are essentially more complicated than linear spaces. (Running ahead, we can explain the source of this complexity: modules are not necessarily projective in the sense of VII.1.3.) There are new phenomena that occur in modules but are unnatural for linear spaces,[2] and these together with the relevant applications, form part of a special discipline — homology. The latter studies this sort of phenomenon in various mathematical structures from a general point of view; in our case we shall talk about "the homology of the algebras of analysis". The introduction to this branch of homology takes place in Chap. VII.

The abstract algebraic theory of modules over rings is reflected in the monographs [46], [83], [78].

---

[1] The persistence with which we emphasize the connection between modules and representations is not accidental: one finds books in which both concepts appear but where their equivalence is kept secret.

[2] For the reader who is not squeamish about the bracketed sections we can specify that what we are referring to is, generally speaking, a negative solution to the typical questions mentioned in Remark 0.2.6.

§1. LEFT MODULES; INITIAL DEFINITIONS AND EXAMPLES

§1.1. Modules in abstract algebra and their morphisms

Definition (VI.1.1). Let A be an algebra and X a linear space. The bi-operator $\dot{m} : A \times X \longrightarrow X$ is called a *left outer multiplication in* X *by the elements of* A, or a *left action* of A on X if, writing $a \cdot x$ instead of $\dot{m}(a, x)$, it satisfies the identity $(ab) \cdot x = a \cdot (b \cdot x)$. The space X itself, furnished with a left outer multiplication $\dot{m}$ (that is, the pair $(X, \dot{m})$; cf. I.1.1) is called a *left module over* A or more briefly, a *left* A*-module*.

Warning. Throughout this and the next section, when we say "module" we always mean just a left module: other versions of the concept of a module will appear only from §3 on. Then we shall use the word "module" in a wider sense.

If X is an A-module, then, by associating to each $a \in A$ the operator $\dot{m}_a : X \longrightarrow X : x \longmapsto a \cdot x$, we obviously get a homomorphism from A into $\mathcal{L}(X)$, that is, a representation of our algebra; we call it the *representation associated with the module* X. Conversely, if we are given a representation T of the algebra A on the linear space X, then by putting $a \cdot x := T_a(x)$; $a \in A$, $x \in X$, we obviously get the structure of an A-module in X — the so-called *module associated with the representation* T. Moreover, it is easy to see that for given A and X, the freezing isomorphism $\mathcal{L}(A \times X, X) \xrightarrow{\simeq} \mathcal{L}(A, \mathcal{L}(X))$ (see, Chap.0 §3.1) realizes a bijection between the set of all left actions of A on X and the set of all representations of A on X.

Definition (VI.1.2). Let A have an identity. Then the A-module X is said to be *unital* if $e \cdot x = x$ for all $x \in X$ (in other words, if the representation associated with this module is a unital homomorphism).

Obviously, an arbitrary A-module X may be identified with a unital module over A_+ if we put $(\lambda e + a) \cdot x := \lambda x + a \cdot x$; $\lambda \in \mathbb{C}$, where A_+ is the unitization of A in the sense of Chap.I §1.2. The A_+-module we get is, in turn, called the unitization of the original A-module.

Let X be an A-module. A subspace Y of X such that $a \cdot x \in Y$ for all $x \in Y$ and $a \in A$ is called a *submodule of* X; this means precisely that Y is an invariant subspace for all the operators of the representation associated with X. Of course Y itself is an A-module with respect to the given action of the algebra A. A submodule of X not coinciding with the whole of X is said to be *proper*.

If Y is a submodule of the A-module X, then the quotient space X/Y itself is obviously an A-module with respect to the outer multiplication, well defined by $a\cdot(x+Y) := a\cdot x + Y$. This module is called the *quotient module of* X *with respect to* Y.

Finally the direct sum and the Cartesian product of any family of A-modules, considered as linear spaces, are themselves A-modules with respect to the obvious coordinatewise operations.

The following maps play the same role for modules as operators do for linear spaces.

Definition (VI.1.3). Let X and Y be A-modules. The map $\varphi : X \longrightarrow Y$ is called a *morphism of* A-*modules* if it is a linear operator and, in addition, $\varphi(a\cdot x) = a\cdot\varphi(x)$ for all $a \in A$, $x \in X$.

It is obvious that, in the case of unital modules, the homogeneity of φ is itself a consequence of the last equation.

The set of A-module morphisms from X to Y is denoted by ${}_{A}h(X, Y)$, where we write ${}_{A}h(X)$ instead of ${}_{A}h(X, X)$. Obviously ${}_{A}h(X, Y)$ is a subspace of $\mathcal{L}(X, Y)$ and ${}_{A}h(X)$ is a subalgebra of $\mathcal{L}(X)$.

If $T^{(1)}$ is the representation associated with X and $T^{(2)}$ is that associated with Y, then a morphism φ is also called an *intertwining operator* (of these representations); in terms of representations this means that $\varphi \circ T^{(1)}_{a} = T^{(2)}_{a} \circ \varphi$ for all $a \in A$.

Of course, the natural embedding of a submodule in a module and the natural projection of a module onto a quotient module are module morphisms. We also note that, if $\varphi : X \longrightarrow Y$ is a morphism, $\operatorname{Ker}\varphi$ is a submodule of X and $\operatorname{Im}\varphi$ is a submodule of Y.

Modules over a fixed algebra A and their morphisms, together with the law of composition of morphisms as maps, form a category denoted by A**-algmod**. This category is additive[3] (watch out! this does not hold for the category of algebras); in particular, the direct sum of two of its objects is, as an object, their Cartesian product. When we talk about *isomorphic modules* we always have in mind isomorphisms in A**-algmod**; obviously, these are precisely the bijective morphisms. Representations associated with isomorphic modules are usually said to be *equivalent*.

When A is unital, the category A**-algmod** contains a complete subcategory consisting of unital modules; the latter is denoted by A**-algunmod**.

---

[3] And even — at the moment we are considering the purely algebraic situation — Abelian; for Abelian categories see, for example, [51], [6].

If M is a subset of the A-module X, then the set $\{a \in A : a \cdot x = 0$ for all $x \in M\}$ is called its *annihilator* and it is denoted by $\mathcal{A}nn_A M$. Obviously this is a left ideal of A and is a bi-ideal if M is a submodule; in particular, $\mathcal{A}nn_A X$ is the kernel of the representation associated with X. If X is such that $\mathcal{A}nn_A X = \{0\}$, then it is called a *faithful* module; if, conversely, $\mathcal{A}nn_A X = X$, then X is called an *annihilator* (or, more rarely, a trivial) module.

Exercise (VI.1.4). Let A be unital. Show that any A-module has a unique decomposition into the direct sum of unital and annihilator A-modules.

Hint. Consider the submodules $\{x : e \cdot x = x\}$ and $\{x : e \cdot x = 0\}$.

A subset M of the A-module X is called a *generating set* and its elements are called the *generators* of this module if $\ell.h.\{a \cdot x : a \in A,\ x \in M\} = X$. If M has this property, it obviously means that any submodule of X that contains M must be X itself.[4] A module having a finite set of generators is called *finitely-generated.*

* * *

The simplest example of a module over A is provided by A itself; another simple example is any algebra containing it, in particular, A_+. Further any left ideal in A or in A_+ is a module over A. In these examples the inner multiplication plays the role of the outer multiplication.

Obviously, the left ideals in A (A_+) are precisely the submodules of the A-modules A (A_+). Concerning the quotient modules of such modules, the two following observations will prove useful to us later on.

Proposition (VI.1.5) (cf. I.3.33). *Let I be a proper left ideal in A. Then there is a bijection between the set of submodules of A/I and the set of left ideals in A containing I which takes every submodule into its preimage under $\mathbf{pr} : A \longrightarrow A/I$. As a consequence I is maximal $\Leftrightarrow$ there are no submodules in A/I except itself and $\{0\}$.* ▶■

Proposition (VI.1.6). *Let $X = A/J$, where J is a left ideal in A. Then*

(I) $\mathcal{A}nn_A X = \{a \in A : ab \in J$ *for all* $b \in A\}$;

(II) *if J is modular, then* $\mathcal{A}nn_A X \subseteq J$.

▶ The identity $a \cdot (b + J) = ab + J$; $a, b \in A$, which holds in X immediately implies (I). Hence, if u is a modular

---

[4] Note that if A is unital, any A-module having a generating set is necessarily unital.

identity for J, then $au \in J$ for any $a \in \mathcal{A}nn_A X$ and therefore $a = au + (a - au) \in J$. ■

We note that the representation associated with the A-module A is, obviously, none other than the regular representation of our algebra.

Another class of elementary examples occurs when we move away from the A-module A in a different direction. First of all we note that every linear space is, of course, a unital module over $\mathbb{C}$ (and a module over $\mathbb{O}$). Concerning unital modules over an arbitrary unital algebra A, then those most like linear spaces are the modules of the form $\bigoplus_{\Lambda} A$ which is what we use to denote the direct sum of a family of copies of the A-module A indexed by the set Λ. In particular the module A^n — the direct sum of n copies of the module A is one of these. Modules of the form $\bigoplus_{\Lambda} A$ and all modules isomorphic to them are said to be *free* (in the purely algebraic sense; for Banach modules a somewhat different definition of "free" is required; see 1.21 later). What was said above about them being like vector spaces is clarified by the following two observations.

Exercise (VI.1.7). Show that a unital A-module X is free $\Longleftrightarrow$ it has a *basis*, that is a family of elements $\{x_\nu;\ \nu \in \Lambda\}$ having either of the following two equivalent properties:

(I) any $x \in X$ can be represented, uniquely, as $x = \sum_{k=1}^{n} a_{\nu_k} \cdot x_{\nu_k}$; $\nu_k \in \Lambda$, $a_{\nu_k} \in A$, $1 \le k \le n$;

(II) for any unital A-module Y and any map between sets $\omega : \{x_\nu\} \longrightarrow Y$ there exists a unique A-module morphism $\varphi : X \longrightarrow Y$ such that $\varphi(x_\nu) = \omega(x_\nu)$ for all $\nu \in \Lambda$.

Proposition (VI.1.8). *A unital A-module X is free $\Longleftrightarrow$ it is isomorphic to an A-module of the form $A \otimes E$, where E is a linear space, with outer multiplication uniquely determined by $a \cdot (b \otimes x) = ab \otimes x$; $a, b \in A$, $x \in E$.*

▸ The module $\bigoplus_{\Lambda} A$ is obviously isomorphic to the module $A \otimes E$, where E is the space of formal linear combinations of the elements from Λ with complex coefficients. Conversely, if we are given the module $A \otimes E$, where E is a linear space with basis $\{x_\nu, \nu \in \Lambda\}$, then this module is obviously isomorphic to $\bigoplus_{\Lambda} A$. ■

If the algebra A is not necessarily unital, then the A-modules of the form $\bigoplus_{\Lambda} A_+$ and all modules isomorphic to them are called *free A-modules* or *free objects in* A-**algmod**. The obvious analogues of Exercise 1.7 and Proposition 1.8 hold for such modules if we replace A by A_+; in particular, they are described, up to an isomorphism, as A-modules of the form $A_+ \otimes E$; $E \in \mathcal{L}in$. We emphasize that, whether unital modules are "free" or not depends, by definition, on which category they are considered to be in; for example, the unital algebra A is free as a unital module (= an object in **A-algunmod**) but is not free as "simply" a module (= an object in **A-algmod**).

Running on ahead, we note that the complexity of modules compared with linear spaces, which we talked about in the introductory part of the chapter, has its origin in the fact that modules over an arbitrary algebra are not necessarily free. The following, in particular, tells us this.

Exercise (VI.1.9). Show that every unital module over a unital algebra A is free (in **A-algunmod**) $\Longleftrightarrow$ A is a skew field.

Hint. If A is a skew field, then the existence of a basis in the A-modules can be proved by a standard argument based on Zorn's lemma. If A is not a skew field, then the module A/I, where I is a maximal left ideal in A (cf. I.3.20), cannot contain a submodule isomorphic to A and hence it has no basis.

Apart from skew fields, there are other algebras for which one can give a clear description of the modules over them.

Example (VI.1.10). Giving the action of the polynomial algebra $\mathcal{P}_n$ on a space X is equivalent to giving a collection of n commuting operators. In fact, if we let $p(S_1, \cdots, S_n)$ denote the result of substituting these operators in place of the independent variables in the expression for $p \in \mathcal{P}_n$, we get $p \cdot x = p(S_1, \cdots, S_n)(x)$ for $x \in X$; moreover the operators $S_1, \cdots, S_n$ turn out to correspond to outer multiplication by the independent variables.

Analogously, giving the action of the group algebra of the group $\mathbb{Z}^n$ on X (see Chap.I §1.1) is equivalent to giving a collection of n invertible commuting operators. The following observation is a generalization of this example. Recall that any homomorphism of G into the group of invertible operators on X is called a *representation of the group G on the linear space X*. Let $\widetilde{T} : s \longmapsto \widetilde{T}_s$ be such a representation. Then for $a = \sum_{k=1}^{n} \lambda_k s_k \in \mathbb{C}(G)$; $s_k \in G$, $\lambda_k \in \mathbb{C}$ and $x \in X$ we put $a \cdot x = \sum_{k=1}^{n} \lambda_k \widetilde{T}_{s_k}(x)$; it is obvious that in this way X is given the structure of a $\mathbb{C}(G)$-module. Conversely, if the $\mathbb{C}(G)$-module X is given, then if we restrict the representation of the algebra $\mathbb{C}(G)$ associated with this module to $G \subset \mathbb{C}(G)$, we obviously get a representation of the group G on X. Thus giving a representation of a group is equivalent to giving a module over its group algebra and studying the first reduces completely to studying the second.

(As group algebras often turn out to have a very nice structure, this information provides an effective method for studying group representations; for the role they play in finite groups, say, see [71], [45] for example.)

§1.2. The modules of analysis; first definitions. Free Banach modules

Definition (VI.1.11). Let A be a Banach algebra. A Banach space X furnished with a bi-operator representing left outer multiplication by the elements A (that is, the structure of a left A-module) is called a *left Banach* A*-module* if this bi-operator is continuous.

As in the case of algebras (cf. Chap.I §2.1) we have selected, in view of its importance, a special case of the following more general definition.

Definition (VI.1.12). Let A be a polynormed algebra. A polynormed space X furnished with a bi-operator representing left outer multiplication by elements from A is called a (left) *polynormed* A*-module* if this operator is separately continuous.

In future, when we talk about a "Banach (polynormed, respectively) A-module" we always mean that A is a Banach (polynormed) algebra.

It is appropriate to consider not only Banach modules and arbitrary polynormed modules, but also a number of intermediate classes, including those of Fréchet modules and jointly polynormed modules (cf. I.2.18). We leave the detailed definitions, which are completely analogous to 1.11 and 1.12, to the reader.

Let us consider what are the concepts that correspond to Banach and polynormed modules in the language of representations. In order to avoid confusion in the terminology, we shall say that a representation of the algebra A (for the moment arbitrary) on a polynormed space X is *topological* if its image consists only of continuous operators. In that case we shall identify our representation with its corestriction to $\mathcal{B}(X) \subseteq \mathcal{L}(X)$ and use the notation $T : A \longrightarrow \mathcal{B}(X)$. Clearly topological representations are precisely those that are associated with those modules that have an outer multiplication that is continuous with respect to the second argument; hence, in particular, the representation associated with a polynormed module is automatically topological.

Proposition (VI.1.13). *Let A be a Banach algebra and let X be a Banach space where, in addition, X is a left A-module in the purely algebraic sense. Then this module is a Banach module $\Longleftrightarrow$ its associated representation is topological and continuous (as a homomorphism from A into $\mathcal{B}(X)$) with respect to the operator norm in $\mathcal{B}(X)$.*

▶ This follows from the fact that, for a Banach module, both the bi-operator representing outer multiplication (Theorem 0.3.4) and the appropriate version of the freezing isomorphism are bounded (see the end of Chap.0 §3.1). ■

We offer the corresponding "polynormed" result as an exercise for the reader.

Exercise (VI.1.14). Let A be a polynormed algebra and let X be a polynormed space where, in addition, X is an A-module in the purely algebraic sense. Check that this is a polynormed module $\Leftrightarrow$ its associated representation is topological and continuous with respect to the strong operator topology in $\mathcal{B}(X)$.

Hint. Use the result of Exercise 0.3.7.

In connection with the concept of a topological representation, recall that we have already met one special class of these: namely *-representations of *-algebras introduced in Chap.IV §4.1. The following definition obviously describes the modules associated with *-representations of star Banach algebras that are continuous with respect to the operator norm in $\mathcal{B}(H)$.

Definition (VI.1.15). Let A be a star Banach algebra. A Hilbert space H with the structure of a Banach A-module is called a *star A-module* if

$$\langle a\cdot x, y\rangle = \langle x, a^*\cdot y\rangle; \quad a\in A, \quad x, y\in H.$$

is satisfied.

* * *

Starting from the appropriate concept for abstract modules (= modules in the purely algebraic sense) on the one hand and for polynormed spaces on the other, one can obviously define a submodule, a quotient module and the Cartesian product of modules for the "algebras of analysis" introduced above. We shall use these concepts freely later. In particular, if the class of Banach modules contains a module, it also contains all its closed submodules and all its quotient modules with respect to closed submodules, and if it contains a finite (but only finite!) family of non-zero modules it also contains their Cartesian product.

Let X be a polynormed A-module. For any subset M of A or A_+ we put $M\cdot X := \overline{\ell.h.}\{a\cdot x : a\in M, x\in X\}$. As we shall see later, the submodule $A\cdot X$, also denoted by X_{es}, plays a very prominent role in a number of questions; it is called the *essential submodule* or the *essential part* of X. A module that coincides with its essential part is said to be *non-degenerate* or (more rarely) *stable*; these include, of course, all unital modules. The representation associated with a non-degenerate module is also said to be *non-degenerate*. Finally a polynormed algebra is said to be *non-degenerate* or *idempotent* if the A-module A is non-degenerate (= the regular representation of this algebra is non-degenerate); obviously this means precisely that $A^2 = A$.

Exercise (VI.1.16). Let A be a Banach algebra with a left b.a.i. $(e_\nu;\ \nu \in \Lambda)$, and let X be a left Banach A-module. Prove that $X_{es} = \{x \in X : x = \lim_\nu e_\nu \cdot x\}$.

We note that under the conditions of this exercise every $x \in X_{es}$ can be represented as $a \cdot y$ for some $a \in A$ and $y \in X_{es}$. This is the so-called the Cohen-Hewitt factorization theorem generalizing Cohen's theorem mentioned earlier (I.2.38); for a proof of this see, for example, [9, §11].[5]

While we are talking about submodules, we shall in passing indicate one of the advantages of star modules in comparison with more general modules.

Proposition (VI.1.17). *The orthogonal complement of any closed submodule of a star module (over a star Banach algebra) is also a submodule.*

▶ Let A, H and H_0 be the given algebra, module and submodule. We take $x \in H_0^\perp$; then, for any $a \in A$ and $y \in H_0$, $a^* \cdot y \in H_0$ implies that $\langle a \cdot x, y\rangle = \langle x, a^* \cdot y\rangle = 0$. Hence $a \cdot x \in H_0^\perp$. ■

* * *

The maps considered compatible with the above structures are those module morphisms that are simultaneously continuous maps or, more briefly, — *continuous morphisms*. These include, in particular, any natural embedding of a submodule in a module and any natural projection of a module onto a quotient module.

The set of continuous morphisms between the polynormed A-modules X and Y is denoted by ${}_A\mathbf{h}(X, Y)$. Obviously ${}_A\mathbf{h}(X, Y)$ is a subspace of $\mathcal{B}(X, Y)$ as well as of ${}_Ah(X, Y)$, and ${}_A\mathbf{h}(X)$ — which we write instead of ${}_A\mathbf{h}(X, X)$ — is a subalgebra of $\mathcal{B}(X)$ and of ${}_Ah(X)$. In the case of Banach modules ${}_A\mathbf{h}(X, Y)$ is obviously a Banach space and ${}_A\mathbf{h}(X)$ is a Banach algebra with respect to the operator norm. Here ${}_A\mathbf{h}(X)$ is, of course, precisely the set of continuous operators on the space X that commute with all the operators of the representation T associated with our module; in future, when we are using the language of representations, we shall write $(\operatorname{Im} T)^!$ or, more briefly, $T^!$ instead of ${}_A\mathbf{h}(X)$ (cf. the notation in Chap.III §2.3).

---

[5] This result was also obtained by Curtis and Figa-Talamanca [314] independently of Hewitt [279].

If we take some class or other of polynormed modules over a fixed algebra as the objects and continuous module morphisms as the morphisms, we get a number of categories. The most important of them for us is the category of left Banach modules over a Banach algebra A, denoted by **A-mod** and its "unital" version which is the category of unital Banach modules over the unital Banach algebra A and is denoted by **A-unmod**. (We note that to within a category isomorphism **$\mathbb{O}$-mod** = **$\mathbb{C}$-unmod** = $\mathcal{B}an$.) We shall meet other categories, for example, the category of all polynormed modules over a polynormed algebra A. When A is an algebra with strongest seminorm system, this last category contains a complete subcategory, naturally identified with **A-algmod**: it is formed from the modules with strongest seminorm system; cf.I.2.17.

It is obvious that all these categories of polynormed modules are additive.[6]

The isomorphisms between objects of the categories under consideration are called *topological isomorphisms* (of polynormed modules); as a consequence of the Banach theorem these include all continuous bijective morphisms of Banach modules. An (iso)morphism between Banach modules which preserves the norm of the elements is said to be *isometric*.

Recall that such terms have already been used by us first for spaces and then for algebras. So when we talk about topological isomorphisms and isometric morphisms, it is necessary, generally speaking, to be precise as to what structures we are talking about; however, this is usually clear from the context.

Finally, definitions and constructions concerning complements (Chap.0 §1.4) can obviously be carried across from spaces to modules. In particular, a (necessarily closed) submodule X_0 of a polynormed module X over a polynormed algebra A is said to be *complemented as a module* or, briefly, is a *complemented module* if there exists another submodule X_1 of X such that the operator $\kappa : X_0 \times X_1 \longrightarrow X : (x, y) \longmapsto x + y$ is a topological isomorphism of polynormed modules. We obviously have the following.

Proposition (VI.1.18). *X_0 is a complemented module* $\Leftrightarrow$ *it has a topological complement* (= a complement in the sense of spaces) *which is a submodule of X.* ▶■

We also note that the criterion in 0.1.14 for complementability, when translated from spaces to modules is as follows:

Proposition (VI.1.19). *A submodule X_0 of a polynormed A-module X is a complemented module* $\Leftrightarrow$ *the natural embedding*

[6] But, in contrast to **A-algmod**, not Abelian; cf. footnote just after Definition VI.1.3.

in $: X_0 \longrightarrow X$ *has a left inverse which is a continuous morphism of* A*-modules* $\Longleftrightarrow$ *the natural projection* $\mathbf{pr}: X \longrightarrow X/X_0$ *has a right inverse which is a continuous morphism of* A*-modules.*

▶ The proof of Proposition 0.1.14 carries over with the obvious modifications. ■

* * *

Of course the simplest examples of modules — such as taking an abstract algebra with outer multiplication reducing to the inner multiplication — obviously have "polynormed" analogues. Thus, for example, a closed left ideal of a Banach algebra A is itself a (left) Banach A-module and so on.

Another important class of polynormed modules is formed by the so-called free modules. We basically concentrate on free Banach modules.

First recall the purely algebraic definition of "free" (in two versions — unital and general) as in §1.1. If we try to carry it across to Banach modules, we come up against the obvious inconvenience that the direct sum $\bigoplus_{\Lambda} E$, where E is a Banach space, has a natural Banach space structure only if Λ is finite. To obtain the "correct" definition giving a fairly wide class of modules one must take another route (which might be suggested by Proposition 1.8).

Again, as in the purely algebraic context, it is necessary to distinguish the unital and the general case; for the moment we deal with the first. Let A be a unital Banach algebra and let E be a Banach space.

Proposition (VI.1.20). *There is the structure of a unital left Banach* A*-module in the Banach space* $A \hat{\otimes} E$*; this is uniquely determined by* $a \cdot (b \otimes x) := ab \otimes x$*;* $a, b \in A$*,* $x \in E$*.*

▶ Given $a \in A$ we consider the continuous bi-operator $\mathcal{R}_a : A \times E \longrightarrow A \hat{\otimes} E : (b, x) \longmapsto ab \otimes x$ and denote the operator associated with it by $\dot{m}_a$. All we have to do is to put $a \cdot u := \dot{m}_a(u)$ for $u \in A \hat{\otimes} E$. The rest is clear. ■

Definition (VI.1.21). Banach A-modules of the form $A \hat{\otimes} E$; $E \in \mathcal{B}an$ and all A-modules topologically isomorphic to these are called *free unital Banach* A*-modules* or the *free objects in* A**-unmod**.

The outstanding role of the modules we have introduced is based on the following characteristic property.

Theorem (VI.1.22). *A Banach* A*-module* X *is free* (as an object in A**-unmod**) $\Longleftrightarrow$ *it has a closed subspace* F *such that for any continuous operator* $\xi : F \longrightarrow Y$*, where* Y *is an arbitrary unital*

Banach A-module, there is a unique continuous Banach A-module morphism $\varphi : X \longrightarrow Y$ *such that* $\varphi|_F = \xi$.

▶ "⇒". Given the module $A \hat{\otimes} E$ as in Definition 1.21, we put $F = \{e \otimes x : x \in E\}$. Let $\xi : F \longrightarrow Y$ be the given operator; then the operator $\varphi : A \hat{\otimes} E \longrightarrow Y$, associated with the bi-operator $(a, x) \longmapsto a \cdot \xi(e \otimes x)$ is, as is easily verified, a Banach A-module morphism that extends ξ and, moreover, it is unique.

"⇐". For a given X and F there is, by hypothesis, a morphism $\varphi : X \longrightarrow A \hat{\otimes} F$ extending the operator $\xi : F \longrightarrow A \hat{\otimes} F : x \longmapsto e \otimes x$. At the same time, by what has already been proved, there exists a morphism $\psi : A \hat{\otimes} F \longrightarrow X$ uniquely determined by $\psi(e \otimes x) := x$; $x \in F$. Hence $\psi \circ \varphi : X \longrightarrow X$ is a morphism that is the identity on F. But, by hypothesis, such a morphism is unique and therefore coincides with 1_X; similarly $\varphi \circ \psi$ is the identity on $A \hat{\otimes} F$. The rest is clear. ■

The subspace F of X having the property indicated in the theorem about "establishing a unique morphism from an operator" is called the *base space* of the free module X. The reader may compare this property with the property of "establishing a morphism from a map" possessed by the basis of a free abstract module in Exercise 1.7(II).

If A is not necessarily unital, then the free objects in A-**mod** (*free "general" A-modules*) are defined as those that are topologically isomorphic to the modules $A_+ \hat{\otimes} E$, $E \in \mathcal{Ban}$ — in other words, as the free unital modules in A_+-**unmod** after identifying this category with A-**mod**. We suggest that the reader formulates and proves the analogue to Theorem 1.22 for such modules.

The concept of "free" naturally generalizes to other categories of polynormed modules besides Banach modules. For example, the free objects in the category of unital left Fréchet modules over a unital Fréchet algebra A (briefly — free unital Fréchet A-modules) are defined as those that are topologically isomorphic to modules of the form $A \hat{\otimes} E$, or, what is equivalent here, $A \overline{\otimes} E$, where $E \in \mathcal{Fr}$. The free objects in the category of all unital left polynormed modules over an (arbitrary unital) polynormed algebra A are defined as those that are topologically isomorphic to the modules $A \otimes_i E$, $E \in \mathcal{PN}$; moreover, if we take the strongest systems of seminorms in A and E we get the purely algebraic definition of a free module in §1.1. It is not difficult to see that such classes of modules have an intrinsic characterization analogous to that established in Theorem 1.22 for Banach unital modules. A fuller discussion of these questions is given in [81, Chap.III].

§1.3. The modules of analysis; typical examples. Remark on the connection between group representations and representations of their group algebras

Without exaggeration it can be said that modern analysis (as also algebra) "is swarming with modules"; one only needs always to call things by their proper names. We now indicate a few typical examples of modules over various algebras with varying degrees of generality.

We begin with function algebras. Let A be a certain function algebra on the set Ω. The first group of examples occurs thus: if Δ is a subset of Ω, then many spaces of functions on Δ are A-modules with respect to pointwise outer multiplication $[a\cdot x](t) := a(t)x(t)$, where x is a function on Δ, $a \in A$, $t \in \Delta$.

Example (VI.1.23). Let Ω be a topological space and let Δ be a subset of Ω. Then $C(\Delta)$ is a polynormed module over $C(\Omega)$ (with the compact-open topology). If Ω is locally compact, then $C_b(\Delta)$ (see Chap.III §1.1) is a Banach $C_0(\Omega)$-module; we note that the essential part of the module $C_b(\Omega)$ is $C_0(\Omega)$. Finally, if Ω and Δ are compact, the previous module becomes the $C(\Omega)$-module $C(\Delta)$; it is not difficult to discover that it is isometrically ismorphic to $C(\Omega)/I$, where $I = \{f \in C(\Omega) : f = 0 \text{ on } \Delta\}$.

The following is constructed in a similar way.

Example (VI.1.24). Let $\mathcal{M}$ be an infinitely smooth manifold and let U be an open subset of $\mathcal{M}$. Then $C^\infty(U)$ is a Fréchet module over the Fréchet algebra $C^\infty(\mathcal{M})$ (see Chap.III §1.3). Some interesting applications of this module are indicated in [203].

It is worth recalling certain Banach modules connected with special spaces Ω such as $[0, 1]$ or $\mathbb{N}$; we name among them the $C[0, 1]$-module $C[0, \frac{1}{2}]$, the $C[0, 1]$-module $L^1[0, 1]$, the l_p-module l_q; $1 \le p < q \le \infty$ [289].

Thanks to the Gel'fand transformation we can, when A is an arbitrary commutative Banach algebra, introduce a whole series of modules similar in structure to those indicated above. In the next example Ω is the spectrum of this algebra.

Example (VI.1.25). The space $C_0(\Omega)$ is a Banach A-module with outer multiplication $[a\cdot f](s) := \hat{a}(s)f(s)$; $s \in \Omega$. The A-module $C_b(\Delta)$ is defined in a similar way for $\Delta \subseteq \Omega$.

The various spaces of vector functions[7] constitute another

---

[7] And also — more generally — cross-sections of vector bundles; see, for example, [4].

source of examples of modules over function algebras.

Example (VI.1.26). Let Ω be a compact space and let E be a Banach space. The space $C(\Omega, E)$ of continuous E-valued functions on Ω with norm $\|\bar{u}\| := \max\{\|\bar{u}(t)\| : t \in \Omega\}$ is a Banach $C(\Omega)$-module with respect to the outer multiplication $[f \cdot \bar{u}](t) := f(t)\bar{u}(t)$; $f \in C(\Omega)$, $\bar{u} \in C(\Omega, E)$, $t \in \Omega$.

Example (VI.1.27). Let U be a domain in $\mathbb{C}^n$, and let $(E; \|\cdot\|_n; n = 1, 2, \cdots)$ be a Fréchet space. Then the space $\mathcal{O}(U, E)$ of holomorphic E-valued functions on U (see Chap.0 §3.5) is a Fréchet $\mathcal{O}(U)$-module with action analogous to that considered in Example 1.26. See [252] and [271] for some important applications of the module $\mathcal{O}(U, E)$; we only note that the class of these modules and those topologically isomorphic to them is precisely the class of free unital Fréchet modules over $\mathcal{O}(U)$ in the sense of the definition given at the end of §1.2 (cf. Exercise 0.3.54).

Modules of smooth vector functions over $C^\infty(\mathcal{M})$ are introduced in a similar way (cf. Example 1.24); we leave the details of the definition to the reader.

... Now let A be a certain Banach operator algebra on a Banach space X (see the beginning of Chap.III §2).

Example (VI.1.28). X itself, taking as outer multiplication the action of an operator on a vector $(a \cdot x := a(x))$, is, of course, a Banach A-module which — for understandable reasons — is called the *natural module over* ("its operator algebra") A.

Exercise (VI.1.29). Let A be $\mathcal{B}(X)$ or $\mathcal{K}(X)$, $x \in X \setminus \{0\}$, $I = \{a \in A;\ a(x) = 0\}$. Show that X, as a natural A-module, is isometrically isomorphic to the quotient module A/I.

Hint. Consider the morphism $A \longrightarrow X : a \longmapsto a(x)$.

Exercise (VI.1.30). Let X be a polynormed space and let $\mathcal{B}(X)$ be furnished with some structure of a polynormed algebra. Check that X is a polynormed $\mathcal{B}(X)$-module (with action $a \cdot x := a(x)$) $\Longleftrightarrow$ the topology in $\mathcal{B}(X)$ is not weaker than the strong operator topology (cf. Exercise 1.14). As a consequence ascertain that the Hilbert space H is a polynormed $\mathcal{B}(H)$-module when $\mathcal{B}(H)$ has the strong and the ultrastrong but not the weak and the ultraweak operator topologies (see Chap.III §2.3).

... Let us turn to modules over the group algebra $L^1(G)$ of a locally compact group G. Besides this algebra itself and the measure algebra ambient to it, there are many other function spaces on G that have the structure of Banach $L^1(G)$-modules, where one takes as outer multiplication some version or other of convolution. Some of these modules will be defined later (Example 3.3) when they will appear more naturally against the background of a certain general construction. We now select a somewhat different example.

Example (VI.1.31). For any G the complex plane $\mathbb{C}$ is a Banach module over $L^1(G)$ with action $f \cdot z := \left(\int_G f(s)ds\right) z; \quad z \in \mathbb{C}$. This $L^1(G)$-module is said to be an *augmentation* module; it is connected with its algebra by the so-called *augmentation morphism* $\tau : L^1(G) \longrightarrow \mathbb{C} : f \longmapsto \int_G f(s)ds$ whose kernel is obviously the *augmentation ideal* I *in* $L^1(G)$ (see III.3.6). This immediately implies that $\mathbb{C}$ is isomorphic in $L^1(G)$-**mod** to the quotient module $L^1(G)/I$. In particular, $\mathbb{C}$ is isomorphic to $l_1(G)/I$ for discrete G where $I = \{f : \sum_{s \in G} f(s) = 0\}$.

One of the most important stimuli to studying modules over group algebras is their close connection with the theory of group representations. This sort of question really belongs to the theory of representations and a detailed account of it goes beyond the context of our book; we recommend to the reader [31, §13], [40, §10], [87, §§21–22] and [53, §29]. Descriptions of all the central results about this connection, in their "Banach" and "Hilbert" versions, deserve at least to be stated. To formulate them it is convenient for us now to cross over from the language of modules to the equivalent language of representations.

Let G be a topological group and let E be a Banach space. A group homomorphism $\check{T} : G \longrightarrow \operatorname{Inv} \mathcal{B}(E)$ (we shall write $\check{T}_s$ instead of $\check{T}(s)$) is called a *bounded representation of* G *on* E if it is continuous with respect to the strong operator topology in $\mathcal{B}(E)$ and its image is bounded in the sense of the operator norm. A bounded representation of the group G in a Hilbert space is said to be *unitary* if all the operators in its image are unitary.

Assertion (VI.1.32) (cf. [30, Proposition 2.1]). *Let* G *be a locally compact group and let* $\check{T}$ *be a bounded representation of* G *on the Banach space* E. *Then there exists a non-degenerate continuous*[8] *representation of the algebra* $L^1(G)$ *on* E, *defined by* $T_f(x) := \int_G f(s)\check{T}_s(x)ds; \quad f \in L^1(G), \quad x \in E$. *Moreover the correspondence* $\check{T} \longmapsto T$ *is a bijection between the set of all bounded representations of the group* G *on* E *and the set of all non-degenerate continuous representations of the algebra* $L^1(G)$ *on* E. *Finally* $\check{T}$ *can be recovered from* T *by using* $\check{T}_s(x) = \lim_\nu T_{s^{-1}(e_\nu)}(x)$, *where* $(e_\nu;\ \nu \in \Lambda)$, *is an arbitrary*

[8] As a Banach algebra homomorphism.

b.a.i. *in* $L^1(G)$ *and* $^{s^{-1}}(\cdot)$ *is the notation for the left shift by* s^{-1}; $s \in G$, $x \in E$.

(For the integrals of vector functions appearing here see, for example, [13].)

Exercise (VI.1.33). Find the (simplified) form that this assertion takes in the case when G is discrete (= when $L^1(G) = l_1(G)$) and give a suitable proof.

Hint. The vector integral in Assertion 1.32 becomes the sum of an absolutely convergent series in E.

For applications, the "Hilbert" special case of Assertion 1.32, enriched by the involvement of another kind of group algebra, is most important.

Assertion (VI.1.34). *Let G be a locally compact group and let $\check{T}$ be a unitary representation of G on a Hilbert space H. Then there exists a non-degenerate *-representation T of the *-algebra $L^1(G)$ on H defined by the equation for T_f given in Assertion* 1.32 *and also* (as a consequence of IV.7.63–66) *a non-degenerate *-representation T' of the *-algebra $C^*(G)$ on H that coincides with T on $L^1(G) \subseteq C^*(G)$. Moreover the correspondence $\check{T} \longmapsto T$ $(\check{T} \longmapsto T')$ is a bijection between the set of all unitary representations of the group G on H and the set of all non-degenerate *-representations of the *-algebra $L^1(G)$ (*-algebra $C^*(G)$, respectively) on H.*

Remark (VI.1.35). Besides the bijections indicated in 1.32 and 1.34, there also exists a natural injection from the bounded representations of the group G into the continuous representations of its measure algebra $M(G)$ (this also comes in "Banach" and "Hilbert" versions; see [30, §2] and for example [31, §13]).[9] If G is, in addition, a Lie group, then its representations are also related, on this occasion by a very interesting method, to those of other group algebras — generally speaking, not Banach — which reflect its additional structure (cf. Chap. III, end of §3.3). For this see, for example, [40, §10.5].

* * *

Recently, a special class of Banach modules over C^*-algebras — the so-called *Hilbert C^*-modules* — have acquired great popularity among specialists in operator theory and algebraic topology. Such an A-module X is by definition furnished with the additional structure of an "A-valued scalar product" — a map $X \times X \longrightarrow A$ with properties similar to the usual ($\mathbb{C}$-valued)

[9] As we saw in Chapter III §3.3, when discussing the characters of $M(G)$, this injection is not usually a bijection.

scalar product, the only difference being that going from λ to $\bar{\lambda}$ in $\mathbb{C}$ is replaced in the corresponding axioms by going from a to a^* in A. In the commutative case $A = C_0(\Omega)$ (IV.7.13) a Hilbert C^*-module over A from a fairly wide class may be identified with a vector bundle over Ω in which the fibres are Hilbert spaces. This leads to the point of view that a general Hilbert C^*-module is the "correct" non-commutative generalization of a vector bundle; such a philosophy (cf. IV.7.17) has turned out to be extremely fruitful in a number of questions. For Hilbert C^*-modules and their applications see [207], [145], [154], [198], [245], [246].

§2. IRREDUCIBILITY

In this section we concentrate on studying "modules with no submodules" — the so-called irreducible modules. These form one of the most important classes of modules and in questions relating to applications in physics it is undoubtedly the most important.

Irreducible modules are nice in many respects. In our account we limit ourselves to what, in our opinion, constitutes their three main features.

Firstly, irreducible representations (= those associated with irreducible modules) being "homomorphisms into the simplest objects" play the same role, when studying non-commutative algebras, as that of characters in the theory of commutative Banach algebras (cf., for example, IV.1.4 and IV.2.14 with Theorem 2.33 and Proposition 2.40 which come later); moreover the two concepts coincide for the latter class of algebras (Proposition 2.12). As a consequence, to regard irreducible representations as the right non-commutative generalization of characters is, in the majority of questions, entirely correct and useful.

Secondly, from the information — apparently very scanty — contained in the definition of irreducible modules, one can make great progress in the study of their structure (with Schur's lemma 2.47 as the "golden key"). Here, a central result which has numerous and important applications is Theorem 2.55 on denseness — a far-reaching generalization of the fact that the image of an irreducible representation on a finite-dimensional space necessarily contains all the operators acting on it.

Finally, topologically irreducible representations (a version of the concept that coincides with the original concept for *-representations of C^*-algebras) play an outstanding role in the theory of involutive algebras whose rudiments were set out in Chap.IV §§4–7. It turns out (see Theorems 2.76 and 2.79) that among the GNS-representations those that are topologically irreducible are just those that correspond to the extreme points of the set of states of the given algebra. This fact enables one to bring into play strong methods of functional analysis (such as the Krein-Milman theorem) and to obtain, with their help, a number of deep results about the structure of star algebras and their representations.

It is worth noting that all this has immediate and important applications to the mathematical tools of modern physics (although this, as it is said by "Mowgli", is really "quite, quite, quite another story" ...).

§2.1. Cyclic and irreducible representations. Definitions and examples

From now on A will always be an arbitrary fixed algebra, unless additional assumptions are explicity stated. The next concept, which is very useful in its own right, leads us on to irreducible modules in real earnest.

Definition (VI.2.1). An element x of an A-module X is called *cyclic* if any $y \in X$ has the form $a \cdot x$ for a certain $a \in A$. A module having a cyclic element is itself called a *cyclic* (or one-generator) *module* and the representation associated with it is called a *cyclic representation.*

It turns out that we are, in essence, already acquainted with cyclic modules:

Proposition (VI.2.2). *An A-module X is cyclic $\Leftrightarrow$ it is isomorphic to the quotient module A/I for a modular left ideal I in A. In more detail, if x is a cyclic element in X, then the kernel of the morphism $\tau_x : A \longrightarrow X : a \longmapsto a \cdot x$ can be taken to be I (that is, we put $I := \mathcal{A}nn_A\{x\}$) and the isomorphism $i : A/I \longrightarrow X$ is the operator induced by this τ_x.*

▶ "$\Leftarrow$". If u is a modular identity for I, then $a(u + I) = a + I$ is satisfied in A/I for all $a \in A$. Hence $u + I$ is a cyclic element in A/I.

"$\Rightarrow$". Obviously τ_x and i are morphisms of A-modules and I is a left ideal in A. Further, x being cyclic implies that τ_x is surjective and hence i is bijective. Finally, any u with $u \cdot x = x$ is a modular identity for I: in fact $(au - a) \cdot x = a \cdot (u \cdot x - x) = 0$ is satisfied for all $a \in A$ and therefore $au - a \in I$. ■

Proposition (VI.2.3). *Under the conditions of Proposition 2.2, suppose that A and X are Banach: then X is cyclic $\Leftrightarrow$ it is topologically isomorphic to A/I, where I, as described above in 2.2, is closed. Moreover the corresponding ideal and isomorphism can be chosen to be the same as in 2.2.*

▶ One only has to add to the proof of Proposition 2.2 (in the part "$\Rightarrow$") the comment that now, by construction, φ and i are continuous. Hence $I = \operatorname{Ker} \tau_x$ is closed and i, by the Banach theorem, is a topological isomorphism. ■

Exercise (VI.2.4). Describe, to within a topological isomorphism, all the cyclic Banach modules over $C(\Omega)$, where Ω is a compact space.

Here finally is our basic definition.

Definition (VI.2.5). An A-module $X \neq \{0\}$ is said to be *irreducible*[10] if it has no submodules other than itself and $\{0\}$ and, in addition, it is not a one-dimensional annihilator (= it is not isomorphic to $\mathbb{C}$ with the zero outer multiplication).

A non-zero algebra consisting of operators is called *irreducible* if its operators have no proper non-zero invariant subspace in common. A representation is said to be *irreducible* if its operators form an irreducible algebra or, what is obviously equivalent, the module associated with this representation is irreducible.

The following is easily deduced from Definition 2.5.

Proposition (VI.2.6). *If an A-module is irreducible then* (I) *it is non-degenerate and* (II) *it contains no element $x \neq 0$ with $a \cdot x = 0$ for all $a \in A$.* ▶■

Exercise (VI.2.7). Show that a representation of an algebra on a finite-dimensional space is irreducible $\Longleftrightarrow$ there is no basis in the latter in terms of which the operators of the representation could be written as matrices of the form

$$\left(\begin{array}{c|c} \cdots & \cdots \\ \hline 0 & \cdots \end{array}\right)$$

with fixed dimension for the blocks and square blocks down the diagonal.

Theorem (VI.2.8). *The following properties of an A-module X are equivalent to its being irreducible*:

(I) *each $x \in X\backslash\{0\}$ is a cyclic element*;

(II) *X is isomorphic to A/I, where I is a maximal modular left ideal in A*;

(III) *$\mathcal{A}nn_A\{x\}$ is a maximal modular left ideal in A for each $x \in X\backslash\{0\}$ and X is isomorphic to $A/\mathcal{A}nn_A\{x\}$.*

Moreover, if X is a Banach module, then "isomorphic" can always be replaced by "topologically isomorphic".

---

[10] In abstract algebra such modules are often called simple. Hence, among other things, the origin of the term "a semisimple algebra" — which is now renamed "a classically semisimple algebra": this is an A which decomposes, as an A-module, into the finite direct sum of simple A-modules (cf. I.3.65–69).

▶ Irreducibility ⇒ (I). By Proposition 2.6(II), the submodule $\{a \cdot x : a \in A\}$ in X is non-zero and therefore coincides with X.

Irreducibility ⇒ (III). In view of Proposition 2.2, the preceding implication implies that all the requirements in (III) are satisfied except perhaps the one requiring $\mathcal{A}nn_A\{x\}$ to be maximal. The latter is implied by Proposition 1.5.

(II) ⇒ irreducibility. This follows from Propositions 1.5 and 2.2.

(I) ⇒ irreducibility and (III) ⇒ (II). Obvious.

Using Proposition 2.3 instead of Proposition 2.2 in these arguments we immediately get the additional result about Banach modules. ∎

Here are some curious applications of this theorem for Banach structures.

Proposition (VI.2.9). *Let A be a Banach algebra and X an irreducible A-module (in the purely algebraic sense). Then there exists a norm in X, unique up to equivalence, that turns it into a Banach module.*

▶ Taking an arbitrary $x \in X \backslash \{0\}$ and using Theorem 2.8(I), we consider the isomorphism $i : A/I \longrightarrow X$ indicated in Proposition 2.2. Since I is maximal (Proposition 1.5) and therefore closed (Proposition I.3.29), then A/I is a Banach module; consequently so also is X with respect to the norm "carried across from A/I by means of i". If $\|\cdot\|$ and $\|\cdot\|'$ are two norms in X having the stated property, then according to Proposition 2.3, i is a topological isomorphism between A/I and each of the Banach A-modules $(X; \|\cdot\|)$ and $(X; \|\cdot\|')$; hence the two norms are equivalent. ∎

Corollary (VI.2.10). *Every purely algebraic isomorphism between irreducible Banach modules is a topological isomorphism.*

Remark (VI.2.11). Later on it turns out that we need not even require in advance that the outer multiplication in this corollary should be continuous (in our modules) with respect to the first variable: we shall see that this is automatically satisfied. But this fact is no longer so simple: see later Theorem 2.62a.

* * *

When studying various classes of algebras (both abstract and polynormed) a typical exercise is to describe, up to an isomorphism, all the irreducible modules over them;[11] moreover in

---

[11] Another typical exercise about decomposing, in some reasonable sense, arbitrary representations into irreducible represention is not considered in this book (cf. what is said at the end of §2.4).

some cases one can succeed in solving the problem completely whereas in others it turns out to be hopeless.

First of all we note that the problem under discussion includes the problem of describing the one-dimensional modules, which, in an equivalent language, is the problem we have repeatedly considered about describing the characters (= one-dimensional representations) of a given algebra. Clearly, by Theorem 2.8, the two problems coincide for those algebras that have all of their maximal modular ideals of codimension 1 (for $\mathcal{P}$ and $C^\infty[0, 1]$ but not for $C(\mathbb{R})$, say; cf. Remark V.3.29).

We pick out, in particular, the case of commutative Banach algebras. In the next proposition A is such an algebra and Ω its spectrum. Fixing $s \in \Omega$, we put $a \cdot z = \hat{a}(s)z$ for $a \in A$, $z \in \mathbb{C}$; in this way $\mathbb{C}$ becomes a Banach A-module denoted by $\mathbb{C}(s)$. Obviously $\mathbb{C}(s) \simeq A/I_s$ and the character associated with this module is χ_s.

Proposition (VI.2.12). *Each irreducible A-module is isomorphic to one, and only one, A-module $\mathbb{C}(s)$ for some $s \in \Omega$. As a consequence, each irreducible representation of a commutative Banach algebra is one of its non-trivial characters.*

▶ Let X be our module. The definition of the spectrum together with Theorem 2.8 imply that X is isomorphic to A/I_s, and hence isomorphic to $\mathbb{C}(s)$, for a certain $s \in \Omega$. All that remains is to note that equivalent one-dimensional representations must simply coincide; therefore $\mathbb{C}(s) \simeq \mathbb{C}(t)$; $s, t \in \Omega_+$ implies $\chi_s = \chi_t$, that is $s = t$. ■

Remark (VI.2.13). We have thus obtained another interpretation of the spectrum of a commutative Banach algebra A (cf. Chap.IV §1.2): it is the set of isomorphism classes of irreducible A-modules.

The irreducible modules have an explicit description for a number of operator algebras — only, as is already clear, not in terms of characters. It turns out that certain algebras may have only one irreducible module, up to an isomorphism.

Example (VI.2.14). Let E be a Banach space. Then E, as the natural module over $\mathcal{K}(E)$ or $\mathcal{B}(E)$ (see 1.28) automatically has the property (I) in Theorem 2.8 and it is therefore irreducible.

Proposition (VI.2.15). *Let H be a Hilbert space. Then every irreducible module over $\mathcal{K}(H)$ is isomorphic to (and if it is a Banach module it is topologically isomorphic to) the natural module.*

(In the language of representations this means that every irreducible representation of the algebra $\mathcal{K}(H)$ is equivalent to the "identity", that is, to the natural embedding of $\mathcal{K}(H)$ in $\mathcal{L}(H)$.)

▶ If we combine Corollary 2.10 and Theorem III.2.5, we see that the given module is isomorphic to $\mathcal{K}(H)/I$, where $I = \{a : a(x_0) = 0\}$ for a certain fixed $x_0 \in H$, $x_0 \neq 0$. But the last module is isomorphic to the natural module: the required isomorphism is generated by the surjective morphism of $\mathcal{K}(H)$-modules $\sigma : \mathcal{K}(H) \longrightarrow H : a \longmapsto a(x_0)$. ■

In the case when H is finite-dimensional, Proposition 2.15 obviously means that every irreducible module over the matrix algebra $\mathcal{M}_n$ is isomorphic to $\mathbb{C}^n$ with the outer multiplication defined as the usual multiplication of column vectors by matrices.

As an obvious consequence of Proposition 2.15, every irreducible $\mathcal{K}(H)$-module is, to within an isomorphism, a star module (in the sense of Definition 1.15). This fact is actually fairly general:

Assertion (VI.2.16) (cf. [31, 2.9.6]). *Let A be a C^*-algebra. Then every irreducible A-module is isomorphic to (and if it is a Banach module, then it is topologically isomorphic to) a star A-module.*[12]

In general, how does one contruct the irreducible modules over C^*-algebras? Of course, in reducing the problem of their description to the case of star modules (that is, to the problem of describing the irreducible *-representations of these algebras), we have made only the first step in studying them. As to the subsequent progress, the general picture at the moment is as follows: the closer, in a well-defined sense, our algebra is to $\mathcal{K}(H)$, the more we know. More precisely, there is a rather clearly indicated class of C^*-algebras having a "well understood" structure of irreducible modules; this is the class of postliminal algebras, already mentioned in Chap.III §2.2. It is not part of our aim to give more detailed information about them; all the same we suppose the reader will be curious to know at least their definition. Now (in contrast to Chap.III) the material we have accumulated enables us to give at least one of a number of equivalent definitions (see, for example, [31, §4] and [58, Ch.VI]).

Definition (VI.2.17). A C^*-algebra is called *liminal* (or a CCR-algebra) if the image of each of its irreducible representations consists of only compact operators. A C^*-algebra is called *postliminal* (or also a GCR-algebra, a C^*-algebra of type I or smooth) if each of its quotient C^*-algebras contains a non-zero closed liminal bi-ideal.

[12] This fact is proved using the correspondence between irreducible representations and pure states, which is the subject of §2.4; however it is more appropriate to mention it now.

Exercise (VI.2.18). Consider $A = C(\Omega, \mathcal{K}(H))$, where Ω is compact (cf. 1.26), as an algebra with pointwise multiplication. Prove that A is a liminal algebra and describe, to within an isomorphism, all irreducible A-modules.

The irreducible modules over non-postliminal C^*-algebras, including in particular those over the ones we know, such as $\mathcal{B}(H)$ "itself", the Calkin algebra and the Fermion algebra, have not been studied anything like as well; in fact we know almost nothing about what these modules look like, except in a few isolated examples. In each of the cases $\mathcal{B}(H)$ and $\mathcal{C}(H)$ there are very many of them: the cardinality of the set of isomorphism classes of irreducible modules over both algebras is equal to 2^c (c is the cardinality of the continuum); see [31, 4.7.22d].

Here is another more specific question originally posed by Naimark. Does there exist a C^*-algebra, other than $\mathcal{K}(H)$ (see Proposition 2.15), over which all irreducible modules are isomorphic? It is known that in the class of separable algebras the answer is negative [200], [226].

Finally, we touch on the question of constructing irreducible modules over group algebras of locally compact groups. First of all, we combine Proposition 2.12 with the result on the structure of the characters of L^1-algebras which was given in Chap.III §3.2 and of which Theorem III.3.13 is a particular case. We immediately get that, if G is commutative all the irreducible $L^1(G)$-modules are one-dimensional and are given by the points of the dual group $\hat{G}$: to each $\tilde{\chi} \in \hat{G}$ there corresponds the module $\mathbb{C}(\tilde{\chi})$ with the action

$$f \cdot z := \left(\int_G f(s)\tilde{\chi}(s)ds \right) z; \quad f \in L^1(G), \quad z \in \mathbb{C}(\tilde{\chi}).$$

Compact groups are another class of groups for which the modules under discussion have a fairly simple structure. We give without proof the basic assertion about this.

Assertion (VI.2.19) (implied, for example, by [54, IV §3]). *Let* G *be compact. Then*

(I) *every irreducible* $L^1(G)$*-module is finite-dimensional and isomorphic to a star module*;

(II) *there exists only a finite number of non-isomorphic irreducible* $L^1(G)$*-modules of a given dimension.*

We note that an analogous result also holds for the C^*-algebras of compact groups;[13] it is not difficult to deduce this from the general properties of enveloping C^*-algebras.

---

[13] This implies, in particular, that such algebras are liminal in the sense of Definition 2.17.

* * *

In the remaining part of the section, when we say "A-module" we mean a polynormed module over a polynormed algebra.

Along with the concept of an irreducible module, which is, as we have seen, of a purely algebraic character, we may also consider a number of other variants of the relevant definitions, in the majority of which we assume that the module has a topology. We limit ourselves to one of these versions, possibly, the most important. The following leads up to it.

Definition (VI.2.20). An element x of the A-module X is said to be *topologically cyclic* if the set $\{a \cdot x : a \in A\}$ is dense in X. A module having a topologically cyclic element and the representation associated with this module are themselves said to be *topologically cyclic.*

Example (VI.2.21). Every GNS-representation is topologically cyclic: the $\hat{e}$ given in Proposition IV.5.14 is a topologically cyclic element.

Definition (VI.2.22). An A-module is said to be *topologically irreducible* if it has no closed submodules other than itself and $\{0\}$, and, in addition, it is not a one-dimensional annihilator. The representation associated with this module is also said to be *topologically irreducible.*

Proposition (VI.2.23). *An A-module is topologically irreducible $\Leftrightarrow$ every non-zero element of A is topologically cyclic.*

▶ The proof is a repetition, with the obvious modifications, of the corresponding argument in Theorem 2.8. ■

Of course "ordinary" irreducibility implies topological irreducibility. The converse is not true:

Exercise (VI.2.24). Let H be a Hilbert space with a countable orthonormal basis $\{e_n\}$, and let $A = \{\varphi \in \mathcal{B}(H) : \|\varphi\|_1 := \sum_{k,l} |\langle \varphi e_k, e_l \rangle| < \infty\}$. Check that $(A; \|\cdot\|_1)$, is a Banach operator algebra and prove that the natural A-module H is topologically irreducible but is not irreducible.

Nevertheless, the two concepts coincide for the most important class of operator algebras:

Assertion (VI.2.25) ("Kadison's theorem"). *Every topologically irreducible Banach module over a C^*-algebra is irreducible.*

For the proof of this fairly deep fact see [31, §2.8].

Another version of the concept under discussion, the so-called n-irreducibility, will be considered in the next section; for other interesting variants — "complete irreducibility", "tensor irreducibility" and so on, see, for example [40, §7.2] and [33, §8].

§2.2. Characterization of the radical and semisimplicity in terms of irreducibility. A structure theorem for non-commutative algebras (and in what way it is deficient)

Let A be, as before, an arbitrary fixed algebra. Recall (§1.1) that the annihilator of any A-module is a bi-ideal in A.

Definition (VI.2.26). An ideal in A (necessarily proper and two-sided) is called *primitive* if it is the annihilator of some irreducible non-singular A-module (= it is the kernel of a certain non-trivial irreducible representation of our algebra).

Proposition (VI.2.27). *Every primitive ideal of a Banach algebra is closed.*

▶ This follows from Proposition 2.9. ■

Proposition (VI.2.28). *Every maximal bi-ideal in A which is modular as a left ideal, is primitive.*

▶ According to Proposition I.3.25, our bi-ideal I is contained in a certain maximal modular left ideal J. Then Proposition 1.6(I) implies that I (being a right ideal) lies in $\mathcal{A}nn_A\, A/J$ and hence (being a maximal bi-ideal) coincides with this annihilator. But A/J, by Theorem 2.8, is an irreducible module; hence I is primitive. ■

Example (VI.2.29). In the algebra $\mathcal{B}(H)$, where H is a separable infinite-dimensional Hilbert space, there are in all two primitive ideals: $\{0\}$ and $\mathcal{K}(H)$. The first of these is primitive since it is the annihilator of the natural $\mathcal{B}(H)$-module H (2.14), and the second since it is a maximal bi-ideal (2.28 and III.2.10). Since these are the only two proper bi-ideals which are closed (III.2.10), there are, by Proposition 2.27, no other primitive ideals.

Primitive ideals are something new for us only in the case of non-commutative algebras:

Proposition (VI.2.30). *The primitive ideals of a commutative algebra are precisely its maximal modular ideals.*

▶ Let $I = \mathcal{A}nn_A A/J$, where J is a maximal modular ideal. Then Proposition 1.6(I) and A being commutative imply that $J \subseteq I$ and Proposition 1.6(II) implies the reverse inclusion; hence $I = J$. Thus the first of the above two classes lies in the second and hence, by Proposition 2.28, they coincide. ■

Remark (VI.2.31). Are the primitive ideals in A primitive ideals in A^{op}? This question (informally speaking, about whether there is the kind of "left-right" symmetry that is inherent in the concept of the radical, say) remains open. Bergman [104] has constructed a suitable counterexample for rings that are not algebras.

We now give, using the concept we have introduced, a new characterization of the radical. We note the following as a preliminary.

Proposition (VI.2.32). *Primitive ideals exist in* A $\Longleftrightarrow$ *irreducible* A*-modules exist* $\Longleftrightarrow$ *maximal modular left ideals exist in* A.

▶ This follows from Definition 2.26 and Theorem 2.8. ■

Theorem (VI.2.33). *The radical of an arbitrary algebra* A *either coincides with the intersection of all its primitive ideals, or, if there are none, coincides with* A *itself.*

▶ Let R be our radical and let R' be the intersection. Proposition 2.32 and Theorem I.3.53 obviously imply that we are justified in limiting ourselves to the case when irreducible A-modules exist.

We take such a module X. By Theorem 2.8, for any $x \in X$; $x \neq 0$, $X \simeq A/J_x$, where $J_x = \mathcal{A}nn_A\{x\}$ is a maximal modular left ideal in A. But then, by Theorem I.3.53, R lies in every J_x and hence also it lies in their intersection — the primitive ideal $\mathcal{A}nn_A X$ Hence, as X is arbitrary, $R \subseteq R'$.

Let us establish the opposite inclusion. By Proposition 2.32, maximal modular left ideals exist in A; let J be one of them. Then by 2.8, A/J is an irreducible module and therefore R', by definition, is contained in $\mathcal{A}nn_A A/J$. But the latter, by Proposition 1.6(II), lies in J. Hence $R' \subseteq J$ and, since J is arbitrary, by Theorem I.3.53, $R' \subseteq R$. ■

This theorem, taking into account Proposition 2.9, immediately implies the following.

Corollary (VI.2.34). (I) *A radical algebra has no non-trivial irreducible representation;*

(II) *for each element* $a \neq 0$ *in a semisimple algebra there exists an irreducible representation* T *such that* $T_a \neq 0$. *Moreover, if our algebra is Banach, then* T *can be chosen to be an irreducible continuous representation on a Banach space.*

The fact indicated in 2.34(II) is often expressed by saying that semisimple algebras have "sufficiently many" irreducible representations (to distinguish the different elements).

Recall (Proposition 2.12) that the irreducible representations of a commutative Banach algebra are its characters. Therefore we can consider Theorem 2.33 to be a non-commutative generalization, for the class of Banach algebras, of the description of the radical in terms of the characters that we obtained earlier (IV.1.4). Do the subsequent results of the theory as set out in Chap.IV §§1-3, and in the first instance does the basic Theorem IV.2.14, admit meaningful non-commutative generalizations? Namely, is it possible to realize the elements of an arbitrary Banach algebra as something like continuous functions with values in a comparatively simply constructed (although, generally speaking, non-commutative) algebra?

Up to now we have succeeded in constructing such a realization by continuous vector functions only for certain special classes of Banach algebras [31, §10]. For general Banach algebras the only result we know is a generalization of the algebraic part of Gel'fand's theorem — the part that remains after forgetting about everything that is connected with the topology on the spectrum. We turn to a presentation of this.

The role that the field $\mathbb{C}$ played in Theorem IV.2.14 is now played by the following class of algebras.

Definition (VI.2.35). An algebra A is said to be *primitive* if it has at least one faithful irreducible A-module (= at least one injective irreducible representation).

Proposition (VI.2.36). (I) *An algebra is primitive* $\Leftrightarrow$ *it is isomorphic to an irreducible operator algebra* $\Leftrightarrow$ *its zero ideal is primitive.*

(II) *A bi-ideal* I *in* A *is primitive* $\Leftrightarrow$ *the quotient algebra* A/I *is primitive.* ▶■

Proposition (VI.2.37). *A commutative primitive algebra is a field.*

▶ By Proposition 2.30, $\{0\}$ is a maximal modular ideal of our algebra. The rest is clear. ■

We introduce into the discussion the set of primitive ideals of the algebra A; this is denoted by $\underline{\Pi}(A)$, or simply by Π, and it is called the *structure set* of this algebra. By Proposition 2.30, the structure set of a commutative Banach algebra is none other than its spectrum. Again, as in that case (see Chap.IV §1.2), it is convenient to think of Π as a point set indexing the primitive ideals and denote the latter by I_s; $s \in \Pi$.

Given $s \in \Pi$ we consider the quotient algebra $A_s = A/I_s$. This algebra is primitive (Proposition 2.36(II)) and it is Banach if A is a Banach algebra (Proposition 2.27).

Finally we take the Cartesian product of the family $\{A_s;\ s \in \Pi\}$, briefly denoted in future by $\bigtimes A_s$, and we introduce the map $\mathcal{G} : A \longrightarrow \bigtimes A_s$ taking $a \in A$ into the "row" $\{a_s = a + I_s \in A_s;\quad s \in \Pi\}$. This $\mathcal{G}$ is the analogue of (more correctly, a substitute for) the Gel'fand transformation. It is convenient to describe its properties using the following terminology and notation.

For an arbitrary family of Banach algebras $\{A_\nu;\ \nu \in \Lambda\}$, we consider the subalgebra $\sum A_\nu$ of its Cartesian product $\bigtimes A_\nu$ that consists of those $b = \{b_\nu;\ \nu \in \Lambda\}$ for which $\|b\|_0 = \sup_\nu \|b_\nu\|$ is finite. (This subalgebra is sometimes called the *complete direct sum of the family* $\{A_\nu\}$.)

Definition (VI.2.38). A subalgebra B of $\bigtimes A_\nu$ is called a *subdirect sum of the family* $\{A_\nu;\ \nu \in \Lambda\}$ if for every ν the ν-th coefficients of the elements of B take all possible values in A_ν. Suppose $\{A_\nu;\ \nu \in \Lambda\}$: is a family of Banach algebras, then if a subdirect sum B of this is a Banach algebra with respect to a certain norm $\|\cdot\|$, it is called a *Banach subdirect sum* of this family provided that $B \subseteq \sum A_\nu$ and, in addition, $\|b\| \geq \|b\|_0$ for all $b \in B$.

Let us return to the map $\mathcal{G}$. We put $\hat{A} = \operatorname{Im}\mathcal{G}$ and, if A is a Banach algebra, we put $\|\check{a}\|^\wedge = \inf\{\|a\|\ ;\ a \in A,\ \mathcal{G}(a) = \check{a}\}$ for $\check{a} \in \hat{A}$.

Theorem (VI.2.39) (cf. IV.2.11–13). *For arbitrary* A

(I) $\mathcal{G} : A \longrightarrow \bigtimes A_s$ *is an algebra homomorphism and it is unital if* A *is unital;*

(II) $\operatorname{Ker}\mathcal{G} = \operatorname{Rad} A$. *As a consequence, if* A *is semisimple, then* $\mathcal{G}$ *is injective;*

(III) $\hat{A}$ *is a subdirect sum of the family* $\{A_s;\ s \in \Pi\}$.

If A *is, in addition, a Banach algebra, then* $(\hat{A}; \|\cdot\|^\wedge)$ *is a Banach subdirect sum of the family* $\{A_s;\ s \in \Pi\}$, *which is isometrically isomorphic to the Banach algebra* $A/\operatorname{Rad} A$.

▶ (II) is a consequence of Theorem 2.33; the other assertions are obvious. ■

Proposition (VI.2.40). (I) *An algebra is semisimple* $\Longleftrightarrow$ *it is isomorphic to a subdirect sum of a family of primitive algebras.*

(II) *A Banach algebra is semisimple* $\Longleftrightarrow$ *it is isomorphic to a Banach subdirect sum of a family of primitive Banach algebras.*

▶ It is sufficient to show that a subdirect sum B of a family of primitive algebras $\{A_\nu;\ \nu \in \Lambda\}$, is semisimple; all the rest is implied by Proposition 2.39. For this we consider the homomorphism $\tau_\nu : B \longrightarrow A_\nu$ that sends an element to its ν-th coordinate and let I_ν be its kernel. By Definition 2.38, τ_ν is surjective; hence $B/I_\nu \simeq A_\nu$ and therefore I_ν is a primitive ideal in B (Proposition 2.36(II)). All that remains is to note that $\bigcap_\nu I_\nu = \{0\}$ and apply Theorem 2.33. ■

In their turn Propositions 2.40(I) and 2.37 imply the following.

Corollary (VI.2.41). *A commutative algebra is semisimple* $\Longleftrightarrow$ *it is isomorphic to a subdirect sum of a family of fields.*[14]

---

[14] If we go outside the class of Banach algebras these are not, generally speaking, the same field; cf. IV.2.14!

The reader probably agrees with us that the above results are less effective than their "models" in Chap.IV §§1–3: there is too much freedom in the concept of a subdirect sum and in the concept of a primitive algebra. (The structure of such algebras is made somewhat clearer in the next section — see Corollary 2.58 — but this does not change the general impression.) Nevertheless, as has been said, these results essentially exhaust what is known, up to now, about the structure of general non-commutative algebras, both abstract and Banach.

┌───────

We have already noted that the spectrum of a commutative Banach algebra is a particular case of the structure set of an arbitrary algebra. However, the latter, generally speaking, does not have a topology which would be the natural generalization of the Gel'fand topology. Nevertheless it does have a certain standard topology generalizing another possible topology on the spectrum — namely the hull-kernel topology mentioned at the end of Chap.IV §1. We offer some information about this topology in the form of exercises.

Let M be a subset of A. Its *hull* is the subset $h(M)$ of $\underline{\Pi} = \underline{\Pi}(A)$ consisting of all primitive ideals containing M.

Exercise (VI.2.42) (cf. IV.1.22–23). Prove that the complements of the hulls of all possible subsets of the algebra A form a topology in $\underline{\Pi}(A)$.

Hint. We always have $h(M) = h(J)$, where J is the minimal bi-ideal containing M. Further, if $\{J_\nu;\ \nu \in \Lambda\}$, is a set of bi-ideals in A, then $\bigcap_\nu h(J_\nu) = h(\ell.h. \bigcup_\nu J_\nu)$. Finally, if J_1 and J_2 are bi-ideals, and $J_3 = \ell.h.\{ab : a \in J_1;\ b \in J_2\}$, then $J_3 \subseteq I$, where I is a primitive ideal, implies either $J_1 \subseteq I$ or $J_2 \subseteq I$. This means that $h(J_1) \cup h(J_2) = h(J_3)$.

As in the case of commutative Banach algebras (cf. Problem IV.1.23), the topology thus constructed is called the *hull-kernel topology* or the *Jacobson topology*. When the structure set is allotted this topology it is called the *structure space* or the *primitive spectrum* of the given algebra.

Exercise (VI.2.43). Describe the structure space of the algebra $\mathcal{B}(H)$, where H is an infinite-dimensional separable Hilbert space.

Answer. $\Pi = \{s, t\}$, where $I_s := \{0\}$, $I_t := \mathcal{K}(H)$ (see Example 2.29), and Π has just three open sets: Π, $\{s\}$ and $\varnothing$.

Exercise (VI.2.44). Describe the structure space of the subalgebra of $C([0, 1], \mathcal{M}_2)$ consisting of all the matrix functions f with $f(1)$ being a diagonal matrix.

Here are a few more general facts.

Exercise (VI.2.45). Prove that (I) the structure spaces of the algebras A and $A/\mathrm{Rad}\,A$ are homeomorphic; (II) the structure space of a unital algebra is quasi-compact (although it is not always Hausdorff); (III) in the structure space of a commutative algebra every one-point subset is closed.

The topology of the structure space provides an appropriate language for expressing many properties of the algebras under consideration. This applies most of all, perhaps, to the postliminal C^*-algebras mentioned earlier (Definition 2.17). These algebras may be obtained, using certain standard constructions, from those C^*-algebras (called algebras with continuous trace), for which the question discussed above about a meaningful analogue of the Gel'fand theorem has been completely solved. Moreover for a C^*-algebra with continuous trace, the continuous vector functions defined on its structure space, which realize the elements of this algebra take values in $\mathcal{K}(H)$; see [128], [31, §10].[15]

On the other hand, it is already clear that for general Banach algebras the structure space is not suitable as the domain of definition of hypothetical vector functions: even in the commutative case the usual functions realizing the elements of the algebra are not necessarily continuous with respect to the hull-kernel topology. The algebra $\mathscr{A}(\overline{\mathbb{D}})$ may be taken as an example (cf. just before Proposition III.1.2).

§2.3. Schur's lemma and its star version (Naimark's theorem). Density theory and its applications - subsequent results on automatic continuity

We begin with a trivial observation.

Proposition (VI.2.46). *Let X and Y be irreducible modules over the algebra A. Then every morphism $\varphi : X \longrightarrow Y$, other than zero, is an isomorphism.*

▶ X being irreducible obviously implies that $\mathrm{Ker}\,\varphi = \{0\}$ and Y being irreducible implies that $\mathrm{Im}\,\varphi = Y$. ■

Recall that an important object — the algebra ${}_Ah(X)$ of the morphisms of this module into itself (§1.1) was defined for every A-module X.

Proposition 2.46 implies the following famous theorem.

---

[15] Actually we are talking about cross-sections of vector bundles in a fairly wide sense, with fibres in $\mathcal{K}(H)$ (generally speaking, for various H).

Theorem (VI.2.47) (Schur's lemma). *If* X *is an irreducible* A*-module, then* ${}_Ah(X)$ *is a skew field.* ▶■[16]

But the reader will, we hope, remember that, under some standard assumptions, there is only one complex algebra that is a skew field.

Theorem (VI.2.48). *Let* X *be an irreducible* A*-module and suppose that at least one of the following conditions is satisfied:*

(I) A *is a Banach algebra with respect to a certain norm, or*

(II) X *is finite-dimensional.*

Then ${}_Ah(X) = \ell.h.\{1\} \simeq \mathbb{C}$.

▶ If (I) is satisfied, then by Proposition 2.9 we are justified in assuming that X is a Banach module; hence, if we combine Proposition 2.46 and Corollary 2.10, we can conclude that ${}_Ah(X)$ consists of continuous morphisms, that is, it coincides, in the notation of §1.2, with the Banach algebra ${}_A\mathbf{h}(X)$. If (II) is satisfied, then since ${}_Ah(X)$ is a finite-dimensional algebra, it is a Banach algebra with respect to any norm. In both cases, by Schur's lemma, ${}_Ah(X)$ is a Banach skew field and the result we need is implied by the Gel'fand-Mazur theorem (II.1.17). ■

In the language of representations the result we have obtained goes as follows:

Theorem (VI.2.48a). *Let* T *be an irreducible representation of the algebra* A *on* X *and moreover suppose either that* A *is a Banach algebra or that* $\dim X < \infty$. *Then* $T^! = \ell.h.\{1\} \simeq \mathbb{C}$ *(in other words, any operator in* X *commuting with all* T_a; $a \in A$, *is a multiple of the identity).* ▶■

In fact, Theorem 2.48 with the condition (II) may be obtained by purely algebraic methods and it is even stronger then:

Exercise (VI.2.49). Show that, instead of (II), it is sufficient to require that the dimension of X is at most countable.

Hint. If for every $\varphi \in {}_Ah(X)$ there is a polynomial p with $p(\varphi) = 0$ (this is automatically so when $\dim X < \infty$), then the result is implied by Theorem 2.47 and the fact that $\mathbb{C}$ is

---

[16] We have before us a classic example of an assertion becoming a rather obvious back number after an adequate language has been found to express it — in this case the language of modules. Of course, at the beginning of the century when Schur made his discovery there was no sign of such a language. The young Emmy Noether was still writing long formulae under the supervision of Jordan and little van der Waerden had no inkling that he would become a mathematician ... .

algebraically closed; cf. the proof of II.1.9. If, however, for a certain φ, $p(\varphi) \neq 0$ for all $p \neq 0$, then, by Theorem 2.47, ${}_A h(X)$ contains a continuum of linearly independent elements $(\varphi - \lambda 1)^{-1}$; $\lambda \in \mathbb{C}$.

The following tells us that additional conditions on A or X are essential.

Example (VI.2.50). Take the algebra of rational functions $\mathcal{R}$. As this is a field, then the $\mathcal{R}$-module $\mathcal{R}$ is irreducible; at the same time the skew field ${}_{\mathcal{R}}h(\mathcal{R})$ is isomorphic to $\mathcal{R}$ itself and, in particular, it is infinite-dimensional.

If we combine Theorem 2.48 and Proposition 2.46, we immediately get the following.

Corollary (VI.2.51). *Let X and Y be irreducible A-modules and moreover suppose either that A is a Banach algebra or that one of the two modules is finite-dimensional. Then the space ${}_A h(X, Y)$ is either trivial or one-dimensional.*

(Translate this corollary as well into the language of representations!)

Remark (VI.2.52). In the theory of representations one considers algebras over an arbitrary field of scalars (and not only over $\mathbb{C}$, as in our book) and this raises the analogous question on the structure of the skew field ${}_A h(X)$, where X is an irreducible module over such an algebra. An interesting version of Theorem 2.48 is obtained when considering the base field $\mathbb{R}$. It turns out that, if analogous conditions are satisfied, then there exist for ${}_A h(X)$ not one but three possibilities: it is, as a real algebra, isomorphic to either $\mathbb{R}$ or $\mathbb{C}$ or to the skew field $\mathbb{H}$ of real quaternions (cf. [9, §25]). Corollary 2.51 is also transformed accordingly: in its "real" version the number $\dim {}_A h(X, Y)$ [17] may take no longer two but four values: 0, 1, 2 or 4 (cf. [40, §8]).

We now show that with the help of Schur's lemma one can make far-reaching deductions about the structure of a wide class of irreducible modules.

Definition (VI.2.53). An A-module X is said to be *n-irreducible* or *n-transitive* $(n \in \mathbb{N})$ if given any linearly independent set $x_1, \cdots, x_m \in X$, with $m \leq n$, then, for an arbitrary set $y_1, \cdots, y_m \in X$, there exists some $a \in A$ such that $a \cdot x_k = y_k$ for all $k = 1, \cdots, m$. The representation

[17] In the theory of representations this number is called the intertwining number of the corresponding irreducible representations.

associated with an n-irreducible module is itself said to be *n-irreducible*.

Of course, if a module is n-irreducible this implies that it is m-irreducible for all $m < n$. In addition, Theorem 2.8(I) implies that 1-irreducibility is none other than ordinary irreducibility.

Lemma (VI.2.54). *An A-module X is n-irreducible $\Longleftrightarrow$ it is 1-irreducible and for any $m \le n$ if $x_1, \cdots, x_m \in X$ is a linearly independent set then there exists $a \in A$ such that $a \cdot x_2 = ... = a \cdot x_m = 0$ and $a \cdot x_1 \neq 0$.*

▶ "$\Rightarrow$" is obvious; we establish "$\Leftarrow$". Let $x_1, \cdots, x_m$ be linearly independent and let $y_1, \cdots, y_m$ be arbitrary. The condition implies that for each k; $1 \le k \le m$, there is $a_k \in A$ with $a_k \cdot x_k \neq 0$ and $a_k \cdot x_l = 0$ for $l \neq k$. Using the 1-irreducibility of X we take, for this same k, some element $b_k \in A$ with $b_k \cdot (a_k \cdot x_k) = y_k$. All that remains is to verify Definition 2.53, putting $a := b_1 a_1 + \cdots + b_m a_m$. ■

The next theorem will later be called the density theorem. In fact, it is a special version, selected as being geometrically the clearest and having applications to analysis, of the well-known "Jacobson-Chevalley density theorem" (see, for example, [83, §2.1.]).[18]

Theorem (VI.2.55). *The following properties of an A-module X are equivalent:*

(I) X *is 1-irreducible and in addition* $_A h(X) = \ell.h.\{1\}$;

(II) X *is 2-irreducible*;

(III) X *is n-irreducible for each* $n = 1, 2, 3, \cdots$.

▶ (I) $\Rightarrow$ (III). By the principle of induction, it is sufficient to show that if (I) holds and X is m-irreducible for a certain $m \in \mathbb{N}$ this implies that X is $(m+1)$-irreducible.

We take linearly independent elements $x_1, \cdots, x_{m+1}$ in X (if there are none, then the assertion is obvious). We consider in A the left ideal $I = \{a : a \cdot x_3 = \cdots = a \cdot x_{m+1} = 0\}$ (when $m = 1$ we put $I := A$). For $k = 1, 2$ we put $\tau_k : I \longrightarrow X : a \longmapsto a \cdot x_k$; X being m-irreducible automatically implies that these are surjective A-module morphisms.

By Lemma 2.54 and the choice of I it is sufficient to find $a \in I$ with $a \cdot x_2 = 0$ and $a \cdot x_1 \neq 0$. We assume that there is no such element; this means that $\operatorname{Ker} \tau_2 \subseteq \operatorname{Ker} \tau_1$. Then there exist an operator φ induced, in the sense of Chap.0 §0.2°, by the operator 1_I and an operator ψ so that the diagram

---

[18] The "actual" density theorem is most reminiscent of the result in Exercise 2.60 later, but it features irreducible modules over arbitrary rings (and not necessarily complex algebras).

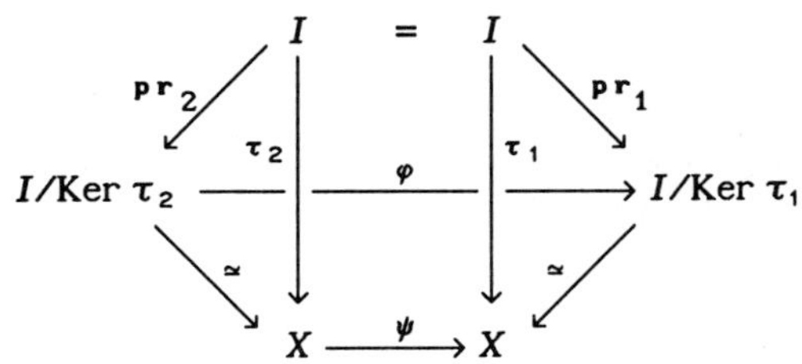

in which $\mathbf{pr}_k$; $k = 1, 2,$ are natural projections, is commutative. Obviously φ and ψ are A-module morphisms and moreover ψ is well defined by $\psi(a \cdot x_2) = a \cdot x_1$; $a \in I$. But then, by (I), $\psi = \lambda 1$ for a certain $\lambda \in \mathbb{C}$; hence $a \cdot z = 0$ is satisfied for $z = \lambda x_2 - x_1$ and all $a \in I$. This means that given the m linearly independent elements $z, x_3, \cdots, x_{m+1}$ there does not exist $a \in A$ with $a \cdot x_3 = \cdots = a \cdot x_{m+1} = 0$ and $a \cdot z \neq 0$. This contradicts the inductive hypothesis about m-irreducibility.

(II) $\Rightarrow$ (I). Our problem is to show that any $\varphi \in {}_A h(X)$ is a multiple of 1. We choose arbitrarily $x_1 \in X \setminus \{0\}$. If $a \cdot x_1 = 0$ is satisfied for some $a \in A$, then also $a \cdot (\varphi(x_1)) = \varphi(a \cdot x_1) = 0$; in view of (II), this means that x_1 and $\varphi(x_1)$ are linearly dependent, that is $(\varphi - \lambda 1)x_1 = 0$ for some $\lambda \in \mathbb{C}$. But then $\varphi - \lambda 1$, since it is not an isomorphism, must be zero by Proposition 2.46.

(III) $\Rightarrow$ (II). This is clear. ∎

If we combine Theorems 2.55 and 2.48 we get the following.

Corollary (VI.2.56). *Let X be an irreducible A-module where either A is Banach or $\dim X < \infty$. Then X is n-irreducible for all $n \in \mathbb{N}$.*

For finite-dimensional modules the result becomes especially clear:

Corollary (VI.2.57). *Let T be an irreducible representation of the algebra A on the finite-dimensional space X. Then the operators T_a; $a \in A$, "fill up" the whole algebra $\mathcal{L}(X)$.*

These two corollaries in turn, taking into account Proposition 2.9, immediately imply the following.

Corollary (VI.2.58). (I) *Let A be a primitive Banach algebra. Then for some Banach space X it is topologically isomorphic to a certain Banach operator algebra $A_0 \subseteq \mathcal{B}(X)$ which has the property that, for any $n \in \mathbb{N}$ and vectors $x_1, \cdots, x_n, y_1, \cdots, y_n$ with $x_1, \cdots, x_n$ being linearly independent, there exists an operator $a \in A_0$ such that $a(x_k) = y_k$ for all $k = 1, 2, \cdots, n$.*

(II) *Any finite-dimensional primitive algebra is isomorphic to $\mathcal{L}(X)$ for a certain finite-dimensional space X.*

Remark (VI.2.59). Example 2.50 together with Theorem 2.55 ((II) $\Rightarrow$ (I)) shows that for at least one module the fact that it is 1-irreducible does not imply that it is 2-irreducible.

Concerning arbitrary 1-irreducible modules, one can establish the following by arguing roughly as for the deduction (I) $\Rightarrow$ (III) in Theorem 2.55.

Exercise (VI.2.60). Let X be an irreducible A-module. Prove that for any $n \in \mathbb{N}$ and for any $x_1, \cdots, x_n \in X$, which are linearly independent over ${}_Ah(X)$ (that is, when considering X as a left linear space over the skew field ${}_Ah(X)$ with the operation $\varphi x := \varphi(x)$) and any $y_1, \cdots, y_n \in X$ there exists $a \in A$ such that $a \cdot x_k = y_k$ for all $k = 1, \cdots, n$.

Finally we explain where the name "density theorem" comes from.

Exercise (VI.2.61). Let T be a representation of the algebra A on the space X. Taking the topology in $\mathcal{L}(X)$ with subbasis the sets $U_{x,y} = \{\varphi : \varphi(x) = y\}$; $x, y \in X$, show that the set $\{T_a : a \in A\}$ is dense in $\mathcal{L}(X)$ with respect to this topology $\Longleftrightarrow$ T is n-irreducible for all $n \in \mathbb{N}$.

Let us now concentrate on Banach algebras. Our next aim is to obtain, using the density theorem, some important results about automatic continuity. First we fulfill the promise made in Remark 2.11.

Theorem (VI.2.62) (Johnson). *Let $T : A \longrightarrow \mathcal{B}(X)$ be a topological representation of a Banach algebra on a Banach[19] space. Then, if T is irreducible, it is continuous.*

Theorem (VI.2.62a) (the same as before but in the language of modules). *Let X be a Banach space which is an irreducible module over the Banach algebra A. Then, if the outer multiplication $(a, x) \longmapsto a \cdot x$ is continuous with respect to the second argument, it is continuous also with respect to the first (and therefore X is a Banach module).*

▶ We prove the required assertion "in the guise of 2.62" for finite-dimensional, and "in the guise of 2.62a" for infinite-dimensional X.

As the primitive ideal $\operatorname{Ker} T$ is closed (Proposition 2.27), $A/\operatorname{Ker} T$ is a Banach algebra. Therefore, if $\dim X < \infty$, the injective operator $T_1 : A/\operatorname{Ker} T \longrightarrow \mathcal{B}(X)$, induced by T, acts between finite-dimensional normed spaces and consequently it is continuous. But then so is $T = T_1 \circ \mathbf{pr}$, where $\mathbf{pr}$ is the automatically continuous natural projection of A onto $A/\operatorname{Ker} T$.

---

[19] In fact Johnson proved this theorem for an arbitrary normed — not necessarily complete — X [130].

We pass to the (fundamental) case when X is infinite-dimensional. Let $x_1, x_2, \cdots$ be an infinite linearly independent system of elements in X of norm 1. We put $I_1 := A$ and $I_n := \{a \in A : a \cdot x_1 = \cdots = a \cdot x_{n-1} = 0\}$ for $n = 2, 3, \cdots$. As $I_n = \bigcap_{m=1}^{n-1} \mathcal{A}nn_A\{x_m\}$ when $n > 1$, Theorem 2.8(III) and I.3.29 imply that all the I_n are closed left ideals in A. Further, we put $\tau_n : I_n \longrightarrow X : a \longmapsto a \cdot x_n$ for $n = 1, 2, \cdots$; obviously these are A-module morphisms.

Is it possible for all these operators τ_n to be discontinuous? If this is so, then by induction it is easy to construct $a_n \in I_n$; $n = 1, 2, \cdots$ such that $\|a_n\| < 2^{-n}$ for all n and $\|a_n \cdot x_n\| > n + \|\sum_{m=1}^{n-1} a_m \cdot x_n\|$ when $n > 1$. Since A is complete, it contains $b := \sum_{n=1}^{\infty} a_n$ and $b_m := \sum_{n=m}^{\infty} a_n$; $m = 2, 3, \cdots$. The relations $\cdots \supseteq I_n \supseteq I_{n+1} \supseteq \cdots$ and the fact that, as we have noted, these ideals are closed, imply that $b_{n+1} \cdot x_n = 0$ for $n = 1, 2, \cdots$. Hence for this particular b we shall have $b \cdot x_n := \sum_{m=1}^{n-1} a_m \cdot x_n + a_n \cdot x_n$ and consequently $\|b \cdot x_n\| \geq \|a_n \cdot x_n\| - \|\sum_{m=1}^{n-1} a_m \cdot x_n\| > n$. As $\|x_n\| = 1$, this contradicts the assumption that the outer multiplication is continuous with respect to the second argument.

So, n has been found such that $\tau_n : I_n \longrightarrow X$ is a continuous operator. By Corollary 2.56, X is n-irreducible and this means that τ_n is surjective. We fix $x \in X$ and take $c \in I_n$ with $\tau_n(c) = x$. Then for each $a \in A$

$$\|a \cdot x\| = \|a \cdot \tau_n(c)\| = \|\tau_n(ac)\| \leq \|\tau_n\| \|ac\| \leq (K\|\tau_n\| \|c\|)\|a\|,$$

where K is the norm of the multiplication bi-operator in A. As x is arbitrary, this bound shows that the outer multiplication in X is continuous (not only with respect to the second but also) with respect to the first argument. ■

In order that the reader estimate the true worth of the next theorem,[20] we recall that in a linear space there may exist, generally speaking, many non-equivalent norms with respect to which it is a Banach space (see, for example, [189]).

---

[20] This theorem has a history: first Gel'fand [113] proved it for commutative algebras using "his" transform. The question of whether the commutativity condition could be discarded remained open for about 20 years, in spite of the effort of a whole host of mathematicians (see the discussion in [64]). Finally in 1967 Johnson (apparently as a thirtieth birthday gift) solved the problem conclusively.

Theorem (VI.2.63) (Johnson) [130]. *Let A be an algebra which is a Banach algebra with respect to both of the norms $\|\cdot\|$ and $\|\cdot\|'$. Then, if A is semisimple, the two norms are equivalent.*

▶ Since the relationship between the two norms is symmetric, our problem is to prove that the identity operator from $(A; \|\cdot\|)$, into $(A; \|\cdot\|')$, is continuous. By the closed graph theorem (Chap.0 §1.2), it is sufficient in turn to establish the following: if the sequence $a_n \in A$ converges to zero in $(A; \|\cdot\|)$, and to a certain element b in $(A; \|\cdot\|')$, then $b = 0$.

Let X be any irreducible A-module which, by Proposition 2.9, we are justified in assuming is a Banach module over $(A; \|\cdot\|)$. Consider the associated continuous representation $T : (A; \|\cdot\|) \longrightarrow \mathcal{B}(X)$; then $\lim_{n\to\infty} T_{a_n} = 0$. But this same T may be considered also as a topological representation of the algebra $(A; \|\cdot\|')$, on the same X, which, as Theorem 2.62 implies, is also continuous; hence $\lim_{n\to\infty} T_{a_n} = T_b$. Consequently, $T_b = 0$, that is $b \in \mathcal{A}nn_A\, X$.

Since X is arbitrary, Theorem 2.33 implies that $b \in \operatorname{Rad} A$. But A is semisimple; thus $b = 0$. ∎

Of course, the theorem we have proved is equivalent to the following assertion: every (purely algebraic) isomorphism between two semisimple Banach algebras is a topological isomorphism. In this form the theorem may be strengthened. For this we need the following proposition.

Proposition (VI.2.64). *The kernel of any surjective homomorphism from a Banach algebra onto a semisimple algebra is closed.*

▶ Let $\kappa : A \longrightarrow B$ be our homomorphism; passing, if required, to κ_+, we are justified in assuming it to be unital. It is sufficient to show that $a = \lim_{n\to\infty} a_n$; $a_n \in \operatorname{Ker} \kappa$, implies $a \in \operatorname{Ker} \kappa$.

For any $b \in A$ we have $ba = \lim_{n\to\infty} ba_n$. Therefore, by Proposition I.2.39(I), for some n the element $c = e_A - (ba - ba_n)$ is invertible in A; consequently, $\kappa(c) = \kappa(e_A) - \kappa(b)\kappa(a) + \kappa(b)\kappa(a_n) = e_B - \kappa(b)\kappa(a)$ is invertible in B. But, by hypothesis, the elements $\kappa(b)$; $b \in A$ run over the whole of B; this means (I.3.51) that $\kappa(a) \in \operatorname{Rad} B$. Hence, since B is semisimple, $\kappa(a) = 0$. ∎

Theorem (VI.2.65) (Johnson). *Every surjective homomorphism from any Banach algebra onto a semisimple Banach algebra is continuous.*

▶ In the commutative diagram below, where κ is our homomorphism, it follows from Proposition 2.64, since $\operatorname{Im} \kappa = B$, that κ_1 is an isomorphism between Banach algebras; therefore by Theorem 2.63, since B is semisimple, κ_1 is continuous. The rest is clear. ∎

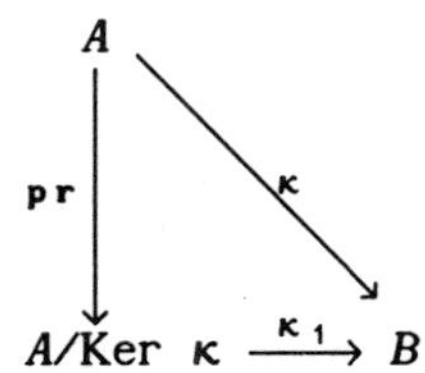

It is interesting that the proof is considerably simpler when A is commutative and moreover — an essential detail! — surjectivity is no longer necessary:

Exercise (VI.2.66). Prove that every homomorphism from a Banach algebra into a semisimple commutative Banach algebra is continuous.

Hint. Using the continuity of the characters of both algebras and Corollary IV.1.4, show that our homomorphism satisfies the conditions of the closed graph theorem.

If B is not commutative, then it is impossible to abandon the requirement of surjectivity in Theorem 2.65. (Construct a discontinuous homomorphism from an algebra with zero multiplication into $\mathcal{M}_2$.) However, it is not known whether one can use only $\overline{\operatorname{Im}\kappa} = B$ in Theorem 2.65 instead of $\operatorname{Im}\kappa = B$. For a discussion of this question and partial results see [72], [142], [205].

Recently Aupetit [205] was able to weaken the conditions of Theorem 2.65 in quite another direction: he proved that surjective operators (not necessarily homomorphisms) between the algebras in Theorem 2.65 are automatically continuous if they do not increase the spectral radius of the elements (cf. II.1.6). It is remarkable that his proof is not, primarily, based on the theory of irreducible modules; it is based on the fact that certain functions connected with the spectral radius are subharmonic.

Having given the deepest positive result about the automatic continuity of homomorphisms — Johnson's Theorem 2.65, it is appropriate to mention the strongest of the available negative results. Up to now, to prove that $\kappa : A \longrightarrow B$ is continuous, we have imposed conditions on B; are there reasonable conditions on A that have the same effect? The situation is that one can construct discontinuous homomorphisms into a Banach algebra fairly easily with almost all specific algebras in the role of A; the methods used, however, are not suitable for $A = C(\Omega)$. The corresponding problem, posed by Kaplansky [150], turned out to be very difficult. It was solved independently by Esterle [301] and Dales [143]; here is their result.

Assertion (VI.2.67). *Let Ω be an infinite compact space. Then if the continuum hypothesis is true* (how do you like that! —

A.H.), *there exists a discontinuous homomorphism from* $C(\Omega)$ *into a certain radical Banach algebra.* *As a consequence* (with the same "if"), *there is in* $C(\Omega)$ *a norm that gives it the structure of a jointly normed algebra but is not equivalent to the uniform norm.*

* * *

We return to Schur's lemma. Has it an analogue for topologically irreducible modules? In a special case, which is important for applications, there is in fact the following.

Theorem (VI.2.68) (Naimark). *If* H *is a star module over a star Banach algebra* A, *then the following properties of* H *are equivalent:*

(I) H *is topologically irreducible*;

(II) ${}_A\mathbf{h}(H) = \ell.h.\{1\} \simeq \mathbb{C}$;

(III) 1 *is the only orthogonal projection in* ${}_A\mathbf{h}(H)$.

(In other words, if $T : A \longrightarrow \mathcal{B}(H)$ is a *-representation of a star Banach algebra, then it is topologically irreducible $\Longleftrightarrow$ any continuous operator on H that commutes with all the T_a; $a \in A$, is a multiple of 1 $\Longleftrightarrow$ any orthogonal projection in H that commutes with all the T_a coincides with 1.)

▶ (I) $\Rightarrow$ (II). Definition 1.15 obviously implies that ${}_A\mathbf{h}(H)$ is a sub-C^*-algebra of $\mathcal{B}(H)$ (and even a von Neumann algebra). Therefore it is sufficient to show that any self-adjoint $\varphi \in {}_A\mathbf{h}(H)$ is a multiple of 1.

Let B be the closure in $\mathcal{B}(H)$ of the set of all polynomials in φ and let Ω be the spectrum of this commutative sub-C^*-algebra of ${}_A\mathbf{h}(H)$. We note that the kernel of any $\psi \in B$ is a closed submodule in H and therefore, if $\psi \neq 0$, it must be trivial since H is topologically irreducible. Consequently, the product of any two non-zero operators from B is non-zero and therefore, since B and $C(\Omega)$ are isomorphic (Theorem IV.7.13), the same holds for the product of any two non-zero continuous functions on Ω. But then, by Urysohn's theorem (Chap.0 §0.3°), Ω is a singleton and therefore $B \simeq C(\Omega) \simeq \mathbb{C}$. The rest is clear.

(II) $\Rightarrow$ (III). This is obvious.

(III) $\Rightarrow$ (I). Let H_1 be a closed submodule of H and let S be the orthogonal projection with $\operatorname{Im} S = H_1$; it is sufficient to show that $S : H \longrightarrow H$ is a morphism of A-modules. By Proposition 1.17, $H_1^\perp$ is also a submodule of H and therefore, for any $a \in A$ and $x \in H$; $x = y + z$, $y \in H_1, z \in H_1^\perp$, we obtain $S(a \cdot x) = S(a \cdot y) + S(a \cdot z) = a \cdot y + 0 = a \cdot S(x)$. The rest is clear. ■

Remark (VI.2.69). Unlike Naimark's theorem, which uses special properties of "star" structures, Schur's lemma itself is not a statement of equivalence: ${}_Ah(X) \simeq \mathbb{C}$, generally speaking, does not imply that X is irreducible. The simplest counterexample is the natural module over the algebra of operators that can be written as upper triangular 2×2-matrices.

Recalling Kadison's deep theorem on C^*-algebras (Assertion 2.25) we indicate another, comparatively simple, case where ordinary irreducibility and topological irreducibility coincide.

Proposition (VI.2.70). *Let A be a commutative star Banach algebra. Then every topologically irreducible star A-module H is one-dimensional (= each topologically irreducible *-representation of this algebra is one of its *-characters).*

▶ Let T be the representation associated with H. Since A is commutative, for any $a \in A$ the operator T_a commutes with all T_b, $b \in A$, and therefore, by Theorem 2.68 and automatic continuity (IV.5.25), it is a multiple of 1. But then all subspaces of H are submodules. The rest is clear. ■

Now let A be an arbitrary Banach algebra and let X be an arbitrary topologically irreducible Banach A-module. Can we assert, as before, that ${}_A\mathbf{h}(X) \simeq \mathbb{C}$? To all who have tackled this question it has become clear that, if this conjecture were true, it would also imply, almost incidentally, a positive solution of the famous "*invariant subspace problem*": does every continuous operator on a Banach space X have a closed invariant subspace (other than X and $\{0\}$)? We note the relevant argument.

We let $l_1(\mathbb{N})$ denote the Banach algebra with underlying space l_1 and the "power series" multiplication $\xi * \eta := \{\zeta_n = \sum_{k+l=n} \xi_k \eta_l\}$, where $\xi = \{\xi_n\}$, $\eta = \{\eta_n\} \in l_1$ and $n = 0, 1, 2, \cdots$ (the required properties are easily verified).

Exercise (VI.2.71). Assume that there exists a continuous operator S in X "with no invariant subspace". Show that this X must have the structure of a topologically irreducible Banach module over $A = l_1(\mathbb{N})$ with ${}_A\mathbf{h}(X) \not\simeq \mathbb{C}$.

Hint. Assuming $\|S\| = 1$, we put $\xi \cdot x := \sum_{n=0}^{\infty} \xi_n S^n(x)$; $\xi = \{\xi_n\} \in l_1$, $x \in X$. The fact that $\delta_1 \cdot x = S(x)$, where $\delta_1 := (0, 1, 0, 0, \cdots)$, implies that the module thus constructed is topologically irreducible. At the same time the operator S, which is clearly not a multiple of the identity, belongs to ${}_A\mathbf{h}(X)$.

All that remains is for us to add that the question posed above remained open until very recently, when the invariant subspace problem itself was answered in the negative by both Enflo [299a] and Read [217]; and so, by Exercise 2.71, this question also has a negative solution. As far as the author knows nobody has discovered to date a simpler example (without using such strong methods) of a topologically irreducible Banach module with ${}_A\mathbf{h}(X) \neq \mathbb{C}$.

└

§2.4. The connection between topologically irreducible *-representations and pure states. Application: a new characterization of the star radical and star semisimplicity

The material in this section borders directly on those concepts and constructions in the theory of involutive algebras that were the subject of Chap.IV §5. No doubt the reader remembers the fundamental role of GNS-representations, which were first defined there, and now may be curious, having become acquainted with irreducible representations and their merits, to know: which of the representations given by GNS-constructions are irreducible and are there many of them? It turns out that, if such a question is slightly touched up, namely if we are interested in topologically irreducible representations, then we get a beautiful and conclusive answer which is the starting point of many applications. (Besides, thanks to Kadison — see Assertion 2.25 — for such an important class as the C^*-algebras the retouched question coincides with the original.)

Let A be a *-algebra. Given a functional f on A let us agree to write $f \geq 0$ if it is positive; for functionals f and g we shall write $f \leq g$ if $g - f \geq 0$. For $f, g \geq 0$ we shall say that g is subordinate to f if $g \leq Cf$ for a certain $C > 0$. The set of all functionals subordinate to a given f is denoted by $\mathrm{Sub}(f)$.

Further, let H be a Hilbert space. When we speak about positive operators on H we have in mind the elements of $\mathcal{B}(H)_{pos}$ and we use the notation $S_1 \leq S_2$; $S_1, S_2 \in \mathcal{B}(H)$ for the order relation defined in Chap.IV §7.5 (with $\mathcal{B}(H)$ as A).

In the next proposition, which plays a key role in what follows, A is a unital star Banach algebra, T is a unital topologically cyclic *-representation of A on H with topologically cyclic vector x; further, $\check{H} := \{T_a x : a \in A\}$ and $f := f_{T,x} : a \longmapsto \langle T_a x, x\rangle$ is the corresponding positive vector functional on A (see IV.5.3). We introduce the notation $\mathrm{Pos}(T)$ for the set of positive operators in $T^!$ [21] and finally,

---

[21] Recall that $T^!$ is, in the language of modules, ${}_A\mathbf{h}(H)$. In general we use the language of representations in this section, which is evidently more convenient here than the language of modules.

if $S \in \mathrm{Pos}(T)$ we let g_S denote the functional on A defined by $g_S(a) := \langle ST_a x, x\rangle$.

Lemma (VI.2.72). *For any* S *in* $\mathrm{Pos}(T)$, g_S *belongs to* $\mathrm{Sub}(f)$.

▸ For any $a \in A$ the operator $Q_a := T_{a*}ST_a$, being positive (IV.7.75), has an algebraic square root (IV.7.27). Hence

$$g_S(a^*a) = \langle ST_{a*a}x, x\rangle = \langle Q_a x, x\rangle = \langle Q_a^{1/2}x, Q_a^{1/2}x\rangle \geq 0.$$

Therefore $g_S \geq 0$. In exactly the same way, if we take $\|S\|1 - S \geq 0$ (see IV.7.77(VI)) instead of S, we get that $g_{\|S\|1-S} = \|S\|f - g_S \geq 0$. ■

Proposition (VI.2.73). *The map* $S \longmapsto g_S$ *is a bijection between the sets* $\mathrm{Pos}(T)$ *and* $\mathrm{Sub}(f)$.

▸ We denote our map by $\alpha : \mathrm{Pos}(T) \longrightarrow \mathrm{Sub}(f)$ and show that it has an inverse.

We take $g \in \mathrm{Sub}(f)$; $0 \leq g \leq Cf$. The "Cauchy–Bunyakovskii inequality" IV.5.5 (applied to g) implies that, for any $a, b \in A$

$$|g(b^*a)|^2 \leq g(a^*a)g(b^*b) \leq C^2 f(a^*a)f(b^*b) = C^2\|T_a x\|^2\|T_b x\|^2.$$

Hence $g(b^*a) = 0$ as soon as $T_a x = 0$ or $T_b x = 0$ and hence — since multiplication is bilinear, we have $g(b_1^*a_1) = g(b^*a)$ whenever $T_{a_1}x = T_a x$ and $T_{b_1}x = T_b x$. This, in its turn, means that the map $\mathcal{R} : \check{H} \times \check{H} \longrightarrow \mathbb{C} : (y, z) \longmapsto g(b^*a)$ is well defined, where $a, b \in A$ are any elements with $T_a x = y$ and $T_b x = z$. Obviously $\mathcal{R}$ is a sesquilinear functional with $\|\mathcal{R}\| \leq C$; therefore by the last theorem in Chap.0 §1.3, there is a uniquely defined $R := R_g \in \mathcal{B}(H)$ such that $\langle Ry, z\rangle = \mathcal{R}(y, z)$ for all $y, z \in \check{H}$, in other words, such that $\langle RT_a x, T_b x\rangle = g(b^*a)$ for all $a, b \in A$. Let us consider what properties it has.

First of all, the last equation implies that, for all $a, b, c \in A$,

$$\langle RT_a(T_b x), T_c x\rangle = \langle RT_{ab}x, T_c x\rangle = g(c^*(ab)) = g((a^*c)^*b) =$$

$$= \langle RT_b x, T_{a*}(T_c x)\rangle = \langle T_a R(T_b x), T_c x\rangle,$$

and hence $\langle RT_a y, z\rangle = \langle T_a Ry, z\rangle$ for all $y, z \in \check{H}$. Since $\check{H}$ is dense in H, this implies that $RT_a = T_a R$ for all $a \in A$; that is, that $R \in T^!$. In addition, $g \geq 0$ implies that, for all $a \in A$, $\langle RT_a x, T_a x\rangle = g(a^*a) \geq 0$. Therefore $\langle Ry, y\rangle \geq 0$ for all $y \in \check{H}$ and therefore for all $y \in H$. By the criterion for an operator to be positive (Chap.0 §1.3), this means that $R \geq 0$.

So $R_g \in \text{Pos}(T)$; and we have thus constructed a map $\beta : \text{Sub}(f) \longrightarrow \text{Pos}(T) : g \longmapsto R_g$. Since T is unital, $\langle R_g T_a x, x\rangle = \langle R_g T_a x, T_e x\rangle = g(e^* a) = g(a)$; hence $g_{(R_g)} = g$ for all $g \in \text{Sub}(f)$. This shows that β is a right inverse to α.

Finally, for any $S \in \text{Pos}(T)$ and $a, b \in A$,

$$\langle ST_a x, T_b x\rangle = \langle T_{b*} ST_a x, x\rangle = \langle ST_{b*a} x, x\rangle = g_S(b^* a)$$

is satisfied. Hence $\langle Sy, z\rangle = \langle R_{(g_S)} y, z\rangle$ for all y, z in $\check{H}$ and, as a consequence, in H. Thus $R_{(g_S)} = S$ and this shows that β is a left inverse to α. The rest is clear. ∎

Exercise (VI.2.74). Prove that, if the map $S \longmapsto g_S$ is extended by linearity to the whole of $T^!$, it realizes an isomorphism of this linear space onto $\ell.h.\text{Sub}(f)$.

Functionals with $\text{Sub}(\cdot)$ as small as possible deserve a special definition.

Definition (VI.2.75). A positive functional f on a *-algebra A is said to be *pure* if $\text{Sub}(f) = \{tf : t \geq 0\}$ (in other words, if $0 \leq g \leq f$ implies that g is proportional to f).

Now everything is ready for the promised criterion of topological irreducibility.

Theorem (VI.2.76) (Segal). *Let A be a star Banach algebra and let f be a positive functional on $A_\times$. Then the* GNS-*representation T^f associated with f is topologically irreducible $\Longleftrightarrow$ f is pure.*

▶ Since $\{T^f_a \hat{e} : a \in A_\times\} = \tilde{H}^f$ is satisfied for $\hat{e} := e + J^f$ (see the end of Chap.IV §5.1), the representation T^f is topologically cyclic. Further, by Proposition IV.5.14(I), $f_{T^f, \hat{e}} = f$. Therefore Lemma 2.72, applied with $T = T^f$, provides a bijection between $\text{Pos}(T^f)$ and $\text{Sub}(f)$, where it is obvious that the set $\{t1 : t \geq 0\}$ maps onto $\{tf : t \geq 0\}$. Hence f is pure $\Longleftrightarrow$ $\text{Pos}(T^f) = \{t1 : t \geq 0\}$. But the last property is equivalent to T^f being topologically irreducible: in one direction this follows obviously from Theorem 2.68 (III) ⇒ (I), and in the other by Theorem 2.68 (I) ⇒ (II). ∎

We show as a first application ("to warm up") that, in the commutative case, pure functionals are well-known to us under another name.

Theorem (VI.2.77). *A state* (see IV.5.1) *f on the commutative star Banach algebra A is pure $\Longleftrightarrow$ it is a *-character of this algebra.*

▶ "⇒". If we combine Theorem 2.76 and Proposition 2.70, we see that T^f is a *-character on A; moreover, as for all

GNS-representations, $\operatorname{Ker} T^f \subseteq J^f$ (IV.5.14(II)).[22] Further, it follows from Corollary IV.5.6(II) (taking b to be the identity) that $J^f \subseteq \operatorname{Ker} f$. Since we are talking about the kernels of functionals, the two inclusions give $\operatorname{Ker} T^f = \operatorname{Ker} f$. Since again $T^f(e) = f(e) = 1$, f must coincide with the *-character T^f.

"$\Leftarrow$". Let χ be a *-character on A. Then $J^\chi = \{a : \chi(a^*a) = 0\} = \{a : |\chi(a)|^2 = 0\} = \operatorname{Ker} \chi$; consequently $\tilde{H}^\chi$ (that is, A/J^χ) is one-dimensional and so H^χ is also one-dimensional. Therefore T^χ is automatically irreducible and all that remains is to apply Theorem 2.76. ■

Exercise (VI.2.78). Get the same result by other means without using Theorem 2.76, but by using in an essential way a property of universal enveloping C^*-algebras (Chap.IV §7.4).

Hint. A state on A determines uniquely a certain state on $C^*(A)$. Hence one can deduce that $f(a) = \int_\Delta \hat{a}(s)d\mu(s)$ for a certain positive measure μ on the subset Δ of *-characters in $\underline{\Omega}(A)$ that can be identified with $\underline{\Omega}(C^*(A))$.

But the main applications of Theorem 2.76 are connected with a remarkable geometric property of the set of pure states. Let A be a unital *-algebra. Recalling the notation $\underline{\mathrm{St}}(A)$ for the set of all states, we let $\underline{\mathrm{PSt}}(A)$ denote the set of its pure states.

Theorem (VI.2.79). $\underline{\mathrm{PSt}}(A)$ *is the set of extreme points* (see the end of Chap.0 §1.2) *of* $\underline{\mathrm{St}}(A)$.

▶ First let f be a pure state and $f = tg + (1-t)h$, where $g, h \in \underline{\mathrm{St}}(A)$, $0 \le t \le 1$. Then $g, h \in \operatorname{Sub}(f)$, and therefore $g, h \in \ell.h.\{f\}$; consequently, since we are talking about states, $f = g = h$. We have shown that f is an extreme point of $\underline{\mathrm{St}}(A)$.

Conversely, let f be an extreme point of $\underline{\mathrm{St}}(A)$ and let g be subordinate to f. Replacing g, if necessary, by a positive multiple, we are justified in assuming that $f = g + h$, where $g, h \ge 0$ and $g, h \ne 0$; the latter, by Proposition IV.5.7(II) means that $g(e), h(e) \ne 0$. We put $t := g(e)$; then $1 - t = h(e)$ and $f = tg_0 + (1-t)h_0$, where $g_0 := t^{-1}g$ and $h_0 := (1-t)^{-1}h$ are states. Hence $g_0 = h_0 = f$ and hence g is proportional to f. We have shown that f is a pure state. ■

Now let A be a unital star Banach algebra. Recall that all the states on A are automatically continuous (IV.5.18) and let us make the obvious observation that $\underline{\mathrm{St}}(A)$ is a convex

---

[22] In the language of modules $\operatorname{Ker} T^f$ is $\mathcal{A}nn_A \tilde{H}^f$ (that is, $\mathcal{A}nn_A A/J^f$); because of this the above inclusion is also a particular case of Proposition 1.6(II).

subset of A^* which is closed in the weak* topology. We are thus in a situation which enables us to apply one of the "strongest tools" in functional analysis — the Krein-Milman theorem (Chap.0 §1.2). The latter, combined with Theorem 2.79, immediately gives the following.

Corollary (VI.2.80). *For any* $f \in \underline{\mathrm{St}}(A)$, $a_1, \cdots, a_n \in A$ *and* $\varepsilon > 0$ *there exist* $f_1, \cdots, f_m \in \underline{\mathrm{PSt}}(A)$ *and numbers* $t_1, \cdots, t_m \geq 0$, $t_1 + \cdots + t_m = 1$ such that $|f(a_k) - \sum_{l=1}^{m} t_l f_l(a_k)| < \varepsilon$ for all $k = 1, \cdots, n$.

This, in its turn, implies the following.

Corollary (VI.2.81). *Let* A *be a star Banach algebra. Then for any* $a \in A$ *its star bound* $B^*(a)$ *is equal to* $\sqrt{\sup\{g(a^*a) : g \in \underline{\mathrm{PSt}}(A_\times)\}}$ (and not just $\sqrt{\sup\{g(a^*a) : g \in \underline{\mathrm{St}}(A_\times)\}}$, as we saw in Chap.IV §5.2).

We now, finally, are in a position to fulfil the promise made in IV.5.30. In the next theorem A is a star Banach algebra, $\mathcal{T} : A \longrightarrow \mathcal{B}(\mathcal{H})$ is its universal representation (IV.5.26) and $\mathcal{T}\,\mathcal{R}ep(A)$ denotes the class of its topologically irreducible *-representations.

Theorem (VI.2.82). $\|\mathcal{T}_a\| = \sup\{\|T_a\| \,;\, T \in \mathcal{T}\,\mathcal{R}ep(A)\}$ *is satisfied for any* $a \in A$.

▶ In the chain of inequalities

$$B^*(a) \leq \sup\{\|T^f_a\| : f \in \underline{\mathrm{PSt}}(A_\times)\} \leq \sup\{\|T_a\| : T \in \mathcal{T}\,\mathcal{R}ep\, A\} \leq$$
$$\leq \sup\{\|T_a\| : T \in \mathcal{R}ep\, A\}$$

the first is implied by Corollary 2.81 and Proposition IV.5.14(III) and the second by Theorem 2.76 and the third is obvious. But both the first and the last of these numbers are equal, by Theorem IV.5.27, to $\|\mathcal{T}_a\|$. ■

This theorem obviously implies a new characterization of the star radical of star Banach algebras.

Theorem (VI.2.83) (cf.2.33). *For any star Banach algebra* A, *-Rad A *consists of all those* $a \in A$ *for which* $T_a = 0$ *for every topologically irreducible* *-*representation* T *of this algebra.* ▶■

Corollary (VI.2.84) (cf. 2.34(II)). *If* A *is a star- semisimple star Banach algebra, then it has sufficiently many topologically irreducible* *-*representations in the sense that, for any* $a \in A\backslash\{0\}$ *there exists a representation* T *in the indicated class such that* $T_a \neq 0$.

The last fact has important applications in harmonic analysis, enabling one to give the "correct" generalization, for non-commutative groups, of the theorem about the sufficiency of the character set (III.3.21).

Let $\check{T}$ be a bounded representation of a locally compact group G on the Banach space E. We say that it is irreducible if the operators $\check{T}_s$; $s \in G$ have no closed invariant subspace in common, except E itself and $\{0\}$.[23] It is easy to see that the irreducible unitary representations of the group G are just those that correspond, under the bijection indicated in Assertion 1.34, to the topologically irreducible *-representations of the algebra $L^1(G)$. This observation, combined with Corollary 2.84 and the fact that, as mentioned earlier (IV.5.37), the algebras $L^1(G)$ are star semisimple, enables one to establish the following assertion.

Assertion (VI.2.85) (Gel'fand-Raikov; see, for example, [87, Theorem 22.12]). *For any two distinct elements* s *and* t *of a locally compact group there exists an irreducible unitary representation* $\check{T}$ *of this group such that* $\check{T}_s \neq \check{T}_t$.

Remark (VI.2.86). If G is compact, then every irreducible representation of G is necessarily finite-dimensional (this result, which is closely connected with Assertion 2.19, is set out in detail in [40, §9]). Thus Assertion 2.85 implies that there are enough homomorphisms into groups of unitary matrices (of various finite dimensions) to distinguish the elements of a compact group.

* * *

Another important group of questions, where the connection between irreducible representations and pure states is applied, is to give a so-called decomposition of the representations of star Banach algebras, primarily C^*-algebras, into (topologically or "simply") irreducible representations. However, a discussion of these questions goes beyond the scope of our book: even to pose them precisely is itself a far from simple matter and demands some tedious work with a number of technical definitions. (We are talking, as a rule, not about a decomposition into a direct sum but about a more refined decomposition into a direct integral by so-called disintegration.) About these problems see, for example, [40, §10], [31, §8], [53, Ch.VIII].

[23] In group theory one says "irreducible" and not "topologically irreducible": the fact is that, in general, one very rarely meets representations of groups, as opposed to algebras, with no invariant subspaces, closed or otherwise.

§3. VARIOUS TYPES OF MODULE, THEIR INTERRELATION AND CERTAIN CONSTRUCTIONS CONCERNING THEM

§3.1. Right modules and bimodules; examples. Reduction of modules of all types to left unital modules

Up to now left modules have been sufficient for us; however, for subsequent work it is indispensable to introduce into the discussion some more variations of this concept.

First of all, if we replace the notation $a \cdot x$ by $x \cdot a$ in Definition 1.1 and require the identity $(x \cdot a) \cdot b = x \cdot (ab)$ to be satisfied, we get the definition of a *right outer multiplication* (or *action*) and then of a *right module* over the given algebra A or of a *right* A*-module.* Obviously, a right action of A is precisely a left action of A^{op} with $a \cdot x$ equal to the previous $x \cdot a$. For a commutative algebra, of course, the two versions of the concept of module coincide.

We introduce one more "composite" concept whose naturalness and usefulness will gradually become apparent in the course of this account.

Definition (VI.3.2). Let A and B be algebras. A linear space X is called a *bimodule which is a left module with respect to* A *and a right module with respect to* B or briefly an A-B*-bimodule* if it is a left A-module and a right B-module and the identity $(a \cdot x) \cdot b = a \cdot (x \cdot b)$ is satisfied.

We shall talk about an "A-bimodule" or "bimodule over A" instead of an "A-A-module".

Such concepts as unital module (and unitization), submodule, quotient module, Cartesian product and also irreducibility can obviously be carried across to right modules and bimodules.

We have defined "pure" modules; definitions of Banach and, more generally, polynormed modules of the types indicated, duplicate 1.11 and 1.12 with obvious modifications. Concepts of the essential part (denoted as before by X_{es}) and a non-degenerate module can be carried across directly to such modules.

By analogy with Definition 1.3, an operator satisfying the identity $\varphi(x \cdot a) = \varphi(x) \cdot a$ is called a *morphism of right* A*-modules.* An operator that is simultaneously a morphism of left A- and right B-modules is called a *morphism of* A-B*-bimodules.* We use the notation $h_A(\cdot, \cdot)$ and $\mathbf{h}_A(\cdot, \cdot)$ for the set of morphisms between right A-modules and ${}_Ah_B(\cdot, \cdot)$ and ${}_A\mathbf{h}_B(\cdot, \cdot)$ for the set of morphisms between A-B-bimodules; as in §1, an "italic" h means that we are talking about arbitrary morphisms and a "bold" $\mathbf{h}$ means that, in the context of polynormed modules, we are talking about continuous morphisms. If we view these morphisms in the light of category theory, we see that each of the categories of left modules introduced earlier has obvious "right-" and "bimodule" versions. In the corresponding notation "A-..." is replaced by "...-A" or "A-...-B": for example, **mod-*A*** is the category of right Banach A-modules, ***A*-mod-*A*** of Banach A-bimodules, ***A*-algunmod-*B*** of pure unital A-B-bimodules and so on. Finally, the concept of an isomorphism in all its variations — algebraic, topological and isometric — obviously extends from left modules to right modules and bimodules.

* * *

Of course, the algebra A itself and any algebra ambient to it, such as A_+, provide examples not only of left but also of right A-modules and also of A-bimodules; every right ideal in A is a right A-module and a bi-ideal is an A-bimodule (pure, Banach or polynormed, according to the circumstances).

The following few observations show that studying left modules itself must sooner or later lead to the appearance on the scene of other types of module as well. We begin with one important construction.

Let A be a Banach algebra, X a left Banach A-module. Then the dual Banach space X^* is obviously a right Banach A-module with action defined by $\langle f\cdot a, x\rangle := \langle f, a\cdot x\rangle$; $a\in A$, $f\in X^*$, $x\in X$. This right A-module is said to be *dual to the left A-module X*. Analogously *the left Banach A-module dual to a given right module* is defined using $\langle a\cdot f, x\rangle := \langle f, x\cdot a\rangle$, while *the Banach B-A-bimodule, dual to a given Banach A-B-bimodule* is defined by using both formulae. A (bi)module of either type dual to any other is called a *dual (bi)module.*

Example (VI.3.3). Let G be a discrete group. Then it is easily verified that the $l_1(G)$-bimodule $l_1(G)^*$ coincides, up to an isometric isomorphism, with $l_\infty(G)$ [24] which is given the outer multiplication $f\cdot g(s) := \sum_{t\in G} f(s^{-1}t)g(t)$ and $g\cdot f(s) := \sum_{t\in G} f(t^{-1})g(t^{-1}s)$, $f\in l_1(G)$, $g\in l_\infty(G)$, $s\in G$.

Exercise (VI.3.4). Let G be $\mathbb{R}$ or $\mathbb{T}$. Indicate the explicit form of the outer multiplications in the $L^1(G)$-bimodule $L^\infty(G) = L^1(G)^*$ and verify that its essential part (as modules of any type) consists of the bounded uniformly continuous functions.

It can be proved that every von Neumann algebra A is a dual A-bimodule; this fact is important for the theory of operator algebras [222].

---

[24] We recall that $l_\infty(G)$ or $c_b(G)$ is the notation for the space of all bounded functions on G with the uniform norm.

The construction of the dual module is generalized by the following.

Proposition (VI.3.5). *Let A and B be Banach algebras, and let $X \in A$-**mod** and $Y \in B$-**mod**. Then $(\mathcal{B}(X, Y); \|\cdot\|_0)$, is a Banach B–A-bimodule with actions $[b\cdot\varphi](x) := b\cdot\varphi(x)$ and $[\varphi\cdot a](x) := \varphi(a\cdot x)$; $\varphi \in \mathcal{B}(X, Y)$.* ▶■

Here is one more source of bimodules that crops up.

Proposition (VI.3.6). *Let A, B and X be the same as in Proposition* 3.5 *except that $Y \in$ **mod**-B. Then $X \hat{\otimes} Y$ is a Banach A–B-bimodule with actions uniquely determined by $a\cdot(x \otimes y) := a\cdot x \otimes y$ and $(x \otimes y)\cdot b := x \otimes y\cdot b$.*

▶ The argument in Proposition 1.20, that enabled us to construct the left action in $A \hat{\otimes} E$ can be carried out with obvious modifications. ■

Example (VI.3.7). Let G_1 and G_2 be locally compact groups. Then the Grothendieck isometric isomorphism $L^1(G_1) \hat{\otimes} L^1(G_2) \xrightarrow{\simeq} L^1(G_1 \times G_2)$ (0.3.36) is "pluralistically" an isomorphism not only of Banach algebras (cf. Chap. III §3.2) but also of Banach $L^1(G_1)$–$L^1(G_2)$-bimodules: in the first of them the outer multiplications are particular cases of those described in Proposition 3.6 and in the second they are given by convolution with respect to the corresponding coordinate (write down the explicit formulae!).

Besides the general constructions indicated, bimodules over Banach function algebras and also over arbitrary commutative Banach algebras often occur when considering functions of two variables.

Example (VI.3.8). Let A_k be a commutative Banach algebra with spectrum Ω_k; $k = 1, 2$. Then $(C_0(\Omega_1 \times \Omega_2); \|\cdot\|_0)$ is a Banach A_1–A_2-bimodule with the operations $[a\cdot u](s, t) := \hat{a}(s)u(s, t)$ and $[u\cdot b](s, t) := u(s, t)\hat{b}(t)$; $s \in \Omega_1$, $t \in \Omega_2$.

This bimodule may be regarded as a two-sided version of the module in Example 1.25. We suggest finding similar versions of the other examples of modules over function algebras considered in §1.3.

Example (VI.3.9). The augmenting $L^1(G)$-module $\mathbb{C}$ from Example 1.31 is also a right $L^1(G)$-module and a $L^1(G)$-bimodule with a right action coinciding with the left action $(z\cdot f \equiv f\cdot z)$.

* * *

The concept of an *n-module* — that is, a left module with respect to one algebra and a right module with respect to n other given algebras — can be introduced analogously to bimodules by requiring that the operators of outer multiplication by elements of distinct algebras commute. Moreover every module may be considered as a module of the more complicated type by enlisting the services of the algebra $\mathbb{O}$ or — for unital modules — $\mathbb{C}$. For example, a left A-module is also an A-$\mathbb{O}$-bimodule and so on.

More interesting is the fact that the apparent diversity of the types of module introduced is deceptive: on the contrary, all these types can be reduced to a single type — left unital modules. We have already noted that ordinary modules reduce to unital modules and right modules reduce to left modules; in the language of category theory this means, in particular, that there are category isomorphisms **mod**-$A \simeq A^{\mathrm{op}}$-**mod** $\simeq A_+^{\mathrm{op}}$-**unmod**. For more complicated modules (we limit ourselves to bimodules) tensor products come to our aid:

Proposition (VI.3.10). *Let A and B be Banach algebras and let X be a Banach A-B-bimodule. There exists a unique method of turning X into a unital left Banach module over $A_+ \hat{\otimes} B_+^{\mathrm{op}}$* (see I.2.26) *so that $(a \otimes b)\cdot x = a \cdot x \cdot b$ for all $a \in A$, $b \in B$, $x \in X$. Moreover any continuous morphism of A-B-bimodules becomes a continuous morphism of left $A_+ \hat{\otimes} B_+^{\mathrm{op}}$-modules. The functor constructed in this way from A-**mod**-B into $A_+ \hat{\otimes} B_+^{\mathrm{op}}$-**unmod** is a category isomorphism.*

▶ We consider X as a unital A_+-B_+-bimodule and having fixed $x \in X$ we take the bi-operator $\dot{\mathcal{R}} : A_+ \times B_+^{\mathrm{op}} \longrightarrow X : (a, b) \longmapsto a \cdot x \cdot b$ and then the operator $m_x : A_+ \hat{\otimes} B_+^{\mathrm{op}} \longrightarrow X$ associated with it. We now take $\dot{m} : (A_+ \hat{\otimes} B_+^{\mathrm{op}}) \times X \longrightarrow X : (u, x) \longmapsto m_x(u)$ as the outer multiplication. The rest is clear. ■

The case $A = B$ (of A-bimodules) deserves, in view of its subsequent special role, a special term.

Definition (VI.3.11). The algebra $A_+ \hat{\otimes} A_+^{\mathrm{op}}$, also denoted by A^{env}, is called *the enveloping algebra* of the Banach algebra A.

We note that $(A^{\mathrm{env}})^{\mathrm{op}}$ naturally identifies with $A_+^{\mathrm{op}} \hat{\otimes} A_+$ and therefore with A^{env} itself. Therefore every Banach A-bimodule may be considered as a left as well as a right module over (one and the same) A^{env}.

If our algebras and modules are unital, then when reducing these modules to left modules there is no further need to adjoin the identity to the corresponding algebras. In particular, for unital A-bimodules, $A \hat{\otimes} A^{\mathrm{op}}$, sometimes denoted by $A^{\mathrm{env}}_{\mathrm{un}}$ plays the role of the algebra A^{env}. We leave the details to the reader.

We direct attention to the left A^{env}-module A_+ and, in the unital case, the left $A^{\mathrm{env}}_{\mathrm{un}}$-module A corresponding to the appropriate A-bimodules. They will play an important role in the study of the homological properties of algebras (see Chap. VII §3.3). Both these modules, obviously, are cyclic.

Of course, all that has been said above about identifying the various types carries across from Banach to other classes of polynormed modules (and of course to abstract modules); it is only necessary to use suitable versions of the tensor product. Thus, for example, for the category of abstract bimodules (over a fixed algebra) an analogue to Proposition 3.10 holds with "$\hat{\otimes}$" replaced by the algebraic tensor product "$\otimes$", for the category of all polynormed bimodules "$\otimes_i$" plays a similar part, and so on. Again we leave the details to the reader (see [81, Chap.II §5.2] about certain categories).

§3.2. Forgetful and replacement operations. Modules of morphisms and tensor products of modules

In this section we describe certain important constructions that enable one to pass from some modules to others. More formally, we shall speak about certain functors connecting the categories introduced above.

Definition (VI.3.12). The functor $\square : A\text{-}\mathbf{mod} \longrightarrow \mathcal{B}an$ that associates to each module its underlying Banach space and to each morphism the same map but considered only as an operator, is called the *forgetting about outer multiplication* or simply the *forgetful functor*.

Definition (VI.3.13). Let $\kappa : A \longrightarrow B$ be a continuous homomorphism of Banach algebras. The functor $\kappa^{\#} : B\text{-}\mathbf{mod} \longrightarrow A\text{-}\mathbf{mod}$, that associates to each B-module X the A-module X_κ with the same underlying space and outer multiplication $a \cdot x := \kappa(a) \cdot x$; $a \in A$, and to each morphism of B-modules the same map but considered as a morphism of A-modules (check the necessary properties!), is called the *replacement functor via the homomorphism* κ.

The functors we have introduced have obvious analogues for other types of module and other classes of underlying spaces. Later on we shall freely use expressions like "A-B-module, considered as a right B-module" having in mind the appropriate forgetful functor $A\text{-}\mathbf{mod}\text{-}B$ into $\mathbf{mod}\text{-}B$.

We note that the forgetful functor in Definition 3.12, selected by us in view of its importance, is a special case of a replacement functor, namely that of replacement via the unique homomorphism from $\mathbb{O}$ into A (in the "unital" version — via the unital homomorphism from $\mathbb{C}$ into A). Further, the module in Example 1.25 is also essentially obtained by replacement: this time from $C(\Omega)$-modules by replacement via $\Gamma : A \longrightarrow C(\Omega)$.

The replacement functor provides a suitable language for considering a whole series of problems in algebra and analysis. In particular, problems about constructing a functional calculus (using various classes of functions; cf. Chap. II §2 and Chap. IV §7.2) may be considered as special cases of problems "about solving $X_\kappa = Y$ for given $\kappa : A \longrightarrow B$ and $Y \in A\text{-}\mathbf{mod}$".

So, for example, if κ is the natural embedding of $\mathcal{P}_n$ in $\mathcal{O}(U)$; $U \subseteq \mathbb{C}^n$, and X is a Banach space that is a polynormed $\mathcal{P}_n$-module, then the corresponding problem, as is easily verified, is precisely the problem about constructing a holomorphic calculus in U for a given system of commuting operators on X (in this case — the outer multiplication operators by the independent variables); see Chap. II §2.2. Having indicated this interpretation, Taylor was able to apply effective methods of homological algebra to the preceding problem; some of his most significant results have already been mentioned, together with references (see *ibid*).

For Taylor's results relating to functional calculi of a general kind, their existence and uniqueness, see [252], [271].

* * *

The next functor — more correctly, a class of functors — deserves a very detailed examination by us.

Recall the co- and contravariant morphism functors $\mathbf{h}_{\mathcal{K}}(\cdot, \cdot)$ defined for general types of categories in Chap. 0 §2.2. We are now interested in the special case $\mathcal{K} := A\text{-}\mathbf{mod}$, where A is a Banach algebra. Here, it is obvious that, as for $\mathcal{K} = \mathcal{B}an$ (see *ibid*), the set $\mathbf{h}_{\mathcal{K}}(X, Y)$ — this is the same as ${}_A\mathbf{h}(X, Y)$ — is a Banach space and the map $\mathbf{h}_{\mathcal{K}}(X, \varphi)$ — this is the same as ${}_A\mathbf{h}(X, \varphi)$ — is a bounded operator. Therefore for a fixed $X \in A\text{-}\mathbf{mod}$ we get a *covariant morphism functor* ${}_A\mathbf{h}(X, ?) : A\text{-}\mathbf{mod} \longrightarrow \mathcal{B}an$ and for a fixed $Y \in A\text{-}\mathbf{mod}$ there is an analogous definition of the *contravariant morphism functor* ${}_A\mathbf{h}(?, Y) : A\text{-}\mathbf{mod} \longrightarrow \mathcal{B}an$ (recall that for $\varphi : X_1 \longrightarrow X_2$ the operator ${}_A\mathbf{h}(\varphi, Y)$ takes $\psi : X_2 \longrightarrow Y$ into $\psi \circ \varphi : X_1 \longrightarrow Y$).

The functors we have considered take values in $\mathcal{B}an$ as though the modular structure were annihilated. There are other functors constructed by a similar principle which, on the contrary, enrich the original structures. In fact, we have already seen in Proposition 3.5 that when $X \in A\text{-}\mathbf{mod}$ and $Y \in B\text{-}\mathbf{mod}$ the space $\mathcal{B}(X, Y)$ (later also denoted by $\mathbf{h}(X, Y)$) has the natural structure of a Banach B-A-bimodule. Therefore, obviously, if we extend this construction to the morphisms, we get a covariant "operator functor" $\mathbf{h}(X, ?) : B\text{-}\mathbf{mod} \longrightarrow B\text{-}\mathbf{mod}\text{-}A$; a contravariant functor $\mathbf{h}(?, Y) : A\text{-}\mathbf{mod} \longrightarrow B\text{-}\mathbf{mod}\text{-}A$ is defined analogously.

If we start from the very beginning using, not one-sided modules, but bimodules, then it is not difficult to discover that there is a construction that generalizes both pairs of functors considered above and with them a whole lot of others. Here, in view of what follows, we have to show a certain precision.

Proposition (VI.3.14). *Let A, B and C be Banach algebras. Then*

(I) *for* $X \in A$**-mod-**B *and* $Y \in A$**-mod-**C, ${}_A\mathbf{h}(X, Y)$ *is a Banach B-C-bimodule with the operations* $[b\cdot\psi](x) := \psi(x\cdot b)$ *and* $[\psi\cdot c](x) := \psi(x)\cdot c$; $b \in B$, $c \in C$, $\psi \in {}_A\mathbf{h}(X, Y)$, $x \in X$. *For fixed X the assignment* $\mathbf{h}(X, ?) : Y \longrightarrow {}_A\mathbf{h}(X, Y)$ *and* $\varphi \longmapsto {}_A\mathbf{h}(X, \varphi)$ *is a covariant functor from* A**-mod-**C *into* B**-mod-**C *(and if Y is fixed,* ${}_A\mathbf{h}(?, Y) : X \longmapsto {}_A\mathbf{h}(X, Y)$ *and* $\varphi \longmapsto {}_A\mathbf{h}(\varphi, Y)$ *is a contravariant functor from* A**-mod-**B *into* B**-mod-**C);

(II) *for* $X \in B$**-mod-**A *and* $Y \in C$**-mod-**A, $\mathbf{h}_A(X, Y)$ *is a Banach C-B-module with the operations* $[\psi\cdot b](x) := \psi(b\cdot x)$ *and* $[c\cdot\psi](x) = c\cdot\psi(x)$; $b \in B$, $c \in C$, $\psi \in \mathbf{h}_A(X, Y)$, $x \in X$. *For fixed X the assignment* $\mathbf{h}_A(X, ?) : Y \longmapsto \mathbf{h}_A(X, Y)$, *and* $\varphi \longmapsto \mathbf{h}_A(X, \varphi)$ *is a covariant functor from* C**-mod-**A *into* C**-mod-**B *(and if Y is fixed* $(\mathbf{h}_A(?, Y) : X \longmapsto \mathbf{h}_A(X, Y)$, *and* $\varphi \longrightarrow \mathbf{h}_A(\varphi, Y)$ *is a contravariant functor from* B**-mod-**A *into* C**-mod-**B). ▶■

Obviously, the first pair of the functors considered earlier was obtained from Proposition 3.14(I) with $\mathbb{O}$ instead of B and C, and the second from 3.14(II) with $\mathbb{O}$ instead of A, A instead of B and B instead of C. We select the case corresponding to $C := \mathbb{O}$ and $B := A$ which is important for the subsequent specialization of Proposition 3.14(I): here we get a covariant functor ${}_A\mathbf{h}(X, ?) : A$**-mod** $\longrightarrow A$**-mod** if we fix $X \in A$**-mod-**A.

Exercise (VI.3.15). Check that, when $X = A_+$, the last functor is naturally equivalent to the identity functor in A**-mod**.

Exercise (VI.3.16). Under the conditions of Proposition 3.14, let $A = B = C = C_0(\Omega)$, where Ω is a locally compact space. Show that in this case the A-bimodule ${}_A\mathbf{h}(A, A)$ is isometrically isomorphic to $C_b(\Omega)$.

It is also known that when $A := \mathcal{K}(H)$ (or $L^1(G)$, respectively) the A-bimodule ${}_A\mathbf{h}(A, A)$ is isometrically isomorphic to $\mathcal{B}(H)$ [88, p.118] (or $M(G)$, respectively, [110]).

We note especially the particular case of the contravariant operator functor $\mathbf{h}(?, Y)$ that occurs when $B := \mathbb{O}$ (that is, B**-mod** $= \mathcal{B}an$) and $Y := \mathbb{C}$. Obviously, this functor associates to each left A-module its dual right A-module in the sense of §3.1. This functor, and also the analogous functors from **mod-**A into A**-mod** and from A**-mod-**B into B**-mod-**A (cf. the same §3.1) will be called the *star functors* and denoted by (*); the simplest case of these — for Banach spaces — has already been considered in Chap. 0 §2.2.

Of course, each of the variants of the morphism functor mentioned up to now has an obvious analogue for unital (bi)modules. In particular, if we take a unital A and regard A as a fixed right A-module, we can consider the operator functor $\mathbf{h}(A, ?) : \mathcal{B}an \longrightarrow A\text{-}\mathbf{mod}$ which in accordance with the general rules takes E into $\mathbf{h}(A, E)$ with the action $[a \cdot \psi](b) := \psi(ba)$. This is the so-called "*co-free functor*"; it turns out to be useful later on.

... In the case of categories of polynormed modules that are more general than the Banach category, the question of what is the appropriate system of seminorms to take as basic in ${}_A\mathbf{h}(X, Y)$ or $\mathbf{h}_A(X, Y)$ (see Proposition 3.14) turns out to be quite delicate and in a number of cases a satisfactory answer cannot, in general, be found. (For a discussion of this and some positive results see [251].) At the same time for a number of applications it is sufficient to know the purely algebraic structure of the morphism sets. Therefore for such categories it makes sense to consider functors with values in some category or other of abstract modules and talk, for example, about ${}_A\mathbf{h}(X, ?) : A\text{-}\mathbf{mod} \longrightarrow \mathcal{L}in$ and so on. We leave the details of the corresponding definitions to the reader.

* * *

Another important class of functors closely connected with the morphism functors is provided by the tensor product construction for modules. Its definition is reminiscent of Definition 0.3.24 and it is a generalization of the latter; however, when one takes into account the outer multiplication, the spaces that arise may turn out to be far smaller than the original spaces.

Let A be an algebra, X a right and Y a left A-module, and let E be a linear space.

Definition (VI.3.17). A bi-operator $\mathcal{R} : X \times Y \longrightarrow E$ is said to be *balanced*[25] if $\mathcal{R}(x \cdot a, y) = \mathcal{R}(x, a \cdot y)$ for all $a \in A$, $x \in X$, $y \in Y$.

Now, unless specially stated, we assume that A, X, Y, and E are all allotted suitable Banach structures.

Definition (VI.3.18). The pair (Θ_A, θ_A), where Θ_A is a Banach space and $\theta_A : X \times Y \longrightarrow \Theta_A$ is a balanced continuous bi-operator, is called a *tensor product of the Banach modules X and Y* if

(I) for every balanced continuous bi-operator $\mathcal{R}$ from $X \times Y$ into (any) Banach space E there exists a unique continuous operator R such that the diagram

---

[25] Another terminology is "A-linear".

$$\begin{array}{ccc} X \times Y & & \\ \theta_A \downarrow & \searrow \mathcal{R} & \\ \Theta_A & \xrightarrow{\;R\;} & E \end{array}$$

is commutative and

(II) in these circumstances $\|R\|_0 = \|\mathcal{R}\|_0$.

So (cf. 0.3.8) the pair (Θ_A, θ_A) has the universality property with respect to the class of all balanced continuous bi-operators from $X \times Y$ into Banach spaces. Obviously, for $A = \mathbb{O}$ and — in the case of unital modules — for $A = \mathbb{C}$, the given definition becomes Definition 0.3.24.

Definition 3.18 implies a uniqueness theorem for (Θ_A, θ_A) whose formulation and proof duplicates, with obvious modifications, what was said in Theorem 0.3.26. We leave the details to the reader.

One of the methods of obtaining an explicit construction for the tensor product of the given modules is as follows. We "forget" outer multiplications in X and Y and consider the space $X \hat{\otimes} Y$. We let N_A denote the closure in $X \hat{\otimes} Y$ of the linear hull of all possible elements of the form $x \cdot a \otimes y - x \otimes a \cdot y$; $a \in A$, $x \in X$, $y \in Y$. Further we put $X \hat{\otimes}_A Y := X \hat{\otimes} Y / N_A$, we let $x \otimes_A y$ denote the coset $x \otimes y + N_A$ and we consider the bi-operator $\vartheta_A : X \times Y \longrightarrow X \hat{\otimes}_A Y$ $: (x, y) \longmapsto x \otimes_A y$. Obviously ϑ_A is balanced.

Elements of the form $x \otimes_A y$ are called *elementary tensors* (in $X \hat{\otimes}_A Y$). Besides the identities analogous to those in $X \otimes Y$ (cf. just before Theorem 0.3.12) there is a new identity $x \cdot a \otimes_A y = x \otimes_A a \cdot y$.

Theorem (VI.3.19) (existence). *The pair* $(X \hat{\otimes}_A Y, \vartheta_A)$ *is the tensor product of the modules* X *and* Y.

▶ If $\mathcal{R} : X \times Y \longrightarrow E$ is balanced, then obviously $N_A \subseteq \operatorname{Ker} R_0$ for the operator $R_0 : X \hat{\otimes} Y \longrightarrow E$ associated with it. Consequently R_0 induces the operator $R : X \hat{\otimes}_A Y \longrightarrow E$. The rest is clear. ■

Proposition (VI.3.20). *The norm of the element* $u \in X \hat{\otimes}_A Y$ *is equal to* $\inf \sum_{k=1}^{\infty} \|x_k\| \|y_k\|$, *where the lower bound is taken over all possible representations* $u = \sum_{k=1}^{\infty} x_k \otimes_A y_k$; $x_k \in X$, $y_k \in Y$, *as an absolutely convergent series.*

▸ This follows immediately from Theorem 0.3.29. ■

We note that in the construction of the module tensor product it is as if the left and right modular structures annihilate each other and one gets "only" a Banach space — rather as when two left, or two right, modular structures are mutually obliterated on forming the morphism space ${}_A\mathbf{h}(\cdot, \cdot)$. On the other hand — we saw this in Proposition 3.6 — the Banach space tensor product of left and right modules has the enriched structure of a bimodule (cf. the bimodule $\mathbf{h}(X, Y)$ in Proposition 3.5). In the case of general bimodules the operation "$\mathbin{\hat{\otimes}_A}$" (again like "$\mathbf{h}(\cdot, \cdot)$"; cf. 3.14) leads to some modular structures being obliterated and to the appearance of others.

Proposition (VI.3.21). *Let A, B and C be Banach algebras, $X \in B$-$\mathbf{mod}$-A and $Y \in A$-$\mathbf{mod}$-C. Then there exists a Banach B-C-bimodule structure in $X \mathbin{\hat{\otimes}_A} Y$, where X is taken as a right and Y as a left A-module, which is uniquely determined by $b \cdot (x \otimes_A y) := b \cdot x \otimes_A y$ and $(x \otimes_A y) \cdot c := x \otimes_A y \cdot c$.*

▸ For $u \in X \mathbin{\hat{\otimes}_A} Y$ and $b \in B$ we define $b \cdot u$ as $\dot{m}_b(u)$, where $\dot{m}_b$ is the operator associated with the balanced bi-operator $(x, y) \longmapsto b \cdot x \otimes_A y$. We define $u \cdot c$ similarly using the bi-operator $(x, y) \longmapsto x \otimes_A y \cdot c$ for the same u and $c \in C$. The rest is clear. ■

Proposition (VI.3.22) ("associativity law", cf. 0.3.43(III)). *Let A, B, C and D be Banach algebras, $X \in A$-$\mathbf{mod}$-B, $Y \in B$-$\mathbf{mod}$-C and $Z \in C$-$\mathbf{mod}$-D. Then we have an isometric isomorphism of A-D-bimodules $(X \mathbin{\hat{\otimes}_B} Y) \mathbin{\hat{\otimes}_C} Z \xrightarrow{\simeq} X \mathbin{\hat{\otimes}_B} (Y \mathbin{\hat{\otimes}_C} Z)$ taking $(x \otimes_B y) \otimes_C z$ into $x \otimes_B (y \otimes_C z)$; $x \in X$, $y \in Y$, $z \in Z$ and defined uniquely by this.* ▸■

Exercise (VI.3.23). Let A be commutative. Show that in this case the spaces ${}_A\mathbf{h}(X, Y)$ and $X \mathbin{\hat{\otimes}_A} Y$, where $X, Y \in A$-$\mathbf{mod} = \mathbf{mod}$-$A$, also have a certain natural Banach A-module structure and moreover the "commutativity law" $X \mathbin{\hat{\otimes}_A} Y \xrightarrow{\simeq} Y \mathbin{\hat{\otimes}_A} X$, generalizing Theorem 0.3.43(II) holds for the second module.

We consider an important example.

Theorem (VI.3.24) (cf. Rieffel [223]). *Let I be a closed right ideal in A_+ having, as a Banach algebra, a left b.a.i. $(e_\nu; \nu \in \Lambda)$. Then for I, as a right A-module, and any $Y \in A$-$\mathbf{mod}$ there exists a topological isomorphism between Banach spaces $\sigma : I \mathbin{\hat{\otimes}_A} Y \longrightarrow I \cdot Y$ uniquely determined by $\sigma(a \otimes_A y) := a \cdot y$; $a \in I$,*

$y \in Y$. *If, however,* I *is, in addition, a bi-ideal, then* σ *is an isomorphism of left Banach* A*-modules* (see 3.21); *in particular, when* $I = A$, *this is an isomorphism between the* A*-modules* $A \hat{\otimes}_A Y$ *and* Y_{es}:

▶ We put $\mathcal{R} : I \times Y \longrightarrow I \cdot Y : (a, y) \longmapsto a \cdot y$; obviously this is a balanced continuous bi-operator. Let $\sigma : I \hat{\otimes}_A Y \longrightarrow I \cdot Y$ be the operator associated with it and let $u \in I \hat{\otimes}_A Y$ be an element of the form $\sum_{k=1}^{n} a_k \otimes_A y_k$; $a_k \in I$, $y_k \in Y$. Then

$$u = \sum_{k=1}^{n} (\lim_{\nu} e_\nu a_k) \otimes_A y_k = \lim_{\nu} (\sum_{k=1}^{n} e_\nu a_k \otimes_A y_k) =$$
$$= \lim_{\nu} (\sum_{k=1}^{n} e_\nu \otimes_A a_k \cdot y_k) = \lim_{\nu} (e_\nu \otimes_A \sum_{k=1}^{n} a_k \cdot y_k) = \lim_{\nu} (e_\nu \otimes_A \sigma(u)).$$

Hence $\|u\| \leq (\sup_{\nu} \|e_\nu\|) \|\sigma(u)\|$, and consequently σ realizes a topological isomorphism between the dense subspaces $\ell.h.\{a \otimes_A y \,;\, a \in I, y \in Y\}$ of $I \hat{\otimes}_A Y$ and $\ell.h.\{a \cdot y : a \in I, y \in Y\}$ of $I \cdot Y$. The rest is clear. ■

Corollary (VI.3.25). (I) *For any* $Y \in A$**-mod** *there exists an isomorphism* $A_+ \hat{\otimes}_A Y \xrightarrow{\simeq} Y$ *in* A**-mod** *and, if* A *and* Y *are unital, then* $A \hat{\otimes}_A Y \xrightarrow{\simeq} Y$ *as well.*

(II) *For any* $X \in$ **mod**$-A$ *there exists an isomorphism* $X \hat{\otimes}_A A_+ \xrightarrow{\simeq} X$ *in* **mod**$-A$ *and, if* A *and* X *are unital, then* $X \hat{\otimes}_A A \xrightarrow{\simeq} X$ *as well.*

(Loosely speaking A_+, and in the unital case A also, is an identity with respect to the operation of tensor multiplication of modules.)

We recommend that the reader prove Corollary 3.25 immediately without basing it on Theorem 3.24.

For what follows it is useful to introduce a special terminology. We say that a left (right) A-module is *reduced* if it is topologically isomorphic to a module of the form $A \hat{\otimes}_A Y$; $Y \in A$**-mod** (or $X \hat{\otimes}_A A$; $X \in$ **mod**$-A$ respectively).

Proposition (VI.3.26). *All unital modules over unital algebras, and more generally, all non-degenerate modules over algebras with b.a.i. are reduced.*

▶ This is obviously implied by Corollary 3.25 and Theorem 3.24. ■

One can give a clear description of tensor multiplication by a cyclic module.

Exercise (VI.3.27). Let I be a closed left ideal in A_+, $X \in$ **mod**-A; construct a topological isomorphism between Banach spaces $X \underset{A}{\hat{\otimes}} (A_+/I) \xrightarrow{\simeq} X/X \cdot I$.

... We return to general concepts.

Proposition (VI.3.28) (cf. 0.3.40). *For any two morphisms* $\varphi : X_1 \longrightarrow X_2$ *in* **mod**-A *and* $\psi : Y_1 \longrightarrow Y_2$ *in* A-**mod** *there exists a unique continuous operator* $\varphi \underset{A}{\hat{\otimes}} \psi : X_1 \underset{A}{\hat{\otimes}} Y_1 \longrightarrow X_2 \underset{A}{\hat{\otimes}} Y_2$ *such that* $(\varphi \underset{A}{\hat{\otimes}} \psi)(x \underset{A}{\otimes} y) = \varphi(x) \underset{A}{\otimes} \psi(y)$ *for all* $x \in X_1$, $y \in Y_1$. *Moreover* $\|\varphi \underset{A}{\hat{\otimes}} \psi\| \leq \|\varphi\| \, \|\psi\|$.

▶ We put $\mathcal{R} : X_1 \times Y_1 \longrightarrow X_2 \underset{A}{\hat{\otimes}} Y_2 : (x, y) \longmapsto \varphi(x) \underset{A}{\otimes} \psi(y)$; obviously this is a balanced bi-operator with norm $\leq \|\varphi\| \, \|\psi\|$. All that remains is to consider the operator associated with $\mathcal{R}$. ■

The operator we have constructed $\varphi \underset{A}{\hat{\otimes}} \psi$ is called the *tensor product of the morphisms* φ *and* ψ. Here is a useful illustration.

Exercise (VI.3.29). Let $A := \mathcal{A}(\bar{\mathbb{D}})$ (see Chap. III §1.4), $I_0 := \{w \in A : w(0) = 0\}$ and $\mathbb{C}(0)$ be the one-dimensional A-module A/I_0 (cf. Proposition 2.12). Show that $\mathbf{in} \underset{A}{\hat{\otimes}} \mathbf{1} : I_0 \underset{A}{\hat{\otimes}} \mathbb{C}(0) \longrightarrow A \underset{A}{\hat{\otimes}} \mathbb{C}(0)$ is the zero operator between one-dimensional spaces.

So the tensor product of two injective, even topologically injective, morphisms is not necessarily injective.[26] On the other hand, it would not be difficult to show that the tensor product of two surjective morphisms is always surjective (cf. 0.3.41).

Now, acting just as we did in defining the functor "$\hat{\otimes}$" in Chap.0. §3.4, we fix $X \in$ **mod**-A and associate to each $Y \in A$-**mod** the Banach space $X \underset{A}{\hat{\otimes}} Y$, and to each morphism $\varphi : Y_1 \longrightarrow Y_2$ in A-**mod** the operator $1_X \underset{A}{\hat{\otimes}} \varphi : X \underset{A}{\hat{\otimes}} Y_1 \longrightarrow X \underset{A}{\hat{\otimes}} Y_2$. Obviously we get a covariant functor from A-**mod** into $\mathcal{B}an$ denoted by $X \underset{A}{\hat{\otimes}} ?$ and called the functor *of tensoring on the left by the module* X. Analogously, for fixed $Y \in A$-**mod** one gets a covariant functor $? \underset{A}{\hat{\otimes}} Y :$ **mod**-$A \longrightarrow \mathcal{B}an$ ("tensoring on the right by the module Y"). Both functors are obviously additive; as a consequence we have the "distributivity" relations

---

[26] A similar phenomenon has already been noted earlier for spaces without an outer multiplication (0.3.42). However, for modules, as it would appear from 3.29, this could be due to far cruder, purely algebraic, reasons.

$$X \mathop{\hat{\otimes}}_{A} (Y_1 \oplus Y_2) \simeq (X \mathop{\hat{\otimes}}_{A} Y_1) \oplus (X \mathop{\hat{\otimes}}_{A} Y_2)$$

and

$$(X_1 \oplus X_2) \mathop{\hat{\otimes}}_{A} Y \simeq (X_1 \mathop{\hat{\otimes}}_{A} Y) \oplus (X_2 \mathop{\hat{\otimes}}_{A} Y).$$

The tensoring operation is connected with the operation of taking the space of morphisms by the following very useful identity.

Theorem (VI.3.30) ("dual associativity law"; cf. 0.3.43(IV)). *Let* $X \in$ **mod**-A, $Y \in A$-**mod**, $E \in \mathcal{B}an$. *Then, if we consider* $\mathbf{h}(Y, E)$ *as a right* A*-module*[27] *there exists a unique isometric isomorphism between Banach spaces*

$$\mathbf{h}(X \mathop{\hat{\otimes}}_{A} Y, E) \xrightarrow{\simeq} \mathbf{h}_A(X, \mathbf{h}(Y, E)),$$

taking $\varphi : X \mathop{\hat{\otimes}}_{A} Y \longrightarrow E$ *into* ψ *such that* $[\psi(x)](y) := \varphi(x \mathop{\otimes}_{A} y)$; $x \in X$, $y \in Y$. *If, in addition,* X *is a* B-A*-bimodule, where* B *is another Banach algebra, then the above map is an isomorphism between the corresponding* (see 3.14(II) and 3.21) *right* B*-modules.*

▶ By Definition 3.18 and the definition of the freezing isomorphism (Chap. 0 §3.1), each of the two spaces is isometrically isomorphic to the subspace of $(\mathcal{B}(X \times Y, E); \|\cdot\|_0)$, consisting of balanced operators. The rest is clear. ■

Exercise (VI.3.31). Using Proposition 3.14 in 3.21 formulate and prove a more general form of the dual associativity law corresponding to the formula ${}_A\mathbf{h}_C(X \mathop{\hat{\otimes}}_{B} Y, Z) \xrightarrow{\simeq} {}_A\mathbf{h}_B(X, \mathbf{h}_C(Y, Z))$.

When talking about tensor products of modules we have, for definiteness, limited ourselves to the case of Banach structures. In a similar way one can introduce, in terms of the universality property with respect to balanced continuous (separately or jointly, according to the sense) bi-operators, tensor products of modules with other classes of underlying spaces. The corresponding definitions duplicate, with obvious modifications, the versions for spaces without an outer multiplication as given at the beginning of Chap. 0 §3.5.

For each kind of tensor product of "the modules of analysis" there is an existence theorem which is proved by producing a suitable explicit construction taking into account the specific character of the given class of modules. Thus, for example, the

[27] Recall that outer multiplication is given by $[\alpha \cdot a](x) := \alpha(a \cdot x)$; see Proposition 3.5 with $\mathbb{C}$ instead of B, Y instead of X and E instead of Y.

injective tensor product of (general) polynormed modules — right by left — over a polynormed algebra can be constructed as $X \otimes_i Y/\ell.h.\{x \cdot a \otimes y - x \otimes a \cdot y\}$. When considering complete Hausdorff structures the appropriate complete Hausdorff injective tensor product can be constructed as the (automatically complete) space $X \bar{\otimes} Y/\overline{\ell.h.}\{x \cdot a \otimes y - x \otimes a \cdot y\}$ if X and Y are Fréchet spaces or, in the general case, as the completion of this quotient space (cf. [90, Chap.IV, Ex.11]) and so on.

More details about the tensor products of polynormed modules are given in [81]; in addition, both there and in [88] there is some more information on the tensor products of Banach modules.

Chapter VII
Problems leading to homology and the elements of homology theory

We have already met the word "homology" and moreover (like "algebra" and "topology") in two senses: as a specific mathematical object (0.4.22) and as a branch of mathematics. This last chapter is devoted to those concepts and results in the theory of "the algebras of analysis" that are just as much part of the subject of homology — or that in each case can be interpreted in typical homological terms. Many of the concepts introduced here, for example projective modules, derivations and extensions, seem at first to be completely unconnected. Gradually, however, some connection between the separate subjects begins to emerge (cf. 2.40) and finally, just before the final curtain, it is shown (§3.3) that they may be considered in the context of a certain unified approach and studied by general methods.

What is the essence of this approach? With an unavoidable hint of crudeness, one could say that the questions studied in this chapter reduce, sometimes directly, sometimes after suitable rethinking, to the consideration of two types of problem in the theory of modules: lifting and extension.[1] We begin with the first of these.

---

[1] From a certain point of view anything that reduces to such a problem in the various "categorically presented" mathematical disciplines also forms part of the subject of homology. But this is only one of the many opinions about a subject with such vague boundaries.

§1. PROJECTIVITY AND OTHER CONCEPTS EXPRESSED IN TERMS OF LIFTING AND EXTENSION PROBLEMS

§1.1. The lifting problem. Projective modules and their connection with free modules

Warning. Everywhere in this section, unless specially stated, A is a fixed unital Banach algebra, the expression "A-module" means a left unital Banach A-module and all morphisms and operators are continuous.

Let P be a certain A-module, let $\sigma : X \longrightarrow Y$ be a certain surjective morphism of A-modules (by the Banach theorem, it can always be assumed that σ is the natural projection of a module onto a quotient module) and let $\varphi : P \longrightarrow Y$ be another morphism of A-modules. In this situation we shall call the map $\psi : P \longrightarrow X$ a *lifting* of φ, or say that it *lifts* φ if the diagram

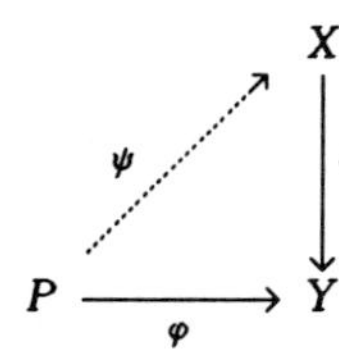

is commutative or, what is the same thing, $\sigma \circ \psi = \varphi$.

Does there exist among the maps lifting φ a (continuous) morphism of A-modules? This typical question in the theory of Banach modules is called the *lifting problem*; if we want to be more precise, we shall talk about the "lifting problem for P given $\sigma : X \longrightarrow Y$ and $\varphi : P \longrightarrow Y$" or, briefly, the "lifting problem (P, σ, φ)". There is an important special case of this that arises when $Y = P$ and $\varphi = \mathbf{1}_P$: does σ have a right inverse morphism (= is it a retraction in A-**unmod**)? This is the so-called *retraction problem* for P given $\sigma : X \longrightarrow P$ or, briefly, the "retraction problem (P, σ)".

Example (VII.1.1). The lifting problem for $P = A$ given any σ and φ has a positive solution (= the required lifting morphism exists): it is sufficient to fix any $x \in X$ with $\sigma(x) = \varphi(e)$ and put $\psi : a \longmapsto a \cdot x$.

Exercise (VII.1.2). Let I be a closed left ideal in A without a right identity (as an algebra). Show that the retraction problem for A/I given $\mathbf{pr} : A \longrightarrow A/I$ has a negative solution.

From among all the surjective module morphisms we select those σ that have a continuous right inverse operator, in other words, for which $\Box(\sigma)$, where $\Box$ is the forgetful functor from A-**unmod** into $\mathcal{B}an$ (cf. VI.3.12), is a retraction; both conditions are obviously equivalent to $\operatorname{Ker}\sigma$ being complemented as a subspace of X (cf. Chap.0 §1.4). Such a σ is called an *admissible epimorphism* (it is automatically an epimorphism in A-**unmod**). The lifting problem (P, σ, φ), in particular the retraction problem (P, σ), is said to be *admissible* if σ is an admissible epimorphism; the majority of lifting problems met in practice are related to, or reduce to, such problems. (So, in the above situation the morphism φ automatically has lifting operators — for example, $\psi_0 \circ \varphi$, where ψ_0 is a right inverse operator for σ and the question is whether there can be found one among them that is a morphism, that is, one that is compatible with the outer multiplication.)

The following definition singles out the modules that are considered to be "homologically best".

Definition (VII.1.3). An A-module P is said to be *projective* if every admissible lifting problem for P has a positive solution.

It is obvious that if $A = \mathbb{C}$ any A-module (= Banach space) is projective.

Remark (VII.1.4). A priori, it is not clear why in this definition one should involve, not all, but only the admissible lifting problems. The advantages of such an approach[2] will be explained gradually in the course of our study (see later, for example, Corollary 1.12, Theorem 1.23 and Corollary 3.3).

Definition 1.3 immediately implies the following.

Proposition (VII.1.5). *If P is projective, then every admissible epimorphism $\sigma : X \longrightarrow P$ is a retraction of A-modules.* ▶■

Proposition (VII.1.6). *A retract* (in A-**unmod**) *of a projective module is itself projective.*

▶ Obviously we need to show that every diagram of the form

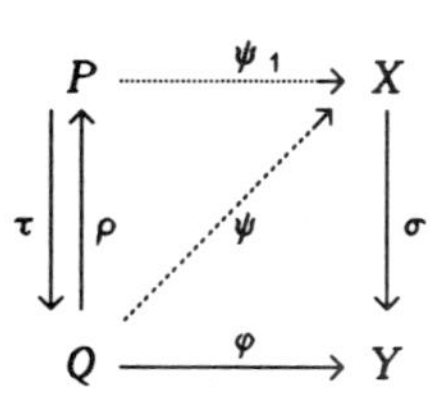

(considered initially without the morphisms depicted by dotted lines), in which P is projective, σ is an admissible epimorphism and $\tau \circ \rho = 1_Q$, may be made into a commutative

[2] Originating from the so-called relative homology; see [277], [296], [51, Ch.IX].

diagram by adding a morphism ψ. For this we note that the lifting problem $(P, \sigma, \varphi \circ \tau)$ is automatically admissible and therefore, by Definition 1.3, the original diagram can be made into a commutative diagram by adding a certain morphism ψ_1. All we have to do now is to put $\psi := \psi_1 \circ \rho$. ▶■

The following proposition, as we shall soon see, provides something more than simply a class of examples.

Proposition (VII.1.7). *A free A-module* (see VI.1.21) *is always projective.*

▶ We consider the admissible lifting problem $(P, \sigma : X \longrightarrow Y, \varphi : P \longrightarrow Y)$[3] in which P is a free module with base space F. If we fix an operator $\rho : Y \longrightarrow X$ with $\sigma \circ \rho = 1_Y$ and in Theorem VI.1.22 take P, X and $(\rho \circ \varphi)|_F$ in place of the X, Y and ξ featured there, we get a morphism of A-modules $\psi : P \longrightarrow X$ such that $\psi|_F = (\rho \circ \varphi)|_F$ and hence $(\sigma \circ \psi)|_F = \varphi|_F$. But, according to the assertion about uniqueness in the same theorem, a morphism from P (to anywhere) is uniquely defined by its values on F. Hence $\sigma \circ \psi = \varphi$. ■

For arbitrary $X \in$ A**-unmod**, we consider the free module $A \hat{\otimes} X$ and put $\pi_X : A \hat{\otimes} X \longrightarrow X : a \otimes x \longmapsto a \cdot x$, that is, we take the operator associated with the bi-operator of outer multiplication. Obviously π_X (denoted simply by π when there is no danger of confusion) is an A-module morphism; it is called the *canonical projection for* X.

Example (VII.1.8). Let A be a (unital) semisimple commutative Banach algebra with spectrum Ω, and let I be a closed ideal of A. Then it is easy to verify that we have a commutative diagram

$$\begin{array}{ccc} A \hat{\otimes} I & \xrightarrow{\ \gamma\ } & C(\Omega \times \Omega) \\ \downarrow{\scriptstyle \pi} & & \downarrow{\scriptstyle \nabla} \\ I & \xrightarrow{\ \Gamma\ } & C(\Omega) \end{array}$$

where the operators are defined by $\Gamma : a \longmapsto \hat{a}(s)$, $\gamma : a \otimes b \longmapsto \hat{a}(s)\hat{b}(t)$ and $\nabla : u \longmapsto f : f(s) := u(s, s)$; $s, t \in \Omega$. Thus, when $\operatorname{Ker} \Gamma = \operatorname{Ker} \gamma = 0$ (and this is so for all the examples of algebras in the given class occurring in this book), the canonical projection for I is the map given by "restricting functions of two variables to the diagonal".

---

[3] Recall that when P is given explicity as $A \hat{\otimes} E$; $E \in \mathcal{B}an$, one can put $F := \{e \otimes x : x \in E\}$.

Proposition (VII.1.9). π_X *is an admissible epimorphism for any* X.

▶ The operator $x \longmapsto e \otimes x$; $x \in X$, is a right inverse to π_X. ■

If, using Banach's theorem, we identify X with $A \hat{\otimes} X/\mathrm{Ker}\,\pi_X$ and use Proposition 1.7, we get the following.

Corollary (VII.1.10). *Every A-module is the quotient module of a projective module where the corresponding natural projection is an admissible epimorphism.*

(This fact is usually expressed by saying that there are sufficiently many projective A-modules.)

Theorem (VII.1.11). *The following properties of an A-module P are equivalent:*

(I) P *is projective;*

(II) P *is a modular direct summand or (what is the same thing; see* VI.1.19) *a retract of a free A-module;*

(III) *the canonical projection* $\pi_P : A \hat{\otimes} P \longrightarrow P$ *is a retraction in* A-**unmod**.

▶ (I) $\Rightarrow$ (III) follows if one considers the retraction problem (P, π_P); (III) $\Rightarrow$ (II) is obvious; (II) $\Rightarrow$ (I) follows from Propositions 1.7 and 1.6. ■

So the projectivity property, although defined at first using the whole class of lifting problems, may be characterized in terms of a single retraction problem: (P, π_P). We also note that the theorem we have proved enables projective modules to be described in purely intrinsic terms:

Corollary (VII.1.12). *An A-module is projective* $\Longleftrightarrow$ *it is, to within a topological isomorphism, a modular direct summand in an A-module of the form* $A \hat{\otimes} E$, $E \in \mathcal{B}an$.

Finally we point out a useful proposition.

Proposition (VII.1.13). *If* P *is projective, then so also is* $P \hat{\otimes} E$ *for any* $E \in \mathcal{B}an$ *(see* VI.3.6 *when* $B = \mathbb{C}$).

▶ If ρ is a right inverse morphism for π_P, then $\rho \hat{\otimes} 1_E$ plays the same role for $\pi_{P \hat{\otimes} E}$. ■

* * *

Up to now we have been talking only about left unital modules. However, by the identification indicated at the end of Chap. VI §3.1 all the definitions and facts about projectivity can be carried over automatically to all other types of Banach module; in future we shall use this freely. In particular, the role that free modules play in A-**unmod**, is taken over in the category A-**mod** by the modules $A_+ \hat{\otimes} E$ (this was mentioned in Chap. VI

§1.2), in **mod**-A it is taken over by $E \hat{\otimes} A_+$, and in A-**mod**-B this role is taken over by $A_+ \hat{\otimes} E \hat{\otimes} B_+$ (everywhere $E \in \mathcal{B}an$), and also by all (bi)modules that are topologically isomorphic to them. We shall also call all these (bi)modules *free* — bearing in mind the relevant category.

As we shall see in §1.2, the retraction problem (P, π_P) lies at the root of the majority of methods for checking whether a given unital module P is projective (cf. 1.11). For "general" modules, it is often convenient to check whether they are projective by way of the following similar problem.

Proposition (VII.1.14). *Let A be (any) Banach algebra, $P \in A$-**mod**. Then, if P is non-degenerate, it is projective $\Longleftrightarrow$ the morphism $\pi_P^0 : A \hat{\otimes} P \longrightarrow P : a \otimes x \longmapsto a \cdot x$* (= the restriction to $A \hat{\otimes} P$ of the canonical projection π_P for P in A_+-**unmod**) *is a retraction of A-modules.*

▶ "$\Leftarrow$". The right inverse to π_P^0 is also a right inverse to π_P.

"$\Rightarrow$". If $P = A \cdot P$, then the image of any morphism from P into $A_+ \hat{\otimes} P$, including, in particular, that of the right inverse to π_P, lies in $A \cdot (A_+ \hat{\otimes} P) = A \hat{\otimes} P$. The rest is clear. ■

From what has been proved, one can see in particular that no misunderstandings arise with the concept of projectivity when one considers unital modules in place of "general" modules:

Corollary (VII.1.15). *Let A be unital. Then P is projective in A-**unmod** $\Longleftrightarrow$ it is projective in A-**mod**.*

The morphism π_P^0 in 1.14 will be called the *canonical morphism for P* and also often denoted simply by π; this does not lead to confusion.

Example (VII.1.16). Let $A = P = L^1(G)$, where G is a locally compact group. Here, as is easily seen, we have a commutative diagram

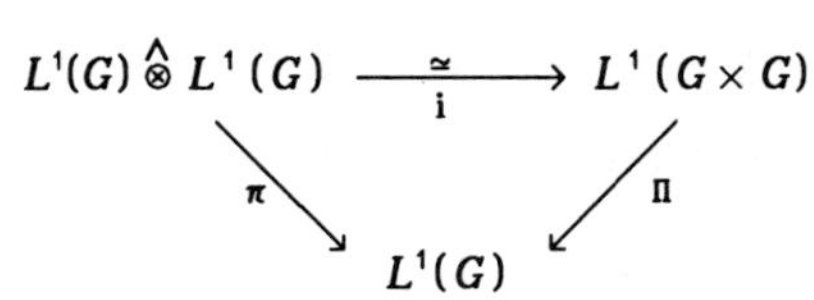

in which i is Grothendieck's isometric isomorphism (see VI.3.7) and $\Pi : u(s, t) \longmapsto f(s) := \int_G u(t, t^{-1}s)dt$; we notice that all three maps are morphisms of Banach $L^1(G)$-modules. Thus, in the case under consideration, π can be identified with the convolution morphism for functions of two variables.

§1.2. Examples of projective modules. The condition that ideals and cyclic modules are projective

We consider what the requirement of projectivity turns out to be in certain classes of modules we have already met.

In this section A is a fixed Banach algebra, initially not assumed to be unital. We begin by considering the A-module A itself and its closed submodules (= closed ideals in A).

Definition (VII.1.16). A is called *projective on the left* if the left A-module A is projective and *hereditary on the left* if, in addition, all closed left ideals in A are projective.

The terms projective and hereditary "on the right" are defined analogously.

Obviously, every algebra having a right identity is projective on the left (cf. §1.1). Another example, this time more concrete, is given by the following exercise.

Exercise (VII.1.17). Show that for any locally compact group G [4] the algebra $L^1(G)$ is projective on the left.

Hint. The convolution morphism Π in Example 1.16 has a right inverse $\rho : f \longmapsto u(s, t) := X(t^{-1})f(st)$, where X is the characteristic function of any compact set in G of Haar measure 1.

Remark (VII.1.18). The question of whether the proper ideals in $L^1(G)$ are projective is more complicated. It can be shown that $L^1(G)$ is hereditary on the left $\Leftrightarrow$ G is compact (cf. [296]; concerning sufficiency cf. later 1.60(II) and 1.81).

We now establish a certain sufficient condition for ideals to be projective which, in a number of cases, gives an effective means of checking this. For an arbitrary set Λ we let $N(\Lambda)$ denote the collection of finite subsets of Λ, directed by inclusion.

Definition (VII.1.19). Let B be a Banach algebra. A bounded family $\{b_\nu;\ \nu \in \Lambda\}$, of elements of B is called a *decomposition of a right a.i. for* B if

(I) the net $(e_\lambda;\ \lambda \in N(\Lambda))$ is a right a.i. in B where, for $\lambda = (\mu_1, \cdots, \mu_n)$, $e_\lambda := \sum_{k=1}^{n} b^2_{\mu_k}$;

(II) there is a $C > 0$ such that $\| \sum_{k=1}^{n} z_k ab_{\mu_k} \| \le C \max_k \| ab_{\mu_k} \|$ for any $a \in B_+$, $\lambda = (\mu_1, \cdots, \mu_n) \in N(\Lambda)$ and complex numbers z_k; $|z_k| \le 1$, $1 \le k \le n$;

---

[4] Or at least for $G = \mathbb{R}$ and $\mathbb{T}$ if the reader does not wish to accept the truth of what was said in Chap.III §3.2.

(III) for any $a \in B$ and $\varepsilon > 0$ the inequality $\|ab_\mu\| \geq \varepsilon$ is satisfied for only a finite set of indices $\mu \in \Lambda$.

Example (VII.1.20). For each $n \in \mathbb{Z}$ we consider the function $f_n(t)$ in the algebra $C_0(\mathbb{R})$ that is equal to $\sqrt{t-(n-1)}$ when $n-1 < t \leq n$, equal to $\sqrt{(n+1)-t}$ when $n < t < n+1$ and zero at the remaining points of $\mathbb{R}$. It easy to verify that $\{f_n;\ n \in \mathbb{Z}\}$, is a decomposition of a right a.i. in $C_0(\mathbb{R})$.

Exercise (VII.1.21) (generalizing Example 1.20). Let Ω be a compact space and let Δ be a closed subset of Ω having a countable defining system of neighbourhoods ("defining" means that any neighbourhood of Δ contains at least one neighbourhood of the system). Show that the Banach algebra $I_\Delta = \{f \in C(\Omega) : f|_\Delta = 0\}$ has a decomposition of a right a.i. and that the condition in the exercise is automatically fulfilled if Ω is metrizable.

Example (VII.1.22). The algebra $\mathcal{K}(H)$, where H is a Hilbert space, also has a decomposition of a right a.i.: to get such a decomposition it is sufficient to consider the family of orthogonal projections onto the vectors of a fixed orthonormal basis.

Theorem (VII.1.23) [268]. *Let I be a closed* (not necessarily proper) *left ideal in the Banach algebra A having, as a Banach algebra, a decomposition of a right a.i. Then I is a projective left A-module.*

▶ According to Proposition 1.14 it is sufficient to construct a right inverse morphism to the canonical morphism $\pi : A \hat{\otimes} I \longrightarrow I$. We take $\{b_\nu;\ \nu \in \Lambda\}$, and C corresponding to $B = I$ (see Definition 1.19) and put $C_1 := \sup_\nu \|b_\nu\|$. We fix some $x \in I$ and for any $\lambda = (\mu_1, \cdots, \mu_n) \in N(\Lambda)$ we put $u_\lambda := \sum_{k=1}^{n} xb_{\mu_k} \otimes b_{\mu_k}$.

Let $\varepsilon > 0$. Then, by 1.19(III), there exists $\lambda_0 \in N(\Lambda)$ such that $\|xb_\nu\| < \varepsilon$ for $\nu \notin \lambda_0$. We take $\lambda', \lambda'' \in N(\Lambda)$ such that $\lambda_0 \subseteq \lambda', \lambda''$; then it is obvious that $u_{\lambda'} - u_{\lambda''}$ can be represented as $\sum_{i=1}^{k} xb_{\mu_i} \otimes b_{\mu_i} - \sum_{j=1}^{m} xb_{\nu_j} \otimes b_{\nu_j}$, where $\mu_i, \nu_j \notin \lambda_0$. Therefore, using 1.19(II) and Proposition 0.3.30, we see that in the last expression the norm of each sum is at most $(C\varepsilon)(CC_1)$. Consequently $\|u_{\lambda'} - u_{\lambda''}\| < 2C^2C_1\varepsilon$.

Since ε is arbitrary, we have shown that the net $(u_\lambda;\ \lambda \in N(\Lambda))$ is a Cauchy net and hence converges in $A \hat{\otimes} I$ (which is complete). We put $\rho(x) := \lim_\lambda u_\lambda$ and $\rho : I \longrightarrow A \hat{\otimes} I : x \longmapsto \rho(x)$. It is easy to verify that ρ is an A-module morphism which is continuous since $\|u_\lambda\| \leq \|x\| C^2 C_1^2$ (guaranteed by the same 1.19(II) and 0.3.30); moreover 1.19(I) implies that $\pi \circ \rho(x) = x$ is satisfied for any $x \in I$. The rest is clear. ■

As it is not difficult to see (cf. 1.20 and 1.21), the theorem we have proved may be used to establish the projectivity of a wide class of ideals in the algebra $C(\Omega)$. In fact, we have the following criterion which we give without proof. We recall that a topological space is said to be *paracompact* if any open cover has an open refinement $\mathfrak{U}$ such that every point of the space has a neighbourhood intersecting only a finite number of sets from $\mathfrak{U}$. The space is said to be *hereditarily paracompact* if any open subset of it is paracompact. According to a well-known theorem of Stone, any metrizable space is paracompact [95, Theorem 5.1.3] and therefore hereditarily paracompact. Conversely, the space of countable ordinals and the complement in $\beta\mathbb{N} = \underline{\Omega}(c_b)$ (see Chap.IV §3.1) of any point in the "Čech growth" $\beta\mathbb{N}\setminus\mathbb{N}$ is not paracompact [95, 5.1.20 together with 3.10.18].

Assertion (VII.1.24). *A closed ideal I in $C(\Omega)$, where Ω is a compact set, is projective $\Leftrightarrow$ the complement of its zero set* (*see* Chap.III §1.1) *is paracompact. Hence the algebra $C(\Omega)$ is hereditary $\Leftrightarrow$ Ω is hereditarily paracompact.*

As another consequence of Theorem 1.23 we get (taking into account Example 1.22) that the algebra $\mathcal{K}(H)$ is projective on the left. In fact every separable C^*-algebra is projective on the left and, furthermore, hereditary on the left; at the same time no infinite-dimensional von Neumann algebra is hereditary on the left (Lykova [193], [195]).

Concerning conditions that are necessary for ideals to be projective, they are usually studied for commutative Banach algebras. In particular, "$\Rightarrow$" in Assertion 1.24 is a special case of the following assertion.

Assertion (VII.1.25) [268]. Let A be a unital commutative Banach algebra with spectrum Ω, and let I be a closed ideal of A. Then, if I is projective, its spectrum Ω_0 (identified with $\Omega\setminus\{s : \hat{x}(s) = 0$ for all $x \in I\}$; cf. IV.1.19) is paracompact.

The proof of this and also of a number of other conditions that are necessary for ideals to be projective (for example, conditions on the norms of certain of their elements), is based on a construction that associates with each projective ideal a certain function of two variables having special properties. One way of representing this is given by the following exercise.

Exercise (VII.1.26). Under the conditions of 1.25 let $\rho : I \longrightarrow A \hat{\otimes} I$ be a fixed right inverse morphism to π_I, let $\gamma : A \hat{\otimes} I \longrightarrow C(\Omega \times \Omega)$ be the operator taking $a \otimes x$ into $\hat{a}(s)\hat{x}(t)$. Show that for any $s \in \Omega_0$ and $t \in \Omega$ the number

$F(s, t) := [\gamma \circ \rho(x)](s, t)$, where $x \in I$ is such that $\hat{x}(s) = 1$, does not depend on the choice of x and that the resulting function $F : \Omega_0 \times \Omega \longrightarrow \mathbb{C}$ is continuous, and equal to zero when $t \notin \Omega_0$ and equal to one when $s = t$.

For applications of the function F, called the *skeleton of the projective ideal* I, see [270] and [81, IV §3].

Let us turn from the submodules of the module A to its quotient modules and to the closely related class of cyclic modules (VI.2.2). First we establish a specific result which is important for applications.

Theorem (VII.1.27) [147]. *Let A be an operator algebra on the Banach space E containing all the finite-dimensional operators (for example $\mathcal{B}(E)$, $\mathcal{K}(E)$ or $\mathcal{N}(E)$). Then the natural A-module E is projective.*

▶ If we fix $x_0 \in E$ and $f \in E^*$ such that $f(x_0) = 1$, we get $\tau : A_\times \longrightarrow E : a \longmapsto a(x_0)$ and $\rho : E \longrightarrow A_\times : y \longmapsto \langle f, \cdot \rangle y$, where the last symbol denotes the one-dimensional operator $x \longmapsto f(x)y$. Obviously τ and ρ are morphisms in A-**mod** and moreover $\tau \circ \rho = 1_E$. Hence E is a retract of the module $A_\times$ and therefore (Theorem 1.11(II)) it is projective. ■

We consider only the simplest class of cyclic modules over group algebras and the result in question is offered in the form of an exercise with a hint.

Exercise (VII.1.28). Show that the augmenting $L^1(G)$-module $\mathbb{C}$ (see VI.1.31) is projective $\Longleftrightarrow$ the group G is compact and that in this case the augmentation ideal is also projective.

Hint. If G is compact, then $\rho : \mathbb{C} \longrightarrow L^1(G) : z \longmapsto f(s) \equiv z$ is a morphism which is right inverse to the augmentation morphism τ and we only have to apply Exercise 1.17. If, however, there exists a morphism ρ that is right inverse to τ (see 1.5), then there exists on G a non-zero integrable function, namely $\rho(1)$ which is invariant under left shifts (= constant almost everywhere).

In the general case, if a quotient module X of the module A is projective, this means that, under certain additional conditions, X can be realized as an ideal in A that is complemented by another ideal.

Proposition (VII.1.29). *Let I be a closed left ideal in the unital algebra A and $X := A/I$. Then for X to be projective it is sufficient and, if I is complemented as a subspace of A, it is also necessary that there exists in A another closed ideal J such that $I \oplus J = A$. Moreover X is isomorphic to J in* A-**mod**.

▶ Sufficiency is implied by Theorem 1.11(II). Concerning necessity, the complementation condition means that the natural projection of A onto X is an admissible epimorphism and therefore (1.5) has a right inverse morphism ρ. All that remains is to put $J := \rho(X)$. ■

The condition that I be complemented is fairly restrictive (cf. what was said in Chap.0 §1.4). It is not known whether it can be discarded in general; however, it can be replaced by the considerably less burdensome requirement that certain A-modules have a "nice geometry" (Selivanov [234]; see also [81, IV §4]). In particular the following holds.

Assertion (VII.1.30). *The condition in* 1.29 *that* I *be complemented in* A *can be replaced by any of the following:*

(I) A *is commutative*;

(II) A *is the unitization of the* L^1*-algebra of a locally compact group*;

(III) A *is a* C^**-algebra and* I *is either maximal* (= X is irreducible), *or it is a bi-ideal.*

The question of how to construct algebras with a sufficiently large stock of projective cyclic modules will be discussed, together with other questions of a similar sort, in §1.4.

Assertion 1.30(I) and Shilov's theorem on idempotents (IV.2.19) leads without special difficulty to the following.

Assertion (VII.1.31). *Let* A *be an (arbitrary) commutative Banach algebra with spectrum* Ω, *and let* I *be a modular closed ideal of* A *with spectrum* Δ. *Then* A/I *is projective* $\Leftrightarrow$ Δ *is closed in* Ω (= Ω is the disconnected union of Δ and $h(I)$). *In particular, the one-dimensional* A*-module* $\mathbb{C}(s)$; $s \in \Omega$, *is projective* $\Leftrightarrow$ s *is an isolated point in* Ω.

Thus if a commutative Banach algebra has projective cyclic modules, this tells us something about the disconnectedness of its spectrum.

§1.3. The extension problem. Injective and flat modules. Flat $L^1(G)$-modules and amenable groups

Now let us leave the lifting problem for the time being and deal with another type of problem that, loosely speaking, is obtained from the first by reversing the arrows.

Warning. Again (cf. §1.1) unless specially stated an "A-module" means a left unital Banach module over a fixed unital Banach algebra A.

Let J be a certain A-module, let $i : Y \longrightarrow X$ be a topologically injective morphism of A-modules, and let $\varphi : Y \longrightarrow J$ be another morphism. In this situation the map $\psi : X \longrightarrow J$ is called an *extension* of φ if the diagram

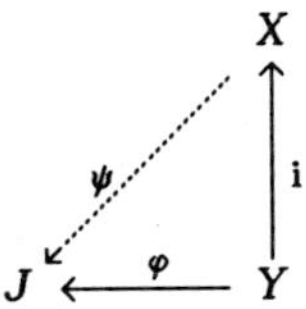

is commutative. (By Banach's theorem Y can be identified with $\operatorname{Im} i$ and it may be assumed that ψ is an ordinary extension of φ in the sense of Chap.0 §0.1°.) The "*extension problem for* J *given* i *and* φ" or, briefly, the "extension problem (J, i, φ)" is the question of whether there exists among the maps extending φ a continuous A-module morphism. When $Y = J$ and $\varphi = 1_J$, we get the question of whether i has a left inverse (= is a coretraction in A-**unmod**). This is called the *coretraction problem* for J given i or, briefly, the coretraction problem (J, i).

Exercise (VII.1.32). Show that any extension problem for $J = A^*$ (dual to $A \in$ **unmod**-A; see Chap.VI §3.1) has a positive solution.

A morphism i that has a continuous left inverse operator is called an *admissible monomorphism*. (Obviously this is a monomorphism in A-**unmod**.) The extension problem (J, i, φ), in particular the coretraction problem (J, i), is said to be *admissible* if i is an admissible monomorphism.

Definition (VII.1.33) (cf. 1.3). An A-module J is said to be *injective* if every admissible extension problem for J (with any given data) has a positive solution.

Obviously, when $A = \mathbb{C}$, all A-modules, that is Banach spaces, are (not only projective; cf. §1.1 but also) injective.

Proposition (VII.1.34) (cf. 1.5). *If* J *is injective, then every admissible monomorphism* $i : J \longrightarrow X$ *is an* A*-module coretraction.*

Further, by "reversing the arrows" in the proof of Proposition 1.6 we obviously get the following.

Proposition (VII.1.35). *A retract (in* A-**unmod**) *of an injective module is itself injective.* ▸■

Let us indicate a class of modules which plays the same role among the injective modules as the class of free modules plays among the projective modules.

Definition (VII.1.36). An A-module is said to be *co-free* if it is topologically isomorphic to an A-module of the form $\mathcal{B}(A, E)$;

$E \in \mathcal{B}an$, with the action $[a \cdot f](b) := f(ba)$; $a, b \in A$, $f \in \mathcal{B}(A, E)$ (in other words, if it belongs to the image of the co-freedom functor in Chap. VI §3.2).

Example (VII.1.37). The $l_\infty(G)$ in Example VI.3.3, considered as a left $l_1(G)$-module, is co-free.

Proposition (VII.1.38) (cf. 1.7). *A co-free A-module is always injective.*

▶ We consider for $J = \mathcal{B}(A, E)$ an admissible extension problem with given $i : Y \longrightarrow X$ and $\varphi : Y \longrightarrow J$. We fix an operator $j : X \longrightarrow Y$ with $j \circ i = 1_Y$ and introduce the map $\psi : X \longrightarrow J = \mathcal{B}(A, E)$ where $[\psi(x)](a) := [\varphi \circ j(a \cdot x)](e)$; $x \in X$, $a \in A$. It is easy to verify that ψ is a continuous A-module morphism such that $\psi \circ i = \varphi$. The rest is clear. ■

Recall the canonical projection π_X from the "projective theory" of §1.1. In the "injective theory" its role is played by the so-called *canonical embedding* $\check{\pi}_X : X \longrightarrow \mathcal{B}(A, X)$ that associates to each $x \in X$ the operator $f : a \longmapsto a \cdot x$. Obviously for any $X \in A$-**unmod**, $\check{\pi}_X$ is an admissible monomorphism (and in particular it is topologically injective). So (cf. 1.10) every A-module is, to within a topological isomorphism, a submodule of a certain injective module, and moreover the corresponding natural embedding is an admissible monomorphism. (There are also "sufficiently many" injective modules — only in a somewhat different sense than for projective modules.)

Finally the following analogue to Theorem 1.11 can be established without difficulty.

Theorem (VII.1.39). *The following properties of the A-module J are equivalent:*

(I) *J is injective;*

(II) *J is a modular direct summand (= retract) of a co-free A-module;*

(III) *the canonical embedding* $\check{\pi}_J : J \longrightarrow \mathcal{B}(A, J)$ *is a coretraction in A-***unmod**. ▶■

* * *

We cancel the "warning" made earlier and go on to consider any type of Banach (bi)module. Earlier (§1.1) the concepts and facts connected with projectivity were carried over from left unital modules to all these types. Now we are obviously justified in acting in exactly the same way with injectivity and in freely using the right- and bimodule versions of 1.33 - 1.39. In particular, this enables us to establish the following connection between projectivity and injectivity.

Proposition (VII.1.40). *Let P be a projective Banach (bi)module (of any type). Then the (bi)module* P^* *(= the* *result of applying to* P *the star functor from* Chap.VI §3.2) *is injective.*

▶ Obviously, it is sufficient to show that $P \in$ **unmod**$-A$ being projective implies $P^* \in A$-**unmod** being injective. In turn it is sufficient, by Theorem 1.11(II), Definition 1.36, Proposition 0.2.7 and the fact that the functor $(^*)$ preserves retractions, to establish this fact when P is free, that is, when it is a module of the form $A \hat{\otimes} E$; $E \in \mathcal{B}an$, with the operation $(a \otimes x) \cdot b := ab \otimes x$. But then if we consider $J := \mathcal{B}(A, E^*) \in A$-**unmod** (cf. 1.36), we see that the dual associativity map $(A \hat{\otimes} E)^* \xrightarrow{\simeq} \mathcal{B}(A, E^*)$ (0.3.43(IV)) is an isomorphism between P^* and J in A-**unmod**. Thus P^* is co-free and hence it is injective (1.38). ■

Exercise (VII.1.41). Extend the result we have obtained from the functor $(^*)$ to any $\mathbf{h}(?, E)$; $E \in \mathcal{B}an$ (cf. Chap.VI §3.2).

The observation made in 1.40 leads up to one of the definitions of a very important class of modules which includes the projective modules.

Theorem (VII.1.42) (cf. [285]). *The following properties of the module* $Z \in A$-**unmod** *are equivalent:*

(I) *for any admissible monomorphism* $i : Y \longrightarrow X$ *in* **unmod**$-A$ *the operator* $i \hat{\otimes}_A 1_Z : Y \hat{\otimes}_A Z \longrightarrow X \hat{\otimes}_A Z$ (see VI.3.28) *is topologically injective*;

(II) *the module* $Z^* \in$ **unmod**$-A$ *is injective.*[5]

▶ We use the well-known fact (which follows easily from the Hahn-Banach and Banach theorems) that an operator between two Banach spaces, in particular $i \hat{\otimes}_A 1_Z$, is topologically injective $\Longleftrightarrow$ its dual is surjective. The dual associativity law (VI.3.30) obviously gives us the commutative diagram

$$\begin{array}{ccc} (X \hat{\otimes}_A Z)^* & \xrightarrow{(i \hat{\otimes}_A 1)^*} & (Y \hat{\otimes}_A Z)^* \\ \downarrow \simeq & & \downarrow \simeq \\ \mathbf{h}_A(X, Z^*) & \xrightarrow{\mathbf{h}_A(i, Z^*)} & \mathbf{h}_A(Y, Z^*) \end{array}$$

where the "vertical arrows" are isometric isomorphisms. Hence $(i \hat{\otimes}_A 1_Z)^*$ being surjective is equivalent to $\mathbf{h}_A(i, Z^*)$ being surjective and the latter is obviously equivalent to the fact that

[5] Of course the word "injective" has completely different meanings in the two parts of the theorem.

the extension problem (Z^*, i, φ) has a positive solution for any $\varphi : Y \longrightarrow Z^*$. The rest is clear. ■

Definition (VII.1.43). The module $Z \in A$-**unmod** is said to be *flat* if it has the (equivalent) properties indicated in Theorem 1.42. A Banach (bi)module of any type is said to be *flat* if the left unital module identified with it (see Chap. VI §3.1) is also flat.

We suggest the reader gives a direct definition of a flat module in **mod**-A and A-**mod**-B (including the relevant versions of Theorem 1.42).

If we rephrase Proposition 1.40, we get the following.

Proposition (VII.1.44). *A projective (bi)module (of any type) is flat.* ▶■

The following gives an important class of flat modules.

Proposition (VII.1.45). *Let I be a closed left ideal in A_+ having a right b.a.i. Then I is flat (in A-**mod** $= A_+$-**unmod**).*

▶ Let $i : Y \longrightarrow X$ be an admissible monomorphism in **mod**-A = **unmod**-A_+. Then if we replace the algebra featuring in Theorem VI.3.24 by its opposite, we get the commutative diagram

$$\begin{array}{ccc} Y \mathbin{\hat{\otimes}_A} I & \xrightarrow{\;i \hat{\otimes}_A 1\;} & X \mathbin{\hat{\otimes}_A} I \\ \downarrow & & \downarrow \\ Y \cdot I & \xrightarrow{\;\;i_0\;\;} & X \cdot I \end{array}$$

where i_0 coincides pointwise with i and the "vertical arrows" are topological isomorphisms (cf. σ in VI.3.24). Since i is topologically injective, then so also is i_0 and hence $i \hat{\otimes}_A 1_I$ is also topologically injective. The rest is clear. ■

Corollary (VII.1.46). *All closed ideals in $C_0(\Omega)$, where Ω is a locally compact space, are flat $C_0(\Omega)$-modules.*

Concerning cyclic modules, the following is "almost a criterion" for them to be flat.

Assertion (VII.1.47). *Let I be a closed left ideal in A_+. Then for A_+/I to be flat it is sufficient and, if $I^\perp := \{f \in A_+^* : f|_I = 0\}$ is complemented in A_+^* (in particular, if I is complemented in A_+), it is also necessary that I have a right b.a.i.*

For the proof and various applications see, for example, [272].

Exercise (VII.1.48). Using the result of Exercise VI.3.29 show that the cyclic $\mathcal{A}(\overline{\mathbb{D}})$-module $\mathbb{C}(0)$ featuring there is not flat.

The question of conditions for modules over L^1-algebras to be flat deserves special attention: it leads to the concept of an amenable group which is important in harmonic analysis.

Let G be a locally compact group and let $L^\infty(G)$ be the Banach space of functions on G that are measurable and essentially bounded with respect to Haar measure; we recall that this is isometrically isomorphic to $L^1(G)^*$ (see, for example, [87, Ch.3 §12]).

The functional $M : L^\infty(G) \longrightarrow \mathbb{C}$ is said to be *left invariant* if $M({}^rf) = M(f)$ for all $f \in L^\infty(G)$ and $r \in G$, and *normalized* if $M(\underline{1}) = 1$ ($\underline{1}$ is the function identically equal to 1).

Definition (VII.1.49). The group G is said to be *amenable* if there exists a continuous left invariant normalized functional on $L^\infty(G)$.

Any compact group is amenable: we can put $M(f) = \int_G f(s)ds$.

It is more complicated to prove that any commutative group is amenable; see [87 §17.5] or [9, 43.5]. On the other hand, no group having a free discrete subgroup with two generators is amenable [25, 1.2.3 and 2.3.2].[6]

The following turns out to be true.

Assertion (VII.1.50). *The augmenting $L^1(G)$-module $\mathbb{C}$ is flat $\Leftrightarrow$ the group G is amenable.*

We shall give the proof only for the case when G is discrete and therefore $L^1(G) = l_1(G)$ and $L^\infty(G) = l_\infty(G)$ (see VI.3.3); the general case is examined, for example, in [81, VII §2.5].

The canonical projection $\pi_{\mathbb{C}}$ for the augmenting $l_1(G)$-module can obviously be identified with the augmentation morphism τ (VI.1.31) and its dual can be identified with the morphism of right $l_1(G)$-modules $\tau^* : \mathbb{C} \longrightarrow l_\infty(G)$ (see VI.3.9 and VI.3.3) that realizes the "embedding of constants" $z \longmapsto f(s) \equiv z$.

---

[6] See also a good new book [61]. For discrete finitely generated groups the question of whether they are amenable is completely solved in terms of the growth — with increasing $n \in \mathbb{N}$ — of the number of different words in n letters that represent the identity element of the group. A criterion of this type was published in 1978 by Grigorchuk [127] and in 1982 by Cohen [163]. Concerning this, it is subtly noted in [61, p.339] that Grigorchuk obtained his result independently of Cohen.

Lemma (VII.1.51). *A continuous functional* $M : l_\infty(G) \longrightarrow \mathbb{C}$ *is left invariant* $\Longleftrightarrow$ *it is a morphism in* **mod**-$l_1(G)$.

▶ Since ${}^s f = f \cdot \delta_s$ (Chap.III §3.1) and $z \cdot \delta_s = z$ whenever $f \in l_\infty(G)$, $z \in \mathbb{C}$, $s \in G$, then M is left invariant $\Longleftrightarrow$ $M(f \cdot \delta_s) = M(f) \cdot \delta_s$ for all f and s. But the last property, by $l_1(G) = \overline{\ell.h.}\{\delta_s : s \in G\}$, is equivalent to M being a morphism. ■

Proposition (VII.1.52). *The following properties of a discrete group are equivalent*:

(I) *the augmenting* $l_1(G)$*-module* $\mathbb{C}$ *is flat*;

(II) *the coretraction problem* $(\mathbb{C}, \tau^*)$ *in* **mod**-$l_1(G)$ *has a positive solution*;

(III) G *is amenable.*

▶ By Proposition 1.34, Theorem 1.39(II) and the fact that $l_\infty(G)$ is "co-free" (cf. 1.37), property (II) is equivalent to $\mathbb{C}$ being injective as a right module, that is, dual to the augmentation module. Hence (II) $\Longleftrightarrow$ (I). Further 1.51 obviously implies that a functional satisfying 1.49 is a left inverse morphism to τ^*. Hence (II) $\Longleftrightarrow$ (III). ■

Together 1.50 and 1.47 provide the following criterion for amenability (first discovered in [30]).

Assertion (VII.1.53). *A locally compact group* G *is amenable* $\Longleftrightarrow$ *the augmentation ideal in* $L^1(G)$ *has a right b.a.i.*

Remark (VII.1.54). If G is amenable, then among the functionals satisfying Definition 1.49 there can also be found a so-called mean functional M_0 such that $\operatorname{ess\,inf} f \le M_0(f) \le \operatorname{ess\,sup} f$ for all real-valued f. (It is easy to construct this by taking the M in Definition 1.49 and putting $M_0(f) := \sup\{\operatorname{Re} M(g) : 0 \le g \le f\}$ for $f \ge 0$.) It is in just this way — in terms of (left invariant) means — that amenable groups are most often defined; moreover many other function spaces can be used instead of $L^\infty(G)$, for example, $C_b(G)$. There exist other approaches to the concept of amenability that are completely different from those in 1.52 and 1.53.[7] Many of them and numerous applications are given in [25], [61].

In fact, if the augmenting $L^1(G)$-module is flat, then all one-sided (left or right) modules over $L^1(G)$, and those over $C^*(G)$ as well, are also flat. In addition, all one-sided modules over many C^*-algebras are flat, including our acquaintances $C_0(\Omega)$, $\mathcal{K}(H)$ and the Fermion algebra. We shall say something about why this should be so in §1.4.

---

[7] One of these has already been mentioned (IV.7.68): G is amenable $\Longleftrightarrow$ its complete and reduced C^*-algebras coincide.

Concerning non-self-adjoint operator algebras, the following assertion, due to Golovin [120], is known.

Assertion (VII.1.55). *Let A be a nest algebra of operators on H* (see Chap.III §2.2). *Then the natural A-module H is flat and injective.*

* * *

From the beginning of the section and up to this point we have talked only about Banach modules. We shall now say a few words about the concepts and facts we are discussing in the context of other categories of polynormed modules. More details of this are in [81].

Concerning the definition of admissible problems and the consequent definitions of projective and injective modules, they carry over word for word to all the categories we considered earlier, including polynormed (bi)modules and Fréchet (bi)modules (of any type). All the results in §1.1, including Theorem 1.11, obviously carry across to projective modules; where the role of the free modules is now played by the free objects in the relevant categories (as recalled in Chap.VI §1.2).

It is even easier to work with certain polynormed function algebras than with Banach algebras: the free modules over them can be described explicitly as vector function modules of a suitable class (cf. VI.1.27). Sometimes this enables us to establish that certain important modules are projective:

Assertion (VII.1.56) (Ogneva [203]). *Let $\mathcal{M}$ be a smooth manifold and let U be an open subset of $\mathcal{M}$. Then the Fréchet $C^\infty(\mathcal{M})$-module $C^\infty(U)$* (see VI.1.24) *is projective.*

(In this situation the canonical projection $\pi_{C^\infty(U)}$ turns out to be precisely — and not "approximately" as in Example 1.8 — the map of restriction to the diagonal $C^\infty(\mathcal{M} \times U) \longrightarrow C^\infty(U) : u(s,t) \longmapsto f(s) := u(s,s)$.)

The situation is far worse when we try to extend the basic results of the "injective" theory (1.38 - 1.42) beyond the context of Banach structures. The reason has already been noted in Chap.VI §3.2: if A and E are not Banach, the space $\mathcal{B}(A,E)$, generally speaking, does not have — unlike $A \hat{\otimes} E$ — a natural polynormed module structure. Therefore the concept of a co-free module and all that has been based on it, loses its sense; in particular, only the first of the two definitions of flatness (1.42), expressed in terms of tensor products, is preserved. (All the same, as Taylor showed, within a certain class of polynormed modules a considerable part of the "injective" theory can be saved; see [251] about this.)

Finally, we recall the purely algebraic situation corresponding, as has already been stressed repeatedly, to algebras and modules with strongest seminorm system on their underlying spaces. Here again we have that symmetry between "projective" and "injective" theories that was lost when transferring from Banach modules to Fréchet modules, say. In addition, since it is obvious that all epi- and monomorphisms are admissible, all the basic definitions and formulations of results can be noticeably simplified — and we arrive at the concepts and facts which were discovered in homological algebra far earlier than in their "Banach" counterpart [38], [51]. Projective "abstract" modules turn out to be direct summands of the modules $\bigoplus_{\Lambda} A$ (see Chap. VI §1.1), injective "abstract" modules turn out to be direct summands of the modules $\mathcal{L}(A, E)$; $E \in \mathcal{L}in$, and the first definition of flatness gives those (and only those) Z for which the functor $? \otimes_{A} Z$ preserves monomorphisms.

We suggest that the reader establishes, as a simple and useful exercise, the remaining details of the corresponding purely algebraic theory.

§1.4. Contractible, biprojective and amenable algebras. Equivalent definitions and general properties

We first establish a few useful connections between projectivity and flatness in the various classes of (bi)modules. In future, everywhere, unless stated otherwise, A is a fixed (not necessarily unital) Banach algebra and all (bi)modules are also Banach.

Proposition (VII.1.57). (I) *Any projective or flat A-bimodule is also projective or flat as a left or a right A-module.*

(II) *If X is a projective left and Y a projective right A-module, then $X \hat{\otimes} Y$ is a projective A-bimodule.*

(III) *If X is a flat left and Y a flat right A-module, then $X \hat{\otimes} Y$ is a flat A-bimodule.*

(IV) *If X is a projective A-bimodule and Y any left A-module, then $X \hat{\otimes}_{A} Y$ is a projective left A-module.*

(V) *If X is a flat A-bimodule and Y any left A-module, then $X \hat{\otimes}_{A} Y$ is a flat left A-module.*

▶ All the required facts are established by much the same arguments by representing a projective module as a retract of a free module (1.11(II)), an injective module as a retract of a co-free module (1.39(II)), and the dual of a flat module as a retract of $\mathcal{B}(A_+, E^*)$; $E \in \mathcal{B}an$ (1.39(III)); moreover extensive application is made of the standard identities for "$\hat{\otimes}$" and "$\mathcal{B}(\cdot, \cdot)$" (0.3.43) or, in the last assertions — for "$\hat{\otimes}_{A}$" and

"$\mathbf{h}_A(\cdot, \cdot)$" (VI.3.22, VI.3.25, VI.3.30). We leave the reader to produce the simplest proofs, we limit ourselves to two.

(III) As X^* is a retract of a right A-module $\mathcal{B}(A_+, E^*) \simeq (A_+ \hat{\otimes} E)^*$, then $(X \hat{\otimes} Y)^* \simeq \mathcal{B}(Y, X^*)$ is a retract of the A-bimodule $\mathcal{B}(Y, (A_+ \hat{\otimes} E)^*) \simeq \mathcal{B}(A_+ \hat{\otimes} E, Y^*)$ and hence a retract of the A-bimodule $\mathcal{B}(A_+ \hat{\otimes} E, \mathcal{B}(A_+, F)) \simeq$ $\simeq \mathcal{B}(A_+ \hat{\otimes} A_+ \hat{\otimes} E, F) \simeq \mathcal{B}(A^{env}, \mathcal{B}(E, F))$, that is, it is the retract of a co-free A-bimodule. The rest is clear.

(V) As X^* is a retract of an A-bimodule $\mathcal{B}(A^{env}, E)$, then $(X \hat{\otimes}_A Y)^* \simeq (Y \hat{\otimes}_{A^{op}} X)^* \simeq \mathbf{h}_{A^{op}}(Y, X^*)$ is a retract of the left A^{op}-module (= right A-module) $\mathbf{h}_{A^{op}}(Y, \mathcal{B}(A_+ \hat{\otimes} A_+^{op}, E)) \simeq$ $\simeq \mathcal{B}(Y \hat{\otimes}_{A^{op}} A_+^{op} \hat{\otimes} A_+, E) \simeq \mathcal{B}(Y \hat{\otimes} A_+, E) \simeq \mathcal{B}(A_+, \mathcal{B}(Y, E))$, that is, it is the retract of a co-free right A-module. The rest is clear. ∎

Within this section we denote the A-bimodule A_+/A by $\mathbb{C}$: obviously this is a one-dimensional space with zero outer multiplications. The natural projection of A_+ onto $\mathbb{C}$ is denoted by σ.

Proposition (VII.1.58). *If $\mathbb{C}$ is projective, injective or flat as a left and as a right A-module, then it is projective, injective or flat as an A-bimodule.*

▶ The assertion about projectivity is implied by 1.57(II) where $\mathbb{C}$ is represented as $\mathbb{C} \hat{\otimes} \mathbb{C}$, the remaining assertions are implied by 1.57(III), the same representation of $\mathbb{C}$ and the identification of $\mathbb{C}^*$ with $\mathbb{C}$ in A-**mod**-A. ∎

We now introduce a few important classes of algebras defined in homological terms (cf. the beginning of the chapter).

Definition (VII.1.59). An algebra A is said to be *biprojective* if the A-bimodule A is projective and *biflat* if this bimodule is flat. A is said to be *contractible* (or cohomologically trivial) if A_+ is biprojective and *amenable* (or homologically trivial) if A_+ is biflat.

(The reason for the names given to the last two classes of algebras will become clear later on; see Assertion 1.86 and Theorem 3.37.)

Modules over the above algebras are especially nice:

Theorem (VII.1.60). (I) *If A is contractible, then all left and right A-modules are projective; if A is merely amenable, then they are all flat.*

(II) *If A is biprojective, then all reduced left and right A-modules are projective; if A is merely biflat, then all these modules are flat. If A is biprojective (flat, respectively) and has a b.a.i., then all non-degenerate left and right*

A-modules are projective (flat). In particular, a biprojective algebra with a b.a.i. is hereditary on the left and on the right.

▶ This is implied by Theorem 1.57((IV) and (V)) (with A_+, A_+^{op}, A or A^{op} in the role of X) and VI.3.24. ■

Theorem 1.60(I) and Proposition 1.58 immediately imply the following.

Proposition (VII.1.61). *If A is contractible, then $\mathbb{C}$ is a projective A-bimodule, and if A is amenable, then $\mathbb{C}$ is an injective and flat A-bimodule.* ▶■

Proposition (VII.1.62). *If A is contractible, then it is biprojective and if A is amenable, then it is biflat.*

▶ Propositions 1.61, 1.5 and 1.34 imply that if we have contractibility (amenability, respectively) the A-morphism σ is a retraction (σ^* is a coretraction, respectively) in A-**mod**-A. Therefore in the first case $A = \operatorname{Ker}\sigma$ is a retract of A_+ and in the second case $A^* \simeq \operatorname{Ker} j$, where j is any morphism which is a left inverse of σ^*, and hence A^* is a retract of A_+^*. The rest is clear. ■

So the interrelation between the four classes of algebras we have introduced is expressed by the following scheme:

$$\textit{contractible} \begin{array}{c} \Longrightarrow \textit{biprojective} \Longrightarrow \\ \Longrightarrow \textit{amenable} \Longrightarrow \end{array} \textit{biflat}$$

We shall gradually confirm that nothing can be added to this without additional assumptions.

Theorem (VII.1.63). *If A is biprojective and unital, then it is contractible; if A is biflat and has a b.a.i., then it is amenable.*

▶ We limit ourselves to proving the second, more complicated assertion. The Banach–Alaoglu theorem (see Chap.0 §1.2) implies that the given b.a.i. has a subnet that is weak* convergent to a certain $\check{e}$ in $A^{**} \supseteq A$. We put $j : A_+^* \longrightarrow \mathbb{C} : f \longmapsto f(e_+) - \check{e}(f|_A)$; it is easy to verify that j is a morphism of A-bimodules which is left inverse to σ^*. Hence, obviously, A_+^* is the direct sum of A^* and $\mathbb{C}$ in A-**mod**-A; therefore, by Theorem 1.39(II), $\mathbb{C}$ is injective in A-**mod** and in **mod**-A, and hence (1.58) it is also injective in A-**mod**-A. Consequently, A_+^* is the direct sum of two injective A-bimodules, and this obviously implies that it is itself injective. ■

Corollary (VII.1.64). *For a unital algebra, being biprojective is equivalent to being contractible and for an algebra with a b.a.i., being biflat is equivalent to being amenable.*

(In particular, we are justified in replacing the "compulsory unitization" A_+ by the "conditional unitization" $A_\times$ in Definition 1.59.)

Exercise (VII.1.65). Show that the converses of both assertions in Theorem 1.63 hold.

Hint. If $\mathbb{C}$ is injective in A-**mod**-A, then A_+^{**} is the direct sum of A^{**} and $\mathbb{C}$ in the same category and hence $\check{e} \in A^{**}$ exists with $a \cdot \check{e} = \check{e} \cdot a = a$ (in A^{**}) for all $a \in A$. As the unit ball of A is weak* dense in the unit ball of A^{**} (see, for example, [27, Theorem V.4.5]), then there exists in A a bounded net which is weak* convergent to $\check{e}$. The latter is a b.a.i. in A.[8]

Our next aim is to show that biprojectivity can be expressed in terms of a single fairly explicit retraction problem and biflatness can be expressed in terms of the coretraction problem "dual to it".

A basic role will be played by the multiplication operator $\pi : A \hat{\otimes} A \longrightarrow A : a \otimes b \longmapsto ab$; it is the canonical morphism for $A \in A$-**mod**. We also let $\pi_+ : A_+ \hat{\otimes} A \longrightarrow A$ and $\pi^+ : A \hat{\otimes} A_+ \longrightarrow A$ denote the canonical projections for $A \in A_+$-**unmod** and $A \in$ **unmod**-A_+ respectively. Obviously π, π_+ and π^+ are morphisms of A-bimodules and π can be considered as a restriction of π_+ and of π^+. We formulate an intermediate proposition.

Proposition (VII.1.66). *The algebra* A *is biprojective* $\Longleftrightarrow$ π_+ *and* π^+ *are retractions in* A-**mod**-A *and it is biflat* $\Longleftrightarrow$ π_+^* *and* π^{+*} *are coretractions in* A-**mod**-A.

▶ "⇒". It is obvious that π_+, π^+, π_+^* and π^{+*} are admissible epi- and monomorphisms[9] and so the first and second statements follow immediately from the the "bimodule" version of Propositions 1.5 and 1.34, respectively.

"⇐". First note that $A_+ \hat{\otimes} A$ is projective in A-**mod** and hence $(A_+ \hat{\otimes} A)^*$ is injective in **mod**-A. If π_+ is a retraction in A-**mod**-A, then it is also a retraction in A-**mod** and so, by Proposition 1.6, A is projective in A-**mod**. Hence, by Proposition 1.57(II), $A \hat{\otimes} A_+$ is a projective A-bimodule and so, if π^+ is also a retraction in A-**mod**-A, A is biprojective by Proposition 1.6 for bimodules. This proves the first statement.

Similarly, if π_+^* is a coretraction, Proposition 1.35 shows that A^* is injective in **mod**-A and so, by Proposition

[8] It is enough to show the existence of a "weak" b.a.i.: see [9, §11].

[9] In this respect π_+ and π^+ are, generally speaking, better than π and π_+^* and π^{+*} are better than π^*.

1.57(III), $A \hat{\otimes} A_+$ is a flat A-bimodule. By the bimodule version of Proposition 1.35, if π^{+*} is also a coretraction in A-**mod**-A, then A is biflat. This proves the second statement. ■

The next step on the road to the promised criterion is of interest in its own right.

Proposition (VII.1.67). *Any biflat, and as a consequence, any biprojective algebra A is idempotent (= coincides with A^2).*

▶ Suppose otherwise; then there exists $f \in A^*$ with $f|_{A^2} = 0$ and $f(a_0) = 1$ for a certain $a_0 \in A$. We extend f to $f_+ \in A_+^*$ by putting $f_+(e_+) = 0$ and we put $u := f_+ \hat{\otimes} f \in (A_+ \hat{\otimes} A)^*$. Then, on the one hand, for all $b \in A_+$, $c \in A$ the obvious equality $f_+(a_0 b) f(c) = f(bc)$ implies that $[u \cdot a_0](b \otimes c) = [\pi_+^* f](b \otimes c)$, that is, $u \cdot a_0 = \pi_+^* f$; on the other hand, for the same b and c, $[a_0 \cdot u](b \otimes c) = f_+(b) f(c a_0) = 0$, that is $a_0 \cdot u = 0$. But by Proposition 1.66 π_+^* has a left inverse ζ in A-**mod**-A; hence $[\zeta(u)](a_0^2) = [\zeta(u) \cdot a_0](a_0) = = [\zeta(u \cdot a_0)](a_0) = [\zeta \circ \pi_+^*(f)](a_0) = f(a_0) = 1$, and at the same time $[\zeta(u)](a_0^2) = [a_0 \cdot \zeta(u)](a_0) = \; = [\zeta(a_0 \cdot u)](a_0) = 0$. A contradiction. ■

Exercise (VII.1.68). Prove directly (by a simpler argument) that biprojective algebras are idempotent.

Here finally is the criterion itself.

Theorem (VII.1.69). (I) *A is biprojective $\Longleftrightarrow$ $\pi : A \hat{\otimes} A \longrightarrow A$ is a retraction in* A-**mod**-A.

(II) *A is biflat $\Longleftrightarrow$ $\pi^* : A^* \longrightarrow (A \hat{\otimes} A)^*$ is a coretraction in* A-**mod**-A.

▶ (I) "⇐". A right inverse to π is also a right inverse to π_+ and to π^+ and therefore A being biprojective is implied by Proposition 1.66.

"⇒". As $A = A \cdot A$ (see 1.67), then the image of any A-bimodule morphism ρ, which is a right inverse to π_+ (see 1.66), lies in $A \cdot (A_+ \hat{\otimes} A) = A \hat{\otimes} A$. All that remains is to consider $\rho|_{A \hat{\otimes} A}$.

(II) We let $j : (A_+ \hat{\otimes} A)^* \longrightarrow (A \hat{\otimes} A)^*$ be the morphism of taking the restriction of functionals (to the space identified with $A \hat{\otimes} A$).

"⇐". Let ζ be a left inverse to π^*; then obviously $\zeta \circ j$ is a left inverse to π_+^* (in A-**mod**-A). Similarly π^{+*} is a coretraction, so Proposition 1.66 implies that A is biflat.

"⇒". Let ζ_+ be a left inverse to π_+^* (in A-**mod**-A). We consider $f \in \operatorname{Ker} j$, in other words, f with $f|_{A \hat{\otimes} A} = 0$. Then $(f \cdot a)(c \otimes b) = f(ac \otimes b) = 0$ for all $a, b \in A$ and $c \in A_+$; therefore $f \cdot a = 0$ and $[\zeta_+(f)](ab) = [\zeta_+(f) \cdot a](b) = = [\zeta_+(f \cdot a)](b) = 0$. Hence $\zeta_+(f)$ is equal to zero on A^2 and hence (by 1.67) it is zero on the whole of A. Consequently ζ_+ is equal to zero on $\operatorname{Ker} j$ and therefore, as j is surjective, it induces some $\zeta : (A \hat{\otimes} A)^* \longrightarrow A^*$. It is easy to verify that the latter is an A-bimodule morphism which is left inverse to π^*. ■

We give a useful application. An element $d \in A \hat{\otimes} A$ is said to be a *diagonal* for A if $a \cdot d = d \cdot a$ for all $a \in A$ and if $\pi(d)$ is an identity element in A. A bounded net $(d_\nu \in A \hat{\otimes} A;\ \nu \in \Lambda)$ is called an *approximate diagonal* for A if $\lim_\nu (a \cdot d_\nu - d_\nu \cdot a) = 0$ for all $a \in A$ and $(\pi(d_\nu);\ \nu \in \Lambda)$ is a b.a.i. in A [133].

Theorem (VII.1.70). *If the algebra has a diagonal then it is contractible and if it has an approximate diagonal, then it is amenable.*

▶ If d is a diagonal, then $\rho : A \longrightarrow A \hat{\otimes} A : a \longmapsto a \cdot d$ is obviously a right inverse to π in **A-mod-A** and so A, by Theorems 1.69(I) and 1.63, is contractible. If there exists an approximate diagonal, then, by the Banach–Alaoglu theorem, it has a certain subnet that converges in $(A \hat{\otimes} A)^{**} \supseteq A \hat{\otimes} A$ to a certain $\check{d}$ and it is easy to verify that $\zeta : (A \hat{\otimes} A)^* \longrightarrow A^* : g \longmapsto f$, where $f(a) := \check{d}(a \cdot g)$, is a left inverse to π^* in **A-mod-A**. Therefore A, by Theorems 1.69(II) and 1.63, is amenable. ■

Proposition (VII.1.71). *Let $\kappa : A \longrightarrow B$ be a continuous homomorphism between Banach algebras having a dense image. Then, if A is contractible or amenable, then so also is B. In particular, a quotient algebra of a contractible (amenable) algebra is itself contractible (amenable).*

▶ The condition obviously implies that for any $u \in A \hat{\otimes} A$ and $a \in A$, $[\kappa \hat{\otimes} \kappa](a \cdot u) = \kappa(a) \cdot [\kappa \hat{\otimes} \kappa](u)$ and $[\kappa \hat{\otimes} \kappa](u \cdot a) = [\kappa \hat{\otimes} \kappa](u) \cdot \kappa(a)$. Therefore it is easy to verify that, if d is a diagonal ($(d_\nu;\ \nu \in \Lambda)$, is an approximate diagonal, respectively) for A, then $[\kappa \hat{\otimes} \kappa]d$ is a diagonal ($([\kappa \hat{\otimes} \kappa](d_\nu);\ \nu \in \Lambda)$, is an approximate diagonal, respectively) for B. All that remains is to apply Theorem 1.70. ■

See [30], [272], [81, IV, §5 and VII §2] for many other constructions that preserve amenability and/or the other properties indicated in Definition 1.59.

In conclusion we gather together what we already know about equivalent definitions of contractibility and amenability and add to this, for the reader's information, a few new facts. In the next assertion I_Δ denotes the kernel of the operator $A_+ \hat{\otimes} A_+ \longrightarrow A_+ : a \otimes b \longmapsto ab$; obviously, this is a left ideal in the enveloping algebra A^{env}.

Assertion (VII.1.72). (I) *A is contractible* $\Leftrightarrow$ *A is biprojective and unital* $\Leftrightarrow$ *A has a diagonal* $\Leftrightarrow$ I_Δ *has a right identity.*

(II) *A is amenable* $\Leftrightarrow$ *A is biflat and has a b.a.i.* $\Leftrightarrow$ *A has an approximate diagonal* $\Leftrightarrow$ I_Δ *has a right b.a.i.*

For a proof of the equivalence of the properties indicated here and also many other criteria, see the sources cited above. One other important criterion for contractibility and amenability will be established later in §3.3.

Practically all the results in this section relating to contractibility and biprojectivity carry over word for word to wider categories of polynormed modules; in particular, for arbitrary polynormed algebras the contractibility criteria indicated in Proposition 1.72 remain in force. (Incidentally, "abstract" contractible algebras were known a long time ago by algebraists under the name of separable algebras; see for example [51, VII §5].) The obstacles already discussed at the end of §1.3 naturally arise when we attempt to extend the results about amenability and biflatness that, as we have seen, use the Banach structure of the dual space in an essential way. However, as far as we know, nobody has yet seriously considered the interesting question of amenability and biflatness beyond the limits of Banach structures.

§1.5. Contractibility, biprojectivity and amenability in specific classes of algebras. Johnson's theorem on amenability (the case of discrete groups)

We now again visit our "zoo" of specific classes of algebras and resume our questionnaire on its inhabitants (cf. introduction to Chap.III). It is not difficult to guess what the new questionnaire to be put to the algebras we have met will look like: are you contractible? Biprojective? Amenable? Or at least biflat? We first discuss the strongest and, as we shall quickly see, the extremely rigid contractibility condition.

Proposition (VII.1.73). *Let* $A = \mathcal{M}_n$ (see Chap.I §1.1) and *let* e_{kl}; $1 \le k, l \le n$, *be the matrix with* 1 *in the "kl-th" place and zero in the others. Then* $d_n := \frac{1}{n}\sum_{k,l=1}^{n} e_{kl} \otimes e_{lk} \in A \hat{\otimes} A$ *is a diagonal for A.*

▶ This follows from the relations $e_{ij}e_{kl} = \delta_k^j e_{il}$, where δ is the Kronecker symbol. ■

Theorem (VII.1.74). *The Cartesian product of a finite number of complete matrix algebras with complex coefficients (generally speaking of different dimension) is a contractible algebra. As a* *consequence, all left and right modules over such an algebra are projective.*

▶ If $A = \bigtimes_{k=1}^{m} A_k$ then $A_k \hat{\otimes} A_k$ can obviously be identified with a subspace of $A \hat{\otimes} A$ and the sum of any set of diagonals for A_k is a diagonal for A. All that remains is to apply Proposition 1.73 and Theorem 1.70. ■

Do other contractible Banach algebras exist besides those just indicated? In view of the presence of the "pathological features of Banach space geometry" (cf. Chap.0 §1.4), this question turns out to be fairly complicated. More explicitly we know the answer to the following question.

Question. Do all, or at least some, of the following (already established) logical arrows have converses: A is isomorphic to the Cartesian product of a finite number of complete matrix algebras ⇒ A is contractible ⇒ A is unital, and all the left Banach A-modules are projective ⇒ A is unital and all the irreducible left Banach A-modules are projective?

A positive answer has been obtained with certain assumptions about A or its irreducible modules having a "nice geometry" (Selivanov [234]; see also [81, IV §§4, 5]). In particular, the following holds.

Assertion (VII.1.75) (cf. 1.30). *Let the unital Banach algebra A be either commutative or a C^*-algebra or the unitization of an L^1-algebra. Then all four conditions on A, indicated in the preceding question, are equivalent.*

Here the commutative case is simplest; we suggest the following to the reader.

Exercise (VII.1.76) [267]. Show that all one-dimensional modules over a unital commutative Banach algebra are projective ⟺ this algebra is isomorphic to the Cartesian product of finitely many copies of $\mathbb{C}$.

It is evident from the proof of Theorem 1.74 that it is true independently of whether it involves Banach, polynormed or, abstract algebras and modules, say. However, the question of whether the converse theorem is true depends essentially on the choice of class of the underlying spaces. The following well-known result (which really started all the preceding discussion) shows how matters stand in pure algebra.

Assertion (VII.1.77) [327], [46]. *Let A be a pure algebra. Then*

(I) *A is contractible ⟺ it is isomorphic to the Cartesian product of finitely many complete matrix algebras with complex coefficients*;

(II) *if A is unital, then all A-modules are projective $\Longleftrightarrow$ all irreducible A-modules are projective $\Longleftrightarrow$ A is isomorphic to the Cartesian product of finitely many complete matrix algebras with coefficients from (generally speaking, different) complex skew fields - in other words, A is classically semisimple*; see I.3.68.

On the other hand, when considering general polynormed structures, the converse of Theorem 1.74 is certainly false.

Assertion (VII.1.78) ([251], for the proof see [81, Theorem IV.5.27]). *Let A be a commutative Arens-Michael algebra. Then it is contractible $\Longleftrightarrow$ all polynormed A-modules are projective $\Longleftrightarrow$ A is topologically isomorphic to $\mathbb{C}^M$* (I.2.16) *for a certain set M.*

The reason for this phenomenon is explained by the following.

Exercise (VII.1.79). Construct a topological isomorphism $\mathbb{C}^M \hat{\otimes} \mathbb{C}^M \xrightarrow{\simeq} \mathbb{C}^{M\times M}$ for which the element that maps into the characteristic function of the "diagonal" $\{(t, t) : t \in M\}$, is a diagonal for $\mathbb{C}^M$.

An interesting example of a contractible algebra that is not an Arens-Michael algebra is $\mathcal{E}(G)^*$ for any compact Lie group G [252, §7]; and therefore c_p, which is topologically isomorphic to $\mathcal{E}(\mathbb{T})^*$, is also an example (see the end of Chap. III §3.3).

We turn from contractible algebras to "simply" biprojective algebras; this condition is also fairly strong but nevertheless allows a far wider class of algebras.

Example (VII.1.80). The algebras c_0 and l_1 (with co-ordinatewise product) are biprojective (and therefore the first of them, in view of the presence of a b.a.i., is even amenable as well; see Theorem 1.63). In fact, given $\xi = \{\xi_n;\ n \in \mathbb{N}\} \in c_0$ (or l_1), consider the series $\sum_{n=1}^{\infty} \xi_n \mathbf{p}^n \otimes \mathbf{p}^n$ in $c_0 \hat{\otimes} c_0$ (or $l_1 \hat{\otimes} l_1$), where $\mathbf{p}^n := (0, \cdots, 0, 1, 0, \cdots)$. Such a series converges: for c_0 this is implied by the bound $\| \sum_{k=m}^{n} \xi_k \mathbf{p}^k \otimes \mathbf{p}^k \| \leq \max\{ |\xi_k| : m \leq k \leq n\}$ (see 0.3.30) and for l_1 it follows immediately by identifying $l_1 \hat{\otimes} l_1$ with l_1^2 (see Chap. 0.3.36). Therefore, if we associate to each ξ the sum of the above series, we get, as is easily verified, a morphism of $c_0(l_1)$-bimodules which is right inverse to the π in Theorem 1.69(I).

The question of which L^1-algebras are biprojective is solved by the following.

Exercise (VII.1.81) [81, IV §5.2]. Show that the algebra $L^1(G)$, where G is a locally compact group, is biprojective $\Leftrightarrow$ G is compact.[10]

Hint. "$\Rightarrow$" is implied by Exercise 1.28 and Theorem 1.60(II). In proving "$\Leftarrow$" it is necessary to consider $\rho : L^1(G) \longrightarrow L^1(G \times G) : f \longmapsto u$, where $u(s, t) := f(st)$. This is an $L^1(G)$-bimodule morphism which is right inverse to the Π in Example 1.16.

Biprojective C^*-algebras are described in the following assertion that is stated here without proof (see [269] and [235]). Let $\{A_\nu;\ \nu \in \Lambda\}$, be a family of Banach algebras. The closed subalgebra of $\sum A_\nu$, consisting of those "rows" $a = \{a_\nu \in A_\nu;\ \nu \in \Lambda\}$ for which, for each $\varepsilon > 0$, $\|a_\nu\| \geq \varepsilon$ is satisfied for only a finite set of indices ν, is called the c_0-sum of this family (see Chap.VI §2.2).

Assertion (VII.1.82). *A C^*-algebra is biprojective $\Leftrightarrow$ it is isometrically isomorphic to the c_0-sum of a certain family of algebras $\{\mathcal{B}(H_\nu);\ \nu \in \Lambda\}$, where all the H_ν are finite-dimensional Hilbert spaces.*

Another important class of biprojective algebras is constructed in the following way [235]. Let E be an arbitrary Banach space. We consider the bi-operator $m : (E \hat{\otimes} E^*) \times (E \hat{\otimes} E^*) \longrightarrow E \hat{\otimes} E^*$ that is (well) defined by $m(x \otimes f, y \otimes g) := \langle f, y\rangle x \otimes g$. Obviously m is a multiplication turning $E \hat{\otimes} E^*$ into a Banach algebra.

Proposition (VII.1.83). *The algebra $A := E \hat{\otimes} E^*$ is biprojective.*

▶ We choose $x_0 \in E$ and $f_0 \in E^*$ arbitrarily with $f_0(x_0) = 1$ and we consider the (well-defined) operator $\rho : A \longrightarrow A \hat{\otimes} A : x \otimes f \longmapsto (x \otimes f_0) \otimes (x_0 \otimes f)$. Elementary calculations show that ρ is an A-bimodule morphism which is right inverse to the π in Theorem 1.69(I). ■

The algebra we have introduced is directly related to the algebra of nuclear operators $\mathcal{N}(E)$: it is connected with the latter by a surjective homomorphism $R : E \hat{\otimes} E^* \longrightarrow \mathcal{N}(E)$, associated with the bi-operator $(x, f) \longmapsto \varphi$, where $\varphi(y) = \langle f, y\rangle x$. Therefore, as a corollary to 1.83, if E is such that R is injective,[11] then $\mathcal{N}(E)$ is a biprojective

---

[10] As usual a reader wishing to limit himself to the specific groups in Chap.III §3.1 is advised only to check that the algebra $L^1(\mathbb{T})$ is biprojective and that $l_1(\mathbb{Z})$ and $L^1(\mathbb{R})$, say, are not.

[11] In other words, if E has the approximation property; see 0.3.39.

algebra. This, by the Schatten-von Neumann theorem formulated earlier (0.3.38), implies the following.

Assertion (VII.1.84). The algebra $\mathcal{N}(H)$, where H is a Hilbert space, is biprojective.

It turns out that Proposition 1.83 provides something more than simply an example of a biprojective algebra: loosely speaking, the algebras featuring there are the bricks from which arbitrary biprojective semisimple Banach algebras with a "nice geometry" are built. For a precise formulation and proof of this result, due to Selivanov, see [235]. It implies, by the way, that the structure spaces (see Chap.VI §2.2) of biprojective algebras of the class under discussion are always discrete and they can have a b.a.i. only under the condition that their irreducible representations are finite-dimensional. In this way — taking into account Exercise 1.81 — one can obtain a proof of both properties for $L^1(G)$ with compact G (cf. VI.2.19).

Finally, the fact established above that c_0 is biprojective is generalized by the following.

Exercise (VII.1.85) [81, IV.5.26]. Show that the algebra $C_0(\Omega)$, where Ω is a locally compact space, is biprojective $\Leftrightarrow$ Ω is discrete.

Hint. If the ρ, featuring in Exercise 1.26, is a bimodule morphism, then the "skeleton" $F(s, t)$ is a (continuous) function equal to one when $s = t$ and zero when $s \neq t$.

* * *

Let us consider what the amenability condition — possibly the most important of those defined in 1.59 — turns out to mean for specific classes of Banach algebras.

Up to now in our account the term "amenable algebra", introduced in 1972 by Johnson [30], has not had any connection with the term "amenable group" (see 1.49) which was already in use in the fifties (Day [140]). In fact the motivation for the first term was a theorem[12] that, in an equivalent formulation, goes as follows.

Assertion (VII.1.86) ("Johnson's theorem on amenability"). *Let G be a locally compact group. Then $L^1(G)$ is an amenable algebra $\Leftrightarrow$ G is an amenable group.*

---

[12] Theorem 2.5 in [30] was proved with the collaboration of Ringrose; see p.33 in the reference cited.

We refer the reader for a proof of this assertion in its general form to [81, Theorem VII.2.35] and we shall only present a proof for the case when G is discrete.

▶ "⇒". If $l_1(G)$ is an amenable algebra, then by Theorem 1.60(I) the augmentation module over it is flat and all we have to do is to apply Proposition 1.52((I) ⇒ (III)).

"⇐". By Theorem 1.69(II) and Example 1.16 our problem is to construct a left inverse to $\Pi^* : l_1(G)^* \longrightarrow l_1(G\times G)^*$ in $l_1(G)$-**mod**-$l_1(G)$. The bimodule $l_1(G)^*$, as we recall, is isometrically isomorphic to the bimodule $l_\infty(G)$ in Example VI.3.3, and it is easy to verify that $l_1(G\times G)^*$ is isometrically isomorphic to $l_\infty(G\times G)$ with the operations

$$[f\cdot u](s,t) := \sum_{r\in G} f(t^{-1}r)u(s,r),$$

$$[u\cdot f](s,t) := \sum_{r\in G} f(r^{-1})u(r^{-1}s,t);$$

where $f\in l_1(G)$, $u\in l_\infty(G\times G)$.

Under these identifications Π^* obviously becomes the morphism of "embedding by constants along diagonals" $l_\infty(G)\longrightarrow l_\infty(G\times G) : g\longmapsto u$, where $u(s,t) := g(st)$, $s,t\in G$.

Let M be a left invariant normalized continuous functional on $l_\infty(G)$. For each $u\in l_\infty(G\times G)$ we consider the vector functions $\overline{v}_1, \overline{v}_2 : G\longrightarrow l_\infty(G)$, where $[\overline{v}_1(s)](t) := u(st, t^{-1})$ and $[\overline{v}_2(s)](t) := u(t, t^{-1}s)$ and then the functions $\zeta_k(u)\in l_\infty(G)$, defined by $[\zeta_k(u)](s) := M[\overline{v}_k(s)]$; $k=1,2$. This obviously defines operators $\zeta_1,\zeta_2 : l_\infty(G\times G)\longrightarrow l_\infty(G)$. For any $u\in l_\infty(G\times G)$ and $r,s,t\in G$ obviously $[u\cdot\delta_r](s,t) = u(rs,t)$; from this it follows easily that $[\zeta_1(u\cdot\delta_r)](s) = M[\overline{v}_1(rs)] = [{}^r\zeta_1(u)](s) = ([\zeta_1(u)]\cdot\delta_r)(s)$. In view of $\overline{l.h.}\{\delta_r : r\in G\} = l_1(G)$, this means that ζ_1 is a morphism in **mod**-$l_1(G)$; an analogous argument shows that ζ_2 is a morphism in $l_1(G)$-**mod**. But for any $s\in G$ the function $\overline{v}_1(s)$ is precisely ${}^s[\overline{v}_2(s)]$; hence, using (only at this point!) that M is left invariant, we get that $\zeta_1=\zeta_2$.

So $\zeta := \zeta_1$ $(=\zeta_2)$ is a morphism in $l_1(G)$-**mod**-$l_1(G)$ and it remains for us to show that $\zeta\Pi^*$ is the identity on $l_\infty(G)$. For this we note that for any $g\in l_\infty(G)$ the value at each $s\in G$ of the vector function $\overline{v}_1$, constructed with respect to $u := \Pi^* g$, is $g(s)$. Therefore, as M is normalized, $[\zeta_1(\Pi^* g)](s) = g(s)$ for all $s\in G$. The rest is clear. ■

Besides being of interest in its own right the result obtained gives an effective method of proving that many other algebras are amenable. In the next theorem, due to Kadison and Ringrose [176, II] and independently to Johnson [30], we take on trust that commutative discrete groups are amenable (cf. what was said in §1.3). For another approach using manipulation with measures see [81].

Theorem (VII.1.87). *The algebra $C_0(\Omega)$, where Ω is a locally compact space, is amenable.*

▶ By Definition 1.59 and Corollary 1.64, it is sufficient to establish that $(C_0(\Omega))_+ = C(\Omega_+)$ is amenable. We put $G := \{f \in C(\Omega_+) : |f(s)| = 1$ for all $s \in \Omega_+\}$; this is a commutative group with respect to multiplication. The definition of the operation and the norm in $l_1(G)$ obviously imply that the homomorphism $\kappa : l_1(G) \longrightarrow C(\Omega_+) : \delta_f \longmapsto f$, $f \in G \subset C(\Omega_+)$ is uniquely defined. Since the group G is amenable, $l_1(G)$ is an amenable algebra by Johnson's theorem and by the Stone-Weierstrass theorem (cf. the end of Chap.IV §6), the image of κ is dense in $C(\Omega_+)$. All that remains is to apply Proposition 1.71. ■

We state without proof a non-trivial converse of this theorem for an important class of Banach algebras.

Assertion (VII.1.88) (Sheinberg [287]). If a uniform Banach algebra A (see Chap.IV §2.2) with spectrum Ω is amenable, then $A = C_0(\Omega)$.

It can also be established by a method similar to that used in 1.87 that the important algebra $\mathcal{K}(H)$ is amenable. [30], [176, II]. However, another argument is more to our liking.

Theorem (VII.1.89). *The algebra $\mathcal{K}(H)$, where H is a separable Hilbert space, is amenable.*

▶ We fix an orthonormal basis in H and let e_{kl}; $k, l \in \mathbb{N}$, denote the operator expressed in terms of this basis as a matrix with one at the "kl-th" place and zeros at the other places. We consider the sequence $d_n := \frac{1}{n} \sum_{k,l=1}^{n} u_{k,l}$ in $\mathcal{K}(H) \hat{\otimes} \mathcal{K}(H)$, where $u_{k,l} := e_{kl} \otimes e_{lk}$ (cf. 1.73). We note that $d_n = \frac{1}{n} \sum_{m=1}^{n} v_m$, where

$$v_m := u_{1,m} + u_{2,m+1} + \cdots + u_{n-m+1,n} + u_{n-m+2,1} + \cdots + u_{n,m-1}.$$

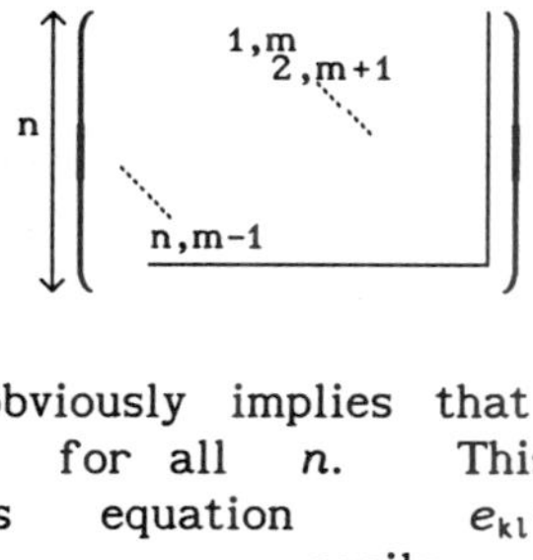

But Proposition 0.3.30 obviously implies that $\|v_m\| \leq 1$ for all m, and hence $\|d_n\| \leq 1$ for all n. This, in turn, with the help of the obvious equation $e_{kl} \cdot d_n = d_n \cdot e_{kl}$ and $\overline{\ell.h.}\{e_{kl} : k, l \in \mathbb{N}\} = \mathcal{K}(H)$, easily implies that

$\lim_{n\to\infty}(a\cdot d_n - d_n\cdot a) = 0$ for any $a \in \mathcal{K}(H)$. Since, furthermore, $\pi(d_n)$ is a standard b.a.i. in $\mathcal{K}(H)$ (I.2.33), this shows that d_n is an approximate diagonal for $\mathcal{K}(H)$. All that remains is to apply Theorem 1.70. ■

Of course Theorems 1.87 and 1.89 are not the only amenability results for the class of C^*-algebras. It is known that all postliminal C^*-algebras are amenable [30], as are all C^*-algebras which are the closure of the union of a family, directed by inclusion, of finite-dimensional sub-C^*-algebras [176, II] (for example, the Fermion algebra). On the other hand, $\mathcal{B}(H)$ and $\mathcal{C}(H)$ are not amenable; as for the reduced C^*-algebra of a discrete group G, it is amenable $\Longleftrightarrow$ G is amenable (Bunce [107]).

Not so long ago a complete description was obtained of the amenable C^*-algebras: it turned out that they are precisely the nuclear C^*-algebras (see the end of Chap.IV §7.4). It was Connes [162] who first established that being amenable implies being nuclear and a few years later Haagerup [264] proved the converse. We warn the reader that in these papers and generally in the majority of papers we have referred to about C^*-algebras (just as in the original proof of Assertion 1.86) an equivalent definition of amenability is used which we going to have to give; see later Theorem 3.37.

§2. DERIVATIONS, AUTOMORPHISMS AND EXTENSIONS; FIRST APPEARANCE OF COCYCLES AND COBOUNDARIES

The questions considered throughout most of this section have an initial statement that looks completely traditional and they are formulated without involving the language of modules; however, modules (as a rule, two-sided) arise naturally in the process of investigating them. We shall see that many typical problems relating to the various concepts under discussion are solved in terms of objects which have a very similar definition — cocycles and (in particular) coboundaries: except that, generally speaking, different problems need cocycles and coboundaries of "different dimensions". More precisely, the necessary information is carried by spaces of cocycles "modulo coboundaries". These spaces, called cohomology groups, will be introduced and studied in §3, with a suitable degree of generality and in close connection with the results of §1.

§2.1. **Derivations and automorphisms**

Definition (VII.2.1). Let A be an algebra. An operator $D: A \longrightarrow A$ is called a *derivation* of this algebra if it satisfies $D(ab) = aD(b) + D(a)b$; $a, b \in A$.

This name, of course, was motivated by the similarity of the above identity and the formula for the derivative of a product of functions in classical analysis. In particular, ordinary differentiation in $C^\infty[0, 1]$ or in $\mathcal{O}(U)$ is also a derivation in the sense just indicated.

Proposition (VII.2.2) ("Leibnitz formula"). *If D is a derivation of the algebra A, then for any $a, b \in A$ and $n \in \mathbb{N}$*

$$D^n(ab) = \sum_{k=0}^{n} \binom{n}{k} D^{n-k}(a)D^k(b)$$

(here and in future an operator to the power zero is, as usual, assumed to be the identity operator).

▶ The standard inductive proof of the classical Leibnitz formula goes across word for word to the case under consideration. ■

The interest in derivations on Banach algebras is, in many respects, due to their close connection with automorphisms (= topological isomorphisms of algebras onto themselves; cf. Chap.I §2.1). The automorphisms of a given algebra A obviously form a subgroup of $\operatorname{Inv} \mathcal{B}(A)$, denoted by $\operatorname{Aut} A$. This group, loosely speaking, shows how rich A is in intrinsic symmetries and therefore it is not surprising that it is the subject of intense study.

Theorem (VII.2.3). *Let D be a continuous derivation of the Banach algebra A. Then the operator e^D* (= the exponential of the element $D \in \mathcal{B}(A)$; see II.2.10) *is an automorphism of this algebra.*

▶ If we use the Leibnitz formula and the obvious properties of all the absolutely convergent double series in A considered below, we get

$$e^D(ab) = \sum_{n=0}^{\infty} \frac{1}{n!} \left(\sum_{k=0}^{n} \binom{n}{k} D^{n-k}(a)D^k(b) \right) =$$

$$= \sum_{l,m=0}^{\infty} \frac{1}{(l+m)!} \binom{l+m}{m} D^l(a)D^m(b) = \sum_{l,m=0}^{\infty} \frac{1}{l!m!} D^l(a)D^m(b) =$$

$$= \left(\sum_{l=0}^{\infty} \frac{D^l(a)}{l!} \right) \left(\sum_{m=0}^{\infty} \frac{D^m(b)}{m!} \right) = e^D(a)e^D(b).$$

Thus e^D is a homomorphism which is continuous because D is and it is invertible, as an element of $\mathcal{B}(A)$, by Proposition II.2.11. The rest is clear. ■

So "where there are derivations there are also automorphisms". We show that, with certain additional assumptions, there is the reverse connection. First we make an observation that is of interest in its own right.

Proposition (VII.2.4). *Let B be a unital Banach algebra and let c be an element of B such that if $d := c - e_B$, then $\underline{r}(d) < 1$. Put $\mathbb{D}_1 := \{z + 1\,;\, z \in \mathbb{D}\}$. Then the element $\log c \in B$* (which is well-defined since $\mathrm{Sp}_B\, c \subset \mathbb{D}_1 \subset \mathbb{C}^{(+)}$; see II.2.21) *is the sum of the series* $\sum_{n=1}^{\infty} \frac{(-1)^{n+1}}{n} d^n$.

▶ The sequence of polynomials $p_n(z) := \sum_{k=1}^{n} \frac{(-1)^{k+1}}{k}(z-1)^k$, considered as being in $\mathcal{O}(\mathbb{D}_1)$ converges to $\log z$. Therefore since the holomorphic calculus for c in $\mathbb{D}_1$ is continuous, it follows that the sequence $p_n(c) = \sum_{k=1}^{n} \frac{(-1)^{k+1}}{k} d^n$ converges to $\log c$ in B. ■

Theorem (VII.2.5) (cf. [146]). *Let A be a Banach algebra and let U be an automorphism of A such that, if $V := U - 1_A$, then $\|V\| < \sqrt{2} - 1$. Then the operator $\log U \in \mathcal{B}(A)$ is a derivation of the algebra A.*

▶ We fix $a, b \in A$; then, according to Proposition 2.4, $\log U(ab) = \sum_{n=1}^{\infty} \frac{(-1)^{n+1}}{n} V^n(ab)$. Since U is a homomorphism, it obviously follows that $V(cd) = (Vc)d + c(Vd) + (Vc)(Vd)$; applying it the required number of times we get that $V^n(ab) = \sum_{k,l=0}^{n} t_{kl}^{(n)}(V^k a)(V^l b)$, where the coefficients $t_{kl}^{(n)}$ are precisely those for which the equation

$$(z_1 + z_2 + z_1 z_2)^n = \sum_{k,l=0}^{\infty} t_{kl}^{(n)} z_1^k z_2^l \tag{*}$$

holds (as polynomials or as functions of two complex variables). Hence

$$\log U(ab) = \sum_{n=1}^{\infty} \frac{(-1)^{n+1}}{n} \left(\sum_{k,l=0}^{\infty} t_{kl}^{(n)}(V^k a)(V^l b) \right). \tag{**}$$

We put $\theta := 2\|V\| + \|V\|^2$; and then (*) implies that $\sum_{k,l=0}^{\infty} t_{kl}^{(n)} \|V\|^{k+l} = \theta^n$. This implies that

$$\sum_{n,k,l}\left\|\frac{(-1)^{n+1}}{n}t_{kl}^{(n)}(V^k a)(V^l b)\right\| \leq \sum_{n=1}^{\infty}\sum_{k,l=0}^{\infty} t_{kl}^{(n)}\|V\|^k\|a\|\,\|V\|^l\|b\| =$$

$$= \sum_{n=1}^{\infty}\theta^n\|a\|\,\|b\|.$$

But the condition on $\|V\|$ is equivalent to $\theta<1$; therefore the corresponding triple series is absolutely convergent in A and so in (**) one can interchange the order of the summation. Thus $\log U(ab)=\sum_{k,l=0}^{\infty}s_{kl}(V^k a)(V^l b)$, where $s_{kl}:=\sum_{n=1}^{\infty}\frac{(-1)^{n+1}}{n}t_{kl}^{(n)}$.

We shall find the numbers s_{kl} explicitly. For this we consider, for sufficiently small $z_1, z_2\in\mathbb{C}$, the series $\sum_{k,l=0}^{\infty}s_{kl}z_1^k z_2^l$. Using the definition of s_{kl} in terms of $t_{kl}^{(n)}$, the absolute convergence of the triple series and then (*), we get that the sum of the above (double) series is

$$\sum_{n=1}^{\infty}\frac{(-1)^{n+1}}{n}(z_1+z_2+z_1z_2)^n = \log(1+z_1+z_2+z_1z_2) =$$

$$= \log[(1+z_1)(1+z_2)] = \log(1+z_1)+\log(1+z_2) =$$

$$= \sum_{k=1}^{\infty}\frac{(-1)^{k+1}}{k}z_1^k + \sum_{l=1}^{\infty}\frac{(-1)^{l+1}}{l}z_2^l .$$

Hence by the uniqueness of the Taylor series of a holomorphic function (of two variables) $s_{kl}=\frac{(-1)^{k+1}}{k}$ when $k>0$ and $l=0$, $s_{kl}=\frac{(-1)^{l+1}}{l}$ when $k=0$ and $l>0$ and $s_{kl}=0$ for all remaining pairs (k,l).

We return to $\log U(ab)$ and see that this is equal to

$$\sum_{k=1}^{\infty}\frac{(-1)^{k+1}}{k}(V^k a)(V^0 b) + \sum_{l=1}^{\infty}\frac{(-1)^{l+1}}{l}(V^0 a)(V^l b),$$

and hence, by Proposition 2.4, coincides with $[\log U(a)]b + a[\log U(b)]$. The rest is clear. ∎

Taking into account Proposition II.2.22(II) we immediately get the following.

Corollary (VII.2.6). *An automorphism U satisfying the conditions of Theorem 2.5 can be represented as e^D, where D is a continuous derivation of the algebra A.*

Proposition (VII.2.7). *There exists for this same U a continuous group homomorphism $t\longmapsto U_t$ from the group $\mathbb{R}$ into* Aut A *with the uniform topology, such that $U_1=U$.*

▸ We put $U_t:=e^{t\log U}$; $t\in\mathbb{R}$. The rest is clear. ∎

The last fact is usually expressed by saying that any automorphism of a Banach algebra that is sufficiently close to the identity, belongs to a certain uniformly continuous one-parameter group of automorphisms of this algebra. (For the "physical meaning" of such groups — they provide a model for the development over time of systems in quantum mechanics — see, for example [10, Chap.III].) For a reader acquainted with Lie groups and algebras one can give an interpretation of this fact: derivations form the Lie algebra of the group of automorphisms.

Exercise (VII.2.8). Suppose that, under the assumptions of Theorem 2.3 (2.5, respectively), we know, in addition, that A is a star algebra and D (respectively U) is a *-operator. Show that then e^D (respectively $\log U$) is also a *-operator.

We note that the condition on U in Theorem 2.5 may be considerably weakened: it is sufficient to require that $\mathrm{Sp}_{\mathcal{B}(A)} U \subset \{z \in \mathbb{C} : -\frac{2}{3}\pi < \arg z < \frac{2}{3}\pi\}$ [134, p.87]. Further weakening is then impossible:

Example (VII.2.9). Let $\xi := e^{\frac{2\pi i}{3}} \in \mathbb{C}$, $\Omega := \{1, \xi, \xi^2\}$ and $A := C(\Omega)$. Then any deriviation of this algebra, since it is obviously zero on the idempotents, is zero. At the same time there do exist automorphisms of the algebra A other than 1 (= the unique automorphism of the form e^D): one such is, for example, $U : f(\eta) \longmapsto g(\eta) := f(\xi\eta)$; $\eta \in \Omega$. Moreover it is easy to see that $\mathrm{Sp}_{\mathcal{B}(A)} U = \Omega$.

The next theorem shows that among Banach algebras only the non-commutative algebras can possess a fairly rich stock of derivations.

Theorem (VII.2.10) (Singer–Wermer). *Let D be a continuous derivation of the commutative Banach algebra A. Then $\mathrm{Im}\, D \subseteq \mathrm{Rad}\, A$. As a consequence, semisimple commutative Banach algebras have no non-trivial continuous derivations.*[13]

▶ By Theorem IV.1.4 it is sufficient to check that $\chi(Da) = 0$ for any $a \in A$ and any character χ on A. With this aim we consider the operators zD and e^{zD} acting on A for any $z \in \mathbb{C}$. As the first of them is a derivation, then the second, by Theorem 2.3, is an automorphism of our algebra. But then the composition $\chi \circ e^{zD}$ is a character and consequently, taking into account Theorem I.2.6, the function $w ; \mathbb{C} \longrightarrow \mathbb{C} : z \longmapsto \chi(e^{zD}(a))$ is bounded. At the same time w, being representable by the series

---

[13] In fact they do not have even discontinuous derivations; see what is said at the end of the section.

$$\chi\left[\left(\sum_{n=0}^{\infty}\frac{(zD)^n}{n!}\right)(a)\right] = \chi(a) + [\chi(Da)]z + \frac{\chi(D^2a)}{2!}z^2 + \cdots$$

is an entire analytic function. Therefore, by Liouville's theorem, $w(z)$ is constant and the coefficients of z^n for $n > 0$ in its Taylor expansion, including $\chi(Da)$, are zero. ∎

Remark (VII.2.11). Such a theorem is no longer true for non-Banach polynormed algebras, or even for metrizable Arens-Michael algebras. An obvious counterexample is ordinary differentiation $f \longmapsto \frac{df}{dt}$ in the algebra $C^\infty[0, 1]$. We get, in particular, another proof that the topology in $C^\infty[0, 1]$ cannot be given by a Banach space norm.

* * *

The following is a natural generalization of Definition 2.1.

Definition (VII.2.12). Let A be an algebra and let X be a bimodule. The operator $D : A \longrightarrow X$ is called a *derivation of* A *with values in* X if it satisfies

$$D(ab) = a\cdot D(b) + D(a)\cdot b; \qquad a, b \in A.$$

Besides derivations of A with values in A itself (as in 2.1), in a lot of questions an important role is played by derivations with values in algebras ambient to A, derivations with values in one-dimensional bimodules (often called point derivations) and some others [168], [328], [218], [142].

Continuous derivations of a polynormed algebra with values in a polynormed bimodule are also called *one-dimensional cocycles* (of A with values in X); obviously they form a subspace of $\mathcal{B}(A, X)$ and this is denoted by $Z^1(A, X)$.

We shall make an important observation. If X is an A-bimodule, then every $x \in X$ generates an operator $A \longrightarrow X : a \longmapsto a\cdot x - x\cdot a$, denoted by $\delta^0 x$; this is obviously a derivation which, in the case of a "polynormed" structure, is automatically continuous. A derivation of this form is called an *inner derivation.*

An automorphism of the algebra A is called an *inner automorphism* if it acts according to the rule $a \longmapsto b^{-1}ab$ for a certain invertible element $b \in A$. Theorem 2.3 is complemented by the following.

Proposition (VII.2.13). *Let* A *be a Banach algebra,* $c \in A$, *and let* $\delta^0 c : A \longrightarrow A$ *be the corresponding inner derivation. Then* $e^{\delta^0 c}$ *is an inner automorphism and moreover* $e^{\delta^0 c}(a) = (e^c)^{-1}ae^c$; $a \in A$.

▶ By induction with respect to $n = 0, 1, \cdots$ it is easy to establish that $(\delta^0 c)^n a = \sum_{k+l=n} \frac{(-1)^k n!}{k!\,l!} c^k a c^l$. Hence, observing that the double series that arises is absolutely convergent in A,

$$e^{\delta^0 c}(a) = \sum_{n=0}^{\infty} \frac{1}{n!} (\delta^0 c)^n a = \left(\sum_{k=0}^{\infty} \frac{(-1)^k}{k!} c^k\right) a \left(\sum_{l=0}^{\infty} \frac{c^l}{l!}\right) =$$
$$= e^{-c} a e^{c} = (e^c)^{-1} a e^c. \blacksquare$$

When A and X are polynormed (in particular, Banach) the inner derivations — in this situation they are then called *one-dimensional coboundaries* of A with values in X — form a subspace of $Z^1(A, X)$ and this is denoted by $B^1(A, X)$; it is obviously the image of the operator $\delta^0 : X \longrightarrow \mathcal{B}(A, X)$ $: x \longmapsto \delta^0 x$. We note that $\operatorname{Ker} \delta^0 = \{x : a \cdot x = x \cdot a$ for all $a \in A\}$; this subspace of the A-bimodule X is also called its *centre* and is denoted by $Z^0(A, X)$; the elements of the centre are called zero-dimensional cocycles. Obviously, to within a linear isometry $B^1(A, X) = X/Z^0(A, X)$.

A typical question in derivation problems is whether there exist so-called outer (= non-inner) derivations of a given algebra with values in a given bimodule; it is these that are most interesting. How abundant they are is indicated by the quotient space $Z^1(A, X)/B^1(A, X)$, denoted by $\mathcal{H}^1(A, X)$ and called the *one-dimensional cohomology group of the* (polynormed) *algebra* A *with coefficients in* X (the meaning of this nomenclature will become clear later in Definition 3.33).

Example (VII.2.14). Let $A = X = \mathcal{K}(H)$. Obviously every $S \in \mathcal{B}(H)$ gives a derivation $D_S : T \longmapsto ST - TS$ in $\mathcal{K}(H)$ and Schur's lemma ensures that $\mathcal{K}(H)^! = \ell.h.\{1\}$ which implies that $D_R = D_S \iff$ the operator $R - S$ is a multiple of the identity. Thus when $S \notin \ell.h.\{\mathcal{K}(H), 1\}$ the derivation D_S is not inner and therefore $\mathcal{H}^1(\mathcal{K}(H), \mathcal{K}(H)) \neq 0$.

Later on we get a number of results about $\mathcal{H}^1(A, X)$ for various A and X when we are armed with the necessary techniques in homology. In particular, it will be shown in (2.41) that $\mathcal{H}^1(\mathcal{K}(H), \mathcal{B}(H)) = 0$, and so there are no other derivations of $\mathcal{K}(H)$ with values in itself, besides those indicated in Example 2.14. (In this example one sees that the study of derivations in the sense of Definition 2.1 naturally leads to the appearance of other bimodules besides A — in the given case of an ambient algebra. In any case this is typical for C^*-algebras; see, say, [58, 8.6].)

Unfortunately we lack the opportunity to dwell in somewhat more detail on the second typical question in the study of derivations — under what conditions are they automatically continuous. A considerable part of the survey by Dales [142] is devoted to this. As outstanding results we note that every derivation of a semisimple Banach algebra A [14] with values in A (Johnson and Sinclair [139]) and every derivation of a C^*-algebra A with values in an arbitrary A-bimodule (Ringrose [221]) is automatically continuous. If we are talking about derivations of a von Neumann algebra A with values in A, then they are, as the important theorem of Sakai [232] and Kadison [174] shows (not only automatically continuous but even) inner; for a new simple proof see [58, 8.6.6]. This fact was established earlier by Kaplansky [152] for $\mathcal{B}(H)$ and for the more general so-called von Neumann algebras of type I.

§2.2. Extensions of algebras and the splitting condition for singular extensions. A few words about perturbations

Definition (VII.2.15). An *extension of a polynormed algebra* A *by a polynormed algebra* I is a polynormed algebra $\mathfrak{A}$ together with a topologically injective homomorphism $i : I \longrightarrow \mathfrak{A}$, the image of which is a bi-ideal in $\mathfrak{A}$, and a topologically surjective (= open) homomorphism $\sigma : \mathfrak{A} \longrightarrow A$ of which this bi-ideal is the kernel.

Thus $\mathfrak{A}$ is a "large" algebra in which there is a certain bi-ideal that can be identified with I and A can be identified with the quotient algebra with respect to this bi-ideal. All this information can be expressed compactly in the language of exact sequences discussed in Chap.0 §4.2. Namely, using Proposition 0.4.12 in an obvious way, we could say that an extension is a triple $(\mathfrak{A}, \sigma, i)$ for which the sequence

$$0 \longleftarrow A \xleftarrow{\sigma} \mathfrak{A} \xleftarrow{\mathrm{i}} I \longleftarrow 0 \qquad (\mathcal{E})$$

is topologically exact. We recall that for Fréchet algebras, and in particular Banach algebras, ordinary (algebraic) exactness is sufficient for the topological exactness of $\mathcal{E}$; see Proposition 0.4.11. Later on we shall often write "the extension $\mathcal{E}$" for brevity; this should not cause any confusion.

[14] Combining this result with the second assertion of the Singer-Werner Theorem 2.10, we see that a semisimple commutative Banach algebra has no non-trivial derivations, continuous or otherwise. Much more recently, Thomas [352] showed that the first assertion of Theorem 2.10 is also true without the assumption of continuity: every derivation of a commutative Banach algebra into itself maps into the radical.

Suppose we know all about A and I; what can be said about $\mathfrak{A}$? Historically this question arose first of all for extensions of semisimple algebras by radical algebras; such extensions are called *radical extensions*. In particular, every algebra $\mathfrak{A}$ provides us with a radical extension

$$0 \longleftarrow \mathfrak{A}/\mathrm{Rad}\,\mathfrak{A} \xleftarrow{\ \mathbf{pr}\ } \mathfrak{A} \xleftarrow{\ \mathbf{in}\ } \mathrm{Rad}\,\mathfrak{A} \longleftarrow 0.$$

Therefore for radical extensions the question posed leads to an attempt to reduce the study of arbitrary algebras — at least in the class of Banach algebras — to the study of semisimple and radical algebras. We referred to these problems in passing in Chap.I §3.4.

What extensions are assumed to be essentially the same?

Definition (VII.2.16). Two extensions $(\mathfrak{A}, \sigma, i)$ and $(\mathfrak{A}_1, \sigma_1, i_1)$ of A by I are said to be *equivalent* if there exists a commutative diagram

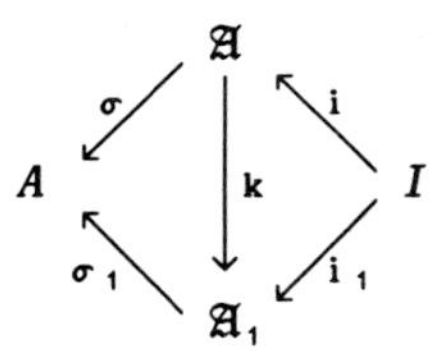

in which k is a topological isomorphism of polynormed algebras.

Exercise (VII.2.17). Show that in the case of Banach algebras (or more generally Fréchet algebras) any continuous homomorphism k making the diagram commutative is a topological isomorphism.

The question of describing all extensions with given "ends" (A and I) turns out, with the exception of a few special cases, to be very difficult. At the same time Wedderburn's theorem (I.3.64) suggests a more tangible question which we naturally deal with in the first instance.

Definition (VII.2.18). The extension $\mathcal{E}$ is said to *split* if there is a (necessarily closed) subalgebra B of $\mathfrak{A}$ such that $\sigma|_B : B \longrightarrow A$ is a topological isomorphism.

Obviously $(\mathfrak{A}, \sigma, i)$ splits $\Longleftrightarrow$ σ has as right inverse some continuous homomorphism $\rho\,; A \longrightarrow \mathfrak{A}$; here ρ is called a *splitting homomorphism* for the given extension. In the case of Fréchet algebras, the existence of a closed subalgebra B of $\mathfrak{A}$ such that $\mathfrak{A} = B \oplus \mathrm{Im}\, i$ is obviously sufficient for splitting.

We pass to one of those few classes of extensions whose study has come to a successful conclusion in an accepted sense.

Definition (VII.2.19). An extension $\mathcal{E}$ is said to be *singular* if $I^2 = \{0\}$ (= multiplication in I is the zero multiplication) and in addition σ has as right inverse a continuous operator $\rho : A \longrightarrow \mathfrak{A}$ (called the *splitting operator* for $\mathcal{E}$).

The existence of a splitting operator is obviously equivalent to $\operatorname{Im} i$ being complemented in $\mathfrak{A}$ (see 0.1.14) or, what is the same thing, $\mathcal{E}$ splitting as a complex in $\mathcal{PN}$ (see 0.4.15). Regarding the condition $I^2 = \{0\}$, this easily implies that I has the structure of a polynormed A-bimodule with operations $a \cdot x := i^{-1}(bi(x))$ and $x \cdot a := i^{-1}(i(x)b)$, where $x \in I$, $a \in A$, and b is any element $\mathfrak{A}$ with $\sigma(b) = a$. If we let X denote I in its new capacity as a bimodule, we shall call the original $\mathcal{E}$ a singular extension of the algebra A *by the* A*-bimodule* X.

Let $\mathcal{E}$ be an extension of the above form in which we shall for simplicity of notation identify X with $\operatorname{Im} i$ (= assume i is a natural embedding). Then, if ρ is a splitting operator, it is obvious that $f_\rho(a, b) := \rho(a)\rho(b) - \rho(ab)$ lies in X for any $a, b \in A$. In this way we have constructed a separately continuous bi-operator $f_\rho \in \mathcal{B}^s(A \times A, X)$ (see Chap.0 §3.1).

We give $\mathcal{B}^s(A \times A, X)$ a new notation $C^2(A, X)$ — it will be more convenient — and we shall call the elements of this space *two-dimensional cochains.* Those cochains f that satisfy

$$a \cdot f(b, c) - f(ab, c) + f(a, bc) - f(a, b) \cdot c = 0, \tag{1}$$

for $a, b, c \in A$, we shall call *two-dimensional cocycles* and those f that have the form $\delta^1 g$ for a certain $g \in \mathcal{B}(A, X)$, where

$$\delta^1 g(a, b) := a \cdot g(b) - g(ab) + g(a) \cdot b, \tag{2}$$

when $a, b \in A$, shall be called *two-dimensional coboundaries.* It is easy to verify that the two-dimensional cocycles form a subspace of $C^2(A, X)$, denoted by $Z^2(A, X)$, and the two-dimensional coboundaries form a subspace of $Z^2(A, X)$, denoted by $B^2(A, X)$. (The only connection with the $Z^1(A, X)$ in §2.1 that we consider for the time being is that the one-dimensional cocycles — they are the derivations — are just those g for which $\delta^1 g \equiv 0$.)

The following clarifies the role of (1) and (2).

Proposition (VII.2.20). *For any splitting operator* ρ, $f_\rho \in Z^2(A, X)$. *If* ρ *and* ρ' *are two such operators, then* $f_\rho - f_{\rho'} \in B^2(A, X)$.

▶ The first assertion is implied by

$$\rho(a)f_\rho(b, c) + f_\rho(a, bc) + \rho(abc) = \rho(a)\rho(b)\rho(c) =$$

$$= f_\rho(a, b)\rho(c) + f_\rho(ab, c) + \rho(abc),$$

taking into account $\sigma\rho = 1_A$ and the definition of the operations in X. The second is implied by the obvious equality $f_\rho - f_{\rho'} = \delta^1 g$ for $g : a \longmapsto \rho(a) - \rho'(a)$. ■

We now introduce into the discussion the quotient space $Z^2(A, X)/B^2(A, X)$ that is denoted by $\mathcal{H}^2(A, X)$ and called the *two-dimensional cohomology group of the algebra* A *with coefficients in* X. Further we let $\mathrm{Ex}(A, X)$ denote the set of equivalence classes of singular extensions of A by X and we let $\theta \in \mathrm{Ex}(A, X)$ denote the class consisting of all split extensions.

Theorem (VII.2.21). *Let* A *and* X *be a polynormed algebra and an* A*-bimodule. Then there exists a bijection between* $\mathrm{Ex}(A, X)$ *and* $\mathcal{H}^2(A, X)$ *taking the equivalence class of the extension with splitting operator* ρ *to the coset* $f_\rho + B^2(A, X) \in \mathcal{H}^2(A, X)$; *moreover it (the bijection) takes the class* θ *to* $0 \in \mathcal{H}^2(A, X)$.

▶ Proposition 2.20 obviously implies that the indicated map — let us denote it by ω — is well defined. Our problem is to construct an inverse map.

Let $f \in Z^2(A, X)$. We introduce the bi-operator $m : \mathfrak{A} \times \mathfrak{A} \longrightarrow \mathfrak{A} : (a + x, b + y) \longmapsto ab + a \cdot y + x \cdot b + f(a, b)$ in $\mathfrak{A} := A \oplus X$; it is easy to verify that m is a multiplication and $(\mathfrak{A}, \mathbf{pr}, \mathbf{in})$ is a singular extension of A by the A-bimodule X. Moreover, if in this construction we replace f by $f + \delta^1 g$, where $g \in \mathcal{B}(A, X)$, we get an equivalent extension: we can take $k : a + x \longmapsto a - g(a) + x$ as the isomorphism in Definition 2.16.

So, the equivalence class of the extension constructed depends only on the coset of f in $\mathcal{H}^2(A, X)$, which enables the map $\omega' : \mathcal{H}^2(A, X) \longrightarrow \mathrm{Ex}(A, X)$ to be well defined. Further it is easy to verify — we leave this to the reader — that $\omega' \circ \omega$ is the identity on $\mathrm{Ex}(A, X)$, $\omega \circ \omega'$ is the identity on $\mathcal{H}^2(A, X)$ and $\omega(\theta) = 0$. ■

Corollary (VII.2.22). *All the singular extensions of* A *by the* A*-bimodule* X *split* $\iff$ $\mathcal{H}^2(A, X) = 0$.

Here are a few exercises of an illustrative character.

Exercise (VII.2.23). Let A be a Banach algebra with a one-sided (say left) b.a.i., $(e_\nu;\ \nu \in \Lambda)$, and let X be a bimodule with zero outer multiplications. Show that $\mathcal{H}^2(A, X) = 0$.

Hint. Given $f \in Z^2(A, X)$ and $a := \sum_{k=1}^{n} b_k c_k;\ b_k, c_k \in A$ it is possible to define $\tilde{g}(a) := \lim_{\nu} f(e_\nu, a) = \sum_{k=1}^{n} f(b_k, c_k) \in X$. The operator $\tilde{g} : A^{[2]} \longrightarrow X$ extends by continuity to $g : A \longrightarrow X$ with $\delta^1 g = -f$.

Exercise (VII.2.24). If we consider $\mathbb{C}$ as an annihilator l_2-bimodule, show that the bifunctional $f(\xi, \eta) := \sum_{n=1}^{\infty} \xi_n \eta_n$; $\xi, \eta \in l_2$, belongs to $Z^2(l_2, \mathbb{C})$ but not to $B^2(l_2, \mathbb{C})$. Show further that there exists a natural linear isomorphism between $\mathcal{H}^2(l_2, \mathbb{C})$ and l_∞/l_2.

Exercise (VII.2.25). Let $A = \{w \in \mathcal{A}(\overline{\mathbb{D}})\ ;\ w(0) = w'(0) = 0\}$. If we consider $\mathbb{C}$ as an annihilator A-bimodule, show that the bifunctional $f(w_1, w_2) := w_1'''(0)w_2'''(0)$; $w_1, w_2 \in A$, belongs to $Z^2(A, \mathbb{C})$ but not to $B^2(A, \mathbb{C})$. The more ambitious reader may like to show that $\mathcal{H}^2(A, \mathbb{C})$ is two-dimensional.

Remark (VII.2.26). The cocycle f in Exercise 2.24 corresponds to the historically first example of a non-splitting radical extension of a Banach algebra which was suggested by Feldman [257] in 1951. In this extension, however, it is not difficult to discover that there exist non-closed subalgebras which are linear complements to the radical (identified with $\mathbb{C}$). As for the extension corresponding to the cocycle f in Exercise 2.25, it has no subalgebras whatsoever which are linear complements to the radical, not even non-closed ones. Check both facts!

The behaviour of the wider class of nilpotent extensions (= those in which $I^n = \{0\}$ for a certain $n \in \mathbb{N}$) depends in many respects on whether the singular extensions split. We give a particular case of the relevant assertion (cf. [81, 1.1.13, 23, 25]).

Proposition (VII.2.27). *Let A be such that $\mathcal{H}^2(A, X) = 0$ for any finite-dimensional X. Then any nilpotent extension of A by a finite-dimensional I splits.*

▶ We limit ourselves to the case $I^3 = \{0\}$: in the general case the argument is based on an obvious induction with respect to the "nilpotency index" n.

We consider $\mathfrak{A}' := \mathfrak{A}/I^2$, $\mathbf{pr} : \mathfrak{A} \longrightarrow \mathfrak{A}'$ and $I' := \mathbf{pr}(I)$. Obviously $\mathfrak{A}'$ is a singular extension of A by I' and therefore (Corollary 2.22) $\mathfrak{A}' = B' \oplus I'$ for a certain closed subalgebra B' of $\mathfrak{A}'$. But $\mathfrak{A}'' := \mathbf{pr}^{-1}(B')$ is a singular extension of A by I^2 and therefore, by the same Corollary 2.22, $\mathfrak{A}'' = B \oplus I^2$ for a closed subalgebra B of $\mathfrak{A}$. Hence, of course $\mathfrak{A} = B \oplus I$. ■

Remark (VII.2.28). When combined with Proposition I.3.59, the fact which we have proved enables Wedderburn's theorem I.3.64 to be reduced to a question of showing that the "groups" $\mathcal{H}^2(A, X)$ are trivial for semisimple finite-dimensional A. We shall return to this question in §3.3.

A certain connection between singular extensions and derivations is provided by the following.

Exercise (VII.2.29). Let $(\mathfrak{A}, \sigma, i)$ be a singular split extension of A by the A-bimodule X. Establish a natural bijection between the set of all closed subalgebras of $\mathfrak{A}$ satisfying Definition 2.18 and the set of all continuous derivations on A with values in X.

Hint. The difference between two splitting homomorphisms takes values in X and is a derivation.

Some further information about extensions is contained in [81] and the references listed there (including [102], [149], [131]).

Remark (VII.2.30). In this book we do not quite reach the theory of C^*-extensions — extensions made up of C^*-algebras and *-homomorphisms (Busby - Brown - Douglas - Philmore, Kasparov, Voiculescu, ...) which has been developing rapidly in recent years. This theory has now become a mathematical discipline in its own right, having a wide range of applications from classical questions in the theory of operators to K-theory; it also has a very direct relation with the "non-commutative algebraic topology" mentioned in Remark IV.7.17. The important particular case where $A = C(\Omega)$ for some metrizable compact space Ω and $I = \mathcal{K}(H)$ has been more fully worked out, and special attention was given to this in the monograph by Douglas [32] in 1980 (see also the survey by Rosenberg [228]). A sample of more recent papers is given in the survey by Kasparov [155].[15]

Another field of application of one- and two-dimensional cohomology groups and also the three-dimensional cohomology groups defined below, is the set of so-called perturbation problems. When discussing some property of a Banach algebra (A, m), where m is a given multiplication, we shall say that it is *stable for small perturbations* or simply *stable* if the algebra (A, m') also has this property for every multiplication m' in A sufficiently close to the original — in the sense of the bi-operator norm $\|m - m'\|$ being small. The algebra A itself is said to be *stable* if for every multiplication m' sufficiently close to the original, the algebras (A, m') and (A, m) are topologically isomorphic. Similarly, by comparing the original and the "perturbed" actions of A (with a fixed multiplication in the algebra) we can define the stability of some property or other of a given A-module (of any type) and also the stability of this A-module itself.

Let X be a Banach A-bimodule. A continuous trilinear operator $f : A \times A \times A \longrightarrow X$ is called a *three-dimensional cocycle* if it satisfies the identity $a \cdot f(b, c, d) - f(ab, c, d) + f(a, bc, d) - f(a, b, cd) + f(a, b, c) \cdot d = 0$, and

[15] See also the new book [8].

a *three-dimensional coboundary* if it has the form $\delta^2 g(a, b, c) := a \cdot g(b, c) - g(ab, c) + g(a, bc) - g(a, b) \cdot c$ for a certain $g \in C^2(A, X)$ (cf. the identity (1) for two-dimensional cocycles); where $a, b, c, d \in A$. It is easy to verify that the space of three-dimensional coboundaries $B^3(A, X)$ is contained in the space of three-dimensional cocycles $Z^3(A, X)$. Having furnished the latter with the tri-linear operator norm, we consider the (seminormed) quotient space $\mathcal{H}^3(A, X) := Z^3(A, X)/B^3(A, X)$; it is called the *three-dimensional cohomology group of* A *with coefficients in* X.

Assertion (VII.2.31) (Johnson [135]; Raeburn and Taylor [230]). *Let* $\mathcal{H}^2(A, A) = 0$ *and* $\mathcal{H}^3(A, A)$ *be Hausdorff (that is,* $B^3(A, A)$ *is closed in* $Z^3(A, A)$.[16] *Then* A *is stable.*

It has also been established [135] that amenability is always a stable property of Banach algebras and, provided that $\mathcal{H}^2(A, A) = 0$, so is commutativity. Concerning the stability of modules, the triviality of certain $\mathcal{H}^k(A, \cdot)$ with $k = 1, 2$ is a sufficient condition for this ([135], [230]), however these results are more conveniently formulated in the language of the spaces $\mathrm{Ext}^k(\cdot, \cdot)$ (see later 3.28).

Besides the papers cited, the papers [175], [166], [168] are devoted to questions of stability, where it is, however, sometimes taken in a somewhat different sense. Many results are summarized in the monograph by Jarosz [96].

§2.3. Admissible complexes and module extensions. Applications to derivations

Warning. In this section when we say "A-module" we always mean a left Banach module over a fixed Banach algebra A. All the results, however, carry across without difficulty to more general polynormed modules with the exception of 2.39 - 2.42.

Let us consider the category A-**mod**; since it is additive it has the concept of a split complex (Definition 0.4.4). At the same time it is one of the categories for which the concept of an exact sequence, or of an exact complex (cf. Chap.0 §4.2) makes sense. Moreover all the assertions formulated for spaces in the section referred to have obvious analogues for modules. In particular, the following is a "modular" version of Theorem 0.4.15.

Theorem (VII.2.32). *The following properties of a (short) complex* $0 \leftarrow X'' \xleftarrow{j} X \xleftarrow{i} X' \leftarrow 0$ $(\underline{\mathcal{S}})$ *in* A-**mod** *are equivalent:*

(I) $\underline{\mathcal{S}}$ *splits (in* A-**mod**);

---

[16] In [230] somewhat more is required, namely $\mathcal{H}^3(A, A) = 0$.

(II) $\underline{\mathcal{S}}$ *is exact and* j *is a retraction (in* A-**mod**);

(III) $\underline{\mathcal{S}}$ *is exact and* i *is a coretraction (in* A-**mod**).

▶ The proof of Theorem 0.4.15 (in the "Banach" case) carries across with obvious modifications. ■

Thus, in A-**mod**, as also in $\mathcal{B}an$, the property of splitting is stronger than the property of being exact; only now — for modules — the difference between the two properties is even more perceptible. Besides the previous examples of exact complexes that do not split "for geometrical reasons" — such as, say, the complex of c_0-modules just before Proposition 0.4.16 — there arise far simpler examples of an essentially algebraic character: such as the complex of $C[0, 1]$-modules $0 \leftarrow C[0, 1]/I_0 \leftarrow$ $\leftarrow C[0, 1] \leftarrow I_0 \leftarrow 0$, where $I_0 = \{f \in C[0, 1] : f(0) = 0\}$. If we want to separate such delicate ("geometric") and crude ("algebraic") reasons for failing to split, we arrive at the definition of an important class of complexes which occupies an intermediate position between exact and split complexes.

Definition (VII.2.33). A complex $\mathcal{X} \in \underline{A\text{-}\mathbf{mod}}$ is said to be *admissible* if it splits as a complex of Banach spaces (= after applying the forgetful functor $\underline{\square} : \underline{A\text{-}\mathbf{mod}} \longrightarrow \underline{\mathcal{B}an}$).

A morphism of A-modules is said to be *admissible* if it is one of the morphisms of a certain admissible complex. Obviously this concept includes as particular cases the admissible epi- and monomorphisms considered earlier in §1. At this point we are nearly ready to give equivalent definitions in the language of complexes for several central concepts in that section.

Proposition (VII.2.34). *An* A-*module* X'' *is projective* (X' *is injective, respectively*) $\Leftrightarrow$ *every admissible short complex* $0 \leftarrow X'' \leftarrow X \leftarrow X' \leftarrow 0$ *splits.*

▶ "$\Rightarrow$" is implied by Proposition 1.5 (1.34, respectively) and Theorem 2.32, "$\Leftarrow$" by a discussion of the complex $0 \leftarrow X'' \xleftarrow{\pi} A_+ \hat{\otimes} X'' \leftarrow \operatorname{Ker} \pi \leftarrow 0$ and Theorem 1.11(III) (or $0 \leftarrow \mathcal{B}(A_+, X')/\operatorname{Im} \check{\pi} \leftarrow \mathcal{B}(A_+, X') \xleftarrow{\check{\pi}} X' \leftarrow 0$ and Theorem 1.39(III)). ■

* * *

Not so long ago we were speaking of extensions of algebras. Extensions of modules are also defined in a similar spirit:

Definition (VII.2.35). An *extension of an* A-*module* X'' *by the* A-*module* X' is a triple (X, j, i) consisting of an A-module and A-module morphisms such that the sequence $0 \leftarrow X'' \xleftarrow{j} X \xleftarrow{i} X' \leftarrow 0$ $(\underline{\mathcal{S}})$ is exact.

By an obvious version of Proposition 0.4.12 for A-**mod**, giving an extension of a module is equivalent to giving an exact sequence "(closed) submodule — module — quotient module". Instead of "the extension (X, j, i)" we shall in future often write "the extension $\underline{\mathscr{S}}$".

Definition (VII.2.36) (cf. 2.16). Two extensions (X, j, i) and (X_1, j_1, i_1) of X'' by X' are said to be *equivalent* if there is a commutative diagram

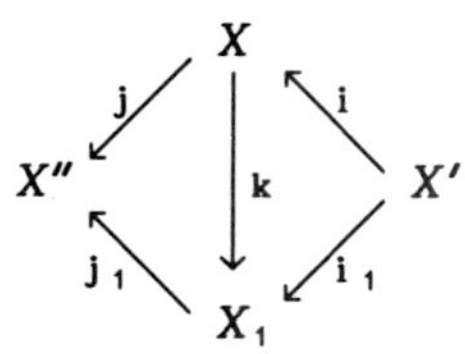

in which k is a topological isomorphism of A-modules.

An extension (of modules) is said to *split* if the corresponding complex $\underline{\mathscr{S}}$ splits. If the latter is admissible (but no more), then the extension is said to be *singular*; in this case an operator, right inverse to j, is called a *splitting operator* (over $\underline{\mathscr{S}}$).

Our next aim is to describe singular extensions of modules to within equivalence as we did for singular extensions of algebras. The set of equivalence classes of singular extensions of X'' by X' is denoted by $\mathrm{Ext}_A(X'', X')$. It contains (as before in $\mathrm{Ex}(A, \cdot)$) a distinguished point — the class of split extensions.

Let $\underline{\mathscr{S}}$ be a certain singular extension; obviously we are justified in identifying $\mathrm{Im}\, i$ with X' and $X/\mathrm{Im}\, i$ with X''. If ρ is a splitting operator, then for any $a \in A$ and $x \in X''$, $f_\rho(a, x) := a\cdot\rho(x) - \rho(a\cdot x)$ lies in X'; thus $f_\rho \in \mathcal{B}(A \times X'', X')$. The following is an obvious analogue of Proposition 2.20 for extensions of modules.

Proposition (VII.2.37). *The bi-operator* f_ρ *satisfies*

$$a\cdot f_\rho(b, x) - f_\rho(ab, x) + f_\rho(a, b\cdot x) = 0, \tag{1}$$

where $a, b \in A$, $x \in X''$. *If* ρ *and* ρ' *are two splitting operators, then* $f := f_\rho - f_{\rho'}$ *satisfies*

$$f(a, x) = a\cdot\varphi(x) - \varphi(a\cdot x), \tag{2}$$

where $a \in A$, $x \in X''$ *for a certain continuous operator* $\varphi : X'' \longrightarrow X'$. *Moreover one can put* $\varphi : x \longmapsto \rho(x) - \rho'(x)$. ▶■

We denote the subspace of $\mathcal{B}(A \times X'', X')$, consisting of bi-operators satisfying (1) (respectively (2) for a certain φ), by $Z(X'', X')$ ($B(X'', X')$, respectively). Obviously the first subspace is closed and contains the second.

Theorem (VII.2.38) (cf. 2.21). *There exists a bijection between the set* $\mathrm{Ext}_A(X'', X')$ *and the quotient space* $Z(X'', X')/B(X'', X')$ *that takes the equivalence class of an extension with a splitting operator* ρ *to the coset of the bi-operator* f_ρ. *This bijection takes the distinguished point to the zero of the quotient space.*

▶ The proof is very similar to that given in 2.21 and differs from it only in being simpler. We leave it to the reader. ∎

Identifying the two sets indicated in Theorem 2.38, we give $\mathrm{Ext}_A(X'', X')$ the structure of a complete seminormed space (with respect to the quotient seminorm of the bi-operator norm in $Z(X'', X')$).

But we now show that the question about constructing module extensions is closely connected with the question discussed in §2.1 about constructing derivations. In the next theorem the space $\mathcal{B}(X'', X')$ is considered to be a Banach A-bimodule (see VI.3.5).

Theorem (VII.2.39). *There exists an isometric isomorphism between the seminormed spaces* $\mathrm{Ext}_A(X'', X')$ and $\mathcal{H}^1(A, \mathcal{B}(X'', X'))$.

▶ We consider the isometric freezing isomorphism $\mathcal{B}(A \times X'', X') \xrightarrow{\simeq} \mathcal{B}(A, \mathcal{B}(X'', X'))$ (see Chap.0 §3.1). Obviously it maps $Z(X'', X')$ onto $Z^1(A, \mathcal{B}(X'', X'))$ and $B(X'', X')$ onto $B^1(A, \mathcal{B}(X'', X'))$. All that remains is to apply Theorem 2.38. ∎

It is now possible to fulfil the promise made at the end of §2.1 and at the same time to demonstrate the first "intrinsic" application of the homology results of §1. From now on until the end of the section A is a Banach operator algebra on the Banach space E where E itself is considered as the natural A-module and $\mathcal{B}(E)$ is considered as a Banach A-bimodule with the ambient algebra operations.

Theorem (VII.2.40). *Let the* A*-module* E *be projective. Then* $\mathcal{H}^1(A, \mathcal{B}(E)) = 0$ *(in other words, every continuous derivation of the algebra* A *with values in* $\mathcal{B}(E)$ *is inner).*

▶ If we combine Theorem 2.39 and Proposition 2.34, we get that $\mathcal{H}^1(A, \mathcal{B}(E, Y)) = \mathrm{Ext}_A(E, Y) = 0$ for any $Y \in A$-**mod**. All that remains is to put Y equal to E itself. ∎

This, taking into account Theorem 1.27, implies the following.

Corollary (VII.2.41) (cf. 2.14). *Let* A *contain all the finite-dimensional operators (for example* A *could be* $\mathcal{B}(E)$, $\mathcal{K}(E)$ or $\mathcal{N}(E)$). *Then* $\mathcal{H}^1(A, \mathcal{B}(E)) = 0$.

Exercise (VII.2.42). Let, further, E be a Hilbert space and A self-adjoint. Show that in this case for $\mathcal{H}^1(A, \mathcal{B}(E)) = 0$ it is sufficient that the A-module E be flat.

Hint. Now $E = E^{**}$ is injective and therefore $\operatorname{Ext}_A(Y, E) = 0$ for all Y.

§3. THE SPACES Ext^n, PREPARATIONS AND APPLICATIONS

In this section the central place is occupied by a construction that associates to two given modules X and Y a certain sequence of linear spaces denoted by $\operatorname{Ext}^n(X, Y)$. Their definition involves complexes of a special kind — so-called projective resolutions — and also involves the n-dimensional homology functor introduced in Chap.0 §4.3. The "Ext" form a universal language in which one can express naturally the majority of the basic concepts in §1, §2 and they provide an effective tool for investigating them.

Warning. Throughout this section, unless explicitly stated otherwise, A is a fixed Banach algebra, not unital as assumed earlier (in view of future examples and applications). The basic working categories are ***A*-mod** (when we talk about modules, as a rule, we are thinking of objects in this category) and the category of chain complexes of Banach A-modules, denoted, according to the general rules in Chap.0 §4.1, by $\underline{\textbf{A-mod}}$.

§3.1. Projective resolutions. The standard (normalized bar-) resolution. Remark on the Koszul resolution

Let X be a certain A-module, generally speaking not projective. We shall attempt to "express it in terms of projective modules" by the following procedure. First using Corollary 1.10, we represent X as a quotient module of some projective module so that the corresponding natural projection is admissible. Then we represent the kernel of this projection, in turn, as the quotient module of another projective module, again with an admissible natural projection; after this we do the same with the kernel of the next and so on. The concept of a resolution that we are talking about provides a convenient "once and for all" way of writing down such a process. Here are the formal definitions.

We agree to identify the module X with a complex in which it itself stands at the zero place and there are zeros at the others. Accordingly, if $\mathcal{X} = \{X, d\} \in \underline{\textbf{A-mod}}$ is such that $X_n = 0$ when $n < 0$, then giving a morphism of complexes $\varepsilon : \mathcal{X} \longrightarrow X$ is equivalent to giving a complex

$0 \leftarrow X \xleftarrow{\varepsilon} X_0 \xleftarrow{d_0} X_1 \xleftarrow{d_1} \cdots$, briefly denoted by $0 \leftarrow X \xleftarrow{\varepsilon} \mathcal{X}$. In both cases we say that we are given a *complex over* X.

Definition (VII.3.1). A complex $\varepsilon : \mathcal{X} \longrightarrow X$ over X is called a *resolution* of the A-module X if the complex $0 \leftarrow X \xleftarrow{\varepsilon} \mathcal{X}$ is admissible (see 2.33). The resolution is said to be *projective* if all the modules in $\mathcal{X}$ are projective.

The *length* of the resolution $\varepsilon : \mathcal{X} \longrightarrow X$ is the smallest $n \geq -1$ such that $X_k = 0$ for $k > n$ or ∞ if there is no such n.

Of course, each module has many projective resolutions. However, essentially they are all the same:

Theorem (VII.3.2) "comparison theorem". *Let* $\varepsilon : \mathcal{P} \longrightarrow X$ *be a projective and* $\varepsilon' : Q \longrightarrow Y$ *an arbitrary resolution of corresponding modules. Then for any morphism* $\varphi : X \longrightarrow Y$ *there exists a morphism of complexes* $\underline{\varphi}$ *such that the diagram*

$$\begin{array}{ccc} \mathcal{P} & \xrightarrow{\varepsilon} & X \\ {\scriptstyle \underline{\varphi}}\downarrow & & \downarrow{\scriptstyle \varphi} \\ Q & \xrightarrow{\varepsilon'} & Y \end{array}$$

is commutative. Any two morphisms with the property indicated are homotopic.

▶ We shall construct by induction the morphisms φ_n; $n = -1, 0, 1, \cdots$. We put $\varphi_{-1} := \varphi$. Now suppose that we have constructed $\varphi_n : P_n \longrightarrow Q_n$; $n \leq k$, so that the diagram

$$\begin{array}{ccccccccc} 0 \leftarrow X & \xleftarrow{\varepsilon} & \cdots & \leftarrow P_{k-1} & \xleftarrow{d_{k-1}} & P_k & \xleftarrow{d_k} & P_{k+1} & \leftarrow \cdots \\ {\scriptstyle \varphi_{-1}}\downarrow & & & {\scriptstyle \varphi_{k-1}}\downarrow & & {\scriptstyle \varphi_k}\downarrow & & {\scriptstyle \varphi_{k+1}}\downarrow & \\ 0 \leftarrow Y & \xleftarrow{\varepsilon'} & \cdots & \leftarrow Q_{k-1} & \xleftarrow{d'_{k-1}} & Q_k & \xleftarrow{d'_k} & Q_{k+1} & \leftarrow \cdots \end{array}$$

is commutative. We put $K := \operatorname{Ker} d'_{k-1} = \operatorname{Im} d'_k$ and $\sigma := d'_k|^K$; the latter, since the lower complex is admissible, is an admissible epimorphism. Further, as $d'_{k-1}\varphi_k d_k = \varphi_{k-1}d_{k-1}d_k = 0$ the morphism $\check{\varphi} := (\varphi_k d_k)|^K$ is well defined. If we consider the lifting problem $(P_{k+1}, \sigma, \check{\varphi})$ that arises and use the fact that P_{k+1} is projective, we may take $\varphi_{k+1} : P_{k+1} \longrightarrow Q_{k+1}$ to be an arbitrarily chosen lifting morphism; obviously it does not disturb the commutativity of the diagram.

All that remains is to put $\underline{\varphi} := \{\varphi_n;\ n \geq 0\}$. The assertion regarding homotopy is proved analogously. ■

If we take two projective resolutions of the same module X and put $\varphi := 1_X$, we immediately get the following.

Corollary (VII.3.3). *Any two projective resolutions of the same module are homotopically equivalent.*

We pass to a description of one specific method of constructing a resolution. For every $n = 0, 1, 2, \cdots$ we consider the functor $\widetilde{\mathbb{B}}_n : \mathcal{B}an \longrightarrow A\text{-}\mathbf{mod}$ acting in the following way: it takes the space E to the free module $A_+ \hat{\otimes} (\underbrace{A \hat{\otimes} \cdots \hat{\otimes} A}_{n} \hat{\otimes} E)$ — the result of applying the *freedom functor* $A_+ \hat{\otimes}\, ?$ to the space included in the brackets — and it takes the operator ψ to the morphism $1_{A_+} \hat{\otimes} 1_A \hat{\otimes} \cdots \hat{\otimes} 1_A \hat{\otimes} \psi$. Further, we put $\mathbb{B}_n := \widetilde{B}_n \circ \square : A\text{-}\mathbf{mod} \longrightarrow A\text{-}\mathbf{mod}$, where $\square$ is the forgetful functor from $A\text{-}\mathbf{mod}$ into $\mathcal{B}an$ (VI.3.12). Obviously all these functors are additive and moreover Proposition 0.4.5 and Definition 2.33 imply the following.

Proposition (VII.3.4). *If the complex* $\mathcal{X} \in \underline{A\text{-}\mathbf{mod}}$ *is admissible, then the complex* $\underline{\mathbb{B}_n}(\mathcal{X}) \in \underline{A\text{-}\mathbf{mod}}$ *splits for every* $n \geq 0$. ▸■

We now fix $X \in A\text{-}\mathbf{mod}$ and take the sequence of modules that arises: $\mathbb{B}_0(X) = A_+ \hat{\otimes} X$, $\mathbb{B}_1(X) = A_+ \hat{\otimes} A \hat{\otimes} X, \cdots, \mathbb{B}_n(x) = A_+ \hat{\otimes} \underbrace{A \hat{\otimes} \cdots \hat{\otimes} A}_{n} \hat{\otimes} X, \cdots$. We consider the operators $d_n : \mathbb{B}_{n+1}(X) \longrightarrow \mathbb{B}_n(X)$ uniquely defined — in view of the universality property of the tensor product — by

$$
\begin{aligned}
d_n(a \otimes a_1 \otimes \cdots \otimes a_{n+1} \otimes x) &= aa_1 \otimes a_2 \otimes \cdots \otimes a_{n+1} \otimes x + \\
&+ \sum_{k=1}^{n} (-1)^k a \otimes a_1 \otimes \cdots \otimes a_k a_{k+1} \otimes \cdots \otimes a_{n+1} \otimes x + \\
&\qquad + (-1)^{n+1} a \otimes a_1 \otimes \cdots \otimes a_n \otimes a_{n+1} \cdot x\,. \qquad (1)
\end{aligned}
$$

Obviously d_n; $n \geq 0$, are A-module morphisms and moreover elementary calculations show that $d_n d_{n+1} = 0$. Thus we have constructed a complex in $A\text{-}\mathbf{mod}$

$$0 \longleftarrow \mathbb{B}_0(X) \xleftarrow{d_0} \mathbb{B}_1(X) \xleftarrow{d_1} \cdots \qquad (\mathbb{B}(X))$$

Proposition (VII.3.5). *The complex over* X

$$0 \longleftarrow X \xleftarrow{\pi} \mathbb{B}_0(X) \xleftarrow{d_0} \mathbb{B}_1(X) \longleftarrow \cdots \qquad (0 \longleftarrow X \longleftarrow \mathbb{B}(X))$$

where $\pi : \mathbb{B}_0(X) = A_+ \hat{\otimes} X \longrightarrow X$ *is the canonical projection, is a projective resolution of the* A*-module* X.

▸ By Proposition 1.7 considered with A_+ instead of A, it is sufficient to show that our complex is admissible. With this aim we consider, for $n = 0, 1, \cdots$, the operators

$s_n : \mathbb{B}_n(X) \longrightarrow \mathbb{B}_{n+1}(X)$ that are well defined by $s_n(a \otimes a_1 \otimes \cdots \otimes a_n \otimes x) = e_+ \otimes a' \otimes a_1 \otimes \cdots \otimes a_n \otimes x$, where $a = \lambda e_+ + a'$ with $\lambda \in \mathbb{C}$ and $a' \in A$. It is easy to verify that these s_n together with $s_{-1} : X \longrightarrow \mathbb{B}_0(X) : x \longmapsto e_+ \otimes x$ provide a contracting homotopy of our complex in $\underline{\mathcal{B}an}$. The rest is clear. ■

Definition (VII.3.6). The complex $0 \leftarrow X \leftarrow \mathbb{B}(X)$ (or, according to the sense, its "truncated part" $\mathbb{B}(X)$) is called the *standard resolution* or the *normalized bar-resolution* of the Banach A-module X.

If $\varphi : X \longrightarrow Y$ is a morphism in **A-mod**, then obviously the family $\mathbb{B}(\varphi) := \{\mathbb{B}_n(\varphi) : \mathbb{B}_n(X) \longrightarrow \mathbb{B}_n(Y); n \geq 0\}$ is a morphism between the complexes $\mathbb{B}(X)$ and $\mathbb{B}(Y)$ in $\underline{A\text{-}\mathbf{mod}}$. If we make the correspondence $X \longmapsto \mathbb{B}(X)$; $\varphi \longmapsto \mathbb{B}(\varphi)$, we get, of course, a covariant functor. It is denoted by $\mathbb{B} : A\text{-}\mathbf{mod} \longrightarrow \underline{A\text{-}\mathbf{mod}}$ and it is called the *standard resolution functor.*

Why is this formal construction, which can be shown to be excessively clumsy, precisely the one that is taken as basic? Let us imagine that someone suggests to the reader, immediately after he becomes acquainted with Definition 3.1, that he should construct a projective resolution of the module X by the simplest method that first comes into his head. We are certain that his resolution would turn out to be an isomorphic copy of the standard resolution. The exact meaning of what has been said is given by the following.

Exercise (VII.3.7) (cf. [81, III.2.8]). Show that the complex $0 \leftarrow X \xleftarrow{\pi} \mathbb{B}(X)$ is isomorphic in $\underline{A\text{-}\mathbf{mod}}$ to the upper row of the commutative diagram

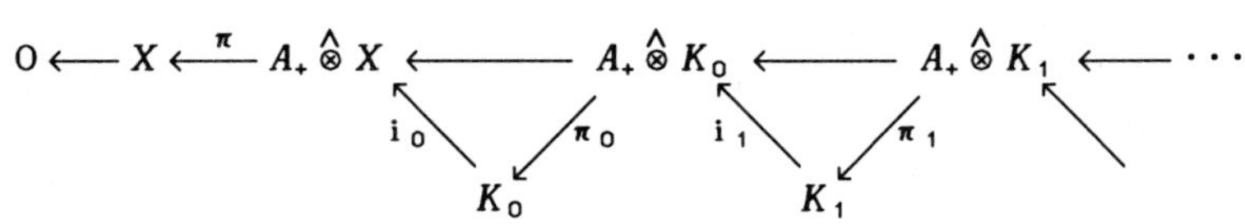

in which all the π_n are canonical projections, the i_n are natural embeddings, $K_0 := \operatorname{Ker} \pi$, and the K_n; $n > 0$, are defined inductively as $\operatorname{Ker} \pi_{n-1}$.

... Up to now, following the warning, we have worked in the category of left modules. But by the already repeatedly applied constructions of Chap. VI §3.1, a Banach module of any type may be considered as an object in **A-mod** for a suitable algebra A (for example, the enveloping algebra if we are talking about A-bimodules). Therefore all the preceding definitions and facts extend automatically to it.

We point out especially the case of the A-bimodule A_+. Its standard resolution as a left A-module has the form

$$0 \longleftarrow A_+ \hat{\otimes} A_+ \longleftarrow A_+ \hat{\otimes} A \hat{\otimes} A_+ \longleftarrow \cdots$$

$$\cdots \longleftarrow A_+ \hat{\otimes} \underbrace{A \hat{\otimes} \cdots \hat{\otimes} A}_{n} \hat{\otimes} A_+. \qquad (\mathbb{B}(A_+))$$

This complex is obviously a projective resolution for A_+ not only in **A-mod** but also in **A-mod-A**. It is called the *standard bimodule resolution*;[17] afterwards in §3.3 it will play a very important role.

The definitions of a projective resolution and the standard resolution carry across to the categories of polynormed modules with various classes of underlying spaces (and of course to categories of abstract modules, where they were invented). For details on this subject and also on certain other versions of the standard resolution, including the so-called non-normalized version, see [251] and [81].

Remark (VII.3.8). Besides the projective resolutions, the "injective coresolutions" of the module X are also useful in a number of questions — these are admissible complexes of the form $0 \longrightarrow X \longrightarrow J_0 \longrightarrow J_1 \longrightarrow \cdots$ with injective J_k. They always exist in the purely algebraic situation where they play just as big a role as the projective resolutions [38], [51], [22], and they also exist for Banach modules [81]. For general polynormed modules the question becomes complicated for the reason mentioned at the end of §1.3 (see Taylor's valuable discussion [251]).

The standard resolution $\mathbb{B}(X)$ is good in that it exists "for all regimes". However, when working with many specific algebras it is often more convenient to use (what "to use" means we shall learn later) other resolutions which are not like the standard one but reflect the peculiarities of the given situation. For example, if A is a hereditary unital algebra with complemented ideals, say, $C[0, 1]$, then every cyclic A-module $X \simeq A/I$ has a projective resolution $0 \longleftarrow X \longleftarrow A \longleftarrow I \longleftarrow 0$ which (since it has length 1) is far simpler than the standard one. If A is the Varopoulos algebra $C(\mathbb{T}) \hat{\otimes} C(\mathbb{T})$, then many A-modules have convenient resolutions of length 2: for example, it is not too difficult to ascertain that $\mathbb{C}(1, 1)$, where $(1, 1) \in \mathbb{T} \times \mathbb{T} = \underline{\Omega}(A)$, has a projective resolution

$$0 \longleftarrow \mathbb{C}(1, 1) \longleftarrow A \longleftarrow (I_1 \hat{\otimes} C(\mathbb{T})) \oplus (C(\mathbb{T}) \hat{\otimes} I_1) \longleftarrow I_1 \hat{\otimes} I_1 \longleftarrow 0.$$

[17] Its name and notation should not lead to error: it is not the standard resolution of the object in A^{env}**-mod** which corresponds to A_+ if we identify A**-mod-**A with A^{env}**-unmod**. We recommend that the reader writes down this last resolution explicitly.

Another resolution of length 2 is often used, called the entwining resolution; see [81, V §2] for more about it and its applications.

The most important of the special resolutions is, perhaps, the *Koszul resolution*: this is a resolution of the A-module X having length n and of the form $0 \leftarrow X \leftarrow \mathcal{K}(P, S_1, \cdots, S_n)$, where P is a projective A-module, $\{S_k \in {}_A\mathbf{h}(P) \colon 1 \le k \le n\}$ is a system of commuting morphisms and $\mathcal{K}(\cdot, \cdot)$ is the notation for the Koszul complex we met in Chap.II §2.2. Such resolutions are typical for modules over many polynormed function algebras, for example, $\mathcal{O}(U)$; $U \subseteq \mathbb{C}^n$ or $C^\infty(U)$; $U \subseteq \mathbb{R}^n$ [252]; for such A there even exists a functor associating with each polynormed A-module its Koszul resolution. We give a precise formulation of one of the relevant assertions.

Assertion (VII.3.9.) *Let* U *be a polydomain in* $\mathbb{C}^n$, *let* X *be a unital Fréchet* $\mathcal{O}(U)$*-module, let* $\mathcal{O}(U, X)$ *be the* $\mathcal{O}(U)$*-module of holomorphic* X*-valued functions on* U (VI.1.27), *and let* S_k *be the operator on* $\mathcal{O}(U, X)$ *taking* $\overline{u}$ *into* $\overline{v}(z_1, \cdots, z_n) := [z_k \cdot \overline{u}](z_1, \cdots, z_n) - z_k(\overline{u}(z_1, \cdots, z_n))$. *Then the complex* $0 \leftarrow X \xleftarrow{\pi} \mathcal{K}(\mathcal{O}(U, X), S_1, \cdots, S_n)$, *where* $\pi : \mathcal{O}(U, X) \longrightarrow X$ *is uniquely defined by the fact that it takes the "one-dimensional" vector function* $w(z)x$ *into* $w \cdot x$; $w \in \mathcal{O}(U)$, $x \in X$,[18] *is a projective resolution of the* $\mathcal{O}(U)$*-module* X.

(This is a crucial fact used in the "homological" proof of Taylor's theorem about multi-operator holomorphic calculi; see II.2.29.)

If A is an infinite-dimensional Banach algebra, it may happen that a Banach A-module does not have a Koszul resolution (Taylor [252, §5]). However, "good" modules may have one; such as, for example, one-dimensional modules over $\mathcal{A}(\overline{\mathbb{D}}^n)$ corresponding to interior points of the polydisc (try to prove it!).

See [165], [38] for more about Koszul resolutions in the context of abstract algebra (where they really arose).

§3.2. Derived functors and Ext^n. One-dimensional Ext in terms of extensions and an obstruction to lifting. Remark on Tor

We shall now speak about additive functors from A-**mod** into fairly good additive categories; in the context of this book it is expedient to concentrate for definiteness on $\mathcal{B}an$. (Though, as

[18] If we identify $\mathcal{O}(U, X)$ with $\mathcal{O}(U) \hat{\otimes} X$, as mentioned at the end of Chap.0 §3.5, this operator becomes the canonical projection for X.

the reader will see, in the majority of applications $\mathcal{L}in$ would suffice.) We recall the functor of the n-dimensional homology $H_n : \underline{\mathcal{B}an} \longrightarrow \langle\mathcal{B}an\rangle$ (Chap.0 §4.3) and also the recently introduced $\mathbb{B} : A\text{-}\mathbf{mod} \longrightarrow \underline{A\text{-}\mathbf{mod}}$.

Definition (VII.3.10). Let $F : A\text{-}\mathbf{mod} \longrightarrow \mathcal{B}an$ be an additive covariant functor. For $n = 0, 1, 2, \cdots$ the covariant functor $H_n \circ \underline{F} \circ \mathbb{B}$ (= the result of the successive application of $\mathbb{B}$, $\underline{F}$ and H_n) is called the *n-th derived functor*[19] of F and is denoted by $F_n : A\text{-}\mathbf{mod} \longrightarrow \langle\mathcal{B}an\rangle$.

Analogously, if we are given a contravariant F, we get the contravariant "n-th derived functor of F", namely $F^n := H^n \circ \overline{F} \circ \mathbb{B}$.

The next two results (the second is offered as an exercise) show that the derived functor may be calculated by taking any projective resolution instead of the standard one.

Theorem (VII.3.11). *Let* $0 \longleftarrow X \longleftarrow \mathcal{P}$ *be a projective resolution of the* A*-module* X. *Then for all* $n \geq 0$ *the seminormed space* $H_n(\underline{F}(\mathcal{P}))$ *coincides, up to a topological isomorphism, with* $F_n(X)$.

▶ By 3.3 and 0.4.5 the complexes $\underline{F}(\mathcal{P})$ and $\underline{F}(\mathbb{B}(X))$ are homotopically equivalent in $\underline{\mathcal{B}an}$. All that remains is to apply Corollary 0.4.24. ■

Exercise (VII.3.12). Let

$$\begin{array}{ccccc} 0 & \longleftarrow & X & \longleftarrow & \mathcal{P} \\ & & \downarrow{\scriptstyle \varphi} & & \downarrow{\scriptstyle \underline{\varphi}} \\ 0 & \longleftarrow & Y & \longleftarrow & \mathcal{Q} \end{array}$$

be a commutative diagram in $\underline{A\text{-}\mathbf{mod}}$ in which horizontals depict projective resolutions. Show that the morphism $F_n(\varphi)$; $n \geq 0$, becomes, after the identifications provided by the preceding theorem, $H_n(\underline{F}(\underline{\varphi})) : H_n(\underline{F}(\mathcal{P})) \longrightarrow H_n(\underline{F}(\mathcal{Q}))$.

Here is the simplest application.

Proposition (VII.3.13). *Let* $P \in A\text{-}\mathbf{mod}$ *be projective. Then* $F_n(P) = 0$ *for* $n > 0$ *and* $F_0(P)$ *is topologically isomorphic to* $F(P)$.

---

[19] More precisely — the n-th projective derived functor, in contrast to its injective analogue; the latter may be defined with the help of the injective version of the standard resolution [81, p.147].

▶ Theorem 3.11 enables us to calculate $F_n(P)$; $n \geq 0$, using the resolution $0 \leftarrow P \xleftarrow{1} P \leftarrow 0$. ■

All three facts (3.11 - 3.13) have obvious analogues for contravariant F.

The next theorem describes the most important property of derived functors and at the same time gives a method of calculation just as effective as 3.11.

Theorem (VII.3.14). *Let* $F : A\text{-}\mathbf{mod} \longrightarrow \mathcal{B}an$ *be a covariant (contravariant, respectively) additive functor, let* $0 \leftarrow X \xleftarrow{j} Y \xleftarrow{i} Z \leftarrow 0$ *be an admissible short complex of A-modules. Then there exist continuous operators* $\xi_n : F_{n+1}(X) \longrightarrow F_n(Z)$ ($\xi^n : F^n(Z) \longrightarrow F^{n+1}(X)$ *respectively*); $n \geq 0$, *such that the sequence*

$$0 \leftarrow F_0(X) \xleftarrow{F_0(j)} F_0(Y) \xleftarrow{F_0(i)} F_0(Z) \xleftarrow{\xi_0} F_1(X) \xleftarrow{F_1(j)} F_1(Y) \leftarrow \cdots$$

$$(0 \longrightarrow F^0(X) \xrightarrow{F^0(j)} F^0(Y) \xrightarrow{F^0(i)} F^0(Z) \xrightarrow{\xi^0} F^1(X) \xrightarrow{F^1(j)} F^1(Y) \longrightarrow \cdots$$

respectively) is exact.

▶ We limit ourselves to covariant F. Propositions 3.4 and 0.4.5 imply that the short complex

$$0 \leftarrow \underline{F}(\mathbb{B}(X)) \xleftarrow{\underline{F}(\mathbb{B}(j))} \underline{F}(\mathbb{B}(Y)) \xleftarrow{\underline{F}(\mathbb{B}(i))} \underline{F}(\mathbb{B}(Z)) \leftarrow 0$$

in $\underline{\mathcal{B}an}$ splits and therefore is automatically exact. Therefore we immediately get the necessary sequence by applying the "basic lemma of homological algebra" (0.4.25). ■

It is easy to verify that the operators ξ_n (ξ^n) are natural; this means that, considered for all admissible short complexes in A-**mod**, they form a natural transformation (see Chap.0 §2.2) between suitably defined functors. We leave the details to the reader.

* * *

For what follows we need another (cf. 2.34) equivalent definition of projective and injective modules — this time in terms of preserving exactness. We recall the morphism functors from A-**mod** into $\mathcal{B}an$ introduced in Chap.VI §3.2. The following can be verified immediately.

Lemma (VII.3.15). *For any A-module X and any short exact complex of A-modules* $0 \leftarrow X_0 \xleftarrow{\sigma} X_1 \xleftarrow{i} X_2 \leftarrow 0$ *the complexes*

$$0 \longleftarrow {}_A\mathbf{h}(X, X_0) \longleftarrow {}_A\mathbf{h}(X, X_1) \longleftarrow {}_A\mathbf{h}(X, X_2) \longleftarrow 0$$

and

$$0 \longrightarrow {}_A\mathbf{h}(X_0, X) \longrightarrow {}_A\mathbf{h}(X_1, X) \longrightarrow {}_A\mathbf{h}(X_2, X) \longrightarrow 0$$

are exact at all terms, except perhaps at ${}_A\mathbf{h}(X, X_0)$ *and* ${}_A\mathbf{h}(X_2, X)$. ▶■

Theorem (VII.3.16). (I) *An A-module X is projective* $\Leftrightarrow$ *the functor* $\underline{{}_A\mathbf{h}(X, ?)}$ *takes each short admissible complex of A-modules into an exact complex* $\Leftrightarrow$ *this functor takes each (arbitrarily long) admissible complex of A-modules into an exact complex.*

(II) *The same but replacing "projective" by "injective" and "*${}_A\mathbf{h}(X, ?)$*" by "*${}_A\mathbf{h}(?, X)$*".*

▶ X being projective means precisely that for any admissible (= figuring in a short admissible complex) epimorphism σ the operator ${}_A\mathbf{h}(X, \sigma)$ is surjective. In conjunction with the lemma, this provides the first logical equivalence; the second we quickly get by going from a "long" complex to its component short complexes (Chap.0 §4.2). This proves (I); (II) is proved similarly. ■

Exercise (VII.3.17). Prove that a left (right, respectively) A-module X is flat $\Leftrightarrow$ the functor $\underline{? \mathbin{\hat{\otimes}_A} X}$ $\underline{(X \mathbin{\hat{\otimes}_A} ?)}$ takes each admissible complex of right (left, respectively) A-modules into an exact complex.

We turn to an important example of a derived functor. We fix $Y \in A$-**mod** and consider the corresponding contravariant morphism functor ${}_A\mathbf{h}(?, Y) : A\text{-}\mathbf{mod} \longrightarrow \mathcal{B}an$.

"Definition" (VII.3.18). We use the notation $\mathrm{Ext}^n_A(X, Y)$ for the n-th derived functor of ${}_A\mathbf{h}(?, Y)$ applied to $X \in A$-**mod** and the contravariant functor itself is denoted by $\mathrm{Ext}^n_A(?, Y) : A\text{-}\mathbf{mod} \longrightarrow \langle\mathcal{B}an\rangle$.

When we refer to the functor we have introduced we shall often use the expression the "n-th" or the "n-dimensional Ext".

Thus, the family of seminormed spaces $\mathrm{Ext}^n_A(X, Y)$; $n \geq 0$, is the cohomology of the complex

$$0 \longrightarrow {}_A\mathbf{h}(\mathbb{B}_0(X), Y) \xrightarrow{{}_A\mathbf{h}(d_0, Y)} {}_A\mathbf{h}(\mathbb{B}_1(X), Y) \longrightarrow \cdots \qquad ({}_A\mathbf{h}(\mathbb{B}(X), Y))$$

If we wish to give the latter a more explicit form, we let $\mathcal{B}^n(A, X, Y)$; $n \geq 0$, denote the space of continuous multilinear operators from $A \times \cdots \times A \times X$ into Y with the norm $\|f\| := \sup\{\|f(a_1, \cdots, a_n, x)\| : \|a_k\|, \|x\| \leq 1\}$ (in particular, $\mathcal{B}^0(A, X, Y)$ is simply $\mathcal{B}(X, Y)$. We consider the complex

$$0 \longrightarrow \mathcal{B}(X, Y) \xrightarrow{\delta^0} \mathcal{B}^1(A, X, Y) \longrightarrow \cdots \longrightarrow \mathcal{B}^n(A, X, Y) \xrightarrow{\delta^n} \cdots, \qquad (\mathcal{B}(A, X, Y))$$

in which $[\delta^n f](a_1, \cdots a_{n+1}, x) :=$

$$= a_1 \cdot f(a_2, \cdots, a_{n+1}, x) + \sum_{k=1}^{n} (-1)^k f(a_1 \cdots, a_k a_{k+1}, \cdots, a_{n+1}, x) + (-1)^{n+1} f(a_1, \cdots, a_n, a_{n+1} \cdot x);$$

for $n > 0$, and $[\delta^0 f](a, x) := a \cdot f(x) - f(a \cdot x)$.

Proposition (VII.3.19). *The family* $\mathrm{Ext}^n_A(X, Y)$; $n \geq 0$, *is to within a topological isomorphism of seminormed spaces the cohomology of the complex* $\mathcal{B}(A, X, Y)$.

▶ For every $n \geq 0$ Theorem VI.1.22, considered with A_+ instead of A, provides a topological isomorphism between ${}_A\mathbf{h}(\mathbb{B}_n(X), Y)$ and $\mathcal{B}(\underbrace{A \hat{\otimes} \cdots \hat{\otimes} A}_{n} \hat{\otimes} X, Y)$; the latter, by the universality property of the tensor product, is isomorphic to $\mathcal{B}^n(A, X, Y)$. The family of corresponding composition maps is obviously an isomorphism between the complexes ${}_A\mathbf{h}(\mathbb{B}(X), Y)$ and $\mathcal{B}(A, X, Y)$. The rest is clear. ■

The observation we have made enables us to recognize in the Ext of "small dimensions" 0 and 1 objects we already know. First of all the obvious equality $\mathrm{Ker}\, \delta^0 = {}_A\mathbf{h}(X, Y)$ implies the following.

Corollary (VII.3.20). *For all* $X, Y \in A$-**mod** *there exists a topological isomorphism between* ${}_A\mathbf{h}(X, Y)$ *and* $\mathrm{Ext}^0_A(X, Y)$ *(the latter is thus Banach). Considered for all* X *with fixed* Y, *these isomorphisms form a natural equivalence between the functors* ${}_A\mathbf{h}(?, Y)$ *and* $\mathrm{Ext}^0_A(?, Y)$.

The case $n = 1$ explains "why Ext is called Ext". Recall the set $\mathrm{Ext}(\cdot, \cdot)$ that corresponds to the admissible extensions of modules and its description as $Z(\cdot, \cdot)/B(\cdot, \cdot)$ (2.38). As $Z(X, Y)$ is no other than $\mathrm{Ker}\, \delta^1$, while $B(X, Y)$ is $\mathrm{Im}\, \delta^0$, we get the following from 3.19.

Corollary (VII.3.21). *For all* $X, Y \in A$-**mod** *the space* $\mathrm{Ext}^1_A(X, Y)$ *coincides, to within a topological isomorphism, with* $\mathrm{Ext}_A(X, Y)$.

It is worth remembering that when $n > 1$ $\mathrm{Ext}^n_A(X, Y)$ may also be interpreted as a set of equivalence classes (reasonably defined) of admissible complexes — only now of the form

$0 \leftarrow X \leftarrow Z_1 \leftarrow \cdots \leftarrow Z_n \leftarrow Y \leftarrow 0$. The corresponding result is stated in the context of abstract algebra in [51, Ch.III].

We emphasize that in the definition of $\mathrm{Ext}_A^n(X, Y)$ the roles of the two modules are not symmetrical; Y is fixed and X "wanders". The same spaces can be obtained also as a result of applying to the variable Y a certain functor defined by fixing X. With this aim we note that on the model of the (covariant) functor ${}_A\mathbf{h}(X, ?)$ we can define the functor ${}_A\mathbf{h}(\mathbb{B}(X), ?) : A\text{-}\mathbf{mod} \longrightarrow \overline{A\text{-}\mathbf{mod}}$ taking Y into the cochain complex ${}_A\mathbf{h}(\mathbb{B}(X), Y)$ and taking the morphism φ into ${}_A\mathbf{h}(\mathbb{B}(X), \varphi) := \{{}_A\mathbf{h}(\mathbb{B}_n(X), \varphi);\ n \geq 0\}$. Then obviously $\mathrm{Ext}_A^n(X, Y)$ is the result of applying to Y the composition of functors $H^n \circ {}_A\mathbf{h}(\mathbb{B}(X), ?)$ which naturally is denoted by $\mathrm{Ext}_A^n(X, ?)$. (We draw attention to the fact that this time we have constructed a covariant functor.) The following obvious proposition is a supplement to Corollary 3.20.

Proposition (VII.3.22). *The functor* $\mathrm{Ext}_A^0(X, ?)$ *considered with values in* $\mathcal{B}an$ *is naturally equivalent to the functor* ${}_A\mathbf{h}(X, ?)$. ▶■

A natural question arises: cannot $\mathrm{Ext}_A^n(X, ?)$ be defined, like $\mathrm{Ext}_A^n(?, Y)$, as a derived functor of a certain functor, say, ${}_A\mathbf{h}(X, ?)$? In the case of the Banach structures under consideration this is in fact so, but the necessary construction is given, not by the projective derived functor — defined in 3.18 — but by the injective derived functor constructed with the help of injective coresolutions (see, for example, the purely algebraic definition in [22]). However, the injective approach to $\mathrm{Ext}_A^n(X, ?)$ unlike that given above, runs into serious difficulties when extending it to polynormed modules which are more general than Banach modules; cf. Remark 3.8.

The reason that the Ext depend on two arguments is connected with their involvement in two exact sequences which are extremely important for their calculation.

Theorem (VII.3.23). *Let* $0 \leftarrow X'' \leftarrow X \leftarrow X' \leftarrow 0$ $(\mathcal{X})$ $(0 \leftarrow Y'' \xleftarrow{\sigma} Y \leftarrow Y' \leftarrow 0$ $(\mathcal{Y})$, *respectively) be a short admissible complex of A-modules. Then for any A-module Y (X, respectively) there exists in* $\langle\mathcal{B}an\rangle$ *an exact sequence*

$$0 \longrightarrow {}_A\mathbf{h}(X'', Y) \longrightarrow {}_A\mathbf{h}(X, Y) \longrightarrow {}_A\mathbf{h}(X', Y) \longrightarrow \mathrm{Ext}_A^1(X'', Y) \longrightarrow \cdots$$

$$\cdots \longrightarrow \mathrm{Ext}_A^n(X', Y) \longrightarrow \mathrm{Ext}_A^{n+1}(X'', Y) \longrightarrow \cdots \qquad (1)$$

(*respectively,*

$$0 \longrightarrow {}_A\mathbf{h}(X, Y') \longrightarrow {}_A\mathbf{h}(X, Y) \longrightarrow {}_A\mathbf{h}(X, Y'') \xrightarrow{\xi} \mathrm{Ext}^1_A(X, Y') \longrightarrow \cdots$$

$$\cdots \longrightarrow \mathrm{Ext}^n_A(X, Y'') \longrightarrow \mathrm{Ext}^{n+1}_A(X, Y') \longrightarrow \cdots) \qquad (2)$$

▶ The first sequence, taking into account Corollary 3.20, is provided by Theorem 3.14 applied to the (contravariant) $F := {}_A\mathbf{h}(?, Y)$. The second, taking into account 3.22, is provided by Theorem 0.4.25 "itself", applied to the sequence of complexes $0 \longleftarrow {}_A\mathbf{h}(\mathbb{B}(X), Y'') \longleftarrow {}_A\mathbf{h}(\mathbb{B}(X), Y) \longleftarrow {}_A\mathbf{h}(\mathbb{B}(X), Y') \longleftarrow 0$ which, by Theorem 3.16, is exact, and by the fact that $\mathbb{B}_n(X)$; $n \geq 0$, is projective. ■

One of the consequences is a new equivalent definition — now in the language of Ext — of the basic classes of modules in §1. We need the following as a preliminary.

Proposition (VII.3.24). *Let* $X \in A\text{-}\mathbf{mod}$, $Y \in \mathbf{mod}\text{-}A$. *Then to within an isomorphism in* $\langle \mathcal{B}an \rangle$, $\mathrm{Ext}^n_A(X, Y^*) = \mathrm{Ext}^n_{A^{\mathrm{op}}}(Y, X^*)$; $n \geq 0$, *where* Y^* *is considered as being in* A-**mod** (see Chap.VI §3.1) *and* Y *and* X^* *in* A^{op}-**mod**.

▶ By Proposition 3.19, it is sufficient to construct an isomorphism between the complexes $\mathcal{B}(A, X, Y^*)$ and $\mathcal{B}(A^{\mathrm{op}}, Y, X^*)$. It is easy to verify that this is given by the family of isomorphisms $\mathcal{B}^n(A, X, Y^*) \xrightarrow{\simeq} \mathcal{B}^n(A^{\mathrm{op}}, Y, X^*)$; $n \geq 0$ that are constructed in an obvious way using the freezing isomorphism $\mathcal{B}(X, Y^*) \xrightarrow{\simeq} (X \hat{\otimes} Y)^* \xrightarrow{\simeq} \mathcal{B}(Y, X^*)$ (see the end of Chap.0 §3.1). ■

Theorem (VII.3.25). *Let* X *be a left Banach* A*-module. Then*

(I) X *is projective* $\Longleftrightarrow$ $\mathrm{Ext}^1_A(X, Y) = 0$ *for all* $Y \in A$-**mod** $\Longleftrightarrow$ $\mathrm{Ext}^n_A(X, Y) = 0$ *for all* $Y \in A$-**mod** *and* $n \geq 1$.

(II) X *is injective* $\Longleftrightarrow$ $\mathrm{Ext}^1_A(Y, X) = 0$ *for all* $Y \in A$-**mod** $\Longleftrightarrow$ $\mathrm{Ext}^n_A(Y, X) = 0$ *for all* $Y \in A$-**mod** *and* $n \geq 1$.

(III) X *is flat* $\Longleftrightarrow$ $\mathrm{Ext}^1_A(X, Y^*) = 0$ *for all* $Y \in \mathbf{mod}\text{-}A$ $\Longleftrightarrow$ $\mathrm{Ext}^n_A(X, Y^*) = 0$ *for all* $Y \in \mathbf{mod}\text{-}A$ *and* $n \geq 1$.

▶ (I) Deducing the third property from the first is a particular case of Proposition 3.13 (in the contravariant form). Deducing the first from the second is provided by Theorem 3.23 (see (2)) together with Theorem 3.16. The rest is clear.

(II) If X is injective then by 3.16 the complex ${}_A\mathbf{h}(\mathbb{B}(Y), X)$ is exact; therefore the first property implies the third. Deducing the first from the second is provided by Theorem 3.23 (see (1)) and the same Theorem 3.16. The rest is clear.

(III) Since X being flat is equivalent to X^* being injective, the result is implied by (II) in conjunction with Proposition 3.24. ■

Theorem 3.23 enables us to give another interpretation (besides Corollary 3.21) of the one-dimensional Ext by connecting it with the question of solving lifting and extension problems. We consider the complex $\mathcal{Y}$ in Theorem 3.23 and we take an arbitrary $X \in A$-**mod** together with $\varphi : X \longrightarrow Y''$ (= an element of ${}_A\mathbf{h}(X, Y'')$). We draw attention to the element $\xi(\varphi) \in \mathrm{Ext}_A^1(X, Y')$ provided by the sequence (2) and we call it the *obstruction to lifting the morphism* φ. Here is why it is so called:

Proposition (VII.3.26). *The lifting problem* (X, σ, φ) *has a positive solution* $\Leftrightarrow$ $\xi(\varphi) = 0$.

▶ Lifting (to a morphism from X into Y) is allowed precisely by those φ which belong to the image of the operator ${}_A\mathbf{h}(X, \sigma)$ and hence, as (2) is exact, to the kernel of ξ. ■

Thus, if X, Y and Y' are A-modules, and moreover Y' is a submodule of Y which is complemented (in $\mathcal{B}an$), then $\mathrm{Ext}_A^1(X, Y)$ can be treated as the set of obstructions to lifting morphisms from X into Y/Y'.

In fact these two interpretations of the one-dimensional Ext are closely connected; the following, in particular, bears witness to this.

Exercise (VII.3.27). Show (in the notation of 3.23) that in identifying the spaces $\mathrm{Ext}_A(Y'', Y')$ and $\mathrm{Ext}_A^1(Y'', Y')$ the equivalence class of the extension $\mathcal{Y}$ becomes the obstruction to lifting the morphism $1_{Y''}$ (to a morphism from Y'' into Y).

Hint. When $X = Y''$ the operator ξ from (2) becomes the connecting operator for $0 \longleftarrow \mathcal{B}(A, Y'', Y'') \longleftarrow \mathcal{B}(A, Y'', Y) \longleftarrow$ $\longleftarrow \mathcal{B}(A, Y'', Y') \longleftarrow 0$. If we apply the general construction for a connecting operator described in the proof of Theorem 0.4.25 to $1_{Y''}$ and choose as its preimage "along the vertical" an operator ρ, right inverse to σ, we arrive at the $f_\rho \in \mathcal{B}(A \times Y'', Y')$ which was constructed, starting from a given extension, in §2.3.

We leave the reader, using the first long exact sequence in Theorem 3.23, to give an interpretation of the elements $\mathrm{Ext}_A^1(\cdot, \cdot)$ as obstructions to extending morphisms and to establish the corresponding analogues to 3.26 and 3.27.

We mention another application of Ext.

Assertion (VII.3.28). *Let* $\mathrm{Ext}_A^1(X, X) = 0$ *be satisfied for* $X \in A$-**mod** *and let* $\mathrm{Ext}_A^2(X, X)$ *be Hausdorff. Then* X *is a stable module* (see the end of §2.2).

* * *

Whereas Ext is defined starting from the morphism functor, in defining the second most important class of derived functors the analogous role is played by the tensor product functor. We fix a right Banach A-module X.

"Definition" (VII.3.29). We use the notation $\mathrm{Tor}_n^A(X, Y)$ for the n-th derived functor of $X \mathop{\hat{\otimes}}_A ? : A\text{-}\mathbf{mod} \longrightarrow \mathcal{B}an$ (see Chap. VI §3.2) applied to Y; the (covariant) functor itself is denoted by $\mathrm{Tor}_n^A(X, ?) : A\text{-}\mathbf{mod} \longrightarrow \langle\mathcal{B}an\rangle$.

The initial distinction between the roles of X and Y in 3.29 may be removed and here — unlike the case for Ext in general (see above) there is complete symmetry in both arguments:

Exercise (VII.3.30). Show that the same spaces $\mathrm{Tor}_n^A(X, Y)$ are the result of applying to X the n-th derived functor of $? \mathop{\hat{\otimes}}_A Y : \mathbf{mod}\text{-}A \longrightarrow \mathcal{B}an$.

Hint. $X \mathop{\hat{\otimes}}_A \mathbb{B}_n(Y) \simeq X \hat{\otimes} \underbrace{A \hat{\otimes} \cdots \hat{\otimes} A}_{n} \hat{\otimes} Y \simeq \mathbb{B}_n(X) \mathop{\hat{\otimes}}_A Y$.

Thus $\mathrm{Tor}_n^A(X, Y)$ can be calculated with the help of projective resolutions of either of these two modules.

If we consider the long exact sequence in Theorem 3.14 for $F = X \mathop{\hat{\otimes}}_A ?$ and $F = ? \mathop{\hat{\otimes}}_A Y$, we get two long exact sequences for Tor that play the same role in their calculation as (1) and (2) in Theorem 3.23 play for Ext. We leave the reader to write out these sequences in an explicit form.

Below we let $\alpha_{X,Y} : \mathrm{Tor}_0^A(X, Y) \longrightarrow X \mathop{\hat{\otimes}}_A Y$ denote the operator induced by the operator $1_X \mathop{\hat{\otimes}}_A \pi_Y : X \mathop{\hat{\otimes}}_A \mathbb{B}_0(Y) \longrightarrow X \mathop{\hat{\otimes}}_A Y$. We note that the latter, when identifying $X \mathop{\hat{\otimes}}_A \mathbb{B}_0(Y) = X \mathop{\hat{\otimes}}_A A_+ \hat{\otimes} Y$ with $X \hat{\otimes} Y$, acts like $x \otimes y \longmapsto x \mathop{\otimes}_A y$.

Exercise (VII.3.31) (cf. 3.25). Show that $X \in \mathbf{mod}\text{-}A$ is flat $\Longleftrightarrow$ $\mathrm{Tor}_1^A(X, Y) = 0$ for all $Y \in A\text{-}\mathbf{mod}$ and $\alpha_{X,Y}$ is a topological isomorphism $\Longleftrightarrow$ $\mathrm{Tor}_n^A(X, Y) = 0$ for all $Y \in A\text{-}\mathbf{mod}$ and $n \geq 0$ and $\alpha_{X,Y}$ is a topological isomorphism. Establish an analogous flatness criterion for $Y \in A\text{-}\mathbf{mod}$.

Remark (VII.3.32). It is not difficult to show that $\alpha_{X,Y}$ is an operator taking the points not separated[20] from zero in $\mathrm{Tor}_0^A(X, Y)$ to zero. It is not, generally speaking, a topological isomorphism; see, for example, [81, Chap. VII §1].

The spaces Tor in the context of modules over the Fréchet algebra $\mathcal{O}(U)$; $U \subseteq \mathbb{C}^n$, play a key role in the second proof of Taylor's theorem about the holomorphic calculus (see Chap. II §2.2).
└

[20] In the sense of the seminorm topology.

§3.3. Cohomology groups and their expression in terms of Ext^n. Applications to derivations and extensions. Cohomological characterization of contractible and amenable algebras

The basic results of this section connect together, to their mutual advantage, the study of lifting and extension problems (§1) and problems involving cocycles and coboundaries (§2).

The following definition provides the Banach analogue of that concept in algebraic homology theory which, at the time, was the starting point of the whole theory. We are talking about the cohomology groups of associative algebras discovered in 1945 by Hochschild [274].

Let A and X be a Banach algebra and a Banach A-bimodule. For $n = 1, 2, \cdots$ we let $C^n(A, X)$ denote the space of continuous n-linear operators from $A \times \cdots \times A$ into X with norm $\|f\| := \sup\{\|f(a_1, \cdots, a_n)\| : \|a_k\| \le 1, \quad 1 \le k \le n\}$; we also put $C^0(A, X) := X$. The elements of $C^n(A, X)$ will be called *n-dimensional cochains.*

For $n > 0$ we consider the continuous operator

$$\delta^n : C^n(A, X) \longrightarrow C^{n+1}(A, X), \quad \text{where} \quad \delta^n f(a_1, \cdots, a_{n+1}) :=$$

$$= a_1 \cdot f(a_2, \cdots, a_{n+1}) + \sum_{k=1}^{n} (-1)^k f(a_1, \cdots, a_k a_{k+1}, \cdots, a_{n+1}) + (-1)^{n+1} f(a_1, \cdots, a_n) \cdot a_{n+1};$$

we also put $\delta^0 : C^0(A, X) \longrightarrow C^1(A, X) : x \longmapsto \delta^0 x$; $\delta^0 x(a) := a \cdot x - x \cdot a$. Since, as is easily verified, $\delta^{n+1}\delta^n = 0$; $n \ge 0$, the sequence

$$0 \longrightarrow C^0(A, X) \xrightarrow{\delta^0} C^1(A, X) \xrightarrow{\delta^1} \cdots \qquad (\mathrm{St}_A(X))$$

is a complex of Banach spaces; it is called the *standard cohomology complex* or the *Hochschild-Kamowitz complex* for A and X. The spaces $\operatorname{Ker} \delta^n$ and $\operatorname{Im} \delta^{n-1}$ are denoted by $Z^n(A, X)$ and $B^n(A, X)$ and their elements are called *n-dimensional cocycles* and *n-dimensional coboundaries* respectively.

Definition (VII.3.33) (Kamowitz [149], Guichardet [118]). The n-dimensional cohomology of the complex $\mathrm{St}_A(X)$ (= the seminormed space $Z^n(A, X)/B^n(A, X)$) is denoted by $\mathcal{H}^n(A, X)$ and is called the *n-dimensional cohomology group of the Banach algebra A with coefficients in the Banach A-bimodule X.*[21]

---

[21] For historical reasons we say cohomology "group" and not "space", which would of course be more correct.

We return to the material of §2 and see that the spaces $\mathcal{H}^1(A, X)$ (of derivations modulo the inner derivations), $\mathcal{H}^2(A, X)$ (of classes of singular extensions)[22] and $\mathcal{H}^0(A, X)$ (= $Z^0(A, X)$, the centre of the bimodule) which were introduced there independently of each other, are now unified. See [276], [192], [222], [283] for some other questions which involve cohomology groups of small dimensions $n \leq 3$. For larger dimensions, as far as we know, no one has yet found reasonably clear interpretations of these spaces. However, the groups of different dimensions are interconnected and a knowledge of some is useful in studying others [134, p.90], [176].

The promised connection with the concepts and results of §1 is provided by the following.

Theorem (VII.3.34). *For A and X as above and any $n = 0, 1, \cdots$, we have, up to a topological isomorphism, $\mathcal{H}^n(A, X) = \mathrm{Ext}^n_{A^{\mathrm{env}}}(A_+, X)$, where A_+ and X in the right hand side are considered as left A^{env}-modules.*

▶ By Theorem 3.11 the Ext in the right hand side of the equation can be calculated with the help of the standard bimodule resolution $\mathbb{B}(A_+)$ (see §3.1), that is, as the cohomology of the complex ${}_A\mathbf{h}_A(\mathbb{B}(A_+), X)$. But this complex, by Theorem VI.1.22 (with A^{env} in the role of A and $\underbrace{A \hat{\otimes} \cdots \hat{\otimes} A}_{n}$ in the role of E) and by the universality property of the n-tuple tensor product, is isomorphic to a complex with terms $C^n(A, X)$ that it is easy to verify will coincide with $\mathrm{St}_A(X)$. The rest is clear. ■

Exercise (VII.3.35). Let A and X be unital. Show that in this case we have, besides Theorem 3.34, the analogous result $\mathcal{H}^n(A, X) = \mathrm{Ext}^n_{A^{\mathrm{env}}_{\mathrm{un}}}(A, X)$.

Exercise (VII.3.36). Introduce the functor $\mathcal{H}^n(A, ?)$ by giving a precise meaning to this notation and amplify Theorem 3.34 by an assertion about the natural equivalence of this functor and $\mathrm{Ext}^n_{A^{\mathrm{env}}}(A_+, ?)$.

An axiomatic description of the cohomology group functor is given, together with some applications, by Craw [172].

So cohomology groups are one of the specializations of Ext. One of the numerous applications of this fact is that it enables one to look afresh at certain classes of algebras introduced in §1.4.

---

[22] And also $\mathcal{H}^3(A, X)$ which is related to perturbations (see the bracketed section of §2.2).

Theorem (VII.3.37). (I) *The algebra* A *is contractible* $\Leftrightarrow$ $\mathcal{H}^1(A, X) = 0$ *for all* $X \in A$**-mod-**A $\Leftrightarrow$ $\mathcal{H}^n(A, X) = 0$ *for all* $X \in A$**-mod-**A *and* $n = 1, 2, \cdots$

(II) *The algebra*[23] A *is amenable* $\Leftrightarrow$ $\mathcal{H}^1(A, X^*) = 0$ *for all* $X \in A$**-mod-**A $\Leftrightarrow$ $\mathcal{H}^n(A, X^*) = 0$ *for all* $X \in A$**-mod-**A *and* $n = 1, 2, \cdots$

▶ If we combine 3.34 and 3.25(I) (3.25(III), respectively), we get that both cohomology conditions in the first (second) assertion are equivalent to the projectivity (flatness) of A_+ as a left A^{env}-module and hence as an A-bimodule. The rest is clear. ■

When applied to the only specific class of contractible Banach algebras we know (see Theorem 1.74 and its discussion) this implies the following corollary.

Corollary (VII.3.38). *Let* A *be the Cartesian product of a finite number of complete matrix algebras. Then* $\mathcal{H}^n(A, X) = 0$ *for all* $X \in A$**-mod-**A *and* $n \geq 1$. *In particular* (see §2.1 and Corollary 2.22) *every derivation of* A *with values in any* X *is an inner derivation and every singular extension splits.*

Remark (VII.3.39). The last result enables us, taking into account what was said in Remark 2.28, to get Wedderburn's theorem as a corollary of the well-known algebraic fact[24] that every finite-dimensional semisimple algebra is a Cartesian product of a finite number of complete matrix algebras.

Regarding applications of the second part of Theorem 3.37, we get the triviality of cohomology groups of positive dimension with coefficients in dual modules for many specific algebras, which were noted earlier as being amenable — $C_0(\Omega)$, $\mathcal{K}(H)$, $L^1(G)$ for amenable $G, \cdots$, together with the standard consequences of this fact about derivations and extensions. In addition, using the isomorphism $\mathcal{B}(H) \simeq \mathcal{N}(H)^*$ (cf. 0.3.45), 3.37(II) implies that, for all $n > 0$, $\mathcal{H}^n(A, \mathcal{B}(H)) = 0$ for any amenable Banach operator algebra on a Hilbert space H. Finally we make an assertion that requires additional arguments.

Assertion (VII.3.40) (Johnson [30]). *Let* A *be a commutative Banach algebra and let* X *be a Banach* A*-bimodule coinciding with its centre. Then* $\mathcal{H}^n(A, X) = 0$ *for* $n = 1$ *and* 2.

It is not known whether the last equation remains true for $n > 2$.

---

[23] This definition of amenability is due to Johnson [30].

[24] See also the bracketed section at the end of Chap.1 §3.

If A is a von Neumann algebra, then in this case there is an interesting special variant of the concept of amenability whose definition is in the spirit of Theorem 3.37(II). Namely, such an A is said to be *ultraweakly amenable* or *amenable as a von Neumann algebra* if every derivation $D: A \longrightarrow X^*$; $X \in A\text{-}\mathbf{mod}\text{-}A$, that is continuous with respect to the ultraweak topology in A and the weak* topology in X^*, is inner. A number of equivalent characterizations have been obtained for these algebras ([181], [162], [298], [306]), of which the most important appears to be "Connes injectivity" [161].[25] See [176], [138] for something about the so-called normal cohomology groups — special versions of the spaces $\mathcal{H}^n(A, \cdot)$ for operator algebras — and for applications of these.

... In one important particular case the cohomology groups can be expressed in terms of Ext without going over to the enveloping algebra.

Theorem (VII.3.41). *Let A be a Banach algebra, $X, Y \in A\text{-}\mathbf{mod}$. Then to within a topological isomorphism, $\mathcal{H}^n(A, \mathcal{B}(X, Y)) = \operatorname{Ext}^n_A(X, Y)$, where $\mathcal{B}(X, Y)$ is considered as an A-bimodule* (see VI.3.5).

▶ There is an obvious way of using the freezing isomorphism to see that the standard cohomology complex for $\mathcal{B}(X, Y)$ coincides, to within an isomorphism in $\underline{\mathcal{B}an}$, with the complex $\mathcal{B}(A, X, Y)$ in Proposition 3.19. The rest is clear. ■

Here is a typical application. Let A be a Banach operator algebra on the Banach space E where the latter is considered as the natural A-module and $\mathcal{B}(E) = \mathcal{B}(E, E)$ as an A-bimodule.

Proposition (VII.3.42). *Let either* (I) *E be projective or* (II) *E be flat and its underlying space be reflexive. Then $\mathcal{H}^n(A, \mathcal{B}(E)) = 0$ for all $n > 0$.*

▶ In case (I) the assertion is implied by Theorems 3.41 and 3.25(I). In case (II), taking into account that the modules E and E^{**} coincide, the assertion is implied by 3.41 and 3.25(III). ■

If we combine 3.42(I) and Theorem 1.27 we get a result that strengthens in an essential way the "one-dimensional" Corollary 2.41 proved earlier.

---

[25] It has recently been shown that a von Neumann algebra A is ultraweakly amenable $\iff$ its "pre-dual" A-bimodule A_* is injective. See [345].

Corollary (VII.3.43) (Kaliman, Selivanov). *Let A contain all finite-dimensional operators. Then $\mathcal{H}^n(A, \mathcal{B}(E)) = 0$ for all $n > 0$.*

If we combine 3.42(II) and Assertion 1.55 we get an assertion whose original proof is connected with the names of Christensen [167], Lance [177] and Nielsen [202].

Assertion (VII.3.44). *Let A be a nest algebra of operators on the Hilbert space H. Then $\mathcal{H}^n(A, \mathcal{B}(H)) = 0$ for all $n > 0$.*

* * *

Besides the cohomology groups of Banach algebras there exists another traditional invariant: the "homology groups" defined by using the so-called standard homology complex (Guichardet [118]). The analogue of Theorem 3.34 for these groups is a theorem expressing them as $\mathrm{Tor}_n^{A^{env}}(X, A_+)$, where X is a coefficient bimodule. For some details about homology groups see [81].

§3.4. Homological dimension. A bound for the homological dimension of biprojective algebras

Definition (VII.3.45). Let X be a left Banach module over the Banach algebra A. The length (cf. §3.1) of its shortest projective resolution is called its *homological dimension* and it is denoted by $\mathrm{dh}_A X$.

In more traditional terms, $\mathrm{dh}_A X$ is obviously the least n for which X can be represented as $P_0/(P_1/\cdots/(P_{n-1}/P_n)\cdots)$, where all the modules are projective and every P_k is complemented as a subspace of P_{k-1}. One can say that $\mathrm{dh}_A X$ measures how far our module is "homologically worse" than being projective. In particular $\mathrm{dh}_A X = 0$ means precisely that X is projective — excluding the case $X = 0$, when we take $\mathrm{dh}_A X = -1$. Thus $\mathrm{dh}_A X$ is an integer ≥ -1 or it is ∞.

(The root of the concept of homological dimension is "Hilbert's theorem on syzygies" [117], [51] in which, talking in modern terms, it was shown that every (abstract) module over the algebra of polynomials in n variables has a projective and even a free resolution of length at most n.)

Proposition (VII.3.46). *Let $\mathrm{dh}_A X < k$. Then for any additive functor $F : A\text{-}\mathbf{mod} \longrightarrow \mathcal{B}an$ we have $F_k(X) = 0$ (or according to the sense, $F^k(X) = 0$).*

▶ This follows from Theorem 3.11. ■

Theorem (VII.3.47). *The following properties of the A-module X are equivalent:*

(I) $\mathrm{dh}_A X \leq n$;

(II) $\mathrm{Ext}_A^{n+1}(X, Y) = 0$ *for any A-module Y;*

(III) *if* $0 \leftarrow X \leftarrow P_0 \leftarrow \cdots \leftarrow P_{n-1} \leftarrow Q \leftarrow 0$ $(\mathcal{P})$ *is a resolution of the module X in which $P_0, \cdots, P_{n-1}$ are projective, then Q is also projective.*

▶ (I) ⇒ (II), (III) ⇒ (I). This is obvious.

(II) ⇒ (III). Let $0 \leftarrow K_{m-1} \leftarrow P_m \leftarrow K_m \leftarrow 0$; $m = 0, \cdots, n-1$, where $K_{-1} := X$ and $K_{n-1} := Q$, be the component (necessarily admissible) complexes of the resolution $(\mathcal{P})$. We consider the long exact sequence for $\mathrm{Ext}(\cdot, Y)$; $Y \in A$-**mod**, provided by Theorem 3.23. Since $\mathrm{Ext}_A^k(P_m, Y) = 0$, $k > 0$ (see 3.25(I)) we get a chain of isomorphisms

$$\mathrm{Ext}_A^1(Q, Y) \simeq \mathrm{Ext}_A^2(K_{n-2}, Y) \simeq \cdots \simeq \mathrm{Ext}_A^n(K_0, Y) \simeq \mathrm{Ext}_A^{n+1}(X, Y) = 0.$$

All that remains, since Y is arbitrary, is to reapply Theorem 3.25(I). ■

If I is a projective left ideal in the left projective algebra A and I is complemented in A as a subspace, but not as a submodule, then obviously $\mathrm{dh}_A A/I = 1$; in particular (see 1.23) the homological dimension of any cyclic module over $C[0, 1]$ is equal to 1.

Exercise (VII.3.48). Show that the homological dimension of the module $\mathbb{C}$ over $A := \mathcal{A}(\overline{\mathbb{D}}^2)$ with the operation $w \cdot z := w(0)z$ is equal to 2.

Hint. Consider the resolution $0 \leftarrow \mathbb{C} \leftarrow A \xleftarrow{d_0} A \oplus A \xleftarrow{d_1} A \leftarrow 0$ with $d_0(w_1, w_2) := z_1 w_1 - z_2 w_2$, $d_1(w) := (z_2 w, z_1 w)$ and ascertain with its help that $\mathrm{Ext}_A^2(\mathbb{C}, \mathbb{C}) \simeq \mathbb{C} \neq 0$.

Exercise (VII.3.49). Let A be commutative and $\mathrm{dh}_A \mathbb{C}(s) \leq n$ for all $s \in \underline{\Omega}(A)$. Show that for any A-module X of finite linear dimension $\mathrm{dh}_A X \leq n$.

Hint. Carry out induction with respect to $\dim X$ using the fact that a family of commuting operators on a finite-dimensional space has a common eigenvector.

Exercise (VII.3.50). Let A be a (linearly) one-dimensional algebra with zero multiplication. Show that $\mathrm{dh}_A A = \infty$.

Hint. Construct a resolution of the form $0 \leftarrow A \leftarrow A_+ \leftarrow \leftarrow A_+ \leftarrow A_+ \leftarrow \cdots$.

The definition of homological dimension and Theorem 3.47 automatically carry across from left modules to the remaining types of (bi)modules.

What are the homological dimensions of the specific modules over specific algebras that we have already met? Best understood are the ideals of function algebras. If $A := C(\Omega)$, where Ω is a compact space, then what was said in Assertion 1.24 means that if Ω is metrizable then $\mathrm{dh}_A I_s = 0$ for all $s \in \Omega$; but, if Ω is more complicated, we meet $s \in \Omega$ with $\mathrm{dh}_A I_s > 0$. Moran [329] has described a space Ω and $s \in \Omega$ with $\mathrm{dh}_A I_s = \infty$ and comparatively recently Krichevets [171] constructed for any $n \in \mathbb{N}$ a compact space Ω such that $\mathrm{dh}_A I_s$; $s \in \Omega$, runs over all values from 0 to n. However, for $c_b \simeq C(\beta\mathbb{N})$ (see Chap. IV §3.1) it is not known what the value of $\mathrm{dh}_A I_s$ is when s is a point in the "Čech growth" $\beta\mathbb{N}\backslash\mathbb{N}$.

For the "multi-dimensional Varopoulos algebra" $A := C(\Omega_1) \hat{\otimes} \cdots \hat{\otimes} C(\Omega_n)$, where the Ω_k; $1 \le k \le n$, are metrizable connected compact spaces, the homological dimension of every maximal ideal is equal to $n-1$.

There are a number of "homologically bad" algebras which are Banach algebras of smooth functions:[26] for example, the homological dimension of any closed ideal in $C^1[a, b]$ is infinite.

For $A := \mathcal{A}(\overline{\mathbb{D}}^n)$ the number $\mathrm{dh}_A I_s$; $s \in \overline{\mathbb{D}}^n$, runs over all values from 0 to $n-1$ depending on the part of the boundary to which s belongs [81, p.289]; in particular, it is 0 for $s \in \mathbb{T} \times \cdots \times \mathbb{T}$. If $s \in \mathbb{D}^n$, then $\mathrm{dh}_A I_s = n-1$, that is, it is equal to $\dim(I_s/I_s^2) - 1$. We note especially a result which shows that the "polydisc algebra" is more than simply an example of this situation, for it connects the question of calculating $\mathrm{dh}_A I_s$ with an old question of whether there are subsets with so-called analytic structure in the spectrum of a commutative Banach algebra (cf. Gleason [119], Read [218]).

Assertion (VII.3.51) (Pugach [212]). *Let A be a commutative Banach algebra, let s be a point of its spectrum and let* $\mathrm{dh}_A I_s = n-1$. *Then* (the linear dimension) $\dim I_s/I_s^2 \le n$ *and moreover when equality holds* (the "non-degenerate" case) s *has a neighbourhood* Δ *in the Gel'fand topology that is an n-dimensional analytic polydisc* (the latter means that there is a homeomorphism $\omega : \mathbb{D}^n \longrightarrow \Delta$ such that for all $a \in A$ the function $\mathbb{D}^n \longrightarrow \mathbb{C} : z \longmapsto \hat{a}(\omega(z))$ is holomorphic).

Considering the augmentation ideal I in $A := L^1(G)$, if G is compact, then, by Exercise 1.28, $\mathrm{dh}_A I = 0$. The case of a non-compact group is a little more complicated; Sheinberg [286] has managed to show that for a wide class of such groups, including the amenable groups, $\mathrm{dh}_A I = \infty$. In general, for any maximal ideal I in the L^1-algebra of a commutative non-compact

[26] But not Fréchet algebras of infinitely smooth functions; see Assertion 3.59 later.

group or (equivalently) the Fourier algebra of a commutative non-discrete group, for example, the Wiener algebra, $\mathrm{dh}_A I = \infty$.

* * *

We return to the general theory. Now, starting from modules, we introduce a few numerical invariants of the algebras themselves.

Definition (VII.3.52). The *left projective global homological dimension* of the algebra A (in future simply the *global dimension*) is the least upper bound of $\mathrm{dh}_A X$ taken over all $X \in A$-**mod**. The *homological bidimension* (in future simply the *bidimension*)[27] of the algebra A is the homological dimension of A_+ as an A-bimodule, that is, $\mathrm{dh}_{A^{\mathrm{env}}} A_+$.

The global dimension is denoted by $\mathrm{dg}\, A$, the bidimension by $\mathrm{db}\, A$. Obviously $\mathrm{dg}\, A = 0 \iff$ all left A-modules are projective and $\mathrm{db}\, A = 0 \iff A$ is contractible. In the unital case there are no new versions of the definitions:

Exercise (VII.3.53). Show that for unital A (I) $\mathrm{dg}\, A$ is also $\sup\{\mathrm{dh}_A X : X \in A\text{-}\mathbf{unmod}\}$ and (II) $\mathrm{db}\, A = \mathrm{dh}_{A^{\mathrm{env}}} A = \mathrm{dh}_{A^{\mathrm{env}}_{\mathrm{un}}} A$.

Theorem (VII.3.54). $\mathrm{dg}\, A \le \mathrm{db}\, A$ *is satisfied for any* A.

▶ Let $\mathrm{db}\, A \le n$ and let X be an arbitrary left A-module. We select from the resolution $\mathbb{B}(A_+)$ the subcomplex

$$0 \longleftarrow A_+ \xleftarrow{\pi} A_+ \hat{\otimes} A_+ \longleftarrow \cdots \longleftarrow A_+ \hat{\otimes} \underbrace{A \hat{\otimes} \cdots \hat{\otimes} A}_{n-1} \hat{\otimes} A_+ \xleftarrow{\mathrm{in}} Q \longleftarrow 0, \qquad (\mathcal{P})$$

where $Q := \mathrm{Ker}\, d_{n-2}$; then, by Theorem 3.47(III), $\mathcal{P}$ is a projective resolution of A_+ in A-**mod**-$A = A^{\mathrm{env}}$-**mod**. We now note that the contracting homotopy for $\mathbb{B}(\cdot)$ indicated in the proof of Proposition 3.5, consists, in the case $\mathbb{B}(A_+)$, of morphisms of right A-modules. Hence the complex $\mathbb{B}(A_+)$, and with it also $\mathcal{P}$, splits in **mod**-A; therefore the functor $?\hat{\otimes}_A X : A\text{-}\mathbf{mod}\text{-}A \longrightarrow A\text{-}\mathbf{mod}$ takes $\mathcal{P}$ into the admissible complex (= splits in $\mathcal{B}an$)

$$0 \longleftarrow A_+ \hat{\otimes}_A X \xleftarrow{\pi\otimes 1} \cdots \longleftarrow (A_+ \hat{\otimes} A \hat{\otimes} \cdots \hat{\otimes} A \hat{\otimes} A_+) \hat{\otimes}_A X \xleftarrow{\mathrm{in}\otimes 1} Q \hat{\otimes}_A X \longleftarrow 0.$$

But the A-module $A_+ \hat{\otimes}_A X$ is isomorphic to X (VI.3.25) and the other terms of the complex, by Proposition 1.57(IV), are

[27] Sometimes: the cohomological dimension (see later 3.55(II)).

projective. In this way we have constructed a projective resolution of an arbitrary $X \in A$-**mod** of length $\le n$. ∎

Both dimensions can be expressed in the language of Ext and cohomology groups:

Theorem (VII.3.55). (I) $\operatorname{dg} A = \min\{n : \operatorname{Ext}_A^{n+1}(X, Y) = 0$ *for all* $X, Y \in A$-**mod**$\} = \min\{n ; \mathcal{H}^{n+1}(A, \mathcal{B}(X, Y)) = 0$ *for all* $X, Y \in A$-**mod**$\}$.

(II) $\operatorname{db} A = \min\{n : \operatorname{Ext}^{n+1}_{A^{\mathrm{env}}}(A_+, X) = 0$ *for all* $X \in A$-**mod**-$A\} = \min\{n : \mathcal{H}^{n+1}(A, X) = 0$ *for all* $X \in A$-**mod**-$A\}$.

▶ (I) is implied by Theorems 3.47(II) and 3.41, (II) by Theorems 3.47(II) and 3.34. ∎

Remark (VII.3.56). Of course 3.55 immediately implies that $\operatorname{dg} A \le \operatorname{db} A$. However, the longer argument in 3.54 has the advantage that it can be carried over to general categories of polynormed modules.

Let us pass on to results about homological dimensions in various classes of algebras. The next theorem was in its time the reason for biprojective algebras being singled out.

Theorem (VII.3.57). *Let* A *be a biprojective Banach algebra. Then* $\operatorname{db} A \le 2$ (that is, $\mathcal{H}^3(A, X) = 0$ *for all* $X \in A$-**mod**-A) *and as a consequence,* $\operatorname{dg} A \le 2$.

▶ We consider in A-**mod**-A the complexes

$$0 \longleftarrow A_+/A \xleftarrow{\mathbf{pr}} A_+ \xleftarrow{\mathbf{in}} A \longleftarrow 0,$$

$$0 \longleftarrow (A_+ \hat{\otimes} A_+)/J \xleftarrow{\mathbf{pr}} (A_+ \hat{\otimes} A_+) \xleftarrow{\mathbf{in}} J \longleftarrow 0$$

$$0 \longleftarrow J \xleftarrow{\zeta} P \xleftarrow{\Delta} A \hat{\otimes} A \longleftarrow 0$$

where

$$J := \ell.h.(A \hat{\otimes} A_+, A_+ \hat{\otimes} A) \subset A_+ \hat{\otimes} A_+, \quad P := (A \hat{\otimes} A_+) \oplus (A_+ \hat{\otimes} A),$$

Δ is the "diagonal embedding" $u \longmapsto (u, u)$ and ζ takes (u_1, u_2) into $u_1 - u_2$. All the complexes are obviously admissible.

Given an A-bimodule X we take the long exact sequences for $\operatorname{Ext}_{A^{\mathrm{env}}}(\cdot, X)$ corresponding to these complexes (the subscript will be omitted in future) and we select from these sections

$$\cdots \longrightarrow \operatorname{Ext}^2(A, X) \longrightarrow \operatorname{Ext}^3(A_+/A, X) \longrightarrow \operatorname{Ext}^3(A_+, X) \longrightarrow \operatorname{Ext}^3(A, X) \longrightarrow \cdots;$$

$$\cdots \longrightarrow \operatorname{Ext}^2(A_+ \hat{\otimes} A_+, X) \longrightarrow \operatorname{Ext}^2(J, X) \longrightarrow \operatorname{Ext}^3((A_+ \hat{\otimes} A_+)/J, X) \longrightarrow$$
$$\longrightarrow \operatorname{Ext}^3(A_+ \hat{\otimes} A_+, X) \longrightarrow \cdots;$$

$$\cdots \longrightarrow \operatorname{Ext}^1(A \hat{\otimes} A, X) \longrightarrow \operatorname{Ext}^2(J, X) \longrightarrow \operatorname{Ext}^2(P, X) \longrightarrow \operatorname{Ext}^2(A \hat{\otimes} A, X) \longrightarrow \cdots$$

Since A, $A_+ \hat{\otimes} A_+$ and $A \hat{\otimes} A$ are projective A-bimodules, all the "edges" of these sections are zero. Further A_+/A and $(A_+ \hat{\otimes} A_+)/J$ are obviously the same — one-dimensional annihilator — A-bimodule. Therefore $\mathrm{Ext}^3(A_+, X) = \mathrm{Ext}^3(A_+/A, X) = \mathrm{Ext}^3((A_+ \hat{\otimes} A_+)/J, X) = \mathrm{Ext}^2(J, X) = \mathrm{Ext}^2(P, X)$. But as A is biprojective, the bimodules $A \hat{\otimes} A_+$ and $A_+ \hat{\otimes} A$ and with them their direct sum P, are projective; this means that $\mathrm{Ext}^2(P, X) = 0$. The rest is clear. ■

Lower bounds for dg and db are, as a rule, more difficult to obtain. The proof of the next fact, about the homological invariants of Banach algebras being non-trivial absorbs quite a lot of different bits of information from functional analysis including for example Theorem 0.1.16 and Assertion 0.3.33.

Assertion (VII.3.58) [270]. *Let* A *be a commutative Banach algebra with infinite spectrum. Then* $\mathrm{dg}\, A > 1$. *As a consequence there exist* $X \in A$**-mod-**A *with* $\mathcal{H}^2(A, X) \neq 0$ *and* (what is equivalent) *a singular extension of the algebra* A *that does not split.*

As a result of combining 3.57 and 3.58, both of the dimensions for biprojective infinite-dimensional Banach function algebras are precisely equal to two; in particular $\mathrm{dg}\, A = \mathrm{db}\, A = 2$ for $A = c_0$ and l_1. One can succeed in demonstrating the same equality for certain non-commutative A also, including $L^1(G)$ and $C^*(G)$ for compact G and also $\mathcal{N}(H)$. If we look at specific modules, we note that $\mathrm{dh}_{c_0} c_b = \mathrm{dh}_{l_1} c_b = 2$ [81, Chap. V §2.2]; this has a "non-commutative analogue" $\mathrm{dh}_{\mathcal{K}(H)} \mathcal{B}(H) = \mathrm{dh}_{\mathcal{N}(H)} \mathcal{B}(H) = 2$.

The proof of the bound $\mathrm{dh}_{c_0} c_b > 1$, which is an important step on the way to the whole of Assertion 3.58, uses in an essential way the fact that c_0 is not complemented in c_b. In certain cases $\mathrm{dh}_A X$ is completely determined by such considerations of Banach-geometry. If A is a biprojective function algebra with a b.a.i., then $\mathrm{dh}_A X = 2$ for $X \in A$**-mod** $\Longleftrightarrow$ the essential part X_{es} has a Banach complement in X (Krichevets [169]).

We now pose, with brief commentaries, a few open questions about the homological dimensions of Banach algebras.

First we consider the class of Banach function algebras. Assertion 3.58 implies that there exist values that are "prohibited" as their dimensions; one such, at least, is 1. On the other hand, it is known that their dimensions may take all even values [233], [170].

Question. In general what is the set of values taken by the homological dimension of Banach function algebras? In particular are there such algebras with $\mathrm{dg}\, A = 3$ and/or $\mathrm{db}\, A = 3$?

The bound $\mathrm{dg}\, A > 1$ (cf. 3.58) has by now been established for various classes of algebras; in particular it holds for all infinite-dimensional *CCR*-algebras (Lykova [194]).

Question. Is it true that $\mathrm{dg}\, A > 1$ or at least $\mathrm{db}\, A > 1$ for all infinite-dimensional semisimple Banach algebras? At least, is this true for infinite-dimensional C^*-algebras?

Finally, our last question, often repeated in the literature [133], [142], [136], [81], concerns two specific algebras and appears, in view of its apparent simplicity, especially "provocative":

Question. What are $\mathrm{dg}\, A$ and $\mathrm{db}\, A$ for $A = C[a, b]$? $A = \mathcal{K}(H)$?

Here the difficulty is that, although these algebras are simply constructed, their tensor powers are by no means simple and these are, evidently, what we have to work with.

... In conclusion — a few words about the dimensions of non-Banach polynormed algebras. Their study was begun by Taylor [251], [252] who considered, among others, several algebras of test functions and generalized functions on domains of $\mathbb{R}^n$ or $\mathbb{C}^n$ including $\mathcal{O}(U)$; $U \subseteq \mathbb{C}^n$, $C^\infty(U)$; $U \subseteq \mathbb{R}^n$ and $\mathcal{E}^*(\mathbb{R}^n)$. Using the Koszul resolution (cf. 3.9), he showed that the homological dimensions of these algebras coincide with the linear dimensions — real or complex, according the sense — of their domain of definition. (Thus, for these algebras dg and db have an intuitive geometric sense and there are no "prohibited values" for them as there are for Banach function algebras.)

As a consequence an analogous result was obtained for certain other traditional algebras of test functions, some of them non-unital [204]. However, until recently it was not clear how things stood with algebras of test functions on manifolds which have no single system of coordinates (such as a two-dimensional sphere); these algebras, as it is not difficult to see, do not have Koszul resolutions. The following was unexpected — at least for the author of this book.

Assertion (VII.3.59) (Ogneva [203]). *Let $\mathcal{M}$ be an arbitrary infinitely smooth manifold. Then the global dimension and bidimension of the Fréchet algebra $C^\infty(\mathcal{M})$ are equal to the topological dimension of $\mathcal{M}$.*

Having formulated this result (based on Assertion 1.56, as we noted before) we take our leave of the reader.

Bibliography

Textbooks, monographs, collections of works

1. *Algebras in analysis*, ed. J.H. Williamson, *Academic Press*, London, (1975).
2. *C^*-algebras and applications to physics*, eds. H. Araki and R.V. Kadison, Springer Lecture Notes in Math., **650**, *Springer*, Berlin, (1978).
3. *Algèbres d'operateurs et leurs applications en physique mathématique*, Mathématique Coll. Int. CNRS, **274**, *Edition du CNRS*, Paris, (1979).
4. M. Atiyah, *Lectures in K-theory, Benjamin*, New York, (1967).
5. *Banach algebras and several complex variables*, eds. F. Greenleaf and D. Gulick, Contemp. Math., **32**, *Amer. Math. Soc.*, Providence, R.I., (1984).
6. H. Bass, *Algebraic K-theory, Benjamin*, New York, (1968).
7. E. Beckenstein, L. Narici and C. Suffel, *Topological algebras, North Holland*, Amsterdam, (1977).
8. B. Blackadar, *K-theory for operator algebras, Springer*, New York, (1986).
9. F.F. Bonsall and J. Duncan, *Complete normed algebras*, Ergebnisse der Mathematik und ihrer Grenzgebiete, **80**, *Springer*, New York, (1973).
10. O. Bratteli and D.W. Robinson, *Operator algebras and quantum statistical mechanics*, I, *Springer*, New York, (1979).
11. I. Bucur and A. Deleanu, *Introduction to the theory of categories and functors*, Pure and Applied Maths. Vol.XIX, *Interscience Publications*, London-New York-Sydney, (1968).
12. N. Bourbaki, *Eléments de mathématique*, Livre II, Algèbre, *Hermann*, Paris, (1964).
13. -----, *Eléments de mathématique*, Livre VI, Intégration, *Hermann*, Paris, (1969).
14. -----, *Eléments de mathématique*, Fascicule XXXII, Théories spectrales, Actualités Sci. Ind., **1332**, *Hermann*, Paris, (1967).
15. L. Waelbroeck, *Topological vector spaces and algebras*, Springer Lecture Notes in Math., **230**, *Springer*, Berlin, (1971).
16. F.H. Vasilescu, *Analytic functional calculus and spectral decompositions, Reidel*, Dordrecht, (1982).

17. T.W. Gamelin, *Uniform algebras, Prentice Hall*, Englewood Cliffs, N.J., (1969).

18. R.C. Gunning and H. Rossi, *Analytic functions of several complex variables, Prentice Hall*, Englewood Cliffs, N.J.,(1965).

19. I.M. Gel'fand, *Lectures in linear algebra*, Nauka, Moscow, (1971).
English translation of 2nd Russian edition: *Interscience*, New York, (1961).

20. -----, D.A. Raikov and G.E. Shilov, *Commutative normed rings, Fizmatgiz*, Moscow, (1960).
English translation: *Chelsea*, New York, (1964).

21. A. Guichardet, *Cohomologie des groupes topologiques et des algèbres de Lie*, Textes Mathématiques **2**, *CEDIC*, Paris, (1980).

22. R. Godement, *Topologie algébraique et théorie des faisceaux*, Actualités Sci. Ind., **1252**, Math. Univ. Strasbourg XIII, *Hermann*, Paris, (1958).

23. K. Hoffman, *Banach spaces of analytic functions, Prentice Hall*, Englewood Cliffs, N.J., (1962).

24. I.C. Gohberg and M.G. Krein, *Introduction to the theory of non-self-adjoint linear operators in Hilbert space, Nauka*, Moscow, (1965). Translated from Russian into English, Translation of Mathematical Monographs, **18**, *Amer. Math. Soc.*, Providence, R.I., (1969).

25. F.P. Greenleaf, *Invariant means on topological groups and their applications, Van Nostrand*, New York, (1969).

26. A. Grothendieck, *Produits tensoriels topologiques et espaces nucléaires*, Mem. Amer. Math. Soc., **16**, **(1955)**.

27. N. Dunford and J.T. Schwartz, *Linear operators*, I: General theory, *Interscience*, New York, **(1958)**; II: Spectral theory. Self-adjoint operators in Hilbert space, *Interscience*, New York, **(1963)**.

28. J. Dauns and K.H. Hoffman, *Representations of rings by sections*, Mem. Amer. Math. Soc., **83**, **(1968)**.

29. N. Jacobson, *Structure of rings*, Amer. Math. Soc. Colloquium Publications, **37**, *Amer. Math. Soc.*, Providence, R.I., (1956).

30. B.E. Johnson, *Cohomology in Banach algebras*, Mem. Amer. Math. Soc., **127**, (1972).

31. J. Dixmier, *Les C^*-algèbres et leurs représentations, Gauthiers-Villars*, Paris, (1964).

32. R.G. Douglas, *C^*algebra extensions and K-homology, Princeton Univ. Press*, Princeton, (1980).

33. D.P. Zelobenko, *Harmonic analysis on semisimple complex Lie groups, Nauka*, Moscow, (1974).

34. -----, *Compact Lie groups and their representations, Nauka*, Moscow, (1970). Translated from Russian, Translations of Mathematical Monographs, **40**, *Amer. Math. Soc.*, Providence, R.I., (1973).

35. ----- and A.I. Shtern, *Representations of Lie groups, Nauka*, Moscow, (1983).
36. W. Żelazko, *Metric generalizations of Banach algebras, Rozprawy Math.*, **47**, Warsaw, (1965).
37. H. Cartan, *Elementary theory of analytic functions of one and several complex variables, Hermann*, Paris and *Addison-Wesley*, Reading, Mass., (1963).
38. ----- and S. Eilenberg, *Homological algebra, Princeton Univ. Press*, Princeton, (1956).
39. J.P. Kahane, *Séries de Fourier absolument convergentes*, Ergebnisse der Mathematik und ihrer Grenzgebiete, **50**, *Springer-Verlag*, Berlin, (1970).
40. A.A. Kirillov, *Elements in the theory of representations, Nauka*, Moscow, (1972).
Translated into English, Die Grundlehren der Mathematischen Wissenschaften, **220**, *Springer-Verlag*, Berlin, (1976).
41. ----- and A.D. Gvishiani, *Theorems and problems in functional analysis, Nauka*, Moscow, (1979).
English translation: *Springer*, New York, (1982).
42. A.N. Kolmogorov and S.V. Fomin, *Elements in the theory of functions and functional analysis, Nauka*, Moscow, (1972).
English translation of first Russian edition: *Graylock Press*, Albany, N.Y., (1961).
43. A.I. Kostrikin and Yu.I. Manin, *Linear algebra and geometry, Izd. Moscow State Univ.*, Moscow, (1980).
English Translation: *Gordon and Breach*, New York, (1989).
44. R.V. Kadison and J.R. Ringrose, *Fundamentals of the theory of operator algebras*, Vol.I. *Academic Press*, London, (1983); Vol.II, *Academic Press*, London, (1986).
45. C.W. Curtis and I. Reiner, *Representation theory of finite groups and associative algebras, Interscience*, New York, (1962).
46. J. Lambek, *Lectures on rings and modules, Blaisdell*, Waltham, Mass., (1966).
47. S. Lang, *Algebra, Addison-Wesley*, Reading, Mass., (1965).
48. L. Loomis, *Introduction to abstract harmonic analysis, Van Nostrand*, New York, (1953).
49. Yu.I. Lyubich, *Introduction to the theory of Banach representations of groups, Izd. Kharkov Univ.*, Kharkov, (1985).
English translation: *Operator Theory: Advances and Applications,* **30**, *Birkhäuser*, Basel, (1988).
50. E.A. Michael, *Locally multiplicatively-convex topological algebras*, Mem. Amer. Math. Soc., **11**, (1952).
51. S. MacLane, *Homology*, Die Grundlehren der Mathematischen Wissenschaften, **114**, *Springer Verlag*, Berlin, (1963).
52. -----, *Categories for the working mathematician, Springer Verlag*, Berlin, (1971).

53. M.A. Naimark, *Normed rings, Nauka*, Moscow, (1968). English translation: *Normed algebras, Wolters-Noordhoff*, Groningen, (1972).
54. ----- [with A.I. Shtern], *Theory of group representations, Nauka*, Moscow, (1976). English translation: *Springer*, New York, (1982).
55. *Certain questions in approximation theory, IL*, Moscow, (1963).
56. *Operator algebras and their connections with topology and ergodic theory*, eds. H. Araki, C.C. Moore, S. Stratila and D. Voiculescu, Lecture Notes in Math., **1132**, *Springer*, Berlin, (1985).
57. *Operator algebras and applications*, ed. R.V. Kadison, Proc. of Symp. in Pure Math. No.38, *Amer. Math. Soc.*, Providence, R.I., (1982).
58. G.K. Pedersen, *C^*-algebras and their automorphism groups, Academic Press*, London, (1979).
59. A. Pietsch, *Nuclear locally convex spaces*, Ergebnisse der Mathematik und ihrer Grenzgebiete, **66**, *Springer-Verlag*, New York, (1972).
60. L.S. Pontryagin, *Continuous groups, Nauka*, Moscow, (1974). English translation of 2nd Russian edition: *Topological groups, Gordon and Breach*, New York, (1966).
61. J.P. Pier, *Amenable locally compact groups, Wiley Intersci. Publ.*, New York, (1984).
62. *Radical Banach algebras and automatic continuity*, eds. J.M. Bachar, W.G. Bade, P.C. Curtis, Jr., H.G. Dales and M.P. Thomas, Lecture Notes in Math., **975**, *Springer*, Berlin, (1983).
63. M. Reed, B. Simon, *Methods of modern mathematical physics*, Vol.I: Functional analysis, *Academic Press*, London, (1972).
64. C.E. Rickart, *General theory of Banach algebras, Van Nostrand*, New York, (1960).
65. A.P. Robertson and W. Robertson, *Topological vector spaces, Cambridge Univ. Press*, Cambridge, (1964).
66. W. Rudin, *Function theory in polydiscs, Benjamin*, New York, (1969).
67. -----, *Functional analysis, McGraw-Hill*, New York, (1973).
68. D. Ruelle, *Statistical mechanics: Rigorous results, Benjamin*, New York, (1969).
69. S. Sakai, *C^*-algebras and W^*-algebras, Springer*, Berlin, (1971).
70. Z. Semadeni, *Banach spaces of continuous functions, PWN*, Warsaw, (1971).
71. J.-P. Serre, *Linear representations of finite groups, Springer*, New York, (1977).
72. A.M. Sinclair, *Automatic continuity of linear operators, Cambridge Univ. Press*, Cambridge, (1976).
73. S. Stratila, *Modular theory in operator algebras, Edit. Acad. Bucarest*, (1981).

74. ----- and L. Zsido, *Lectures on von Neumann algebras, Edit. Acad. Bucarest*, (1979).
75. M. Takesaki, *Theory of operator algebras*, I, *Springer*, Berlin, (1979).
76. F. Treves, *Topological vector spaces, distributions and kernels, Academic Press*, London, (1967).
77. J.L. Walsh, *Interpolation and approximation by rational functions in the complex domain*, Amer. Math. Soc. Colloquium Publications, Vol.20, *Amer. Math. Soc.*, Providence, R.I., (1960).
78. C. Faith, *Algebra: rings, modules and categories* I, Die Grundlehren der Mathematischen Wissenschaften, **190**, *Springer-Verlag*, New York, (1973); II, **191**, (1976).
79. P.R. Halmos, *A Hilbert space problem book, Van Nostrand*, Princeton, N.J., (1967).
80. F. Hausdorff, *Set theory*, *ONTI*, Moscow, (1937).
English translation of 3rd German edition: *Chelsea*, New York, (1957).
81. A.Ya. Helemskii, *The homology of Banach and topological algebras, Kluwer Academic Press*, Dordrecht, (1989).
82. L. Hörmander, *An introduction to complex analysis in several variables, Van Nostrand*, New York, (1966).
83. I.N. Herstein, *Non-commutative rings*, Mathematical Association of America, *Wiley*, New York, (1968).
84. E. Hille and R.S. Phillips, *Functional analysis and semigroups*, Amer. Math. Soc. Colloq. Publ., **31**, *Amer. Math. Soc.*,Providence, R.I., (1957).
85. S.S. Khoruzhii, *Introduction to algebraic quantum field theory, Nauka*, Moscow, (1986).
86. T. Husain, *Multiplicative functionals on topological algebras*, Research Notes in Mathematics, **85**, *Pitman*, London, (1983).
87. E. Hewitt and K.A. Ross, *Abstract harmonic analysis* I: Structure of topological groups. Integration theory, group representations, Die Grundlehren der Mathematischen Wissenschaften, **115**, *Springer-Verlag*, Berlin, (1963); II, **152**, (1970).
88. J. Cigler, V. Losert and P. Michor, *Banach modules and functors on categories of Banach spaces, Marcel Dekker*, New York, (1979).
89. B.V. Shabat, *Introduction to complex analysis*, 2 vols., *Nauka*, Moscow, (1985).
90. H. Schaefer, *Topological vector spaces, Springer-Verlag*, New York, (1971).
91. G.E. Shilov, *Mathematical analysis. Finite-dimensional linear spaces, Nauka*, Moscow, (1969).
English translation: *Linear algebra, Prentice Hall*, Englewood Cliffs, N.J., (1971).
92. R. Schatten, *A theory of cross-spaces, Princeton Univ. Press*, (1950).

93. R. Edwards, *Fourier series: a modern introduction*, 2 vols, *Holt, Rinehart and Winston*, New York, (1967).
94. -----, *Functional analysis, Holt, Rinehart and Winston*, New York, (1965).
95. R. Engelking, *General topology, PWN*, Warsaw, (1977).
96. K. Jarosz, *Perturbations of Banach algebras*, Lecture Notes in Math., **1120**, *Springer*, Berlin, (1985).

Articles in journals and collections of papers

97. V.V. Anshelevich and M.Sh. Gold'shtein, *Operator algebras in statistical mechanics and the non-commutative probability theory*, Current problems in mathematics: Newest results, **27**, Akad. Nauk SSSR, (1985), 191–228.
98. H. Araki and G.A. Elliott, *On the definition of C^*-algebras*, Publ. Res. Inst. Math. Sci. **9** (1973/74), 93–112.
99. R. Arens, *A generalization of normed rings*, Pacific J. Math. **2** (1952), 455–471.
100. -----, *The group of invertible elements of a commutative Banach algebra*, Studia Math. (Ser. Specjalna) Zeszyt **1** (1963), 21–23.
101. -----, *The adjoint of a bilinear operation*, Proc. Amer. Math. Soc. **2** (1951), 839–848.
102. W.G. Bade and P.C. Curtis, *The Wedderburn decomposition of commutative Banach algebras*, Amer. J. Math. **82** (1960), 851–866.
103. -----, ----- and H.G. Dales, *Amenability and weak amenability for Beurling and Lipschitz algebras*, Proc. London Math. Soc. **55** (1987), 359–377.
104. G.M. Bergman, *A ring primitive on the right but not on the left*, Proc. Amer. Math. Soc. **15** (1964), 473–475.
105. F.F. Bonsall and J. Duncan, *Numerical ranges* II, London Math. Soc. Lecture Note Series, **10**, Cambridge, (1973).
106. J.W. Bunce, *Characterisations of amenable and strongly amenable C^*-algebras*, Pacific J. Math. **43** (1972), 563–572.
107. -----, *Finite operators and amenable C^*-algebras*, Proc. Amer. Math. Soc. **56** (1976), 145–151.
108. L. Waelbroeck, *The holomorphic functional calculus and non-Banach algebras* [1, pp.187–251].
109. N.Th. Varopoulos, *Tensor algebras and harmonic analysis*, Acta Math. **119** (1967), 51–112.
110. J.G. Wendel, *Left centralizers and isomorphisms of group algebras*, Pacific J. Math. **2** (1952), 251–261.
111. J.H. Williamson, *A theorem on algebras of measures on topological groups*, Proc. Edinburgh Math. Soc. **11** (1958/59), 195–206.
112. I.M. Gel'fand, *Normed rings*, Dokl. Akad. Nauk SSSR **23** (1939), 430–432.
113. -----, *Normierte Ringe*, Mat. Sb. **9** (51:1) (1941), 3–23.

114. ----- and M.A. Naimark, *On the embedding of normed rings into the ring of operators in Hilbert space*, Mat. Sb. **12** (54:2) (1943), 197–213.

115. ----- and G. Shilov, *Über verschiedene Methoden der Einführung der Topologie in die Menge der maximalen Ideale eines normierten Ringes*, Mat. Sb. **9** (51:1) (1941), 25–39.

116. F. Gilfeather, A. Hopenwasser and D.R. Larson, *Reflexive algebras with finite width lattices: tensor products, cohomology, compact perturbations*, J. Funct. Anal. **55** (1984), 176–199.

117. D. Hilbert, *Über die Theorie der algebraischen Formen*, Math. Ann. **36** (1890), 473–534.

118. A. Guichardet, *Sur l'homologie et la cohomologie des algèbres de Banach*, C.R. Acad. Sci. Paris, Sér. A, **262** (1966), 38–41.

119. A.M. Gleason, *Finitely generated ideals in Banach algebras*, J. Math. Mech. **13** (1964), 125–132.

120. Yu.O. Golovin, *Homological properties of Hilbert modules over nested operator algebras*, Mat. Zametki **41**:6 (1987), 769–775.
= Math. Notes 41 (1987), 433–437.

121. E.A. Gorin, *Certain characteristic properties of C(X), Theory of functions*, Funct. analysis and its appl. **14** (1971), 186–195.

122. -----, *Spectral stability of certain Banach algebras*, Funct. anal. i ego Pril. **8** (1974), 73–74.
= Funct. analysis and its appl. **8** (1974), 158–159.

123. -----, *Several remarks in connection Gel'fand's theorems on the group of invertible elements of a Banach algebra*, Funct. anal. i ego Pril. **12** (1978), 70–71.
= Funct. Analysis and its appl. **12** (1978), 54–55.

124. -----, *On the research of G.E. Shilov in the theory of commutative Banach algebras and their subsequent development*, Usp. Mat. Nauk **33**:4 (1978), 169–188.
= Russian Math. Surveys, **33**:4 (1978), 193–217.

125. ----- and V.M. Zolotarevskii, *Maximal invariant subalgebras in algebras with involution*, Mat. Sb. **85** (1971), 373–387.

126. ----- and V.Ya. Lin, *On separable polynomials over commutative Banach algebras*, Dokl. Akad. Nauk SSSR **218** (1974), 505–508.
= Soviet Mathematics **15** (1974), 1357–1361.

127. R.I. Grigorchuk, *Symmetric random walks on discrete groups*, Usp. Mat. Nauk **32**:6 (1977), 217–218; Extended presentation, *Multi-component random systems*, *Nauka*, Moscow, (1978).
= Advances in probability and related topics No.6, Dekker, New York, (1980), pp.285–325.

128. J. Dixmier and A. Douady, *Champs continus d'espaces hilbertiens et de C^*-algèbres*, Bull. Soc. Math. Fr. **91** (1963), 227–284.

129. B.E. Johnson, *An introduction to the theory of centralizers*, Proc. London Math. Soc. **14** (1964), 299–320.

130. -----, *The uniqueness of the (complete) norm topology*, Bull. Amer. Math. Soc. **73** (1967), 537–539.
131. -----, *The Wedderburn decomposition of Banach algebras with finite-dimensional radical*, Amer. J. Math. **90** (1968), 866–876.
132. -----, *Continuity of derivations on commutative algebras*, Amer. J. Math. **91** (1969), 1–10.
133. -----, *Approximate diagonals and cohomology of certain annihilator Banach algebras*, Amer. J. Math. **94** (1972), 685–698.
134. -----, *Introduction to cohomology in Banach algebras*, [1, pp.84–100].
135. -----, *Perturbations of Banach algebras*, Proc. London Math. Soc. **34** (1977), 439–458.
136. -----, *Low-dimensional cohomology of Banach algebras*, [57, part 2, pp.253–259].
137. -----, *Continuity of generalized homomorphisms*, Bull. London Math. Soc., **19** (1987), 67–71.
138. -----, R.V. Kadison and J.R. Ringrose, *Cohomology of operator algebras* III, *Reduction to normal cohomology*, Bull. Soc. Math. France **100** (1972), 73–96.
139. ----- and A.M. Sinclair, *Continuity of derivations and a problem of Kaplansky*, Amer. J. Math. **90** (1968), 1067–1073.
140. M.M. Day, *Means on semigroups and groups*, Bull. Amer. Math. Soc. **55** (1949), 1054–1055.
141. H.G. Dales, *The uniqueness of the functional calculus*, Proc. London Math. Soc. **27** (1973), 638–648.
142. -----, *Automatic continuity: a survey*, Bull. London Math. Soc. **10** (1978), 129–183.
143. -----, *A discontinuous homomorphism from* $C(X)$, Amer. J. Math. **101** (1979), 647–734.
144. ----- and J.P. McClure, *Higher point derivations on commutative Banach algebras*, Part I, J. Funct. Anal. **26** (1977), 166–189; Part II, J. London Math. Soc **16** (1977), 313–325.
145. M. Dupré and P. Fillmore, *Triviality theorems for Hilbert modules, Topics in modern operator theory* (5th Int. Conf. on Operator theory), *Birkhäuser*, Basel, (1981), p.71–79.
146. G. Zeller-Meier, *Sur les automorphismes des algèbres de Banach*, C.R. Acad. Sci., Paris, Sér. A, **264** (1967), 1131–1132.
147. Sh.I. Kaliman and Yu.V. Selivanov, *The cohomology of operator algebras*, Vest. Mosk. Gos. Univ. Ser.1 Mat. Mekh. **29**(5) (1974), 24–27.
148. E. Kallin, *A non-local function algebra*, Proc. Nat. Acad. Sci. **49** (1963), 821–824.
149. H. Kamowitz, *Cohomology groups of commutative Banach algebras*, Trans. Amer. Math. Soc. **102** (1962), 352–372.
150. I. Kaplansky, *Maximal fields with valuations*, Duke Math. J. 9 (1942), 303–321

151. -----, *Normed algebras*, Duke Math. J. **16** (1949), 399-418.
152. -----, *Modules over operator algebras*, Amer. J. Math. **75** (1953), 839-858.
153. L. Carleson, *Interpolations by bounded analytic functions and the corona problem*, Ann. Math. **76** (1962), 547-559.
154. G.G. Kasparov, *Hilbert C^*-modules: theorems of Steinspring and Voiculescu*, J. Operator Theory **4** (1980),133-150.
155. -----, *Operator K-theory and its applications*, Current problems in mathematics. Newest results. **27**, Akad. Nauk SSSR, Moscow, (1985), pp.3-32.
156. Y. Katznelson, *Sur les fonctions opérant sur l'algèbre des séries de Fourier absolument convergentes*, C.R. Acad. Sci. Paris **247** (1958), 404-406.
157. S. Kwapien and A. Pelczynski, *The main triangle projection in matrix spaces and its applications*, Studia Math. **34** (1970), 43-68.
158. J.L. Kelley and R.L. Vaught, *The positive cone in Banach algebras*, Trans. Amer. Math. Soc. **74** (1953), 44-55.
159. P.C. Curtis, *Derivations of commutative Banach algebras*, Bull. Amer. Math. Soc. **67** (1961), 271-273.
160. R. Kiehl and J.L. Verdier, *Ein einfacher Beweis des Kohärensatzes von Grauert*, Math. Ann. **195** (1971), 24-50.
161. A. Connes, *Classification of injective factors*, Ann. Math. **104** (1976), 73-115.
162. -----, *On the cohomology of operator algebras*, J. Funct. Anal. **28** (1978), 248-253.
163. J.M. Cohen, *Cogrowth and amenability of discrete groups*, J. Funct. Anal. **48** (1982), 301-309.
164. P.J. Cohen, *On a conjecture of Littlewood and idempotent measures*, Amer. J. Math. **82** (1960), 191-212.
165. J.-L. Koszul, *Homologie et cohomologie des algèbres de Lie*, Bull. Soc. Math. France **78** (1950), 65-127.
166. E. Christensen, *Perturbations of operator algebras*, Part I, Invent. Math. **43** (1977), 1-13; Part II, Indiana Univ. Math. J. **26** (1977), 891-904.
167. -----, *Derivations of nest algebras*, Math. Ann. **229** (1977), 155-161.
168. -----, *Derivations and their relation to perturbations of operator algebras*, [57, part 2, pp.261-274].
169. A.N. Krichevets, *On the connection between the homological properties of some Banach modules and questions of the geometry of Banach spaces*, Vest. Mosk. Univ. Ser. Math. Mekh. 2 (1981), 55-58.
= Moscow Univ. Math. Bull. **36** (1981), 68-72.
170. -----, *Calculation of the global dimension of tensor products of Banach algebras and a generalization of a theorem of Phillips*, Matem. Zametki **31** (1982), 187-202.
= Math. Notes **31** (1982), 95-104.
171. -----, *The homological dimension of* $C(\Omega)$, No.9012 - B.86 Viniti, (1986).

172. I.G. Craw, *Axiomatic cohomology for Banach modules*, Proc. Amer. Math. Soc. **38** (1973), 68–74.

173. R.V. Kadison, *Irreducible operator algebras*, Proc. Nat. Acad. Sci. USA **43** (1957), 273–276.

174. -----, *Derivations of operator algebras*, Ann. Math. **83** (1966), 280–293.

175. ----- and D. Kastler, *Perturbations of von Neumann algebras*. I, Stability of type, Amer. J. Math. **94** (1972), 38–54.

176. ----- and J.R. Ringrose, *Cohomology of operator algebras*, Part I, *Type I von Neumann algebras*, Acta Math **126** (1971), 227–243; Part II, *Extended cobounding and the hyperfinite case*, Ark. Mat. **9** (1971), 55–63.

177. E.C. Lance, *Cohomology and perturbations of nest algebras*, Proc. London Math. Soc. **43** (1981), 334–356.

178. -----, Tensor products and nuclear C^*-algebras [57, part 1, pp.379–400].

179. -----, *Some problems and results on reflexive algebras*, [56, pp.324–330].

180. A.T. Lau, *Extremely amenable algebras*, Pacific J. Math. **33** (1970), 329–336.

181. -----, *Characterisations of amenable Banach algebras*, Proc. Amer. Math. Soc. **70** (1978), 156–160.

182. -----, *Analysis on a class of Banach algebras with applications to harmonic analysis on locally compact groups and semigroups*, Fund. Math. **118** (1983), 161–175.

183. ----- and A.L. Paterson, *The exact cardinality of the set of topological left invariant means on an amenable locally compact group*, Proc. Amer. Math. Soc., **98** (1986), 75–80.

184. G. Lassner, *Topological algebras of operators*, Rep. Math. Phys. **3** (1972), 279–293.

185. -----, Über *Realisierungen gewisser* **-Algebren*, Math. Nachr. **52** (1972), 161–166.

186. -----, *Topological algebras and their applications in quantum statistics*, Wiss. Z. KMU, Leipzig, Math.-Nat. R. **30** (1981), 572–595.

187. R. Levi, *Notes on the Taylor joint spectrum of commuting operators*, Spectral Theory, Banach Center Publ. No.8, PWN, Warsaw, (1982), pp.321–332.

188. -----, *Cohomology invariants of systems of essentially commuting operators*, Funct. Anal. i ego pril. **17** (1983), 79–80.
= Funct. anal. and its applications **17** (1983), 229–230.

189. H. Löwig, Über *die Dimension linearer Räume*, Stud. Math. **5** (1934), 18–23.

190. F. Löffler and V. Timmermann, *The Calkin representation for a certain class of algebras of unbounded operators*, Rev. Roumaine Math. Pures Appl. **31** (1986), 891–903.

191. J. Lindenstrauss and L. Tzafriri, *On the complemented subspaces problem*, Isr. J. Math. 9 (1971), 263–269.

192. G. Lindblad, *Dissipative operators and cohomology of operator algebras*, Lett. Math. Phys. **1** (1976), 219–224.

193. Z.A. Lykova, *Homology characteristics of operator algebras*, Vest. Mos. Gos. Univ. Ser. Mat. Mekh. **1** (1986), 8–13.
= Moscow Univ. Math. Bull. **41** (1986), 10–15.

194. -----, *A lower bound on the global homological dimension of infinite-dimensional CCR-algegras*, Usp. Mat. Nauk **41**:1 (1986), 197–198.
= Russian Math. Surveys **41**:1 (1986), 233–234.

195. -----, *Connection between homology characteristics of a C^*-algebra and its commutative sub-C^*-algebras*, to appear in VINITI No.8477 B, (1985).

196. A. Markoe, *Maximal ideals of Stein algebras*, ed. W. Zame, Conference on Complex Analysis, Buffalo State Univ. New York, 1973.

197. B. Mitiagin, S. Rolewicz and W. Żelazko, *Entire functions in B_0-algebras*, Stud. Math. **21** (1962), 291–306.

198. A.S. Mishchenko, *Banach algebras, pseudo-differential operators and their application to K-theory*, Usp. Mat. Nauk **34**:6 (1979), 67–79.
= Russian Math. Surveys, **34**:6 (1979), 77–91.

199. B. Malgrange, *Ideals of differentiable functions*, Oxford University Press, London, (1967).

200. M.A. Naimark, *On a problem of the theory of rings with involution*, Usp. Mat. Nauk **6** (1951), 160–164.

201. -----, *A continuous analogue of Schur's lemma*, Dokl. Akad. Nauk SSSR **98** (1954), 185–188.
Extended presentation: Izv. Akad. Nauk SSSR Ser. Mat. **20** (1956), 3–16 = Amer. Math. Soc. Translations (2) **9** (1958), 217–231.

202. J.P. Nielsen, *Cohomology of some non-self-adjoint operator algebras*, Math. Scand. **47** (1980), 150–156.

203. O.S. Ogneva, *Equality of the homological dimension of the Fréchet algebra of smooth functions on a manifold with the dimension of the manifold*, Funct. Anal. i ego Pril. **20** (1986), 92–93.
= Funct. anal. and its appl., **20** (1986), 248–250.

204. ----- and A.Ya. Helemskii, *Homological dimensions of certain algebras of fundamental (test) functions*, Mat. Zametki **35** (1984), 177–187.
= Math. Notes **35** (1984), 93–99.

205. B. Aupetit, *Automatic continuity conditions for a linear mapping from a Banach algebra onto a semisimple Banach algebra*, [62, pp.313–316].

206. R.T. Powers, *Self-adjoint algebras of unbounded operators*, Part I, Comm. Math. Phys. **21** (1971), 85–124; Part II, Trans. Amer. Math. Soc. **187** (1974), 261–293.

207. W.L. Paschke, *Inner product modules over B^*-algebras*, Trans. Amer. Math. Soc. **182** (1973), 443–468.

208. A. Pelczynski and Cz. Bessaga, *Some aspects of the present theory of Banach spaces*, Stefan Banach Oeuvres Vol.2, PWN, Warsaw, (1979).

209. G. Pisier, *Contre-exemple à une conjecture de Grothendieck*, C.R. Acad. Sci., Paris, Sér.1, **293** (1981), 681–683.

210. L.I. Pugach, *Projective and flat ideals of function algebras and their connection with analytic structure*, Mat. Zametki **31** (1982), 223–229.
= Math. Notes **31** (1982), 114–117.

211. -----, *On the homological dimension of Banach algebras of smooth functions*, Usp. Mat. Nauk **37**:4 (1982), 175–176.
= Russian Math. Surveys, **37**:4 (1982), 135–136.

212. -----, *Homological properties of function algebras and analytic polydiscs in their spaces of maximal ideals*, Rev. Roumaine Math. pure et appl. **31** (1986), 347-356.

213. M. Putinar, *Functional calculus with sections of an analytic space*, J. Oper. Theory **4** (1980), 297–306.

213a. -----, *The superposition property for Taylor's functional calculus*, J. Oper. Theory **7** (1982), 149–155.

214. -----, *Functional calculus and the Gelfand transformation*, Stud. Math. **79** (1984), 83–86.

215. -----, *On analytic modules: softness and quasi-coherence*, *Complex analysis and applications* (Conf. Proc., Varna, 1985), Bulgarian Acad. Sci., (1986), pp.534–547.

216. G. Racher, *On amenable and compact groups*, Monatshefte für Math. **92** (1981), 305–311.

217. C.J. Read, *A solution to the invariant subspace problem*, Bull. London Math. Soc. **16** (1984), 337–401.

218. T.T. Read, *The powers of a maximal ideal in a Banach algebra and analytic structure*, Trans. Amer. Math. Soc. **161** (1971), 235–248.

219. C.E. Rickart, *An elementary proof of a fundamental theorem in the theory of Banach algebras*, Mich. Math. J. **5** (1958), 75–78.

220. J.R. Ringrose, *On some algebras of operators*, Part I, Proc. London Math. Soc. **15** (1965), 61–83; Part II, Proc. London Math. Soc. **16** (1966), 385–402.

221. -----, *Automatic continuity of derivations of operator algebras*, J. London Math. Soc. **5** (1972), 432–438.

222. -----, *Cohomology theory for operator algebras*, [57, part 2, pp.229–252].

223. M.A. Rieffel, *Induced Banach representations of Banach algebras and locally compact groups*, J. Funct. Anal. **1** (1967), 443–491.

224. -----, *Induced representations of C^*-algebras*, Adv. Math. **13** (1974), 176–257.

225. D.A. Raikov, *On the theory of normed rings with involution*, Dokl. Akad. Nauk SSSR **54** (1946), 387–390.

226. A. Rosenberg, *The number of irreducible representations of simple rings with no minimal ideals*, Amer. J. Math. **75** (1953), 523–530.

227. J. Rosenberg, *The role of K-theory in non-commutative algebraic topology, Operator algebras and K-theory*, Contemp. Math. No.10, Amer. Math. Soc., Providence R.I., (1982), pp.155–182.

228. -----, *Homological invariants of extensions of C^*-algebras*, [57, part 1, pp.35–75].

229. H.L. Royden, *Function algebras*, Bull. Amer. Math. Soc. **69** (1963), 281–298.

230. I. Raeburn and J.L. Taylor, *Hochschild cohomology and perturbations of Banach algebras*, J. Funct. Anal. **25** (1977), 258–266.

231. S. Sakai, *A characterisation of W^*-algebras*, Pacific J. Math. **6** (1956), 763–773.

232. -----, *Derivations of W^*-algebras*, Ann. Math. **83** (1966), 273–279.

233. Yu.V. Selivanov, *The values assumed by the global dimension in certain classes of Banach algebras*, Vest. Mosk. Univ. Ser.1 Mat. Mekh. **30**:1 (1975), 37–42.
= Moscow Univ. Math. Bull. **30** (1975), 30–34.

234. -----, *Projectivity of certain Banach modules and the structure of Banach algebras*, Izv. Vyss. Ucebn. Zaved Matematika **188** (1978), 110–116.
= Soviet Math. (Iz.VUZ) **22** (1978), 88–93.

235. -----, *Biprojective Banach algebras*, Izv. Akad. Nauk SSSR Ser. Mat. **43** (1979), 1159–1174.
= Math. USSR-Izv. **13** (1979).

236. I.E. Segal, *The group algebra of a locally compact group*, Trans. Amer. Math. Soc. **61** (1947), 69–105.

237. -----, *Irreducible representations of operator algebras*, Bull. Amer. Math. Soc. **53** (1947), 73–88.

238. -----, *Two-sided ideals in operator algebras*, Ann. Math. **50** (1949), 856–865.

239. Z. Slodkowski and W. Żelasko, *On joint spectra of commuting families of operators*, Stud. Math. **50** (1974), 127–148.

240. M. Takesaki, *On the cross-norm of the direct product of C^*-algebras*, Tohoku Math. J. **16** (1964), 111–122.

241. W. Timmermann, *Ideals in algebras of unbounded operators* I, Math. Nachr. **92** (1979), 99–110.

242. -----, *States on algebras of unbounded operators*, JINR, Preprint E5-85-91, Dubna, (1985).

243. B. Timmermann and W. Timmermann, *On ultrastrong and ultraweak topologies on algebras of unbounded operators*, JINR, Comm. E2-10242, Dubna, (1976).

244. A. Tikhonov, *Über die topologische Erweiterung von Räumen*, Math. Ann. **102** (1930), 544–561.

245. E.V. Troitskii, *Contractibility of the complete general linear group of the C^*-Hilbert module $l_2(A)$*, Funct. Anal. i ego Pril. **20** (1986), 58–64.
= Funct. anal. and its appl., **20** (1986), 301–307.

246. V.A. Trofimov, *Reflexive and self-dual Hilbert modules over certain C^*-algebras*, Usp. Mat. Nauk **42**:2 (1987), 247–248.
= Russian Math. Surveys **42**:2 (1987), 303–304.

247. J.L. Taylor, *The structure of convolution measure algebras*, Trans. Amer. Math. Soc. **119** (1965), 150–166.

248. -----, *The cohomology of the spectrum of a measure algebra*, Acta Math. **126** (1971), 195–225.

249. -----, *A joint spectrum for several commuting operators*, J. Funct. Anal. **6** (1970), 172–191.

250. -----, *The analytic-functional calculus for several commuting operators*, Acta Math. **125** (1970), 1–38.

251. -----, *Homology and cohomology for topological algebras*, Adv. Math. **9** (1972), 137–182.

252. -----, *A general framework for a multi-operator functional calculus*, Adv. Math. **9** (1972), 183–252.

253. -----, *Functions of several non-commuting variables*, Bull. Amer. Math. Soc. **79** (1973), 1–34.

254. -----, *Banach algebras and topology*, [1, pp.118–186].

255. A.S. Feinstein, *The joint spectra of a family of linear operators*, Izv. Akad. Nauk AzSSR, Ser. fiz.-tekh. **5** (1978), 11–15.

256. -----, *The joint essential spectrum of a family of linear operators*, Funct. Anal. i ego Pril. **14** (1980), 83–84.
= Funct. anal. and its appl., **14** (1980), 152–153.

257. C. Feldman, *The Wedderburn principal theorem in Banach algebras*, Proc. Amer. Math. Soc. **2** (1951), 771–777.

258. R.S. Phillips, *On linear transformations*, Trans. Amer. Math. Soc. **48** (1940), 516–541.

259. J. Phillips and I. Raeburn, *Automorphisms of C^*-algebras and second Čech cohomology*, Indiana Univ. Math. J. **29** (1980), 799–822.

260. ----- and ------, *Central cohomology of C^*-algebras*, J. London Math. Soc. **28** (1983), 363–375.

261. O. Forster, *Uniqueness of topology in Stein algebras*, ed. F. Birtel, Function algebras, *Scott, Foresman*, Glenview, Illinois, (1966), pp.157–163.

262. M. Fukamiya, *On a theorem of Gelfand and Naimark and the B^*-algebra*, Kumamoto J. of Sci. Ser.1, **1** (1952), 17–22.

263. U. Haagerup, *Injectivity and decomposition of completely bounded maps*, [56, pp.170–222].

264. -----, *All nuclear C^*-algebras are amenable*, Inv. Math. **74** (1983), 305–319.

265. R.E. Harte, *Spectral mapping theorems*, Proc. Roy. Irish Acad. Sect.A, **72** (1972), 89–107.

266. -----, *A spectral mapping theorem for holomorphic functions*, Math. Zeit. **154** (1977), 67–69.

267. A.Ya. Helemskii, *The homological dimension of normed modules over Banach algebras*, Mat. Sb. **81**(123) (1970), 430–444.
= Math. USSR-Sb. **10** (1970), 399–411.

268. -----, *A description of the relatively projective ideals in the algebras* $C(\Omega)$, Dokl. Akad. Nauk SSSR **195** (1970), 1286–1289.
= Soviet Math. Dokl. **11** (1970), 1680–1683.

269. -----, *A certain method of computing and estimating the global homological dimension of Banach algebras*, Mat. Sb. **87** (1972), 122–135.
= Math. USSR-Sb. **16** (1972), 125–138.

270. -----, *Smallest values assumed by the global homological dimension of Banach function algebras*, Trudy sem. Petrovsk. **3** (1978), 223–242.
= Amer. Math. Soc. Transl. (2) **124** (1984), 75–96.

271. -----, *Homological methods in Taylor's holomorphic calculus of several operators in a Banach space*, Usp. Mat. Nauk **36**:1 (1981), 127–172.
= Russian Math. Surveys **36**:1 (1981), 139–192.

272. -----, *Flat Banach modules and amenable algebras*, Trudy Moskov Mat. Obshch. **47** (1984), 179–218.

273. W. Hodges, *Krull implies Zorn*, J. London Math. Soc. **19** (1979), 285–287.

274. G. Hochschild, *On the cohomology groups of an associative algebra*, Ann. Math. **46** (1945), 58–67.

275. -----, *On the cohomology theory for associative algebras*, Ann. Math. **47** (1946), 568–579.

276. -----, *Cohomology and representations of associative algebras*, Duke Math. J. **14** (1947), 921–948.

277. -----, *Relative homological algebra*, Trans. Amer. Math. Soc. **82** (1956), 246–269.

278. E. Hewitt, *Measure algebras on locally compact groups: a case history in functional analysis*, Stud. Math. (Ser. Specjalna) Zeszyt **1** (1963), 41–52.

279. -----, *The ranges of certain convolution operators*, Math. Scand. **15** (1964), 147–155.

280. S.K. Tsui, *Extensions of derivations from* C^**-algebras*, Semesterbericht Funktionalysis **8** Tubingen (1985), pp.187–200.

281. E. Čech, *On bicompact spaces*, Ann. Math. **38** (1937), 823–844.

282. M.D. Choi and E.G. Effros, *Nuclear* C^**-algebras and the approximation property*, Amer. J. Math. **100** (1978), 61–79.

283. ----- and -----, *Lifting problems and the cohomology of* C^**-algebras*, Can. J. Math. **29** (1977), 1092–1111.

284. A. Szankowski, $\mathcal{B}(H)$ *does not have the approximation property*, Acta Math. **147** (1981), 89–108.

285. M.V. Sheinberg, *Homological properties of closed ideals that have a bounded approximate unit*, Vest. Mos. Univ. ser.1 mat. mekh. **27**:4 (1972), 39–45.
= Moscow Univ. Math. Bull. **27** (1972), 103–108.

286. -----, *On the relative homological dimension of group algebras of locally compact groups*, Izv. Akad. Nauk SSSR ser. mat. **37** (1973), 308–318.
= Math. USSR-Izv. **7** (1973), 307–317.

287. -----, *A characterisation of the algebra* $C(\Omega)$ *in terms of cohomology groups*, Usp. Mat. Nauk **32** (1977), 203–204.

288. G.E. Shilov, *Sur la théorie des idéaux dans les anneaux normés de fonctions*, Dokl. Akad. Nauk SSSR **27** (1940), 900–903.

289. -----, *On the extension of maximal ideals*, Dokl. Akad. Nauk SSSR **29** (1940), 83–85.

290. -----, *On regular normed rings*, Trav. Inst. Math. Stekloff, **21** (1947).

291. -----, *On the decomposition of a commutative normed ring into a direct sum of ideals*, Mat. Sb. **32** (1953), 353–364.
= American Math. Soc. Translations, (2) **1** (1955), 37–48.

292. K. Schmüdgen, *Lokal multiplikativ konvexe* Op^**-algebren*, Math. Nachr. **85** (1978), 161-170.

293. -----, *The order structure of topological* **-algebras of unbounded operators*, Rep. Math. Phys. **7** (1975), 215–227.

294. Yu.A. Shreider, *The structure of maximal ideals in rings of measures with convolution*, Mat. Sb. **27** (1950), 297–318.
= American Math. Soc. Translations (1) **8** (1962), 365–391.

295. S. Eilenberg and S. MacLane, *General theory of natural equivalences*, Trans. Amer. Math. Soc. **58** (1945), 231–294.

296. S. Eilenberg and J.C. Moore, *Foundations of relative homological algebra*, Mem. Amer. Math. Soc. **55**, (1965).

297. P. Eymard, *L'algèbre de Fourier d'un groupe localement compact*, Bull. Soc. Math. France **92** (1964), 181–236.

298. G.A. Elliott, *On approximately finite-dimensional von Neumann algebras* II, Can. Math. Bull. **21** (1978), 415–418.

299. P. Enflo, *A counterexample to the approximation problem in Banach spaces*, Acta Math. **130** (1973), 308–317.

299a. -----, *On the invariant subspace problem for Banach spaces*, Acta Math. **158** (1987), 213–313.

300. J.A. Erdos, *Non-self-adjoint operator algebras*, Proc. Roy. Irish Acad. Sect.A, **81** (1981), 127–145.

301. J. Esterle, *Injection de semi-groupes divisibles dans des algèbres de convolution et construction d'homomorphismes discontinus de* $C(K)$, Proc. London Math. Soc. **36** (1978), 59–85.

302. -----, *Elements for a classification of commutative radical Banach algebras*, [62, pp.4–65].

303. -----, *Quasi-multipliers, representations of* H^∞ *and the closed ideal problem for commutative Banach algebras*, [62, pp.66–162].

304. E.G. Effros, *Aspects of non-commutative order*, [2, pp.1–40].

305. -----, *Aspects of non-commutative geometry*, [3, pp.135–156].

306. -----, *Amenability and virtual diagonals for von Neumann algebras*, J. Funct. Anal. **78** (1988), 137–153.

307. -----, *Advances in quantized functional analysis*, Proc. of the ICM, Berkeley, Calif. 1986, Vol.2, pp.906–916.
308. ----- and E.C. Lance, *Tensor products of operator algebras*, Adv. Math. **25** (1977), 1-34.
309. ----- and A. Kishimoto, *Module maps and Hochschild-Johnson cohomology*, Indiana Univ. Math. J. **36** (1987), 257–276.

List of additional references

310. E. Albrecht, *On joint spectra*, Stud. Math. **64** (1979), 263–271.
311. W. Arveson, *Ten lectures on operator algebras*, Regional Conference Series in Math. **55**, *Amer. Math. Soc.*,Providence, (1984).
312. W. Beer, *On Morita equivalence of nuclear C^*-algebras*, J. Pure and Appl. Algebra **26** (1982), 249–267.
313. D. Voiculescu, *Some extensions of quasi-triangularity* II, Revue Roum. Math. Pures et Appl. **18** (1973), 1439–1456.
314. P.C. Curtis, A. Figa-Talamanca, *Factorisation theorems for Banach algebras*, ed. F. Birtel, Function Algebras, *Scott, Foresman*, Glenview, Illinois, (1966), pp.169–185.
315. R.E. Curto, *Connections between Harte and Taylor spectra*, Rev. Roumaine Math. Pures Appl. **31** (1986), 203–215.
316. A.J. Lazar, S.K. Tsui and S. Wright, *A cohomological characterisation of finite-dimensional C^*-algebras*, J. Operator Theory **14** (1985), 239–247.
317. M. Grosser, *Bidualräume und Vervollständigungen von Banachmoduln*, Lecture Notes in Math., **717**, *Springer*, Berlin, (1979).
318. M. Putinar, *A spectral mapping theorem*, Preprint INCREST No.3, (1981).
319. ------, *Uniqueness of Taylor's functional calculus*, Proc. Amer. Math. Soc., **89** (1983), 647–650.
320. -----, *Spectral theory and sheaf theory*, Part I, Operator theory: advances and applications Vol.11, *Birkhäuser*, Basel, (1983), 283–297; Part II, Math. Z. **192** (1986), 473–490.
321. J.L. Taylor, *Measure algebras*, Regional Conference Series in Math., **16**, *Amer. Math. Soc.*, Providence, (1972).
322. S. Shirali and J.W. Ford, *Symmetry in complex involutory Banach algebras* II, Duke Math. J. **37** (1970), 275–280.
323. M. Takesaki, *Structure of factors and automorphism groups*, Regional Conference Series in Math., **51**, *Amer. Math. Soc.*, Providence, (1983).
324. E.C. Lance, *On nuclear C^*-algebras*, J. Funct. Anal. **12** (1973), 157–176.
325. A.I. Shtern, Nuclear *C^*-algebras and related questions*, Current problems in Mathematics. Newest Results. **26**, Akad. Nauk SSSR, Moscow, (1985), p.107–126.

326. G.R. Allan, H.G. Dales and J.P. McClure, *Pseudo-Banach algebras*, Stud. Math. **40** (1971), 55–69.
327. F. de Meyer and E. Ingraham, *Separable algebras over commutative rings*, Lecture Notes in Math., **181**, *Springer*, Berlin, (1971).
328. J. Wermer, *Bounded point derivations on certain Banach algebras*, J. Funct. Anal. **1** (1967), 28–36.
329. W. Moran, *The global dimension of* $C(X)$, J. London Math. Soc. **17** (1978), 321–329.
330. E. Albrecht and H. Zassenhaus, *Maximal Abelian quasi-nilpotent operator algebras*, Preprint.
331. O. Bratteli, *Derivations, dissipations and group actions on* C^**-algebras*, Lecture Notes in Math., **1229**, *Springer*, Berlin, (1986).
332. M. Wodzicki, *The long exact sequence in cyclic homology associated with an extension of algebras*, C.R. Acad. Sci., Paris Sér.1, **306** (1988), 399–403.
333. N. Groenbaek, *A characterisation of weakly amenable Banach algebras*, Studia Math., **94** (1989), 149–162.
334. H.G. Dales and R.J. Loy, *Prime ideals in algebras of continuous functions*, Proc. Amer. Math. Soc., **98** (1986), 426–430.
335. A.M. Karyaev and N.V.Yakovlev, *Automatic complementability of some classes of annihilator ideals in Banach algebras*, Vest. Mos. Gos. Univ. Ser. 1, Mat. Mekh. (1990), 95–98.
336. P.C. Curtis and R.J. Loy, *The structure of amenable Banach algebras*, J. London Math. Soc., **40** (1989), 89–104.
337. A. Kyriazis, *On topological algebra bundles*, to appear.
338. E. Christensen and A.M. Sinclair, *On the vanishing of* $H^n(A, A^*)$ *for certain* C^**-algebras*, Pacific J. Math., **137** (1989), 55–63.
339. E. Christensen, E.G. Effros and A.M. Sinclair, *Completely bounded multilinear maps and* C^**-algebraic cohomology*, Invent. Math. **90** (1987), 279–296.
340. A. Mallios, *Topological algebras: Selected topics*, *North Holland*, Amsterdam, (1986).
341. A.N. Plichko, *Automatic continuity, bases and radicals in metrizable algebras*, in the press.
342. T.A. Sarymsakov, Sh.A. Ayupov, J. Hadjiev and V.I. Chilin, *Ordered algebras*, *Fan*, Tashkent, (1983).
343. M. Fragoulopoulou, *An introduction to the representation theory of topological* **-algebras*, Schriftenreihe Math. Inst. Univ. Münster, Ser. 2, **48** (1988).
344. -----, *Tensor products of locally* C^**-algebras and applications*, Math. Inst. Univ. Athens, preprint.
345. A.Ya. Helemskii, *The homological background to Connes amenability: injectivity of the pre-dual bimodule*, Mat. Sb. **180** (1989), 1680–1690.

346. -----, *Topological homology: algebras with various conditions of homological triviality*, Lecture Notes in Math., *Springer*, Berlin, to appear.

347. S. Popa and F. Radulescu, *Derivations of von Neumann algebras into the compact ideal space of a semifinite algebra*, Duke Math. J., **57** (1988), 485-518.

348. V.S. Yakovlev, *On the representation of some commutative nuclear *-algebras*, Ukr. Mat. Jurn. **39**:2 (1987), 235-243.
= Ukrainian Math. J. **39** (1987), 317-319.

349. Yu.M. Berezanskii, G. Lassner and V.S. Yakovlev, *Decomposition of positive functionals on commutative nuclear *-algebras*, Ukr. Mat. Jurn. **39** :5 (1987), 638-641.
= Ukrainian Math. J. **39** (1987), 521-523.

350. N.V. Yakovlev, *Examples of Banach algebras with a radical that cannot be complemented as a Banach space*, Usp. Mat. Nauk **44**:5 (1989), 185-186.
= Russian Math. Surveys **44**:5 (1989), 224-225.

351 V. Ptak, *On the spectral radius in Banach algebras with involution*, Bull. London Math. Soc., **2** (1970), 327-334.

352 M.P. Thomas, *The image of a derivation is contained in the radical*, Annals of Math. **128** (1988), 435-460.

Index

Index of notation